Electric Ut
A Guide to the Competitive Era

By Peter Fox-Penner

Public Utilities Reports, Inc.
Vienna, Virginia
1-800-368-5001

1997

First Printing, April 1997

ISBN 0-910325-67-7
Library of Congress Catalog Card No. 97-066610
Printed in the United States of America

Preface

A number of forces beyond the recent interest in restructuring have given rise to this book. Years before the present tumult, I became fascinated with the interaction of regulation and competition. My early experiences in the environmental movement, state government, and later at the University of Chicago only heightened this interest.

Sometime in 1990, it dawned on me that my favorite restructuring text, *Markets for Power*, was in need of a serious update. In addition, I have often wished for a modern reference text on utility issues. Len Hyman's book and many specialized texts notwithstanding, there is no book that explains electric power technology, regulation, and economics in a thorough but non-mathematical fashion.

I was at work on a primitive manuscript in 1993 when I was asked to join the Department of Energy, partly to work on electricity issues. During my years at the Department, electric restructuring grew from a fascination for policy wonks like me into a topic of cocktail conversation in much of the developed world.

My years at DOE led me to believe that the most useful contribution I could make would be to write a readable, well-referenced, comprehensive survey of the topic. In doing so, I realized that I would likely satisfy no one: Colleagues would yawn at the absence of new answers, busy leaders would search in vain for a crisp bottom line, and practitioners of all sorts would find too little or too much of the particular things they were looking for.

I have tried to anticipate some of these concerns by creating a main text that can be skimmed or read without reference to endnotes, text boxes, or appendices. The latter material is for dedicated readers, researchers, and others who are interested in greater detail. I have also tried to include numerous subheads, road maps, and a good index to facilitate use as a reference.

The debts I have incurred while writing this are extraordinary, and begin and end with my family. Not yet three, my daughter thinks that writing a book means coloring on the pages of an already-printed work. Periodically, she would select a book from our bookshelf and give it to me to write on as I left for the office. My debt to my wife (a far better writer/economist than I) cannot be expressed in words.

Beyond this, almost every one of my friends and colleagues has helped me with advice, faxed articles, and much more. I thank all of them—a long list that includes Dan Adamson, Rick Adcock, Bob Anderson, Shimon Awerbuch, Vicky Bailey, Greg Basheda, Doug Bauer, Mark Bernstein, Rosina Bierbaum, Carl Blumstein, Doug Bohi, Dave Boomsma, Ken Bossong, Clark Bullard, Bill Burton, Dallas Burtraw, Nancy Brockaway, Marc Chupka, Gigi Coe, Roger Colton, Greg Conlon, Rich Cowart, David Donniger, Dave Dworzak, Bruce Edelston, Mike Einhorn, Joel Eisenberg, James Fama, Susan Farkus, Rich

Felak, Chris Fitzgerald, Mike Foley, Jonathan Frenzen, Dirk Forrister, Bill Gaffney, Jeff Genzer, Rita Glenn, Larry Golub, Diane Grueneich, George Hall, Jeff Hall, Bruce Hannon, Ray Hartman, Hal Harvey, Bruce Hedman, Ned Helme, Steve Henderson, Alan Hoffman, Bill Hogan, Elgie Holstein, Mary Ellen Hopkins, William Hughes, Renz Jennings, Pam Johnson, Henry Kelly, Jessie Knight, David Kovner, Bill LeBlanc, Murray Liebman, Armory Lovins, Mary Ann Luntz, Larry Mansuetti, Jan McFarland, Betsy Moler, Gary Nakarado, David Nemtzow, Ren Orans, Karen Palmer, Jerry Pfeffer, Kevin Porter, Harvey Reiter, the Regulatory Assistance Project, Lauren Renshaw, Art Rosenfeld, Harvey Salgo, Rich Scheer, Marius Schwartz, Kyle Simpson, Scott Sklar, Pablo Spiller, Peter Spinney, Bonnie Suchman, Rich Tabors, Mike Toman, Susan Tomaski, Jean VanVlandren, Ed Vine, Gordon Wagner, Wesley Warren, Tom Welch, and Barbara Williams.

A few special thanks are in order. I am grateful to Jim Bailey, Marty Lobel, Susan Johnson, Thomas Whitehill, and Denise Benoit, who worked remarkably well on an extremely tough schedule to move from manuscript to book. I also thank my friend and colleague James Hoecker for agreeing to write the foreword. Kathy Correll, Deborah Bailey, and Nicolle Lewis valiantly prepared the manuscript, and many hours of near-thankless research assistance was provided by David Kovner, Bruce Campbell, Greg Pettis, Richard Sweet, and Maureen Smith.

I have precious memories of my restructuring colleagues at the Department of Energy, led by Secretary Hazel O'Leary, Charlie Curtis, Bob Nordhaus, Sue Tierney, Doug Smith, Dave Meyer, Val Jensen, and Diane Pirkey. I feel likewise blessed to have inherited warm, superbly talented colleagues at The Brattle Group, led by Larry Kolbe. I could not have done this without their support, and I appreciate it very much.

Finally, a remarkable group of experts reviewed and edited various parts of this book, under extreme time pressure: Merribel Ayres, Richard Ayres, Les Baxter, Matthew Brown, Ralph Cavanagh, Roger Colton, Joel Eisenberg, Joe Eto, Bob Gee, Pat Giordano, Chuck Goldman, Frank Graves, Phil Hanser, Bill Hermann, Eric Hirst, Bob Niekum, Mike Oldak, Dave Owens, Hannes Pfeffenberger, Dave Penn, and Diana West. I cannot thank them enough for the many immeasurable improvements they made to this work. This notwithstanding, the opinions in this book are my own (and not those of The Brattle Group) and I remain solely responsible for errors and omissions.

This volume represents an attempt to distill the issue of electric restructuring into a single, comprehensive work. Many restructuring issues are highly fact-specific and require individualized study in order to reach appropriately specific conclusions.

Over the course of our history, the American people have gradually entrusted more and more of what might be provided by governments to the self-correcting forces of the marketplace. Markets for essential services are rarely free of government influence, but proper policies can help them retain their fundamentally beneficial character. It is my hope that electric restructuring is a large and fascinating step along this path.

Peter Fox-Penner
Washington, D.C.
February 1997

Foreword

Most readers intent on digesting the contents of this volume will come to it already persuaded that the U.S. electric power system is destined to change permanently and profoundly in the next few years. That does not necessarily mean they know why so vital and (by most global measures) so successful an industry is about to be reinvented. After all, domestic electric utilities, largely investor-owned vertically integrated monopolies, have furnished relatively inexpensive and highly reliable electric service for several generations. Their historical record of efficiency, fuel diversity, financial stability, and quality of service is remarkable. Why, then, restructure the power business?

The first, deceptively significant, answer is "because we can." The movement to restructure evidences a high degree of confidence that the electric system can be bent to new paradigms that can be made to work in practice. This confidence in the efficacy of change derives, in part at least, from the stability of our political institutions, which permits indulgence of the American appetite for making the best better even when large amounts of money are at risk and regulatory and economic control is placed in doubt. In addition, the industry and the American economy currently enjoy relative security, if not tranquility—witness the soaring stock market, low inflation, modest unemployment, and stable energy, fuel, and utility stock prices. Perhaps most importantly, American command of electric power production and delivery technologies, as well as bold new information technologies, provides tremendous operational capabilities and, with them, great flexibility. In that regard, the past triumphs of electrical engineering have helped seal the fate of the old regime. In other words, if one were empowered to choose precisely when to begin reshaping so basic an industry, with all the attendant risks, a more propitious time could scarcely be selected.

Part competitive necessity, part scholarly enterprise, part politics, part corporate generalship, and part mob psychology, electric restructuring has enlisted everything from the most formidable analyses to the most subjective policy judgments and insupportable projections. Some proponents of change want merely to equip the power business to withstand the competitive and economic challenges of the next century and to enable it to deliver the diverse services customers will then demand. For others, the pursuit of more competition and less regulation or the adoption of a single institutional or operational model of the market is an end in itself. In any event, the temptation to be prescriptive about the brave new world is indeed great.

T. S. Kuhn once observed (about revolutions in science) that when a fundamental shift in a predominant mode of thinking—a so-called "paradigm"—happens, "it is rather as if the professional community had been suddenly transported to another planet where familiar objects are seen in a different light and are joined by unfamiliar ones as well." We are unquestionably viewing this industry in a new light while trying to divine the practical long-term consequence of "unfamiliar objects" like the independent system operator, the power exchange, the regional transmission group, the futures contract, and the Btu marketplace. It is from such disparate ingredients that public and private policymakers are today hoping to construct a coherent and workable new model for future regulated and unregulated power markets.

The second, and more clinically satisfying, explanation of why the power industry must change relates to its peculiar pathologies, or what Alfred Kahn has called its "technological and institutional failure, not its successes." Until the early 1990s, electric rates had risen steadily for three decades, instead of declining steadily as they had for 20 years after World War II. The economic maladies of the 1970s and 1980s took a serious toll—double digit inflation, high interest rates, long plant construction times, the reversal of economies of scale, fuel crises, the chronic and enduring problems of nuclear generation, flat levels of demand for power, and excess generation capacity. In many jurisdictions, average embedded utility costs and revenue requirements came to exceed the marginal cost of building new plants, a disparity made more glaring by the advent of new gas-fired electric generation technology. These unfortunate developments produced widening differences in rates between utilities and among jurisdictions. Although invisible to most consumers, such market dysfunctions meant that many existing utility arrangements simply became unsustainable.

Restructuring now seems irresistible. Transmission open access arrived in 1996. Mergers and acquisitions are increasing dramatically. And, state retail competition proceedings have suddenly proliferated. But, even converts to competition may legitimately wonder what utilities are on their way to becoming. Dr. Fox-Penner has performed a valuable service in this regard. His book is not a definitive history of restructuring or even a predictor of its end results; it is too soon for that. Rather, in a single volume, the author makes us confront both the industry's successes and its failures. He examines issues as diverse as stranded investment, environmental quality, and reliability. He identifies the "subtle, complex, and indeterminate" impact that wholesale and retail competition, decontrol of the generation sector, emergence of regional issues and institutions, customer choice, industry consolidation and realignments, and associated developments in information technologies are likely to have on the public welfare. In my estimation, most readers will conclude from this presentation that, even as opportunities for greater efficiency and better services multiply in the restructured world, the task of protecting the consumer and the public interest will become more, not less, complicated. This is true largely because responsibility for ensuring a socially beneficial as well as a profitable industry will be more widely shared among federal, state, and even regional forums and among the units of an increasingly disaggregated, and partly deregulated, industry.

While I may not share Dr. Fox-Penner's skepticism about whether competition portends real financial benefits for residential consumers, I concede that the potential for unfavorable outcomes remains great. Competition may reduce rates in many jurisdictions, but it may also jeopardize existing low rates in others. Inattention and poor policy choices may adversely affect reliability, conservation, power supply diversity, efficiency, and even universal service. As this book suggests, however, abject opposition to competition or restructuring is too late to be meaningful and, in any case, offers precious few solutions to such problems.

The success of restructuring will not depend upon mere faith in competition, deference to paradigm shifts, or confidence that regulators will correct market excesses. It will instead require dialogue, study, prudence, experimentation, courage, and a healthy respect for what this important industry has already accomplished in service to the American consumer. All those helping to forge a new electric power business for the 21st century should take time to grapple with the full breadth of the issues set forth in this excellent volume.

James Hoecker
Washington, D.C.
February 1997

Note: Dr. Hoecker is a member of the Federal Energy Regulatory Commission.

Introduction

Electricity has been the lifeblood of the 20th century and will power the information revolution of the 21st century. The "grid" that connects every American home and office is subject to disturbances matched in scale only by the largest natural disasters and wars. When cities go dark—even for a few hours—security issues rise to the level of a major emergency, and the demographic profile of a region changes in a night.

Early this century, electricity transformed the cities of the developed world from dim, mechanical smoke-filled centers to tantalizing beacons of entertainment and industry. Between 1900 and 1925, almost every home in Chicago was connected to central station power and 30 million electrical lamps were installed.[1] A few decades later, rural electrification and federal power projects brought the revolution to rural villages and farms across America. These changes forever decoupled human existence from rhythm of the sun.

Today, this revolution continues to sweep across the developing nations of the world. Demand for urban electricity is skyrocketing in areas such as Asia, which is expected to double its energy consumption in less than ten years.[2] According to the British journal *The Economist*, the developing world will add as much new electrical capacity in the next 25 years as the rest of the world put in place during the past nine decades.[3] In rural villages, electric power, often from decentralized, renewable sources, contributes to vast improvements in sanitation, health care, and literacy. Speaking of the intimate connection between rural electricity services and social and political change, one international banker bluntly observed: "In developing countries, make no mistake: electricity is democracy."[4]

In America, the larger cities entered the age of electricity via a patchwork of newly-established small electric power companies that served overlapping areas at free-market prices. Economies of scale in ever-larger plants drove electricity prices down, and new electric devices transformed life at work and home.

In Europe, where state ownership had a long tradition, most power companies developed as government entities. In the United States, advocates of private ownership with government regulation won out in the lucrative urban markets. When these private firms did not find it sufficiently profitable to expand into rural areas, the federal government stepped in. As a result, the United States developed a unique industry structure: vertically integrated, regulated, private utilities in dense areas operating alongside public and cooperative utilities in rural areas (the latter first supplied predominantly by large federal generating plants).

By most measures, this structure served well for its first 50 years. In the United States, the variety of electric devices in households, stores, offices, and factories has mushroomed beyond Thomas Edison's wildest dreams. The average use of electric power grew sevenfold between 1940 and 1970, and has increased by an additional 25% since then.[5] The proportion of households with electric service rose to 98.8% in 1956, after which time the U.S. Census Bureau ceased publishing this information.[6] Inflation-adjusted electricity prices have fallen, risen, and fallen again in real terms, ending the past 50 years about where they were in 1925.[7] And while the industry remains a major source of air, water, and other pollutants, technology and regulations have enabled it to reduce power plant emissions by factors of hundreds to thousands since 1970.[8]

As the 20th century progresses to the 21st century, it is clear that the organization of the industry will change. The natural monopoly rationale for regulation is no longer widely accepted. In the United Kingdom, Argentina, Spain, Norway, New Zealand, and parts of the United States and Canada, all or parts of the power industry have been deregulated. Although many technical features of the industries differ, the ongoing deregulation of the telecommunications and natural gas industries also have a growing influence on the power sector policies.

Will this deregulation (or "restructuring") usher in exotic new energy services, such as the ability to purchase only environmentally-superior "green" energy? Or are individual consumers in for little more than aggravation and confusion? Does deregulation threaten to singlehandedly overturn the United States' commitment to stabilize greenhouse gases, jeopardizing the progress forged in the historic Rio accords? Will restructuring result in decades of upheaval, bankruptcies, and billion-dollar losses, including a government takeover of all remaining nuclear power plants?

This volume explores the options for introducing utility deregulation, the likely impacts of increased competition, and the many remaining questions of structure and regulation, such as:

- Does fair and effective competition necessitate the elimination of all vertical integration in the utility industry?

- If transmission and distribution remain regulated natural monopolies, what form of regulation is best for them?

- What are the likely benefits and impacts of allowing individual consumers to choose their own electricity suppliers—so called "customer choice," "direct access" or "retail wheeling?"

Generally there are no simple answers for these questions. Indeed, in many instances, our information and understanding is so sketchy that we can scarcely illustrate the general magnitude of the costs, benefits, or risks of alternative policies. The most important objective of this work is therefore to reinforce the fact that each of these issues involves tradeoffs between important economic and policy considerations.

OLD WINE IN NEW BOTTLES

Electric utility deregulation has been the subject of hundreds, if not thousands, of recent speeches and publications. In modern times, these works date back at least to Leonard Weiss' 1975 Brookings study of "the possibilities for both more competition and for various types of antitrust action" in electric utilities. A 1976 University of Michigan compilation examined restructuring alternatives ranging from universal government ownership to complete deregulation.[9]

Sounding 15 years ahead of its time, the *Wall Street Journal* reported in 1981 that, "in academia, the idea of deregulating electric utilities has been debated for years. But in the past several months, it has attracted serious attention from policymakers in government and industry."[10] In 1982, the Edison Electric Institute examined the determinants of utility industry structure as well as seven "alternative models" of a deregulated industry. This was followed shortly by works by Plummer (1983), Poole (1985), Moorhouse (1986), Smith (1988), and many others.[11]

Many of these works relied on a collection of views and facts to suggest that more competition is desirable.[12] Others argued for a particular new structure, such as spot pricing, reconfiguring the industry into regional power companies, or adherence to the model posed by natural gas deregulation.

Among these "early" studies, the 1983 work of MIT economists Paul Joskow and Richard Schmalensee, *Markets for Power,* stands as the most ambitious attempt to reconcile broad questions of industry structure with accumulated data and experience. Like the 1982 EEI study (and, to a lesser extent, the Michigan study), and Moorhouse, Joskow and Schmalensee began by surveying the technological and economic features of the industry. Drawing heavily on the distinction between ownership (i.e., integration) and contracts, they examined four alternative industry structures. In contrast to many other works before and since, Joskow and Schmalensee expressed many doubts concerning the advantages of competition, and emphasized the importance of a number of conditions that will be re-examined in chapters that follow.

Joskow and Schmalensee completed their analysis more than ten years ago. Their timing could not have been more ironic: 1983 marked the point at which deregulated, deintegrated generating plants began to make their first appearance in the industry.[13] Joskow (1989) returned to chronicle the state of the industry. In spite of a number of theoretical and practical problems, he concluded that "the competitive genie is [now] out of the bottle," and "the time is ripe for regulatory change."[14]

Changes in the industry since that time have fractured the literature into several diverging bodies of work, many of which are growing too fast to survey adequately. Important works have examined the restructuring phenomenon generally;[15] others focus on its impact on individual states and regions[16] or the public interest aspects of the industry.[17] International[18] and interindustry[19] comparisons are more prevalent, and many works have focused on governance of the generation, transmission, or distribution (retail) sector. Many of these will be discussed in later chapters.

A GUIDE TO THE BOOK

This book is organized into three parts.

Part I (Chapters 1–5) is an introduction to the technology, economics, and organization of the electricity industry. This material is essential background material for those not already familiar with the industry; readers familiar with the industry may wish to skim or skip much of this material.

Chapter 1 begins with a review of the economic and political crosscurrents that have given rise to electricity restructuring. This tapestry includes deregulation in other industries and nations, the antinuclear and environmental movements, economic research into the weaknesses of traditional regulation, and global management trends toward deintegrated business structures.

Chapters 2 and 3 examine the technology and basic economics of electric power systems. These two chapters explain the operation of power systems on the two essential time horizons: the short run (the next few minutes through the next few days) and the long run (the next several years). Briefly, the short-run chapter examines power plant control, system reliability, and the interdependence of all producers on a power network; Chapter 3 takes a longer-term look at the industry's methods of planning and capacity expansion.

Chapters 4 and 5 complete the introductory materials by examining the legal, regulatory, and business aspects of the industry. In keeping with the economics of industrial organization, Chapter 4 examines the economic and legal forces that help determine the industry's organization—the size, composition, product mixes, and legal and regulatory rules governing utilities.

Finally, Chapter 5 details the economic, technical, and regulatory changes that have unleashed the wave of restructuring that now engulfs the industry.

Part II (Chapters 6 to 10) is devoted to economic and regulatory issues that are at the root of the original economic regulation of the industry—what we economists call the "private good" aspects of the industry. Chapter 6 begins by explaining the evolution of competition in the electric industry from fringe competition in generation to full-scale competition under open access. In addition to helping illustrate the difficulties raised by the limited introduction of competition into a largely-regulated industry in the United States, this chapter introduces the concepts of open access transmission, spot and congestion pricing, and other essential building-blocks for the new industry.

Mirroring the distinctions introduced in Chapters 2 and 3, Chapters 7 and 8 explain the mechanics of competitive power markets in the short run and the long run, respectively. Short-run competition centers on several models of deintegrated generation and transmission, including the so-called "poolco" and "bilateral" models, as well as the emerging concept of an "independent system operator" or ISO. Long-term issues (Chapter 8) include the entry of new firms and capacity expansion by existing firms in a competitive environment, as well as the integration of long-term planning for all parts of the system and the preservation of effective competition.

Chapters 9 and 10 turn from the bulk power system, or large-scale competitive issues, to retail customer issues, traditionally regulated by state utility commissions. Chapter 9 begins with a detailed consideration of the processes by which retail competition occurs in the industry and the challenging legal, regulatory, and business arrangements that are necessary to enable and safeguard retail choice. Chapter 10 examines several variations on retail competition in a framework designed to illustrate the relative benefits and tradeoffs inherent in retail restructuring alternatives. This chapter also examines the prospects for the emergence of new bundles of energy and communications services that are likely to become far more prevalent in the restructured power marketplace.

Part III (Chapters 11 to 17) examines the origins, economics, and regulatory issues associated with a disparate but important collection of public interest topics that have and will continue to confront the industry. This collection of public interest topics is not comprehensive; there are many other interesting public aspects of the power industry we cannot treat in sufficient detail.[20]

Chapter 11 begins with an examination of universal service and programs that are designed to serve low-income customers. In all developed countries, electric power is considered an "economic necessity"—a private good so important that life, health, or livelihood may depend on access at reasonable prices. Whether one considers this the enshrinement of a public good known as universal access or government redistribution of private wealth, these programs evidence a widespread public belief in a role for government as a (private) service guarantor.

A second, widely recognized public aspect of electric power is its impact on the environment (Chapter 12). The quality of the environment is a well-recognized public good. When the production of electricity impacts the environment, it is said to have a *negative externality* and it is generally (though not universally) accepted that government action is necessary to restore economic optimality.

Chapter 13 examines utility industry efforts to use electricity with greater efficiency, or so-called Demand-Side Management (DSM) programs. Chapter 14 continues with two closely related public policy topics—renewable energy and fuel diversity. This chapter also includes a brief discussion of energy security.

Chapter 15 examines the issue of power system reliability, which has the attributes of a public good due to the technical features of the grid. Except in limited circumstances, bulk power reliability is a good shared by all, and my enjoyment of reliability does not consume or diminish yours. Moreover, reliable electric service is important for the provision of other (non-power) public goods such as national defense.

Finally, Chapter 16 examines electric utility capital costs that have been incurred under regulation and may not be recoverable after restructuring, or so-called stranded costs. While these costs are in not formal public goods, the manner in which our legal and regulatory systems treat legally-mandated investments is certainly a matter of broad economic and political relevance.

Chapter 17 concludes the volume with a brief summary of the major policy issues, possibilities, and uncertainties inherent in the vast undertaking of electric restructuring.

NOTES

1 Platt, 1991, ch. 9.

2 Energy Information Administration, 1995, pp. 15–20

3 "Asia Delivers an Electric Shock." *The Economist,* October 28, 1995, p. 77.

4 Testimony of Brooks Browne, President, Environmental Enterprises Assistance Fund, before the United States Department of Energy Hearings for the 1995 National Energy Plan, Austin, TX., 8/12/94. Much the same fervor was expressed in the United States during the era of rural electrification. American historian Jordan Schwarz writes that the drive for electrification in the U.S. northwest "bordered on a secular religion" during the 1920s and 1930s. (Schwarz, 1994, p. 49)

5 U.S. Census Bureau, *Historical Statistics of the United States from Colonial Times to the Present* (1974), S108-119; EIA(1995c) Tables 5 and 6.

6 U.S. Census, op cit. Probably because many of our larger homes have more than one residential service, the number of U.S. residential electric customers today exceeds the number of residential households counted by the census.

7 U.S. Census Bureau, *Historical Statistics of the United States from Colonial Times to the Present, S108-119; Statistical Abstract,* 1993, Series 961.

8 A comparison of "existing" coal boilers and advanced combined cycle gas technology with selective catalytic reduction (a control for nitrogen oxides) indicates that emissions per kWh are reduced by more than 100-fold for NOx, a factor of one thousand for particulates, and a factor of two for carbon dioxide. (Pace, 1990, Tables ES-1, ES-3). The Environmental Protection Agency reports that, between 1980 and 1994, total electric utility emissions of SO2 dropped from 9.4 million tons to 7.4 million tons, as total power generation rose 42%. During the same period, Nox emissions were approximately constant (EPA, 1995; EIA, 1993b, 1995b).

9 Shaker, Steffy, and McCracken, 1976.

10 Emshwiller, John. "Debate Heats Up on the Merits of Deregulating Utilities," *Wall Street Journal,* June 2, 1981, p. B1.

11 See Plummer (1983), Poole (1985), Moorhouse (1986), Smith (1988). Among many others, also see Berlin, Cicchetti and Gillen (1974), Cohen (1977), Alexander (1981), U.S. Department of Energy (1987), Price Waterhouse (1987), Bohn, et al. (1984), Pierce (1986), Council on Economic Regulation (1988), Naill and Belanger (1989), Kleeman (1989), Corey (1989), Gordon (1990), Studness (1991).

12 In contrast, EEI produced a study in 1986 focused exclusively on legal and regulatory aspects of utility deregulation (EEI, 1976).

13 At the time *MFP* was published, only 12 of the 50 U.S. states had finalized policies that allowed such generation (Fox-Penner, 1990a).

14 Joskow (1989) p. 206.

15 Public Utilities Reports (1994), Joskow (1989), Economic Report of the President (1995), Energy Information Administration (1993b), Tenenbaum, Lock, and Barker (1992), Gegax and Nowotny (1993), and Kahn and Gilbert (1993), Dasovich, Meyer, and Coe (1993), Tonn and Schalffhauser (1994), O'Connor (1994), Moskowitz, et al (1993), Stelzer (1995), Hogan (1996).

16 Public Service Commission of Wisconsin (1995), CONEG (1995), Northwest Power Planning Council (1995), Public Service Commission of New York (1995).

17 Hamrin, et al (1994), Flavin and Lessen (1994), Cavanaugh (1994), Moskowitz and Foy (1994), Lovins (1994).

18 For example, Gilbert and Kahn (1996), Holmes (1992a,b), and International Energy Agency (1991).

19 See Chapter 1, *infra.*

20 For example, utilities contribute very substantial tax revenues to federal, state, and local governments; in many rural counties utility taxes or payments in lieu of taxes are the largest single source of local government revenue. Four towns in Massachusetts report that utilities pay 25 to 56% of all property taxes collected in these towns. (Deloitte and Touche, LLP (1996) p. 24. Also see "Report Cautions Policymakers: Think About Tax Issues while Restructuring." *Electric Utility Week*, October 28, 1996.) Because the present tax code is structured around the integrated regulated industry, tax payments from the power industry will probably drop significantly absent altered tax legislation or regulation.

Another important topic we will not treat is the impact of restructuring on industry research and development. In all industries, research and development has positive externalities, taking it outside the realm of private-good economics. When the goods produced by the industry are public or regulated, as will be the case with transmission and distribution and perhaps other power industry goods, ordinary private-market R&D incentives are likely to prove particularly inadequate. In a sign that these problems were already starting to affect the industry, Congressman George Brown, Jr. recently released figures from the U.S. General Accounting Office that indicate a 33% decline in electric industry total R&D outlays over the past three years. U.S. General Accounting Office, "Federal Research: Changes in Electricity-Related R&D Funding," GAO/RCED 96-203, Sept. 1996. We touch on these issues in the context of our discussions of the environment, renewable energy, and reliability, but the topic is worthy of more attention than we can provide. Briefly, the most extensive discussion of the potential impact of restructuring on utility R&D is probably the proceedings of the California Public Utilities Commission's (CPUC's) restructuring proceeding, R.94-04-031/I.94-04-032 and an associated proceeding convened by the California Energy Commission (CEC) in late 1994 and early 1995. The CPUC proceeding resulted in the creation of an R&D working group and a proposed R&D surcharge on all California sales applying only to R&D for power-industry goods considered to be public in nature. Report of the RD&D Working Group to the California Public Utilities Commission, Decision 95-12-063, September 6, 1996. The CEC proceeding resulted in a special report of the Research and Development Committee of the CEC, June 1995.

Part I

INTRODUCTION TO THE ELECTRIC UTILITY INDUSTRY

Chapter 1

Origins of the Modern Deregulation Debate

"The system of government-franchised monopolies is the most extreme form of government interference in the free market that exists in America today."

—Congressman Dan Schafer

Recent interest in electric restructuring stems from a mixture of political and economic forces. First, there has been a steady erosion of the view that electric utilities are natural monopolies. Second, much research suggests that strict government regulation is inefficient or even perverse, and can be replaced by better alternatives. These views have been furthered by the apparent success of deregulation or the removal of state ownership from other industries all over the world as well as the philosophical shift in the United States toward less government. Third, parts of the public interest community, such as some in the environmental movement, have come to embrace utility deregulation—although many others in these groups remain opposed or indifferent. Finally, several management trends in the nonutility industries (e.g., the increased use of computers to customize product manufacturing) have captured the fancy of many regulated utilities. The use of some of these methods has convinced some utility executives that competition to sell power to consumers is exciting and inevitable.

The move to deregulate utilities cannot be attributed to any one of these factors alone, nor do they necessarily tell us much about how utility restructuring should occur. However, all of these forces play important roles in the evolving utility restructuring debate.

ARE UTILITIES NATURAL MONOPOLIES?

The economic justification for electric utility regulation arose out of the assertion that electric utilities were "natural monopolies." According to the original theory, a natural monopoly means that a single firm is the lowest-cost means of supplying a single area with a product or service loosely known as electric power.[1] The name for this economic condition, which was never quite correct, is *economies of scale*, or more precisely, *declining average costs*. In plain English, this means that the total costs of supplying all customers are lowest with a single provider.

If a market is a natural monopoly, the theory goes, competition to serve the market will lead to excessive instability. This follows somewhat intuitively from the fact that "economies of scale" reduce costs as output increases. In a competitive race for increased market share, rivalry will result in price wars that leave only a single surviving monopoly anyway. Instead, why not avoid the turmoil, grant an exclusive license to sell, and require the monopolist to sell without discrimination or excess profits?

3

This is precisely the argument Samuel Insull, former secretary to Thomas Edison and first president of the Commonwealth Edison Company of Chicago, used to introduce the idea of rate regulation to his industry in 1898.[2] Insull borrowed a concept first applied to the railroads following the Civil War and later applied to the early natural gas industry. (Due to the work of economists such as Richard Ely, Edmund James, and Henry Carter Adams, the idea entered the mainstream of the economics profession in the 1880s.[3])

As the electric power industry grew, it divided into three main stages or segments: generation, transmission, and distribution (see Figure 1-1). Briefly, generators make electric power out of other forms of energy; transmission lines transport the power to near the final user; and distribution lines also transport power, but they change the power from high voltages and large quantities to low voltages and smaller wires more suited to homes and offices—much like the difference between wholesalers and retailers.

Originally, most utilities performed all three functions, and for most Americans this is still true today. Most American ratepayers pay a single power bill to one company that owns enough generators to supply most of its own customers, and owns most of the large and small power lines in its area. The central organizational feature of utility restructuring is to change this pattern dramatically by separating power plant operation from the ownersip and operation of transmission or distribution systems.

Some economists argued from the start that electric utilities were not really natural monopolies; a minority of economists continue to hold this view. In the mainstream, however, there was little disagreement that all three segments of the industry, and thus the industry as a whole, were a natural monopoly during its first few decades. By the 1970s, many argued that economies of scale in power generation had come to an end, while transmission and distribution remained natural monopolies. Hence, the simple reasoning went, generation need no longer be regulated, although regulation of the other two stages of the industry should continue.[4] (See Box 1-1.)

Most observers now believe that the generation stage of the industry has lost enough of its economies of scale to qualify for deregulation. Because most of the large utilities in the United States own and operate all three segments, a complex question arises: Should we continue to regulate separate pieces of utilities and deregulate other parts, or should utilities be forced to divide into several smaller, independent companies. Because two-thirds of the industry (transmission and distribution) will remain regulated, the term *restructuring* is used more commonly than deregulation to describe this overall change.

LEARNING FROM CHANGES IN OTHER UTILITY INDUSTRIES

In other utility industries—notably telecommunications—technological change and other factors have prompted a reexamination of the need for price and entry regulation. Some of this was triggered by the fact that the original theory of natural monopoly had two severely limiting features. First, the regulated firm is assumed to sell only a single good, whereas most firms are multiproduct. Second, natural monopoly cost conditions must be the result of using whatever production technologies are available to the industry. How can the inevitable effects of technical change be incorporated by firms in the industry, particularly when this change appears to remove the natural monopoly characteristics that gave rise to regulation?

Figure 1-1. A Simple Electric System

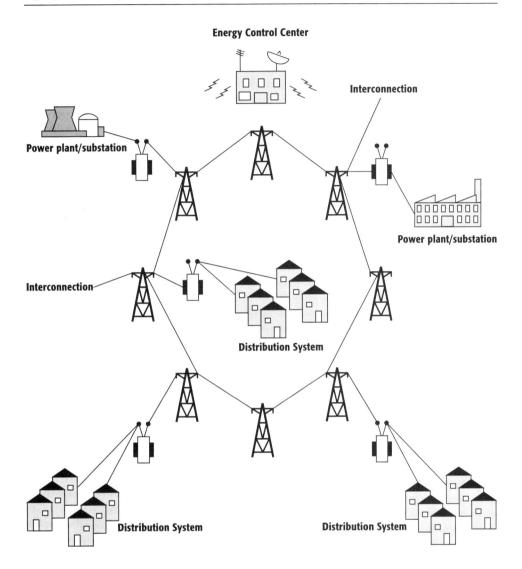

In modern electric power systems, generating stations continuously feed electric energy into a web of transmission lines (loosely referred to as "the grid") at very high voltages. Near the ultimate consumers, power is transformed to a lower voltage and routed to homes and businesses via a low-voltage, localized set of transmission lines called the distribution system.

The capital plant required to create electric power is often called the generation stage or segment of the industry. The power lines and associated hardware used for transmission comprise the transmission segment, and the distribution system and much of the marketing, accounting, and other managerial functions make up the distribution segment.

The controversy over the extent to which these segments can be considered distinct economic activities arises out of the fact that all three segments must be operated in extremely close coordination to avoid power failures. Planning and construction of new facilities in one stage must also account for planning and construction in other stages.

Box 1-1. Questioning Regulation's Effectiveness

As early as the 1930s, some economists voiced doubts about the natural monopoly justification for regulation. Economists Burton Behling and Horace Gray argued that utility monopolies were not natural, but rather the product of "skillful and unscrupulous manipulators."[1] Later and with wider currency, University of Chicago Nobel laureate George Stigler proposed that regulation was established solely to protect the profits of regulated firms, not to protect against monopolistic behavior.[2] Stigler's student and later colleague, Professor Sam Peltzman, countered with a more widely-accepted theory that regulation provided some consumer protection as it simultaneously kept the industry comfortably shielded from competition.[3]

This hardly settled the issue. Economist Walter Primeaux noted that "the economic literature concerning natural monopoly presents no standard definition of the theory, and various economists specify different conditions for such a monopoly to exist." Table 1.1 is a list of justifications compiled by Primeaux for electric utility regulation.

Table 1.1. Primeaux's Classification of the Reasons Why "Natural Monopolies" Have Become Regulated

Attributes of natural monopoly dependent on economies of scale for their existence or implementation:	Attributes of natural monopoly not dependent on economies of scale for their existence or implementation:
Economies of scale in production characterize a natural monopoly. The economies result in persistently decreasing long-run average total costs, which means that a firm's costs continue to fall as output is increased.	The item supplied by a natural monopoly is a necessity. Duplication of facilities in a natural monopoly causes inconvenience to customers.
Firms in a natural monopoly are characterized by having relatively high fixed costs.	Natural monopolies supply articles or conveniences that are used at the place where produced and in connection with the plant or machinery supplying the output.
A single producer in an industry characterized as a natural monopoly operates at lower costs than if two or more firms serve the market.	Natural monopolies may result from a special limitation of raw materials.
The nature of the business in a natural monopoly is such that a large number of competing plants is impossible.	Natural monopolies may arise from secrecy.
Higher customer prices will result if more than one firm serves a market in an industry characterized as a natural monopoly.	
Price differences are a customer attraction in a natural monopoly.	

SOURCE: Primeaux, 1986.

Stigler and Friedland (1962), Moore (1970) and Jarrell (1978), among others, produced research that was intended to demonstrate that regulation has not been effective at policing monopoly firms.[4] The emphasis in these studies has been on the behavior of regulatory commissions and the difficulty of policing the regulated firms, not on cost conditions. These studies concluded that the justification for regulation was never particularly strong and that regulation itself does not accomplish its intended purpose.[5]

One of the best-known findings from this era is the assertion that regulated utilities have the incentive to "gold-plate" their facilities—i.e., to buy more capital equipment, or to pay more for the capital they do buy, than a competitive firm. This theory is often referred to as the "Averch Johnson (AJ) effect," after the economists who discovered it.[6] Although most observers think that there is some truth to the idea, the size and importance of the effect is hotly disputed in the industry and among scholars.[7]

In any case, later chapters examine how difficult it is to design near-perfect regulatory schemes. If the original designers of regulation were going to err, it may have been wise to place the tendency to err on the side of overbuilding in a nation hungry for larger, cheaper, and vastly more widespread power supplies.

By the 1970s, the influence of those who doubted the purpose of regulation and those who doubted that it worked as intended combined to focus substantial attention on utility deregulation. One of the most popular alternatives was the suggestion by UCLA economics professor Harold Demsetz that a periodic auction of the exclusive right to sell electricity in a certain area—an idea called "franchise bidding"—was superior to rate regulation.[8] During this period, many other important critiques of conventional utility regulation were also published.[9]

---

1 See Behling (1938) and Gray (1940).

2 See George Stigler (1971, 1972)

3 For example, see the compilations produced by Shaker, Steffy and McCracken (1976), Poole (1985), Moorhouse (1986), and the article by Gordon (1990).

4 Kahn (1988), vol. II. pp. 108–109, discusses disagreements with the conclusions reached in these works. Work of this nature began at least as early as Mother and Crawford (1933). More recently, see Ryan (1981), Cornell and Webbink (1985), Navarro (1986), and Pacer (1986). For surveys, see Joskow and Noll (1981) and Joskow and Rose (1991).

5 For a summary as of 1981, see Joskow and Noll in Fromm (1981). More recent broad surveys are Noll (1989) and Joskow and Rose (1989).

6 Economist Stanislaw Wellisz is credited with discovering the effect independently at the same time as Averch and Johnson, so the effect is more properly known as the Averch-Johnson-Wellisz effect ("AJW"). See Averch and Johnson (1962) and Wellisz (1963).

7 See, for example, Bailey (1973), Shepherd and Gies (1974), Dayan (1975), and the discussion on pp. 49 59 in vol. II of Kahn (1971). For an extensive discussion, see Bonbright, Danielsen, and Kamerschen (1989).

8 As Spulber (1989) points out, the idea of franchise bidding dates back to the English economist E. Chadwick in 1859. Also see Thomas Hazlett, "Private Contracting versus Public Regulation as a Solution to the Natural Monopoly Problem," in Poole (1985). For a mathematical analysis, see Spulber (1989), p. 252–264. For critiques of the idea, see Williamson (1976) and Schultz (1979).

9 Some of the major works of this era include Crew and Rowley (1987), Bailey and Baumol (1984), Posner (1974), Jordan (1972), and Hilton (1972). See also Kahn (1990).

Box 1-2. Subadditive Costs: Discovery and Example

Subadditivity was "discovered" by a group of economists associated with what was then Bell Laboratories, Princeton University, and several other universities and labs. Wharton Professor Elizabeth Bailey described the discovery in a publication:[1]

> One of the fundamental insights on which this book is based is the finding that it is subadditivity of costs, and not scale economies, that determines when society can be served more economically by a monopoly firm.
>
> The insight was attained by a roundabout process. The first contributor was William Baumol, who in July 1970 devised a formal "burden test" for the prices charged by a multiproduct firm.... Baumol's aim was to determine whether or not, at a given set of prices, some one of a firm's products received cross subsidies from the consumers of the firm's other products.
>
> Two years later—in the summer of 1972—while examining this test, it occurred to Edward Zajac and Gerald Faulhaber, both of Bell Laboratories, that Baumol's insight on the nature of cross subsidy might be deceptively simple. Faulhaber, in particular, tried to persuade himself that if two products of a firm passed the burden test individually, they would pass it jointly. He began using game-theoretic tools to explore this hypothesis.
>
> Faulhaber posited a situation in which three towns desire water supplies in given quantities. The cost of supplying any one of the towns by itself is $300, of supplying any two of them via one facility is $400, and of all three towns jointly is $660. He pointed out that, from the point of view of society, it was cheaper for only one company to supply all three towns because $660 is less than the cost of supply by three separate plants (3 x $300 = $900), or of supply of any two towns by one plant with the other town supplied by a second plant ($400 + $300 = $700). However, there are then no fixed prices at which a single firm can keep the market entirely for itself. For example, at a price per town of $660/3 = $220, it is cheaper for two of the towns to split off and supply themselves (at a cost of $400/2 = $200 apiece) than for them to join the third town as customers of the single supplier. Thus, Faulhaber proved that there can be a situation in which there is a natural monopoly in the sense of subadditivity of production costs, but in which monopoly supply may be difficult to preserve in the free marketplace.

---

1   This passage is from the foreword of Baumol, Panzar, and Willig (1982).

Figure 1-2. Traditional Economies of Scale vs. Subadditive Costs

Economics typically picture a natural monopoly market by showing that the average cost per unit sold is declining over the range of total market demand. Notice that the total cost curve slopes up, but not too steeply.

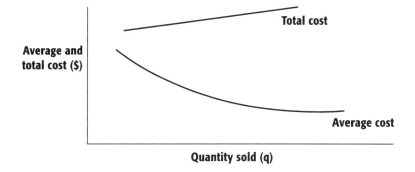

The following cost curves don't look like those of a traditional natural monopoly. However, these cost functions are all subadditive and therefore "natural monopolies."

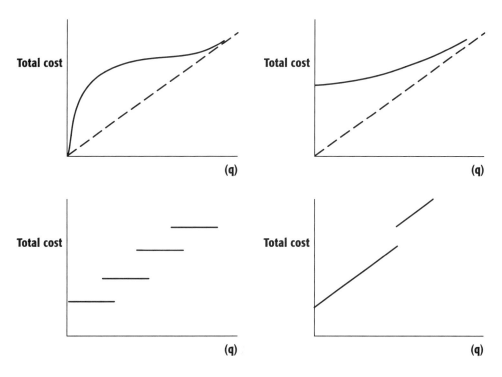

Source: Based on Sharkey (1982)

These questions produced a landmark discovery that resolved some of Primeaux's complaints concerning the vague conditions under which natural monopolies were asserted to exist. Researchers found that a new cost condition called *subadditive* costs—rather than declining average costs or "economies of scale"—was the proper condition establishing a "natural monopoly."[5]

The name may seem forbidding, but cost subadditivity is really quite simple. An industry has subadditive costs if it is cheaper for one firm to supply all of one product to one marketplace. Costs need not be declining as output goes up for any firms in the market, which was the old requirement for natural monopoly. Box 1-2 describes this concept in greater detail, along with a numerical example.[6]

It is suspected that power generation no longer has economies of scale, but what about subadditivity? Is a firm that owns two generators less costly than two firms that own one each? Interestingly, the few researchers who have studied this question have concluded that generation may still be subadditive today, even though scale economies have flattened out.[7] Other evidence concerning the growth of average plant and unit sizes is also inconclusive. As a result, there is a lack of formal economic proof of the fact that the era of natural monopoly is over in the generation portion of the industry. Nonetheless, there is widespread agreement that, in the generation stage of the industry, competition is preferable to price regulation. Except perhaps among economists, it seems that this belief has as much to do with the considerations reviewed in this chapter as it does with whether power generation costs are subadditive. Later chapters consider the risks posed by deregulating generation if this portion of the industry is indeed subadditive.

ANALOGIES BETWEEN OTHER UTILITY INDUSTRIES AND ELECTRIC POWER

Many formerly regulated industries, including trucking, airlines, natural gas, cable TV, and telephones, have been freed of some price or entry regulation. According to much of today's talk about electricity, it is being deregulated largely because most everything else in the United States already has been deregulated. "This is the last great monopoly," one member of Congress thundered recently as he argued for the rapid introduction of deregulation legislation. Thomas Bliley, Chairman of the U.S. House of Representatives Commerce Committee, recently called the nation's experience with gas deregulation "Exhibit A" for the need for similar change in the electric power industry.[8]

Meanwhile, it is a fact of life that the large overlap between lawyers, economists, policy analysts, and legislators involved in natural gas, telecommunications, electric power, and other industries often prompts interindustry comparisons.[9] These comparisons are often easily grasped, but they are not always useful or relevant. To explore this further, interindustry comparisons are made in Chapter 2, after electric power technology is examined in more detail.

Box 1-3. A Capsule History of the Electric Utility Industry

The electric power industry was born in 1882, when Thomas Edison demonstrated that a series of electric lights could be powered from a power generator in another nearby building. Edison established a number of lighting companies that sold light—not electricity—in many large eastern cities. These firms grew into the electric utilities of the 1930s.

By the early 1900s, many small electric light companies served urban centers. Chicago, for example, had more than 47 companies offering service to the same parts of the city. As the use of electricity grew, it became apparent that the lowest-cost means of supplying a single urban area was through a single power network.

Some cities established government-owned systems—a tradition that continues today—as others opted to give exclusive service rights (franchises) to single private companies. This "franchised monopoly" eventually submitted to regulation of prices and the conditions of service at the hands of state public utility commissions that were established in almost every state between 1907 and 1922.

World War I greatly increased the use of electricity for national defense purposes, and the federal government began to assert an interest in the power industry. In 1920, the government passed the Federal Water Power Act, which gave the federal government exclusive rights to develop hydroelectric facilities.

Utility plants became larger and larger, reaching sizes of 90 megawatts (MW) by 1928. New technology also steadily increased the ability to ship larger and larger amounts of power over longer distances, prompting utility systems to expand and interconnect more with neighboring systems. Many of the important technical breakthroughs in these areas occurred in the late 1920s through the 1940s, including the development of high-voltage transmission line, computers capable of analyzing power systems, and remote switching and control apparatus. M.I.T. Professor Vannevar Bush, soon to become famous as the nation's apostle of science as the "endless frontier" of human progress, greatly advanced the science of analog computation when, in 1931, he created the first successful computer simulation of a large power system.

As the industry grew, several firms bought up scores of smaller firms, creating holding company empires that extended across many states. By 1929, seven utility holding companies controlled 60% of the power generated in the United States. The extreme financial leverage used by these holding companies caused rapid and severe losses in the crash of 1929 and provoked great public outcry. In 1935, Congress responded by enacting the Public Utility Holding Company Act and the Federal Power Act. The former required utilities to divest themselves of elaborate holdings and prevented utilities from investing in non-utility businesses; the latter caused sales of electricity across state lines to become regulated by the Federal Power Commission.[1] This established two tiers of rate regulation for large utilities: Operations within states were regulated by state commissions, whereas sales between utility systems in neighboring states were controlled by the FPC.

continued on next page

> **Box 1-3. A Capsule History of the Electric Utility Industry** *(continued)*
>
> Private electric utilities chose not to build power lines in many remote areas of the United States, believing that the cost of connection outweighed the gains from increased sales.[2] Out of a concern that rural economic development and the quality of life would be impeded by an absence of electricity, the federal government established power authorities (e.g., the Tennessee Valley Authority (established in 1933) and the Bonneville Power Administration), and passed laws that enabled the establishment of rural electric cooperatives.
>
> Between 1940 and 1970, this combination of public and investor-owned utilities expanded its total sales more than tenfold, from 118 to 1400 billion kWh (average consumption per customer grew by almost this multiple), eventually reaching almost every American home and business. As plant size and transmission capability increased, inflation-adjusted prices fell.
>
> After the Second World War, the Atomic Energy Commission began to investigate the use of nuclear energy in power plants. In 1957, the federal government absolved private nuclear manufacturers of most of the potential costs of nuclear accidents, paving the way for commercial nuclear power. The first few plants were small demonstration units built with government subsidies, but by 1963 General Electric had sold a 620 MW plant to Jersey Central Power and Light with no direct federal assistance.
>
> In 1965, the Northeast Coast was hit with the nation's largest, most sustained blackout. Nevertheless, the 1960s were the pinnacle of the utility industry's growth, financial performance, and public esteem.
>
> ---
>
> 1   See Chapter 4 for additional discussion.
>
> 2   This is a topic of some dispute. Compare Vennard (1968) and Tollefson (1968), for example

THE 1970s WATERSHED

For decades following the onset of regulation, electric utilities experienced large, steady sales growth and declining prices. By 1920, electric utility capacity had grown from 1 million kilowatts (kW) to over 20 million kW—about enough to meet the power demands of 20 million households and businesses. Between 1925 and 1970, the industry quadrupled the number of customers, but increased plant capacity thirteenfold and sales by a factor of 25. In short, the U.S. economy underwent a sustained period of electrification that enabled the electric power industry to sell ever more power per customer and make far more intensive use of its plants in 1970 than it was able to do in 1925.[10] Box 1-3 is a capsule history of some of the more important industry milestones during this period.

The industry's economic performance was equally astonishing during this period of electrification. Between 1906 and 1970, the average price of power to residences declined from ten cents per kilowatt-hour to about 2.6 cents—*even before adjustment for inflation*. Of course, declining costs and prices as output increases are just what the theory of natural

Figure 1-3. Major Events in the History of the Electric Utility Industry

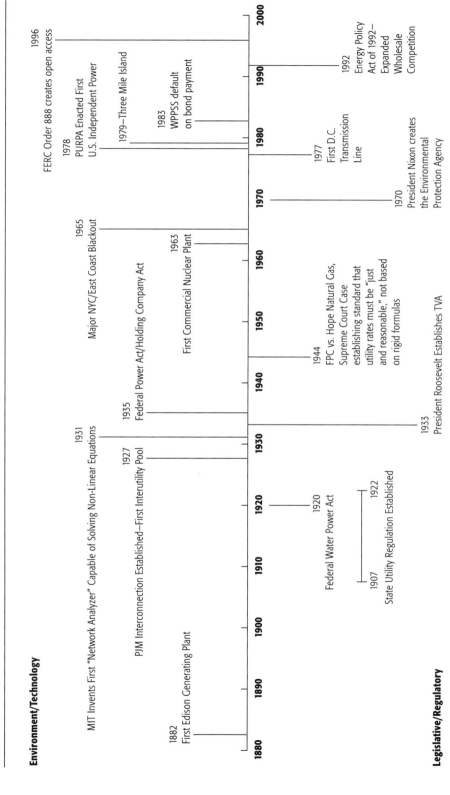

13

monopoly predicts, and for over 50 years the industry did not disappoint.[11] Remarkably, few utilities had ever needed to ask their regulators for significant rate increases—some, never had to ask at all.

In the 1970s, these patterns abruptly changed.[12] Due to the cumulative effects of inflation, oil price shocks, and other factors, interest rates rose to their highest levels ever during the 1970s, reaching a peak in 1981. The price of fuels used in power plants also sky-rocketed. Between 1971 and 1983, the price that utilities paid for oil grew by a factor of 10, natural gas prices grew 700%, and the price of coal quadrupled. Productivity increases in the industry were not nearly large enough to offset these cost increases. Some ana-lysts believe that the onset of environmental regulation of utilities contributed to reduced productivity growth.[13]

These cost increases were first exacerbated and later overwhelmed by increases in the costs of building new power plants, especially nuclear plants. To get a sense of the enor-mity of these construction cost increases, consider that the average cost per kilowatt of capacity for new nuclear power plants finished between 1968 and 1971 was $161/kilo-watt. For plants finished between 1979 and 1984, average costs rose almost tenfold, to $1,373/kW. And the 20 most expensive nuclear plants in the United States ranged in cost from $1,607/kW to $5,810/kW.[14]

With some exceptions, cost increases prompted solely by increased fuel prices, interest rates, and even environmental compliance were reviewed by regulators and then approved. Utilities had little control over the cost of their fuel or capital. Indeed, regulators allowed many utilities to make automatic adjustments to their rates based on their costs of fuel, so that utilities would not be stuck paying for very high fuel prices while they went through the process of receiving a rate increase, which takes months or even years.[15]

Rate increases prompted by escalating construction costs between 1975 and 1990 were not accepted easily by regulators or the public. Unlike fuel price spikes, which utilities did not control (and, in any case, seemed to be short-lived), the cost of a power plant is at least somewhat within the control of the utility that is building it.[16] Like a mortgage on a house, increases in the original cost lead to much higher ongoing payments in order to pay off the principal and interest associated with a major purchase. A 1986 study estimated that 18 utilities that were still building nuclear plants were likely to have first-year rate increases of 25% or more—some as large as 80%.[17] A divisive national debate raged over the causes of the huge increases in the cost of nuclear power and orders for new nuclear plants ceased abruptly in 1978.[18]

Public outcry over large, rapid rate increases (known as "rate shocks") ultimately caused regulators to "disallow" recovery of approximately $10 billion of the costs of nuclear plants.[19] Investor-owned utilities were forced to pay for these "writeoffs" out of their investors' pockets, prompting the first utility bankruptcies since the first part of the cen-tury. In the state of Washington, cost increases in a series of nuclear plants caused the Washington Public Power Supply System to default on payments on its bonds—the largest default on tax-exempt bonds in U.S. history. These financial problems caused utility stocks to fall in value to levels below the book value of their assets, further weak-ening their ability to raise capital for plant completion and other projects.

These developments occurred at the same time that utilities were first facing generation competition from independent power producers. The introduction of competition at the federal level occurred in the Public Utilities Regulatory Policy Act of 1978 (PURPA). This statute was enacted both to increase energy efficiency by stimulating highly-efficient power production and to increase competition into generation. Speaking to support the legislation in 1977, Senator Charles Percy of Illinois said:[20]

> In the last five to ten years the economies of scale in the utility industry have ended. To expand production of electricity costs up to twice as much as the average cost of power...we may be at the beginning of a new era in electric power production.

PURPA responded to this "new era" by creating the legal means of owning and operating a power plant exempt from price regulation or a "Qualifying Facility" or QF. Rules later set by the Federal Energy Regulatory Commission required QFs either to cogenerate heat and power in one plant, making it more efficient than a typical power plant, or to use renewable energy sources to generate the power.[21] PURPA allowed QFs to sell power at wholesale to utilities that would then blend the power with their own regulated generation and sell the blend to retail customers.

Worried that regulated utilities would not want to buy from competitors, PURPA *required* electric utilities to buy from any QFs in their territories. The price the utility paid was originally required to be the utility's own "avoided cost," i.e., the cost the utility would have had to incur if it built another power plant or purchased power from someone else.

These purchase requirements were introduced just when utility plants were encountering some of their largest cost overruns, and some energy policy experts were pointing out that energy conservation was often cheaper than new power plant construction.

One of the changes made to regulation—a change history will view as transitional—was to increase upfront review of new utility plants. Prior to the 1970s, public utility commissions paid relatively little attention to the particular type of plant a utility planned to install next. In many states, utilities became subject to elaborate "integrated resource planning" rules that removed most of the utility's independence in choosing what type of investment to make. The new planning regime—still in existence in some states today, but gradually giving way to unplanned, competitive generation—led to contentious and cumbersome regulatory approvals, and was widely disliked by utilities and many others.

Together, these developments promoted the first sense of a broadscale shift in the utility industry. For the first time, industry members and observers alike began to agree that the structure and regulation of the industry was not working right. Some observers chose to characterize these developments as a breakdown of the "regulatory compact," i.e., the broad agreement that utilities will be allowed to earn a fair return on their investment if they meet their obligations as a regulated entity.[22] Unfortunately, neither the simple natural monopoly model nor the legislation and case law establishing regulation provide much guidance as to what to do when utility costs and rates are increasing, rather than decreasing as one would expect in a natural monopoly.

TODAY'S PICTURE

The combined effect of these pressures has already done much to cure the ills of the late 1970s and early 1980s. Prudent disallowances have largely become a thing of the past and utility rates are again low in real terms. However, the perceived success of independent power generation, coupled with widespread dissatisfaction with utility regulatory processes, escalated the pressure to increase competition in the industry. The Energy Policy Act of 1992 (EPAct) expanded wholesale competition,[23] laying the groundwork for the Federal Energy Regulatory Commission's "MegaNOPR" rulemaking,[24] which made the entire FERC-jurisdictional transmission "open-access." State utility regulators have also enacted significant reforms—most prominently, California's ongoing restructuring proceeding and more recent proposals in almost every state.[25]

Other nations have experimented with partial deregulation of their power sectors. In Norway, the United Kingdom, New Zealand, Argentina, and some other nations, generation and transmission plants that were previously owned by government agencies have been auctioned off or "privatized." In some cases, this made the industry resemble its U.S. counterpart, where utilities are all private, but regulated, companies. Some nations have leapfrogged the U.S. model, moving all the way from state ownership to a competitive generation market in one enormous transition. Many observers view U.S. restructuring as playing catch-up to the more ambitious electricity deregulation schemes around the world.[26]

THE ENVIRONMENTAL COMMUNITY AND UTILITY DEREGULATION

During the 1970s, the environmental community reacted to changes in the utility industry in several ways. Represented most visibly in the work of Amory Lovins, this group argued that the electric utility industry was the most important example of an energy infrastructure that had become too centralized and too reliant on large-scale technologies. Lovins and his contemporaries took note of the trend toward ever-larger generating units (which had not yet reversed itself) and predictions of larger plants to come. These authors argued that the United States could obtain energy in a less expensive and less harmful manner if it pursued a "soft path,"or an energy system with less electrification, less centralization, and smaller-scale energy plants.[27]

Lovins' work did not use conventional economic arguments, but rather a compelling mixture of thermodynamics and polemics. Rather than examining power industry costs, as in most previous research, Lovins examined the uses to which electricity was put in the United States (cooking, heating, and so on). Lovins found that electricity was being used to do work that could be done more cheaply by solar power and other forms of energy, at least under certain conditions.[28]

These views dovetailed with the growing worldwide opposition to nuclear power plants, which are among the largest and most capital-intensive utility plants ever built. Although the specific economic and environmental problems with nuclear power were technology-specific, the soft energy movement associated the sheer size of these units with their adverse environmental and political aspects. With regard to power plants, the motto of the movement echoed the title of a popular broadside: *Small is Beautiful*.[29]

The confluence of the soft energy movement and the economic critics of utility regulation made for a fascinating evolution of soft path concerns into traditional regulatory policy. Without much apparent analysis, the environmental movement drew the conclusion that decentralization and the "soft path" were consonant with reduced utility regulation and greater competition in utility markets. Lovins' own evolution typifies the transition. In *Soft Energy Paths*, the seminal work, there is little or no mention of utility industry competition or deregulation. Five years later, he mentioned competition for energy conservation and new cogeneration plants, although not in particularly forceful terms. By 1993, Lovins is the purveyor of a marketing intelligence service on energy efficiency products called *Competitek*, maintaining that "economic glasnost and competition are crucial."[30]

Recent research on the environmental impacts of greater utility competition has convinced the environmental community that the story is not so simple. Nevertheless, a substantial portion of the community still believes that the long-run impacts are positive.[31]

POST-INDUSTRIAL CAPITALISM AND THE MANAGERIAL REVOLUTION

In dozens of competitive industries, new information technology, management practices, and the globalization of trade are causing momentous changes in the way companies think of themselves and manage their operations. As markets become global, firms are forced to grow or form global alliances that are fashioned to balance regional preferences with worldwide economies of scope and scale. Operating manufacturing or service operations on a global scale poses tremendous challenges and requires unprecedented flexibility and agility in logistical and operational management as well as great financial strength.

The decline in information costs and advances in control technologies have also allowed producers to move away from the mass production approach of the industrial revolution toward customized or "virtual" production. "The closer a corporation gets to cost-effective instantaneous production of mass-customized goods and services, the more competitive and successful it will be," write management consultants William Davidow and Michael S. Malone.[32] (It is a bit humorous to talk to the electric power industry about instantaneous production, because electricity is a product that has always been produced instantaneously on demand for the customer). The techniques of flexible production, lean (or "just-in-time") production, and many new management approaches to maximizing quality and customer product choices all represent global trends toward de-homogenizing a firm's products and services—even as the area served by the firm increases.

These developments have not escaped the attention of executives and planners in the electric power sector. To many in the industry, deregulation will allow the power industry to fit in with these trends in unregulated industries. Fereidoon Sioshansi, a senior researcher working at a U.S. utility, writes:[33]

> Thanks to advances in the technologies of communication and automation, it is becoming cheaper and easier to collect, manipulate, transmit and store information. These advances had enabled service and manufacturing industries to lower their supply costs, to serve their customers better and to differentiate

their products. The same developments have also introduced new competitive pressures in historically noncompetitive industries such as banking, financial services, and transportation and telecommunications.

So far, the penetration of these new technologies into electricity supply has been insignificant. Electricity is being generated and delivered to customers in much the same manner as it was 100 years ago. For years there has been talk of supply and demand integration, distribution automation, real-time pricing, sophisticated load management devices, priority of service schemes and the like. But for the majority of electric utilities, these concepts have remained only ideas. Why has the electric power industry remained relatively unaffected by the information revolution that has affected different industries at different times and in different ways?

The fact is that these developments *are* being absorbed by utility managers and policymakers—perhaps unevenly, but at a rapid pace. Utility consultant Vinod Dar likens the change in the utility industry to the transition from mainframe computing dominated by IBM to decentralized computing provided by Intel, Compaq, Apple, Microsoft, and Novell—"stranding" prior investments in mainframes.[34] One small example of the changes enabled by better and cheaper information technologies can be found in the U.S. Federal Energy Regulatory Commission's (FERC's) reduced regulation of natural gas pipelines. In order to facilitate the effectiveness of unregulated sales of gas pipeline capacity, the FERC requires pipelines to post the availability of unused pipeline capacity on electronic bulletin boards (EBBs). These bulletin boards enable all market participants to instantaneously have identical information about pipeline capacity–a possibility that simply did not exist at a reasonable cost even ten years ago.

As information technology becomes cheaper and more powerful, it is certain that ever-improving information systems will enable new modes of competition and coordination in the power industry. Even without it, however, other industries will continue to present utilities with new models of product customization, decentralized management, and flexible production.

CONCLUSION

The forces reviewed in this chapter—the deregulation trend in other industries, worldwide experience with partial electric deregulation, industry financial problems, a loss of faith in regulation, the environmental movement, and the post-industrial managerial revolution—all set the stage for a more intensive examination of electric power restructuring. Although these forces help us understand *why* restructuring is occurring, they do little to explain the most appropriate new structure for the industry.

The following chapter explores the power industry's unique technology and its implications for industry structure and reorganization—recognizing that technology is changing more quickly, and less predictably, than ever before. A firm understanding of the way power systems work will help to make it easier to explore several new models of a less-regulated industry.

Chapter 1

NOTES

1 Bonbright (1961) contains a classic formulation.

2 Platt (1991), p. 125.

3 The first volume of the proceedings of the American Economic Association, now known as the *American Economics Review*, contained three articles on the subject of natural monopoly regulation. Citing E. Benjamin Andrew, "The Economic Law of Monopoly," *Journal of Social Science* 26(1890):1–12 and Arthur T. Hadley, "Private Monopolies and Private Rights," *Quarterly Journal of Economics* 1(1887):28–44, historian Harold L. Platt asserts that, by 1890, the natural monopoly hypothesis had "gained wide currency." (Platt, 1991, p. 310). For additional historical research, see Thomas Hazlett, "The Curious Evolution of Natural Monopoly Theory," in Poole (1985).

Political factors far beyond the industry's cost characteristics also contributed to the establishment of regulation. Between 1890 and 1910, small electric companies proliferated in American cities and obtained numerous franchises (permission to lay wires in or above streets). In some cities, the awarding of franchises was rife with corruption and political influence. A combination of civic reformers, advocates of municipal ownership of utilities, and state officials eager to check a source of municipal political power ultimately settled on regulation by state regulatory body as a means of curbing corruption and stabilizing the industry. See Platt (1991) and Hirsh (1995) for more information. For a discussion of arguments that regulation simply existed to protect the industry, see the next subsection in this chapter.

4 The economic evidence reviewed in Chapter 4 suggests that economies of scale remain a relevent consideration. It is true that technological change has flattened economies of scale in individual power plants, and the industry trend is no longer to build ever-larger generators. However, it is still common to put many medium-size units at one site, and there may also be economies of scale in generation facility ownership.

5 A rich new body of work arose from this insight and eventually went on to extend the new theory to include multiproduct natural monopolies, sustainable prices, contestable markets, and other results. See Sharkey (1982) and Baumol, Panzar and Willig (1982).

6 Mathematically, the definition of subaddivitity is as follows: If $C(x)$ means "the cost of one firm providing all of customer x's demand," then an industry is subadditive if $C(a+b) < C(a) + C(b)$.

7 For example, Gegax and Nowotny (1993) argue that there is no proof that the generation portion of the power industry fails subadditivity.

8 Mary Driscoll, "Customers Choice Tops Bliley's 1997 Agenda." *Energy Daily*, September 5, 1996, p.1.

9 The list of works relying on analogies is almost endless. To cite a few select examples, Pierce (1991) argues that the deregulation of natural gas provides the ideal model for deregulating electric power; Bobbish (1992) argues essentially the opposite. Santa and Sikora (1994) provide a tremendously detailed comparison of the similarities and differences in the regulatory laws governing electric and gas utilities. Carl Danner (1993) argues that "where telecommunications regulation has been may be where natural gas should go. Also see Backus and Baylis (1996), Weiner (1996) and Hartman and Tabors (1995).

10 Edison Electric Institute (1974).

11 Edison Electric Institute (1974), p.165.

12 Joskow (1989) pp. 149–163 and Hyman (1992) pp. 151–154 review the economic evidence. Heimann (1991) provides a useful, very brief summary. Munson (1984) provides an environmentalist perspective. The U.S. General Accounting Office (1984) provides a useful contrast to these somewhat downcast evaluations.

13 Environmental regulation of electric utilities was not new, but escalated following the establishment of the Environmental Protection Agency in 1970. See R.J. Gordon (1982), Gollop and Roberts (1983) and Joskow and Rose (1989) for discussions of the impacts of environmental regulation on utility productivity.

14 Joskow (1989) p. 151 and Charles River Associates, et al (1986). An earlier, very influential work in this area is Komanoff (1981).

15 These mechanisms are called *fuel adjustment clauses* and are a part of a utility's official published price or tariff, as set by local regulators. See Kelly, Simmons, and Prior (1979) for a complete discussion of fuel cost issues.

16 The causes of cost overruns at nuclear and large coal plants during this period were the subject of great controversy. The greatest acrimony arose between those who argued that nuclear construction in the United States was hamstrung by inefficient and overzealous regulation and those who believed that nuclear power generation was a fundamentally unsound economic idea exacerbated by cost-of-service regulation. For a more complete discussion, see, among many excellent works on the subject, the Office of Technology Assessment (1983), Jones (1989), Morgan (1993), and Rhodes (1993). For specific work on the economics of plant construction, see Komanoff (1981), the Office of Technology Assessment (1983), and Crowley (1981).

17 Charles River Associates (1980) Table 2-3.

18 An accident that disabled the Three Mile Island nuclear plant in Pennsylvania further discouraged nuclear power investment. For an excellent overview, see the Office of Technology Assessment (1984).

19 Joskow (1989) p. 161.

20 See Fox-Penner (1990) p. 520.

21 Codified pursuant to PURPA in 18 CFR 292, Sections 201 and 210.

22 See Swidler (1991), p. 14.

23 Federal Energy Regulatory Commission (1989, 1993); Tenenbaum and Henderson (1991).

24 Promoting Wholesale Competition Through Open Access Non-Discriminatory Transmission Services by Public Utilities, Docket No. RM95-8-000. See Chapter 6.

25 Order Instituting Rulemaking on the Commission's Proposed Policies Governing Restructuring California's Electric Services Industry and Reforming Regulation, R.94-04-031, 1994, and Rhode Island Utility Restructuring Act of 1996 (H-8124). See Chapters 7 and 10–16.

26 Recent international reviews include Gilbert and Kahn (1996) and Cicchetti and Sepetys (1996). Appendix 7A briefly reviews electrricity regulation in the United Kingdom.

27 In addition to *Small is Beautiful* (Schumacher, 1973), Lovins (1976, 1977), Wilson Clark (1975) and Sant (1980) are the seminal works.

28 Lovins' assertions that soft energy forms, including energy efficiency, could provide energy services more cheaply than conventional energy were the subject of bitter disputes. As an example, Asbury and Webb (1980) wrote, "The number of situations in which they [soft path options] actually do reduce costs is unknown and depends upon application-specific performance requirements and prices." As discussed in a later chapter, the growth of utility energy conservation programs has greatly altered the form (and reduced the ferocity) of the dispute. Other elements of the debate, such as the extent to which the U.S. government has subsidized various energy technologies, remain unresolved.

29 See Note 27, *supra.*

30 Lovins (1977); "Saving Gigabucks With Negawatts," Rocky Mountain Institute, Old Snowmass, CO, 26 November 1984; Oddvar Lind, "If I Were the World's Energy Minister," (interview with Lovins), *Our Common Future*, Norwegian Ministry of the Environment, Oslo, Norway, May 8, 1990. Other authors follow a similar course. In "The Greening of American Energy Policy," Weiss and Salzman (1989) note that "The utility industry is becoming increasingly competitive as growing numbers of small generators and energy efficiency suppliers offer their services in the marketplace." (p. 714) A 1982 report by the Edison Electric Institute provides one of the few relatively complete descriptions of the connection between the soft energy movement and deregulation. Edison Electric Institute (1982) p. VI-5 ff.

31 See Chapter 13 for a more complete examination of the environmental and energy policy aspects of restructuring.

32 Davidow and Malone (1992).

33 Sioshansi (1990) p. 17.

34 Vinod Dar, *Public Utilities Fortnightly*, April 1, 1995

Chapter 2

How Power Systems Work

To understand electric restructuring, it is essential to know a little about how power systems operate. In particular, it is useful to understand the relationship between generation, transmission, and distribution; what creates reliable electricity and how the system provides it; and how the economics of power systems makes each utility dependent on its neighbors.

Electricity is generated by transforming stored energy fuel, hydropower, or sunlight into electric power. The transformation process is, somewhat misleadingly called generation and it occurs in power generating plants or generators. The transmission system delivers power from generators to local distribution systems. All electric power networks that are larger than a few square blocks use alternating current (AC), which flows back and forth 60 times a second (60 cycles/second or *hertz*).[1] Most transmission systems use AC because it is easy to change voltage up or down by letting power flow through a *transformer* (a passive, self-operating device). Inexpensive and reliable transformation is useful because, at higher voltages, less power is lost in the wires while it is transmitted across the grid.

The cheapest way to supply power to many customers is to transform power up to very high voltages at the boundary of the power plant and then transform it back down at the other end of the grid, as close as possible to the actual user of the power.[2] Power plants generate voltages of 4,600 to 20,000 volts; transformers "step up" the power to transmission lines of 27,000 to 765,000 volts, and then reduce voltage in a series of transformers at the other end of the system. In the United States, most high-voltage transmission lines are owned by investor-owned utilities and state and federal power authorities.[3]

The transmission system feeds *substations* (essentially transformers) that reduce voltage and spread the power from each transmission line to many successively smaller *distribution* lines, culminating at the retail user. A large factory may choose to take power from the system after just one substation transformer reduces voltage from (say) 230,000 volts to 34,500 volts.[4] Next door to the factory, an apartment building may take power after another transformer reduces voltage to 13,800 volts and a nearby home takes power after it has been further transformed down to 115 or 230 volts. Figure 2-1 shows a prototypical path from generator to customer.[5]

From the standpoint of accountants and engineers, the distribution system is usually considered to begin where voltage is reduced to 37,000 volts. However, from the standpoint of economic function, distribution involves delivering power to the retail customer, while transmission involves getting it there—much like the traditional distinction

Figure 2-1. The Transmission and Distribution System

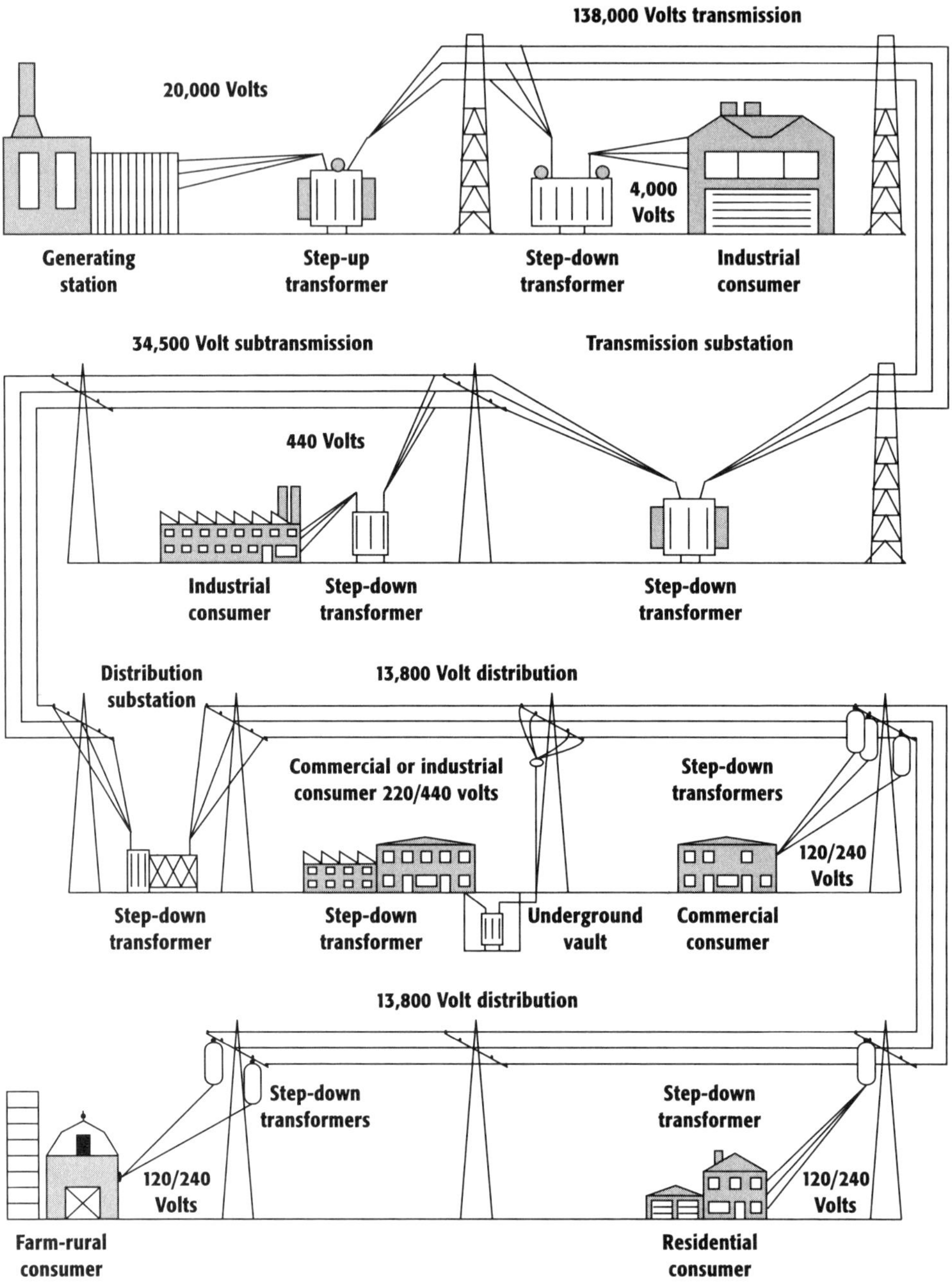

138,000 Volts transmission

20,000 Volts

4,000 Volts

| Generating station | Step-up transformer | Step-down transformer | Industrial consumer |

34,500 Volt subtransmission

Transmission substation

440 Volts

| Industrial consumer | Step-down transformer | Step-down transformer |

Distribution substation

13,800 Volt distribution

Commercial or industrial consumer 220/440 volts

Step-down transformers

120/240 Volts

| Step-down transformer | Step-down transformer | Underground vault | Commercial consumer |

13,800 Volt distribution

Step-down transformers

Step-down transformer

120/240 Volts

120/240 Volts

| Farm-rural consumer | | Residential consumer |

Typical electrical supply from generator to customer showing transformer applications and typical operating voltages.

Source: Pansini (1992)

between wholesale and retail trade. The distinction can be important because, as shown in Figure 2-1, some large industries take power right from the transmission lines, before it is delivered to the lower-voltage parts of the system. For example, the Energy Information Administration defines the distribution system as "the portion of an electric system that is dedicated to delivering electric energy to an end user."[6]

For accounting purposes, utility companies and regulators always divide utility activities, revenues, and assets into three main segments.[7] In 1992, the net book value of all U.S. generating plants represented 54% of the industry's total assets; transmission and distribution plant accounted for 11% and 25%, respectively.[8] These figures illustrate that generating plants are by far the most expensive part of the system. However, the closer the power system is examined, the more difficult it is to discuss any one of the three traditional segments in isolation.

INTERDEPENDENCE AND TIME HORIZONS

One of the defining features of the electric power industry is the high level of interdependence between its three traditional stages. The extent of this interdependence, while not without parallel, is unusual enough to make the industry unique in this respect, and to raise structure and policy questions that are extraordinarily complex. It is easiest to understand these questions by thinking in terms of the time horizon that industry managers use to make operation and management decisions.

When discussing industry characteristics, it is common for economists to distinguish between the short run (or term) and the long run (term). The traditional distinction is that, in the short run, all firms are pretty much stuck with their existing capital plants, whereas in the long run they can add or retire these "fixed assets."[9]

Conversely, it is in the long run that firms in competitive markets adjust their capital so as to produce profit-maximizing amounts of products. In the power industry, the long run is used to ensure that each firm has enough production capacity to serve all demand for power at published prices—i.e., that the market clears without shortages. Because power demand has been growing in most years since the birth of the industry, most of the industry's long-run planning procedures center on projecting additional capacity needs and planning to make them available in time for the demand.[10]

The long- and short-run concepts work well in the electric utility industry, except that within the short run there are two distinct and important time horizons: immediate (the next few minutes and hours) and the next few days and weeks.[11] These are different than the traditional definition of short term, which is usually a calendar quarter or a year. For reasons that will become apparent below, the two short-run time frames go by the names *operations control* (immediate) and *operations planning*. These short-run time frames interact with long-run planning and change. It is easier to understand the time frames if they are examined one at a time.[12]

In the immediate time frame, power system operations are driven by several technical considerations that are required to keep the power grid from collapsing.[13] The first consideration is the simple conservation of energy principle which dictates that the sum of electrical generation from all generators must equal the total amount of power used by all customers *at every moment in time*. If generation were to exceed total use, energy would build up somewhere in the system. Because electric power networks can store only a little bit of power within them, a supply-demand mismatch quickly overloads the weakest parts of the system. (Conversely, if generation is too low, lights will dim or go dark.)

The need for coordinated system operation stems from more than a simple energy balance. Because the system uses alternating current, every generating plant must be in precise synchronization in order to keep the network at the same frequency and maintain voltage. Imagine, for example, a radio station with two transmitters and antennae, each set to slightly different frequencies. Anyone with a radio within reach of both antennae would find it very difficult to tune the station in—one signal would interfere with the other.

The power system is something like this, but it is stretched across vast parts of the United States, and there is the additional need to keep an instant supply-demand balance. Perhaps a better analogy is to imagine a gigantic collection of large pitchforks scattered across a region, all manufactured to give the same note. The pitchforks must all vibrate in perfect synchronization and continuously adjust their volume so that the overall sound is uniform at every point. Energy supplied to each pitchfork must be changed continuously because listeners are always shifting, wind and weather affect the pitchforks, and pitchforks need servicing and occasionally break down. If one pitchfork loses frequency, it must be removed from service immediately and the system instantly readjusted, or else the whole pitchfork network might suddenly fall silent.

To complete the analogy, one final wrinkle needs to be added. Obviously, when a pitchfork's sound starts to drop off, energy is applied to it to get it going again at the proper frequency and volume. Analogously, power must reach every part of the electric power system at the same frequency and voltage as the rest of the system. However, over long distances, transmission lines *store* tiny amounts of power in them. These stored amounts, called *reactive power* or *reactive losses*, usually have the unfortunate effect of lowering the voltage at the other end of a transmission line. However, if the voltage does not stay within a safe tolerance band, electric appliances and motors won't work right, and the entire system can collapse. So, by remote control, the proper amount of reactive power must be supplied to keep the voltage from drooping in certain spots.

The challenge of keeping a power system in supply-demand balance, synchronized, and voltage-supported is made difficult by the fact that the electric transmission system generally does not allow power to be directed down a specific path from one generator to one consumer. The transmission system is more like a large water pool into which electricity flows from all generators. All users take from this pool, and the system is adjusted so that the total water flowing into the pool equals the total water being withdrawn by all users at every moment.[14]

As might be expected, it is not possible to direct the flow of water from any one producer into the pool and then to any one pipe. Like water flowing within a pool, electricity in the transmission system flows across the system of wires via the paths of least resistance. These paths change moment by moment. Trying to predict the flow of electrons is akin to putting a drop of ink into a water pipe flowing into a pool, and then trying to predict how the ink drop will diffuse into the pool, and which combination of outflow pipes will eventually contain ink. Most likely, the ink will end up widely diffused, and many, many outflow pipes will carry a tiny bit of the ink that came out of the originating pipe.

In the utility industry, it has been common to assume, for purposes of making legal transmission contracts, that power flows along a certain set of lines in the transmission network. The assumed lines are called the "contract path." Although the contract path is selected as the most likely main path for the power, major portions sometimes flow on other lines, and are referred to as *parallel, loop,* or *inadvertent* flow.[15]

If power flows across all lines in a network, and several utilities' plants and lines are all connected, how is it possible to make sure that one company's flow does not take up another company's line? The answer is that it is not possible to prevent this, except in a few unusual, especially severe circumstances. As a result, power from one utility *regularly* flows on the lines of others. (See Figure 2-2)

Figure 2-2. An Example of Loop Flow

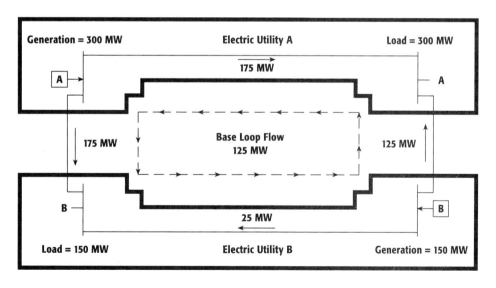

In this example, both electric utility A and B are generating power to supply their own loads. The actual transmission paths, however, are such that utility A supplies power to the load of utility B and vice versa. This is not necessarily an undesirable situation. Utilities in similar situations to A and B have designed their transmission system to operate in this manner in order to reduce the investment in transmission.

Source: Casazza (1996)

Figure 2-3. Hierarchy of Action Centers

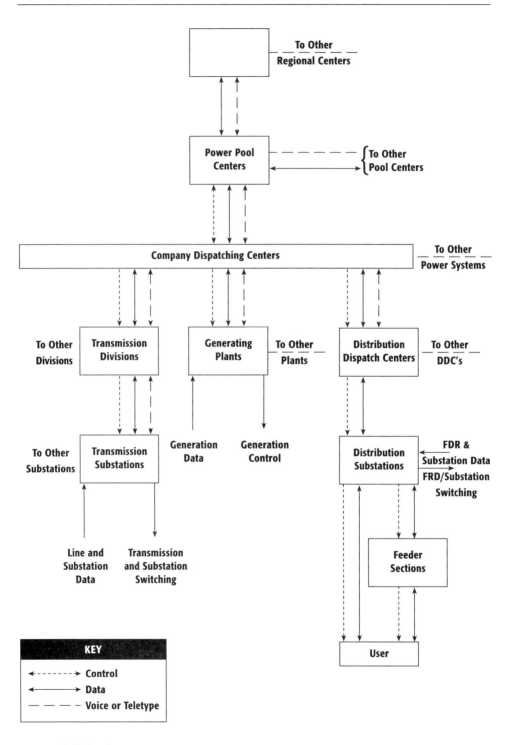

Source: Rustbakke (1983)

Figure 2-3a. Hierarchy of Immediate Utility Control Centers

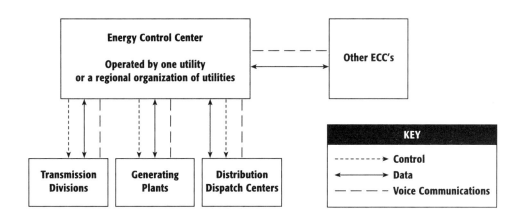

Interconnected utilities also face the problem of inadvertant exchanges of electric power. Inadvertent flows mean that one utility's operations can adversely affect the operation of all others with little predictability or warning. Many of the operating and planning procedures developed by the industry exist because transmission systems have this unique interdependence. As will be discussed later, the industry's ways of coping with these flows is changing as part of restructuring.

CONTROL AREAS AND BULK POWER RELIABILITY

To keep electric power systems in each area of the nation in momentary balance, the electric power network has been engineered into "control areas." There are about 133 control areas in the United States, ranging in size from 100 MW to 30,000 MW.[16] Each area has an *energy control center* (ECC) that is staffed continuously by system controllers who monitor the entire system in their region. Each ECC has instantaneous control over most of the generators, power lines, transformers, and other large parts of the system in its region; its job is to keep the area's system in balance and to respond to emergencies and other events that destabilize the grid.[17] Because there are about 3,500 utilities in the United States and only 133 control areas, it follows that many utilities do not have moment-to-moment control over their generators and lines. Figure 2-3b shows the location of the energy control centers (EECs) in North America. Each control Center monitors all major plants and lines in its area and keeps in touch with all adjacent ECCs. A rough schematic of the controls exercised by ECCs is shown in Figure 2-3a.

To counteract voltage drops caused by reactive losses, system operators remotely operate equipment (similar to small generators) at the far ends of the transmission systems, usually at substations. These generators "support" the voltage drops by measures such as counteracting reactive losses. Because reactive power is measured in units called "VARs," reactive power support is sometimes called *VAR support*.[18] Note that VAR support cannot be sent down the lines from faraway plants. By its nature, it must be supplied at the other end of the power system from the generators.

Figure 2-3b. North American Interconnected Control Areas

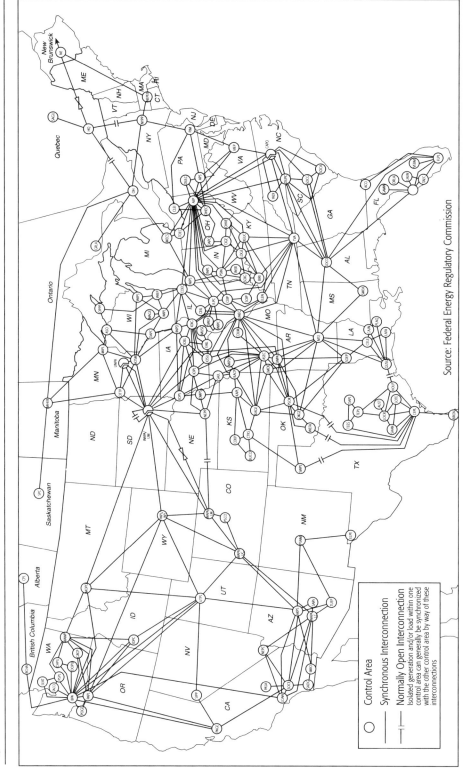

Source: Federal Energy Regulatory Commission

Figure 2-3b. North American Interconnected Control Areas *(continued)*

Controlling Organization and Center Location

1. AEP	American Electric Power Service Corp.	Canton, OH
2. APS	Allegheny Power Service Corp.	Charleroi, PA
3. ARPS	Arizona Public Service Co.	Phoenix, AZ
4. ASEC	Associated Electric Cooperative, Inc.	Springfield, MO
5. AUST	Austin Electric Department	Austin, TX
6. BIRI	Big Rivers Rural Electric Coop, Corp.	Henderson, KY
7. BPA	Bonneville Power Administration	Vancouver, WA
8. BROV	Brownsville Public Utility Board	Brownsville, TX
9. CAJN	Cajun Electric Power Cooperative	New Roads, LA
10. CAPO	Carolina Power & Light Co.	Raleigh, NC
11. CEIL	Central Illinois Light Co.	Peoria, IL
12. CEIP	Central Illinois Public Service	Springfield, IL
13. CEKP	Central Kansas Power Co.	Hayes, KA
14. CELE	Central Louisiana Electric Co.	Alexandria, LA
15. CEPL	Central Power & Light Co.	Corpus Christi, TX
16. CTKS	Central Telephone & Utilities Corp./ Western Power Div.	Great Bend, KA
17. CHPU	Chelan County PUD #1	Wenatchee, WA
18. CIGE	Cincinnati Gas & Electric Co.	Cincinnati, OH
19. CLEI	Cleveland Electric Illuminating Co.	Brecksville, OH
20. COLM	Columbia Water & Light Dept.	Columbia, MO
21. COEC	Commonwealth Edison Co.	Lombard, IL
22. DAPC	Dairyland Power Cooperative	La Crosse, WI
23. DAPL	Dallas Power Cooperative	Dallas, TX
24. DAPO	Dayton Power & Light Co.	Dayton, OH
25. DOPU	Doublas County PUD #1	East Wenatchee, WA
26. DUPC	Duke Power Co.	Charlotte, NC
27. DULC	Duquesne Light Co.	Pittsburgh, PA
28. EAIL	Eastern Iowa Light & Power Coop.	Wilton, IA
29. ELPE	El Paso Electric Co.	El Paso, TX
30. ELEN	Electric Energy Inc.	Joppa, IL
31. EMDE	Empire District Electric Co.	Joplin, MO
32. FLPC	Florida Power Corp.	St. Petersburg, FL
33. FLPL	Florida Power & Light Co.	Miami, FL
34. GAMW	Gainsville Regional Utilities	Gainsville, FL
35. GRCP	Grant County PUD #2	Ephrata, WA
36. GUSU	Gulf States Utilities Co.	Beaumont, TX
37. HOLP	Houston Lighting & Power Co.	Houston, TX
38. IDPC	Idaho Power Co.	Boise, ID
39. ILPC	Illinois Power Co.	Decatur, IL
40. IMID	Imperial Irrigation District	Imperial, CA
41. INSR	Indiana Statewide Rec., Inc.– Hoosier Energy Div.	Bloomington, IN
42. INPL	Indianapolis Power & Light Co.	Indianapolis, IN
43. INPD	Interstate Power Co.	Dubuque, IA
44. IOEL	Iowa Electric Light & Power Co.	Cedar Rapids, IA
45. IOIG	Iowa-Illinois Gas & Electric Co.	Davenport, IA
46. IOPL	Iowa Power & Light Co.	Des Moines, IA
47. IOPS	Iowa Public Service Co	Souix City, IA
48. IOSU	Iowa Southern Utilities Co.	Centerville, IA
49. JACO	Jacksonville Electric Authority	Jacksonville, FL
50. KACP	Kansas City Power & Light Co.	Kansas City, MO
51. KAGE	Kansas Gas & Electric Co.	Wichita, KS
52. KAPL	Kansas Power & Light Co.	Topeka, KS
53. KEUC	Kentucky Utilities Co.	Lexington, KY
54. LAFA	Lafayette Utilities System	Lafayette, LA
55. LALW	Lakeland Dept. of Electric & Water Utilities	Lakeland, FL
56. LASD	Lake Superior District Power Co.	Ashland, WI
57. LOAN	Los Angeles Dept. of Water & Power	Los Angeles, CA
58. LOCR	Lower Colorado River Authority	Austin, TX
59. LOGE	Lousville Gas & Electric Co.	Louisville, KY
60. MAGE	Madison Gas & Electric Co.	Madison, WI
61. MECS	Michigan Electric Coordinated System	Ann Arbor, MI
62. MIPL	Minnesota Power & Light Co.	Duluth, MN
63. MIPU	Missouri Public Service Co.	Jefferson City, MO
64. MOPO	Montana Power Co.	Butte, MT
65. MSS	Middle South Services, Inc.	Pine Bluff, AR
66. MUSC	Muscatine Water & Electric Plants	Muscatine, IA
67. NEPC	Nevada Power Co.	Las Vegas, NV
68. NEPEX	New England Power Exchange	W. Springfield, MA
69. NEPP	Nebraska Public Power District	Hastings, NE
70. NIPS	Northern Indiana Public Service Co.	Hammond, IN
71. NOSM	Northern States Power Co.	Minneapolis, MN
72. NYPP	New York Power Pool	Guilderland, NY

73. OHEC	Ohio Edison Co.	Akron, OH
74. OHVE	Ohio Valley Electric Corp.	Piketon, OH
75. OKGE	Oklahoma Gas & Electric Co.	Oklahoma City, OK
76. OMPS	Omaha Public Power District	Omaha, NE
77. ORLA	Orlando Utilities Commission	Orlando, FL
78. OTTP	Otter Tail Power Co.	Fergus Falls, MN
79. PAGE	Pacific Gas & Electric Co.	San Francisco, CA
80. PAPL	Pacific Power & Light Co.	Portland, OR
81. PAPL/WY	Pacific Power & Co.–Wyoming	Glenrock, WY
82. PASA	Pasadena Light & Power Dept.	Pasadena, CA
83. PJM	Pennsylvania-New Jersey-Maryland Interconnection	Valley Forge, PA
84. PLAQ	Plaquemine Light Dept.	Plaquemine, LA
85. POGE	Portland General Electric Co.	Portland, OR
86. PSCO	Public Service Co. of Colorado	Denver, CO
87. PSIN	Public Service Co. of Indiana	Plainfield, IN
88. PSNM	Public Service Co. of New Mexico	Albuquerque, NM
89. PSOK	Public Service Co. of Oklahoma	Tulsa, OK
90. PSPL	Puget Sound Power & Light Co.	Redmond, WA
91. SAJL	Saint Joseph Light & Power Co.	Saint Joseph, MO
92. SARV	Salt River Project	Phoenix, AZ
93. SAAN	City Public Service Board of San Antonio	San Antonio, TX
94. SADG	San Diego Gas & Electric Co.	San Diego, CA
95. SEAT	Seattle City Light	Seattle, WA
96. SIPP	Sierra Pacific Power Co.	Reno, NV
97. SMEA	South Mississippi Electric Power Assn.	Hattiesburg, MS
98. SOCA	South Carolina Public Service Auth.	Moncks Corner, SC
99. SOLE	Southern California Edison Co.	Rosemead, CA
100. SOCG	South Carolina Electric & Gas Co.	Columbia, SC
101. SOCO	Southern Services, Inc.	Birmingham, AL
102. SOEP	Southwestern Electric Power Co.	Shreveport, LA
103. SOIG	Southern Indiana Gas & Electric Co.	Evansville, IN
104. SOIP	Southern Illinois Power Coop	Marion, IL
105. SOTE	South Texas Electric Coop. Inc.	Hays, KS
106. SUCO	Sunflower Electric Coop	Hays, KS
107. SWLP	Springfield Water, Power & Light	Springfield IL
108. SWPS	Southwestern Public Service Co.	Amarillo, TX
109. SWPA	Southwestern Power Administration	Springfield, MO
110. TACO	Tacoma City Light	Tacoma, WA
111. TAEC	Tampa Electric Co.	Tampa, FL
112. TEES	Texas Electric Services Co.	Fort Worth, TX
113. TEPL	Texas Power & Light Co.	Dallas, TX
114. TOEC	Toledo Edison Co.	Toledo, OH
115. TUEP	Tucson Electric Power Co.	Tucson, AZ
116. TVA	Tennessee Valley Authority	Chattanooga, TN
117. UNEC	Union Electric Co.	St. Louis, MO
118. UNPA	United Power Association	Elk River, MN
119. UPPP	Upper Peninsula Power Co.	Houghton MN
120. UTPL	Utah Power & Light Co.	Salt Lake City, UT
121. VESM	Vero Beach Municipal Utilities	Vero Beach, FL
122. VIEP	Virginia Electric Power Co.	Richmond, VA
123. WAPA/LC	Western Area Pwr. Adm., Lower Colorado	Phoenix, AZ
124. WAPA/LM	Western Area Pwr. Adm., Lower Missouri	Loveland, CO
125. WAPA/UC	Western Area Pwr. Adm., Upper Colorado	Montrose, CO
126. WAPA/UM	Western Area Pwr. Adm., Upper Missouri	Watertown, SC
127. WAMP	Washington Water Power Co.	Spokane, WA
128. WEFA	Western Farmers Electric Coop.	Anadarko, OK
129. WIEP	Wisconsin Electric Power Co.	Milwaukee, WI
130. WIPC	Western Illinois Power Cooperative	Jacksonville, IL
131. WIPL	Wisconsin Power & Light Co.	Madison, WI
132. WIPS	Wisconsin Public Service Co.	Green Bay, WI
133. YADI	Yadkin, Inc.	Badin, NC

Canadian-Mexican Control Areas

1. SCHA	S.C. Hydro & Power Authority	Vancouver, BC
2. CPL	Calgary Power Ltd.	Alberta
3. GRLA	Great Lakes Power Company, Ltd.	Sault St. Mark, ONT
4. HQ	Quebec Hydro-Electric Commision	Montreal, Quebec
5. MHEB	Manitoba Hydro-Electric Board	Winnipeg, Manitoba
6. NB	New Brunswick Electric Power Comm.	Fredericton, NB
7. NSPC	Nova Scotia Power Corp.	Halifax, NS
8. OH	Ontario Hydro	Toronto, ONT
9. SAGU	Saguenay Power System	Quebec
10. SPC	Saskatchewan Power Corp.	Regina, SASK

Suppose the system is humming along in balance, and then a large transmission line suddenly fails. Neither generation nor demand has changed, so the system is still in energy balance, but suddenly hundreds of megawatts of power that were traveling on the failed line have rearranged themselves according to the new paths of least resistance across the altered network topology. In some cases, the new flow pattern will quickly overload some other lines, since they were not set up to handle their original flow plus the flow from the broken line. Transmission lines have fuses or circuit breakers to protect them against overloads, so the newly-overloaded lines will automatically shut themselves off. This, in turn, may overload and shut off additional lines, and the process may continue if remedial actions are not taken.

Within a matter of seconds, this process can sequentially cause dozens of power lines and power plants to shut themselves off in response to one "disturbance," "contingency," or "initiating event." The pattern of shutdowns will depend on exact electrical conditions all over the network at that moment, and can easily cascade, line by line and plant by plant, across thousands of miles of the system. To reduce cascading failures, power systems are designed to automatically disconnect the rest of the system from the area with a problem. Even when this works correctly, however, many plants and lines can go down over a very wide area, and the disconnected areas still have to rebalance lines and loads, which often requires self-imposed outages.[19]

Interestingly, most of the power outages experienced at home or work have nothing to do with this kind of large-scale failure. Surveys show that about 80% of all outages are caused by a fallen distribution line or a nearby transformer failure, affecting only a few hundred customers. To the area controllers, this outage is so small that it barely affects the energy balance in the control area. In most cases, automatic compensators on power plants and substations can automatically readjust the system a tiny bit, and no other part of the system feels the loss. Thus, maintaining large scale or bulk power reliability is substantially different than maintaining distribution system reliability. One relies on control strategies to respond, rebalance, and restore; the other depends on sending trucks out to repair broken parts.

Reliability events surrounding a severe storm around Washington, D.C. illustrate these points. A total of 129,000 customers lost power due to a combination of transmission line outages and many downed distribution lines. In the first 12 hours following the storm, the utility concentrated on restoring service to all substations that had lost power. "Once the substations were back up," a power company spokesman said, "you could see a pattern of outages that tracked pretty much along the worst of the storm track." About 150 large distribution lines had been knocked out along with hundreds of more smaller lines to individual homes. Because the utility does not remotely monitor the status of every distribution line, the utility had to rely on the 68,000 calls it received from customers who lost service.[20]

Although the operators of power systems have gotten very good at keeping the system balanced and operational in the face of unforseen events, there are plenty of occasions when outages occur. It is sometimes simply impossible for controllers to readjust the system fast enough to prevent significant loss of service.[21] As an example, on December

14, 1994, a major cascading outage caused a power loss for over 1.7 million customers scattered across eleven western states and two Canadian provinces. In this disturbance, it took exactly one minute and 25 seconds between the time that the first line failed in Southern Wyoming and the sequence of shutdown lines and plants made it to the Diablo Canyon nuclear plant near Los Angeles, almost 1,000 miles away. It took only a few more minutes to restore about 40% of the customers, while total restoration of the western grid took about four hours.[22]

SAFEGUARDING AGAINST LARGE BLACKOUTS

To facilitate their response to emergencies, system operators have learned to take a number of precautionary measures. Many of these measures are long term in nature—that is, they have to do with how the system is designed, planned, and constructed. For example, it is not possible to keep the system balanced, or to restore balance after a major disturbance, if there is not enough power capacity to supply all demand. These long-term issues are examined in the next chapter.

With respect to immediate operation, system operators have learned to protect against widespread outages by using computer programs that simulate the operation of the power network in an area. Much as in a computer game or a flight simulator, operations planners can practice responding to the outages of plants and lines in their specific area. These exercises help show operators what sort of precautions they must take to avoid blackouts. For example, a simulation might show that if one particular line goes out, the entire system will cascade into failure if a particular nearby plant is not operating, but will be correctable if the plant is running. System operators will change the dispatch to energize the plant purely as a precaution. A simulation that tells operators that the system is operating so that no instant failures of any likely combination of plants or lines will cause the entire system to go down is known as a *secure dispatch*.[23]

Depending on their size and design, power plants take anywhere from several minutes to several days to get up to full power generation. Starting from a full off position, no power plant can get energized and synchronized fast enough to help prevent a blackout. As a first line of defense, operators keep some power plants running and synchronized with the system, but with their outputs set very low. These "spinning reserves" can generate additional power within a matter of seconds, and can be programmed to kick in automatically when needed. Other power plants that can be started quickly ("non-spinning reserves") are kept on standby, and can start in a matter of minutes.[24]

To facilitate these immediate-protection measures, utilities in most developed systems provide planning information to the system controller days and weeks ahead of time. This is the second of the short-term time frames, so called "operations planning." Depending on the specific protocols, utilities in a control area furnish the system controllers advance information concerning their plants, expected system loads, and other expected conditions. This information helps system operators run adjusted simulations that accurately reflect the exact status of plants and lines in the system at that moment. The technical name for informing system controllers which plants will be available a week or month ahead is called *scheduling* or *unit commitment*.[25]

System controllers use this advance information to plan most of the "settings" they use to control the system a week or a day in advance. When actual system conditions conform to the week- and day-ahead forecasts, the controllers have much less to adjust rapidly and usually less to worry about. The goal of the operations planning time frame is to make the immediate time frame work better—to allow more planning and reflection than immediate control requirements can accommodate.

THE ECONOMICS OF INTERDEPENDENCE AND COORDINATION

The extent of the power system's short-run physical interdependence is remarkable, if not entirely unique. No other large, multi-stage industry is required to keep every single producer in a region—whether or not it is owned by the same company—in immediate synchronization with all other producers. Some colleagues like to call the North American power system "the largest machine on Earth."

Why do producers put up with the bother of having to coordinate every one of their units with those of other plants in their area? Simply because this is substantially cheaper *and* more reliable than building a system that is isolated from all others. "In short," wrote the authors of the 1964 National Power Survey, "interconnection is the coordinating medium that makes possible the most efficient use of facilities in any area or region."[26]

The early pioneers of the electric utility industry built individual systems, with no more than a handful of small plants matched to one small customer area. Between 1893 and 1898, the legendary utility magnate Samuel Insull found that the major economic implication of both short- and long-run interdependence (the latter is examined in the next chapter) was that it was cheaper to connect all generators and all customers in one region, rather than trying and match one generator to a customer or customer group. The industry also quickly found out that interconnection and central control increased reliability dramatically.[27]

Interconnection and central dispatch increase reliability by giving the operators of the grid more ability to adjust or restore the system after a failure. Because one operator has immediate control over most of the system, she can react to the situation with many possible responses, making it more likely that the outage can be minimized. A typical control center has control over dozens of power stations and hundreds of high-voltage lines, and it monitors virtually every plant and line in its area. (It does not, however, monitor most substations or distribution lines; this function is left to each utility).

If this does not sound significant, consider this example. One utility, Union Electric, recently studied the amount of additional power plant capacity it would need to maintain reliable service if it was not interconnected to its neighboring utility. The answer was an additional 1,300 MW, 16% more generating plant than it requires when interconnected.[28] Remembering that power plants are much more expensive per megawatt than transmission capacity, interconnecting sounds like a good deal—even if it means information-sharing and some loss of immediate control.[29]

Lower Operating Costs. In addition to providing reliablity, another benefit of interconnection is that operations costs are lower when the system is working normally. These cost savings arise in part out of the fact that customer uses of electricity on the grid are constantly shifting—some are going up while others go down. By aggregating many loads, simple mathematics dictates that the aggregate load will be much smoother than each individual load, and steady and more frequent loads are cheaper to supply than varying or infrequent ones.[30]

Economics of "Merit-Order Dispatch". Power plants have somewhat different variable costs—costs that depend directly on the amount of power produced in a given period. Insull found that these differences meant that a single controller of many plants could control ("dispatch") the system to save the entire system money. Recognizing that total electricity generated must equal demand at every moment, a utility can instruct its system operators to operate whatever combination of available plants at whatever levels yield the lowest aggregate total cost of operations, adjusting the system so as to ensure that the network is reliable. This operating procedure is called *merit order* or *economic dispatch,* and it is the norm in most of the world.[31]

The economic benefits of load aggregation and economic dispatch go well beyond the simple (though sound) idea of using the cheaper plant first. Appendix 2A describes what are usually called the economic benefits of interconnection in greater detail.

Interestingly, the U.S. industry has created institutions and protocols that contain a significant division between centralized control for reliability purposes and central dispatch for reducing costs. Control over all major utility plants for bulk power reliability purposes is universal—all utilities submit to their area controller in one way or another because there is no other way to run the grid right now. Regional reliability councils, (discussed below) which include all utilities with bulk power resources, are chartered specifically to promulgate standards for reliable grid operation, but *not* to study the economics of central dispatch or interconnection, nor for studying or forming cost-reducing power pools.[32]

The main focus of most economic coordination and dispatching is not the control area, but rather each utility by itself. Many utilities obtain the economic benefits of interconnection by trading with neighbors, and many others are in some form of power pool. Except in a handful of power pools, however, each utility retains its own responsibility for maximizing the economic benefits of interconnection.

This means that the dispatch of plants in most areas is something of a two-tier process. First, each utility in the control area sets out its own least-cost dispatch. Area controllers then look at the whole area and make sure this collection of system settings results in a secure system. System controllers will override a utility's least-cost dispatch only if, when combined with other conditions and dispatches in the region, their experience and computer simulations tell them that the total control area is vulnerable to a large failure.[33]

The idea of merit dispatch can apply to any collection of power plants, not just the collection owned by a single utility. Often, several utilities in one or more control areas agree to centrally dispatch all of their plants in merit order, i.e. to act as if they are one utility system for the purpose of lowering the costs of producing power. Multi-utility agreements to "share" resources so as to reduce all participants' costs are variously called coordination, interconnection, brokering, or power pooling agreements.

Multilateral Agreements and Power Pools. The contracts described thus far are bilateral (involving only buyer and seller), but there are also important multilateral associations and contracts in the industry, an important example of which is a power pooling agreement. Power pools are well-described by economist Scott Herriott:[34]

> Power pools are partnerships among corporately unaffiliated electric utility companies, formed to realize cost savings through mutual cooperation. There are many ways in which cooperation can reduce the collective costs of the utilities in a region, including joint system planning and emergency energy exchange, but one aspect of cooperation is common to all power pools. Within a geographical region and at any given time, some utilities typically experience higher short-run marginal costs than others owing to differences of plant mix or instantaneous loads. When the firms with low marginal costs increase their production and sell this "economy energy" to the higher cost firms, the poolwide cost of generation is decreased.

The division between the institutions and agreements governing interconnected operations for reliability purposes and those designed to realize economic interconnection benefits is the result of the long evolution from small isolated systems ito a large national grid, including the evolution of American regulatory and economic law and numerous politico-regulatory considerations. Ultimately, an examination of these questions leads to many of the most fundamental issues underlying the present restructuring—a subject that is considered in detail in Chapter 4.

RELIABILITY: INSTITUTIONS AND FACTS

As noted above, distribution system outages and bulk power outages are quite different. Distribution outages are usually isolated physical disconnections of a small part of the system, due either to a disruptive event (storm, or construction project) or to the failure of a nearby transformer. Bulk power outages can also be caused either by disruptive events or component failures, but these failures are in large plants or lines and often lead to larger outages.

Between 1970 and 1979, there were about 60 bulk power system outages per year. Each of these outages lost power to an average of 110,000 customers. Of these, 50% were restored within an hour, and almost all were fully restored in seven hours or less.

Distribution failures are literally thousands of times more common. A 1979 Department of Energy survey found an average of 81,000 outages per utility each year for the eight sampled utilities. Compared to bulk power system outages, distribution failures affect far fewer customers (an average of 156 customers per outage) and vary widely in duration,

Box 2-1. Power Pooling Agreements

Formally speaking, interconnection means agreeing to physically connect two companies' transmission lines. However, because flows will tend to go from one system to another unpredictably once connected, interconnection agreements must contain an agreement to sell or buy power, depending on which way it may be flowing. Interconnection points are always metered to measure this, and flows can often be controlled or at least modified by system controllers. Hence, an interconnect gives system controllers more freedom to adjust the system if reliability concerns call for it, and it also gives the interconnected systems the opportunity to buy or sell surplus power (sometimes inadvertently) across the interconnect.

Pooling and brokering agreements are more formal, multiparty agreements to change power. Gross and Balu (1996, p. 1) explain pooling as follows: Pooling refers to a formalized agreement between interconnected companies to utilize their power systems so far as to achieve specific common goals. Utilization may entail operations as well as planning. Specific common goals are typically reliability and economic oriented. The broad scope of objects in the information pools include: (i) to take advantage of resource diversity and load diversity; (ii) to exploit economies of scale in generating units; (iii) to harness competition in the electric generating markets; (iv) to coordinate maintenance plans; and (v) to improve overall interconnected operations (input/output).

Some of the salient characteristics and aspects of power pools are:

  i. ease of entry and exit, and nature of membership (voluntary/involuntary);

  ii. membership (buyers and sellers);

  iii. freedom to undertake transactions between parties;

  iv. costing/pricing mechanism;

  v. horizon and nature of services;

  vi. level of centralization and control; and

  viii. existence and nature of settlement systems.

Pooling arrangements are essential pre-commitments to buy and sell all surplus power to and from pool members which can be exchanged with benefits to both parties—i.e., the buyer can save money by buying and the seller can earn a profit. The pool "automatically" receives cost and information from all members and matches buyers and sellers according to the pool rules. The "higher" the pool, the more automatic and comprehensive the sales and purchases.

A "brokering" system is like a pool, but is less automatic. Members buy and sellers post bids for sale and "ask" for the purchase on a "bulletin board" and the broker puts buyers and sellers together. The decision as to whether to sell or shop can be case-by-case in a brokering system, and there may be more flexibility in setting prices than in a pool.

from one or two minutes to many hours. However, due to the sheer number of interruptions, if reliability is measured simply by the number of minutes an average customer is without power, Figure 2-4 shows that distribution system outages dwarf bulk power outages in importance.

Historically, because distribution system outages seemed to matter most, the industry treated reliability as an individual-utility issue. The main controllable factors that affect distribution system reliability are the quality and design of the system's components and distribution system maintenance. Hence, a utility's reliability could best be improved by focusing on things that are largely within the control of the local utility, including the speed of response to the ubiquitous "a tree hit my power line" phone calls from darkened customers.

In the United States, the notion that bulk power reliability could be taken for granted ended abruptly at 5:16 p.m. on November 9, 1965, when a break in a line between northern New York and Canada cascaded into an outage that affected thirty million people over eight states along the Atlantic Coast. The sequence of failures took 12 minutes to occur and 13 hours to restore, in part because the utility industry had never experienced so large and complex a failure in modern times. With thousands of office building workers trapped in elevators, evening subway trains and traffic stalled by an absence of power, an unfamiliar chaos enveloped the East Coast for an entire night and etched the importance of reliability deep into the modern American psyche.[35]

In the years following this outage, the electric utility industry, government agencies, and others debated the best way to prevent (or at least minimize) bulk power failures. President Johnson commissioned a report by the head of the Federal Power Commission, a project that took two years to complete. Among other things, the report concluded that there could be "no absolute assurance" that similar events would not recur.[36]

Reliability Councils. Eventually, those who favored greater government regulation of reliability and those who felt that the industry could provide reliability through voluntary cooperation compromised on a plan to create the North American Reliability Council (NERC) and ten regional councils. Membership would be "voluntary," but all utilities that owned generation or transmission would join. These councils would each perform studies of their systems, engineer them to prevent large outages, and put in place voluntary standards for system design and operation which would reduce the chance of major outages.

The regional councils have two primary functions. First, they promulgate voluntary system planning and operating criteria that are intended to ensure that each utility that has generation or transmission assets builds and operates them in a way that allows system controllers to preserve bulk power reliability. For example, one reliability council's system operating criteria read as follows:[37]

1. Each system should provide sufficient transmission capacity within its system to serve its load and meet its transmission obligation to others without unduly relying on or without imposing an undue degradation of liability on any other system, unless pursuant to prior agreement with the system(s) so affected.

Figure 2-4. Most of the Periods of Interrupted Electric Service are Due to Distribution System Failures

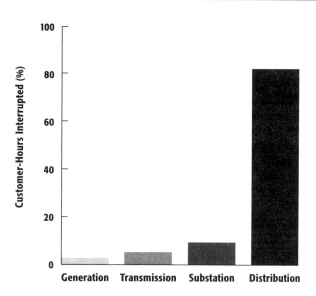

Based on a four-year sample of electric utilities in the United States, the U.S. Department of Energy confirmed that the most frequent and usually the longest outages were caused by breaks in distribution lines rather than failures of substances or large plants or lines.

Source: DOE, 1981, p. 44

2. Each system should provide sufficient transmission capacity, by ownership or agreement, for scheduled power transfers between its system. In transferring such power, there should be no undue degradation of reliability on any system not a party to the transfer.

3. Each system should conduct the necessary studies to demonstrate that its power transfers, in the absence of the loop flow, will not unduly rely on or impose an undue degradation of reliability on parallel transmission capacity of any other system, except pursuant to prior agreement with affected system(s).

Each system should plan its system with adequate transfer capacity so that its power transfer will not have an undue loop flow impact on other systems, and so that planned schedules do not depend on opposing loop flow to keep actual flows within the path transfer capability. A system adding facilities should recognize that the addition itself could result in a component of loop flow that should be accommodated.

Loop flow is an inherent characteristic of interconnected AC systems and the mere presence of loop flow on circuits other than those of the transfer path is not necessarily an indication of a problem in planning or in scheduling practices.

4. Each system should provide, by ownership or agreement, sufficient reactive capacity and voltage control facilities to satisfy the requirements of its own system without imposing an undue burden upon other systems.

These criteria continue on and become much more technical, eventually specifying the maximum permissible impact each utility system can have on its connected neighbors as a result of one, two, or three sudden plant or line outages. For example, the criteria specify that the loss of one plant or line in system A must not affect system B's voltage by more than 25%, and the voltage drop must not exceed 20% for more than a third of a second. Frequency must not dip more than 0.4 cycles/second, except for more than a tenth of a second.[38]

The second function of the reliability councils is to serve as arenas for conducting joint studies of grid reliability. In effect, these joint studies determine whether utilities are living up to the criteria they have mutually agreed to live up to. A Department of Energy report provides a typical description of this activity:[39]

> Utilities study emergency transfer capability [ability of lines to carry power following an outage] for both planning and operating purposes, the former ranging ten years or more in advance and the latter usually one-half year in advance. Extreme disturbance studies are also made to complement transfer studies. The activity of the industry in these kinds of studies is demonstrated by the many reports published by utilities. For example, in the SERC region, 47 intra-and interregional load flow studies were conducted from 1974 to 1978. Of these, seven included stability studies and 13 included inertial analyses. The purpose of 34 of these studies was to assess reliability; the others were for planning and operating purposes.

Until now, the voluntary approach to bulk power reliability has worked fairly well. As discussed in Chapter 15, however, this approach is almost certainly insufficient for an industry that is substantially deregulated.

ANALOGIES TO OTHER DEREGULATED INDUSTRIES—REPRISE

Returning to the earlier discussion of interindustry analogies, many policymakers and industry figures make reference to similarities between electric restructuring and changes in the telecommunications or natural gas industries. Some go beyond these industries, drawing comparisons to airlines, railroads, banks, and many other examples. Testifying in favor of electric power deregulation in 1996, Enron CEO Kenneth T. Lay introduced his testimony with a slide showing the history of reduced regulation of 12 formerly-regulated activities, starting with stock brokerage fees in 1975 and culminating with telecommunications in 1996 and electric power immediately following.[40] Many others have joined this debate in one way or another,[41] including a few who suggest that interdependence in power markets is not unusual or important.[42]

Table 2.1 compares some of the broad, relevant attributes of these three industries—the original public utilities. The table begins with two simple measures of industry size, annual revenues, and capital intensity. Measured in annual revenues, electric utilities are the largest utility sector by 20%. They are also still the most capital-intensive sector, although capital intensity has been dropping and may ultimately resemble that of other industries.

The third row of the table reviews some of the fundamental differences between the three utility production technologies. Natural gas is a storable commodity. It is stored in large fields near gas wells and in large urban areas; the pipeline system itself can be used to store gas simply by increasing its pressure. Because gas can be directed down a single set of pipes, the system is far more resilient than the electric power grid and far less vulnerable to rapid, large scale interruptions.

With regard to telecommunications, messages are routinely stored today and momentary interruptions ("busy signals") are common. Calls must be routed along specific paths, and when portions of the long-distance system go down, calls can often be rerouted on other circuits.

At least for now, electric power is the least directable and most delicate of the three utility technologies. It cannot be stored economically, and there is very little "give" in the system which allows imbalances between supply and demand.

The fourth row of the table reviews the stages of production in the three industries. All three industries have something resembling a stage that "generates," "transmits," and then "distributes" (or sells) the product at retail. In the natural gas industry, natural gas is produced in wells and gathering systems, transmitted in large interstate pipes, and distributed by local gas utilities. In telecommunications, "generators" are those who create the content of communications, everything from individual callers to information services like Dow Jones. Long-distance and local distributions are somewhat analogous to electric transmission and distribution, respectively.

The legal and economic histories of regulation in the natural gas, telecommunications, and electric power industries are all quite different. All three industries have three major stages: one for the generation of the item to be delivered, one for long-distance transport, and one for local delivery. The federal court decree of 1982, which separated long-distance and local telephone service and began long-distance deregulation, constituted the beginning of deregulation of the transmission stage of the industry. The recently-enacted Telecommunications Act of 1996 continues the deregulation of long-distance and introduces federally-sanctioned local area deregulation.[43]

There is little disagreement that new technologies that reduced the cost of long-distance transmission networks, and may now reduce the cost of local networks, were critical elements in the deregulation of telecommunications. This stands in stark contrast to the electricity sector, where technical change has made generation potentially competitive, but has not yet threatened the natural monopoly status of transmission or distribution. Hence, at the most elemental level, deregulation in the telecommunications and electric industries stems from a very different sort of technical change.

Table 2.1. A Comparison of Some of the Main Features of the Natural Gas, Telephone, and Electric Power Industries

Feature	Discussion
Key Technology Attributes	**Natural Gas:** Storable in large fields; can be directed down specific paths; network is less vulnerable to interruptions and imbalances between supply and demand.
	Telephones: Busy signal—network can refuse service temporarily; calls can be directed or rerouted.
	Electric Power: Not economically storable; cannot be readily directed down specific routes; supply-demand imbalance must be maintained; low tolerance for interruption disturbances, and disturbances can instantly spread to larger and larger areas.
Industry Stages and Their Regulatory Status	**Natural Gas:** Producers were deregulated in 1978; pipelines are federally regulated; local distributors are state-regulated.
	Telephones: Content origination never regulated; "transmission" (long-distance) largely deregulated in 1983; local distribution recently deregulated under certain conditions.
	Electric Power: All three stages regulated; federal government regulates sales between utilities and transmission rates; state regulators regulate plant construction, retail rates, distribution.
Other Regulatory Differences	**Natural Gas:** Federal regulators can give interstate pipelines permission to build and exercise eminent domain.
	Telephones: Federal regulators place extensive requirements on local and long-distance providers.
	Electric Power: Only state governments give permission to build new power plants or lines; no federal power of eminent domain.
Approximate Industry Size	**Natural Gas:** $63.83 billion revenues; $138.1 billion assets (1994).[1]
	Telephones: $94.44 billion revenues; $210.5 billion assets (local carriers, 1994).[2]
	Electric Power: $202.71 billion revenues; $550 billion assets.[3]
"Capital Intensity"[4]	**Investment Per Dollar of Revenue**

	Average Industry	Natural Gas	Telephones	Electric Power
	0.6	2.2	2.2	2.7

Table 2.1. A Comparison of Some of the Main Features of the Natural Gas, Telephone, and Electric Power Industries *(continued)*

Feature	Discussion
Peacetime Public Safety Aspects of Industry	**Natural Gas:** Loss of service during periods of extreme weather can be life-threatening.
	Telephones: Loss of service removes most immediate access to police and other emergency services.
	Electric Power: Depending on the size of the area experiencing the disturbance, economic costs (spoiled food, ruined computers) can be very large. Public safety faces widespread threats if large areas remain dark for long.

November 7, 1996

1   Investor-Owned Gas Utilities. Source: 1994 Gas Facts Tables 12-1, 13-1.

2   Regional Carriers and Independent Companies. Source: FCC Data Book, 1994, pp. 9–10.

3   Total Electric Power Industry. Source: Energy Information Administration, 1994.

4   DOE (1979) p. 18. In a much more recent analysis, Schreiber (1996) shows that pretax interest coverage and funds from operations as a percent of interest follow roughly this pattern.

In natural gas, the production of gas from wells ("generation") was the first subject of federal price decontrol. The transportation segment, natural gas pipelines, remains price- and entry-regulated, and most local distribution is subject to traditional cost-of-service regulation or regulated open-access. This situation is closely analogous to electric utilities, where deregulation began in the generation stage and transmission and distribution stages are evolving toward open-access, common carrier models.[44]

In other respects, however, the deregulation of telecommunication and electricity share some similarities. One of the most complex regulatory aspects of telecomm, and a source of great consumer benefit, is the growth of new services that may be added to basic phone service (e.g., call forwarding and message waiting). Some observers argue that retail electric deregulation will create the possibility of new kinds of electric services, including combinations of telecommunications and electric power. The provision of these services will raise concerns similar to those experienced over the last several years with regard to telecommunications.[45] In particular, concerns about the maintenance of universal service in telecommunications are mirrored in concerns about universal service and residential ratepayer equity in general in electricity restructuring.

The final row of the table illustrates the broad-scale costs of outages to communities and regions for each of the utility industries. Although phone and natural gas outages can be serious, the community and national security implications of electricity outages are unparalleled by any other industry in the nation. Every aspect of what is known as state-provided security, from the control of the millions of criminals in American prisons to the safety of the country's nuclear arsenal, depends on the ready availability of electric power.

Table 2.1 and this discussion illustrate the fact that the technical attributes of the three utility systems have many similarities and many dissimilarities. As a result, highly specific analogies (and conclusions therefrom) generally cannot be regarded as reliable without considerable additional analysis. MIT Professors Paul Joskow and Richard Schmalensee put it nicely:[46]

> Sound policy cannot be made on the basis of casual analogies…Currently, electric power is supplied by complex and highly developed systems with unusual technical characteristics. These make it likely that reliance on an economist's instinct, developed through countless examples drawn from agriculture and manufacturing, will produce incorrect conclusions.

On the other hand, analogies are both inevitable and very useful learning tools; attempting to avoid or deny them is pointless and harmful. The key is to use them wisely.

INTERDEPENDENCE: DOWN BUT A WAYS FROM OUT

Despite the massive technological changes the industry has undergone—including the plateau in economies of scale for generators—interdependence continues to be the central technical feature of the power industry throughout the developed world. And this will continue as as long as it is cheaper for most people to get their electricity from the grid than to supply themselves independent of the grid.

What about the thousands of industrial plants, hospitals, and even apartment buildings that own small power plants and generate all or some of their power? There are certainly many situations in which these plants reduce the total cost of service, but in most cases, this is because the generator can also use the waste heat from the plant rather than pay for it separately, or it can get fuel or capital more cheaply than the utility. Even then, these plants become a seamless part of the total utility grid. Indeed, they become almost as interdependent with the rest of the power system as the local utility's own plants. The area's control center usually has the right to turn them up, down, or off to keep the system in balance.[47]

Producer interdependence will be with us for another several decades or so, but there is no question that it is in decline. Technologies are developing which will allow consumers to make electric power in their own homes or offices, possibly as cheaply as they can get it from the grid. This will cause the electric power system to become more decentralized than it is today.[48] However, as long as independent producers are connected to the grid for backup power (when a consumer's backyard unit fails, for example), a plethora of scattered little generators will interdepend with each other in much the same way that fewer, much larger plants do today. Small, cheap generators (including fuel cells) alone will not end interdependence—indeed, without cheap storage, interdependence might become even more important and more complicated.

Cheap local electricity storage will end interdependence for good—and this is just a matter of time. When electricity can be stored inexpensively at each customers' site, electricity will become a commodity much like fuel oil or gasoline. There is little interdependence between the producers of petroleum products because their products are storable, and each firm can produce and deliver when and where it chooses. Indeed, it is hardly noticeable when one gas station runs out of gas or a large petroleum pipeline goes down. Drivers remember to keep their tanks full enough to get them where they need to go, and so it will be when homeowners fill the electricity tanks in their back yards.

The challenge of analyzing industry structure and setting public policy in the face of this epochal technical change will be daunting. At some point in the distant but foreseeable future, the industry's central technology will flip-flop, and the grid will change from the immediate coordination and delivery link to traditional storable commodity delivery role. The shift will be complicated by the fact that this may happen over a long transition period, first in some places in United States and the world and later in others. The central control and coordination function of the grid is pretty much all-or-nothing.

Although it might be ideal to restructure the electric utility industry *today* to accommodate the eventual death of strong interdependence, doing so would require a bit more foresight than even the best among us are blessed with. With the exception of a look toward the distant future in the final chapters that follow, this book examines utility industry structure in what remains of the era of interdependence.

Appendix 2A

Economic Benefits of Interconnected Operation

1. Load Diversity and Aggregation

Almost no electricity customers are required to plan their electric use in advance. As a result, each customer's use of electricity varies widely from moment to moment—zooming up or down in a matter of seconds. Because the grid must remain in immediate balance, this could present a huge problem of prediction and readiness to serve. However, if the loads of many customers are added together, they tend to "average out," creating a smoother pattern of peaks and valleys.

Samuel Insull and his contemporaries discovered that the smoother nature of aggregate loads ("load diversity") makes it cheaper to supply aggregate loads. In a 1914 example used to highlight this point, Insull calculated that 193 customers in one apartment building had individual maximum daily demands of 92 watts, but when combined their maximum was actually registered as only 29 kW. For the entire Commonwealth Edison system, approximately the same ratio held.[49] As long as electric generating capacity remains significantly more expensive than transmission and distribution capacity, the total cost of serving aggregate loads is cheaper (under normal conditions) than the costs of serving loads individually.

2. *The Savings from Economic Dispatch–An Example*

If one is attempting to serve an aggregation of loads from one group of power plants, economic dispatch of the power plants is the lowest-cost way to do it. This is an intuitive result for one company—obviously, one would want to use one's cheapest plants first. However, this result applies to several utilities that pool their generation, *even if pooling causes one utility's costs to go up*. The key is that all utilities in the pool must combine their savings and their costs, i.e., transfer money from utilities whose costs went down to those whose costs went up, so that after each accounting period every utility has reduced its costs to levels lower than it could achieve by going it alone.

Scenario	Cost of Operating A's Plants	Cost of Operating B's Plants	Total Cost of Operating System
Utility A and B try to match own-loads	$175	$225	$400
Utilities use least-cost "central dispatch"	$180	$210	$390
Savings from centralized least-cost dispatch	($5)	$15	$10

Utility A has two plants, A1 and A2; Utility B has plants B1, B2, and B3. All plants are connected to each other and to both utilities' distribution system (and therefore all customers) via the transmission system. The entire system is controlled or *dispatched* by the area controller.

If each utility simply measured the total power demanded by all of its customers, and set the output of its plants to this level, the system could not work or would be too expensive to operate. Because electricity does not flow in a directed manner, portions of the transmission system may become overloaded and fail. In this case, the controller would have to revise some system outputs and settings, attempting to keep each utility responsible for its total load.

To prevent this, the area controller measures the total power demanded by both utility A and utility B's customers. It then uses computer calculations to estimate the combination of power plant settings that will supply this level of power at the lowest aggregate cost, while still ensuring that the total system will operate safely and reliably.

3. *Other Benefits of Interconnection*

Interconnection allows for cost reductions and increased reliability other than those achieved by the simple phenomena of load aggregation and coordinated dispatch. Simply put, interconnection makes it less costly to achieve a given level of bulk power reliability than operating independently. These benefits are described by the Federal Energy Regulatory Commission (1981, p. 15, 19) as follows:

1. Economies of scale—savings from the construction of larger generating units and higher capacity transmission facilities.

2. System reliability—benefits from maintaining a suitable reliability level with less generation reserves and from improved reliability through coordinated transmission planning.

3. Operating reserve—savings in operating reserve, both spinning and nonspinning, resulting from sharing the reserve required to cover contingencies.

4. Installed reserve—savings due to installed capacity reductions resulting from coordinated capacity between system.

5. Staggered construction—savings from scheduling of new facilities to minimize periods and amounts of over or under capacity.

6. Economy energy interchange—savings from buying energy that is less expensive than the alternative available to the purchasing utility. The most highly developed form of this is central dispatch of two or more systems.

7. Load diversity—savings from reduced capacity requirements that result when interconnected systems reach peak load at differing times—seasonally, weekly, daily, and hourly.

8. Maintenance coordination—savings from optimum scheduling of outages for maintenance by two or more systems to minimize the use of high cost replacement power and energy.

9. Maximizing hydroelectric utilization—savings from coordinated scheduling and dispatching of hydroelectric facilities of two or more systems to optimize the peaking power and energy available from these facilities.

10. Diversity of errors—benefit from the compensating effect of independent decisions by separate managements in limiting the consequences of error in forecasting, generating mix and construction scheduling.

11. Siting flexibility—opportunities for a greater choice of power plant sites in the combined areas of coordinating systems.

12. Resource diversity—better ability to substitute alternative energy sources during shortages, through the greater fuel-by-wire capabilities of coordinating systems.

13. Maximum transmission utilization—economies from fuller use of the transmission investment through coordinated system planning, coordinated daily scheduling of generating units, coordinated scheduling of generation and transmission outages, and real-time monitoring of transmission status from a single control center.

14. Emergency response—improved ability to avoid loss of load or minimize its duration during emergencies, as a result of centralized control of the operations.

15. Utility planning and operating quality—improved level of proficiency resulting from information exchanges and joint analyses with other utilities.

A 1980 study of reliability by the Department of Energy noted that interconnection also increased total system energy efficiency and reduced utility financial risks.[50]

Note that items 1, 2, 5, and 15 on this list are realized through coordinated *planning* as well as coordinated operation. This highlights a link between the long-term or planning time frame and the short-term operating time frame that is explored further in Chapter 4. Briefly, long-term planning may enable cost reductions during emergency or extreme conditions of system operation. These cost savings do not occur every day, and cannot be determined without ex-post analysis of problem situations, but they have proven to be real.

Finally, it should be obvious that the economic benefits of interconnection are not cost-less. Interconnection requires increased investment in bulk power transmission facilities and communications and control hardware and personnel.[51] As a general principle, the total economic costs and benefits of greater interconnection should be the basis for deciding the optimal amount of interconnectedness in the U.S. power system. This is another question concerning industry structure which is discussed in later chapters.

Continuing this chapter's focus on the short-run time frame, FERC (1981, p. 19–20) includes this discussion of the magnitude of the benefits of coordinated operations.[52]

Chapter 2

NOTES

1   Most non-U.S. systems use 50 hertz A.C. power; the frequency difference has little impact on system operation. Some very high capacity transmission lines use direct current, or DC, which flows in one direction from higher to lower voltage. To use DC transmission, AC must be converted to DC at one end of the line and back into AC at the other. This is economical only over special very high-capacity long lines in the U.S., such as from the large hydropower plants in the Pacific Northwest down to power-hungry Los Angeles. DC links also work well for crossing under waterways, such as across the English Channel. For an engineer's discussion of DC transmission, see Rustbakke (1983). For an interesting historic note on the development and standardization of AC, see Hughes (1983) ch. 5.

2   This is why most homes and businesses are only a few dozen feet from the ubiquitous grey cylindrical cans that hang at the top of many power poles or the large green metal boxes that sit outside many apartments, subdivisions, and offices. In both cases, these are the last of several transformers that reduce the voltage from the transmission lines down to 240 or 120 volts, a level that is useful and relatively safe for household applications.

3   See Chapter 5.

4   According to Rustbakke (1983), about 1% of all electric customers are served by tapping into subtransmission or transmission lines directly, i.e. taking service at higher voltage levels. This 1% of the customer base consists of very large users, such as steel mills, and accounts for much more than 1% of all U.S. electricity use.

5   For additional information on transmission and distribution system design, see Pansini (1992) and Powel (1964).

6 EIA (1995b). In a recent epochal step toward complete industry restructuring, the Federal Energy Regulatory Commission established a new means of separating transmission and distribution. The new method uses a seven-question test to determine whether a particular part of the system should be classified transmission or distribution. *Federal Energy Regulatory Commission, Order 888,* April 25, 1996. See Chapters 6 and 7.

7 In other words, if the asset is a generator or related hardware, it is part of generation. If it is a transmission line of 22,000 volts or higher it is probably transmission; any lines of lower voltage, and well as some substations (low-voltage transformers) are distribution.

8 EIA (1993). The remaining 10% consists of administrative plant, such as office buildings, and other assets.

9 For a couple, see Baumol and Blinder (1991) p. 551

10 The fact that power prices are, in simplified terms, set equal to the utility's average cost per unit produced makes this kind of future planning much easier. One can look at projected demand, determine the amount of capacity that must be added, compute its cost and the new system average cost, and then make sure that average cost equals the price assumed in the demand projection. See Fox-Penner (1988) and Hyman (1992).

11 Utility engineers call the activities conducted in the immediate time frame *operations control*, while *operations planning* occurs in the few weeks preceding each day.

12 This is tradition in power systems planning texts (e.g., Rustbakke (1983) and is discussed in many other works, such as Itíc, et al (1996) and Fernando and Kleindorfer (1996).

13 This is a simplified explanation of the short-run operation of electric power systems. For a more detailed but non-technical discussion, see Fox- Penner (1990c). For an old but very readable description of system operation, see Federal Power Commission (1964) ch. 11.

14 This analogy is elaborated on in Fox-Penner (1990), Parsini (1992), and Casazza (1993), among many other works. This water pool analogy for the transmission system is less accurate than the pitchfork analogy, which reflects the more complex need to simultaneously balance energy, frequency, and reactive power.

15 Utilities readily admitted that use of the contract path was a convenient fiction crafted to make contracts easier to write. Agreeing on a contract path saved all parties to the transaction the trouble of running studies to see exactly whose lines were carrying how much power, which is a complicated exercise. As competition and third-party use of the transmission system has increased, use of the contract path tradition has caused increasing tension. One important aspect of industry restructuring will be to discard this convention in favor of entirely different ways of setting prices for transmission.

16 The entire U.S. bulk power network is actually three separate grids, one in the Eastern United States, one in the West, and one in Texas. These grids are connected with D.C. links that allow for much more control than exists within the three networks (and until 1967, East and West were not connected at all). The "Eastern Interconnect" has 99 control areas, the "Western Interconnect" has about 34, and the "Texas Interconnection" (more commonly referred to by its reliability council name ERCOT) has about ten. See FERC (1981) p. 26, U.S. Department of Energy (1979) v. I. p. 16 and (1981) p. 14.

17 For a clear and concise description of ECC functions, see FERC (1981) p. 26.

18 "VAR" stands for "Volt-Ampere Reactive," and is a unit of reactive power. See Fox-Penner (1900c). The following utility engineer's description of the process of system control is interesting for its emphasis on operator judgement as well as for its technical detail:

"The control of voltage and reactive power are inseparable. It is a complex task and the approach to it is still based as much on operator experience as on any defined method. The present operating procedure is to monitor and control a number of buses in the system. The dispatcher controls these voltages based on operating practices and off-line load-flow studies. He employs the generator bus voltages for control, in addition to the controlled capacitors and reactors…A methodical, often long, search procedure…is required before an acceptable set of variables can be found to satisfy the desired conditions." (Rustbakke, 1983, p. 268).

Note that generator VAR requirements are determined not only by the VAR support needed for the transmission system, but also by the VARs consumed by end-users as part of their electric power load.

19 For a useful discussion of the mechanics of outages, see U.S. Department of Energy (1981) p. 39. For additional discussion, including load-shedding to prevent cascading, see FERC (1981) p. 31.

20 *Washington Post*, June 30, 1996, p. 33

21 Even more remarkably, system operators often cannot look back at some outages and determine exactly how they occurred. In some cases the cause and the sequence of events is clear; in others, the failure was so fast and so large that there is no way to trace the flow of power during the event. Kirby, Hirst, and Vancouvering (1995).

22 See the Western Systems Coordinating Council, System Disturbance Report, December 14, 1994. To restore service when an outage like this occurs, operators bring the system back on line in parts, matching each new generator back on line to a new group of customers, keeping the system balanced. This is exactly why everyone's lights do not go back on at once following a significant outage.

23 The detailed criteria for declaring a dispatch secure vary from system to system. In some cases, operators study all possible single outages and protect against them ("first contingency"). In others, they routinely look at more than one simultaneous outage. Many more continency scenarios are examined for planning purposes than during the ongoing course of operations. See Fox-Penner (1990c).

24 These are just two examples of a number of protective measures system operators employ. Additional measures include operating reserves (plants on cold standby), selection of generating units based on their electrical ability to survive disturbances, emergency operating procedures, and transmission operating protocols. For additional background, see the Department of Energy (1979) v. II p. 29ff and ch. 10 and DOE (1981) p. 19–20 and 34ff, Rustbakke (1983) and Fox-Penner (1990).

25 See U.S. Department of Energy (1979) Chapter 10 and Rustbakke (1983) Chapter 14.

26 Federal Power Commission (1964) p. 27.

27 Utility pioneer Samuel Insull demonstrated the economic aspects of interconnection to a skeptical industry with the "Lake County experiment" in 1922. Insull purchased the isolated generating plants from 22 small towns in the county, each of which lit the town only during the evenings, as well as from a few farmers in the county who owned small power plants. Investing a comparatively large sum in a transmission system that interconnected these facilities, Insull improved sales and service enough to more than justify the investment and turn the industry's attention toward what was then called "central station power." See Platt (1991) and U.S.D.O.E. (1980) VII. p. 395.

28 David Whitely, Union Electric Company, Presentation to EEI Power Systems Planning and Operations School, December 4, 1995.

29 Many other studies have computed similar substantial cost. Savings from levels of interconnection at least as high, on average, as the present system. For a few examples, see Baldwin (1996) and the Department of Energy (1979) v. II p. 408.

30 This depends only on the presence of significant fixed costs relative to variable costs, and does not require economies of scale or subadditivity. As long as power delivery requires a significant up-front financial commitment, this capital cost must be amortized from revenue derived from the sale of the product of the capital. The larger the aggregate sales derived from the capital, the lower the amortization collection required per unit sold.

31 Strictly speaking, economic dispatch means dispatching a single utility using least-cost or merit order, whereas the term "central dispatch" is sometimes used to denote two or more utilities agreeing to dispatch all of their units in pooled merit order. For an excellent concise discussion of economic dispatch, see Federal Energy Regulatory Commission (1981) p. 27.

Economic dispatch is a key link between the short and long run in the utility industry. Long-term planning (Chapter 3) determines the set of plants from which the dispatcher will be able to choose. In the short run, under normal conditions the dispatcher chooses from among available plants the cheapest combination.

32 In the words of a 1981 report by the Federal Energy Regulatory Commission (FERC), "A conceptual distinction can be drawn between coordination for economy and coordination for reliability, although in practice the distinction may be more difficult to maintain. The object of coordination for economy is to improve the productivity of the resources employed in the generation and transmission of electricity. The object of coordination for reliability is to apply the combined resources of several systems to a contingency on any one." (FERC, 1981, p. 2).

The Department of Energy (1979, p. 272) explains this point as follows:

"The Reliability Councils are typically advisory in that only overview reliability guidelines and reporting standards are provided. Consequently, the responsibility of the Reliability Councils are typically limited in scope and subject matter to these things associated with broad overview of utility studies verifying general conformity.

The Reliability Councils generally do not consider the economics of proposed facility additions although it is difficult if not impossible to separate the economics from the other aspects of the planning decision, as stated by T.J. Nigel, Senior Executive Vice President of AEP, "Economics is fundamental to be placing process. Cost enter into most, if not all trade-off decisions in both planning and operating an electric utility."

33 Even in this situation, the solution often involves taking precautionary measures that do not involve redispatching power plants, but rather putting different units on standby, adjusting power line settings, and so on. In theory, the dispatch that the utility gave to area controllers was lowest cost for that utility, and *any* changes in the system by controllers will raise production costs. Also ideally, system controllers will choose to change the original dispatch in the least costly manner possible, including a possible redispatch of plants. However, in most cases area controllers are not responsible for finding the lowest-cost approach to maintaining reliability in the immediate time frame.

34 Herriott (1985). For a good recent summary, see Hunt and Shuttleworth (1996), p. 39ff.

35 Pop culture commemorated the incident with a musical comedy starring Doris Day, *Where Were You When the Lights Went Out?* See Richard Munson (1985), p. 119.

36 Ibid, p. 120.

37 Western Systems Coordinating Council, "Reliability Criteria for Transmission System Planning," April, 1995.

38 A concise description of NERC's operating committee activities can be found in FERC (1981) p. 32.

39 DOE (1979) v.II p.184.

40 Testimony of Kenneth T. Lay before the Subcommittee on Energy and Power, Committee on Commerce, U.S. House of Representatives, May 15, 1996.

41 For example, Pierce (1991) argues that the deregulation of natural gas provides the ideal model for deregulating electric power; Bobbish (1992) argues essentially the opposite. Santa and Sikora (1994) provide a tremendously detailed comparison of the similarities and differences in the regulatory laws governing electric and gas utilities; Tye and Graves (1996) draw similar parallels in a more formal economic treatment. Carl Danner (1993) argues that "where telecommunications regulation has been may be where natural gas should go."

42 See, for example, McGuire (1996).

43 Amending the Communications Act of 1934 [47 U.S.C. § 151 *et seq.*]

44 See Charles River Associates (1985) for an extensive discussion of the parallels and differences between gas and electric common carriage.

45 The 1996 Economic Report of the President discusses similarities between electric power and telecommunications regulatory policy issues in additional detail.

46 Joskow and Schmalensee (1983) p. 8–9. University of California economist Bart McGuire argues just the opposite—that there is nothing whatsoever that is unique about electric power from the standpoint of deregulation. (See McGuire (1996)).

In "Estimating the Costs and Benefits of Utility Regulation," written in 1974, Schmalensee concludes that estimates of the economic benefits of deregulation are highly dependent on the particular technologies and industry form present before and after deregulation. "Given the current state of the art and its likely development," Schmalensee concluded, "it seems reasonable to suggest that if a control mechanism could be devised that reasonably certainly would never lead to *worse* performance than would a hands-off policy and that had a better than even chance of improving performance, that mechanism should replace utility regulation across the board." (Schmalensee, 1974, p. 63).

47 This has been an area of great disagreement among nonutility power plants and utility systems. Briefly, utilities want the right to control these plants in order to provide for the safe operation of the network. However, many independent power plants earn revenue only if they are on, while utility rates are not based on which power plants are on or off at any moment. Consequently, independent power plants want to be turned down or off only when it is absolutely necessary. The control centers—generally operated by the local utility—do not care how much the independent plant makes, and might even wish to harm the IPP, viewing it as a competitor. The terms and conditions setting forth the rules for utility control of IPPs are therefore complex and contentious, and have led to the pressure to make the regional controllers independent of the local utility, so as to be unbiased towards any plant or owner group. See United States Department of Energy (1986), Spiewak (1987) [and subsequent editions], and Zweifel and Beck (1987).

48 For an introduction to the distributed utility, see Weinberg, Iannucci, and Reading (1991) and the Electric Power Research Institute (1992).

49 Hughes (1983) p. 217.

50 U.S. Department of Energy (1980) p. 45.

51 U.S.D.O.E. (1980) ch. 6 discusses these costs in greater detail.

52 Additional discussions of these benefits are found in DOE (1979) [generally, but see esp. v. II p. 272 and p. 406ff] and Federal Power Commission (1964), esp. ch. 12.

Chapter 3

Power System Long-Term Planning

In the utility industry, long-term planning results in the addition or retirement of plants, lines, and other system parts. This planning must be done far enough in advance so that operators are sure of having enough capacity on hand to dispatch the system safely in ordinary times and emergencies. Planners look out several years into the future and work to build systems that future dispatchers can operate reliably at the lowest possible cost.

As noted in the previous chapter, the condition of having enough capacity on hand to provide reliable operation is known in the power business as *adequacy*. The interlocking concepts of security and adequacy map into the more common business concepts of short-and long-term time horizons, respectively. Adequacy refers to the long-term availability of sufficient capacity to serve; security hinges on the ability to preserve the immediate operation of the grid without sudden outages.[1]

Because new power capacity takes time to plan and install, and future loads are always uncertain, adequacy cannot be guaranteed. Traditionally, the industry has dealt with adequacy by designing systems that would be adequate for all but the most extremely high and unusual load levels. A common standard used as a planning criterion by utilities is that sufficient capacity should be on hand to serve all loads at all times except for the highest 2.4 hours of each year (a "loss of load expectation" of 1 day in ten years).[2] A crude but still more common criteria is to maintain a minimum stock of generation and transmission capacity above the highest predicted peak load—a margin of "reserve capacity" or "reserve margin" that is usually around 15 to 18% of expected peak demand.

Competition will change the industry's long-term planning methods—probably quite dramatically. In order to appreciate how these changes might occur, it is important to review how planning has been practiced in the industry.

TRADITIONAL SYSTEM PLANNING

The first requirement of system operation is to balance energy into and out of the system at every instant. The level of energy flowing in one instant is called power and it is measured in watts (kilowatts, megawatts, etc).[3] Units of power are the universal measure of electrical capacity or size.

As discussed earlier, interconnecting many customers creates an aggregate load that is lower and smoother than the sum of each individual demand. To meet the essential condition that total generation instantly equals total load, the power system in a given area must have a capacity as large as the highest moment of total use, or *peak demand*.[4] In simple terms, long-term planning is looking into the future, guessing what peak demand will be one or more years from now, and building more plants and lines[5] to make sure the system can deliver this peak demand—all at the lowest total life-cycle cost.[6]

Traditional long-term planning is a multi-step, iterative process. Figure 3-1 shows the sequence of steps that a traditional utility goes through in a typical annual planning exercise.[7] The process begins (upper left of figure) with an estimate of future demands at every moment over the next 10 or 20 years, especially peak demand. If this step is not roughly correct, planners will build either too little future capacity, degrading adequacy and therefore reliability, or too much, leading to excess capacity.

A major complication in demand forecasts is predicting future prices. This is required because future demand will depend on the cost of electricity. In spite of this difficulty, utilities were able to make fairly accurate long-term forecasts during the decades in which their prices were stable or falling. Starting in the 1970s, price and demand prediction became much more difficult, and some planners made large and costly forecasting mistakes. Appendix 3A discusses price and load forecasting in more detail.

Armed with what she hopes is an accurate forecast of future demand, the planner next experiments with a number of options for adding and retiring plants and lines (right side, Figure 3-1). System planners use a number of special computer programs to examine expansion options. One kind of program is much like the one used by system operators to test any actual or projected system to ensure that it would be electrically stable. Another kind simulates the future economic dispatch of the system to ensure that variable costs are as low as possible.

In this part of the process, traditional planners have been able to study a number of different combinations of plant size, location, and fuel type, as well as many different grid reinforcement strategies. Generation and transmission planning must be done iteratively because any new plant will change flows on the transmission system and may necessitate adding or changing transmission lines.[8] The tradeoffs between more generation and transmission are important and complex. A Federal Energy Regulatory Commission report elaborates:[9]

> Essentially, there is a tradeoff between adding generating reserve capacity at $200 to $300 per kilowatt for peaking units versus installing additional transmission facilities (say, 500 kV at $300,000/mile) to achieve an equivalent reliability benefit to the system. In general, transmission up to some distance can be added much more cheaply than generation, although...transmission only provides access to reserve capacity; it cannot substitute for it.

If a new plant is needed, the specific choice of what, where, when is determined by the size of the expected capacity shortfall (after lower-cost conservation measures are employed), the cost of fuel alternatives available at the plant site, and the expected time profile of plant operation. A variety of plant configurations can be chosen, each representing a combination of fixed (construction) costs and variable (operating) costs. There are often tradeoffs: A plant can be built which is initially expensive, but runs cheaply and reliably for a long time, or one can be built with a shorter expected life, lower construction costs,

Figure 3-1. Simplified Stages of Traditional Electric Utility Generation and Transmission Planning Process

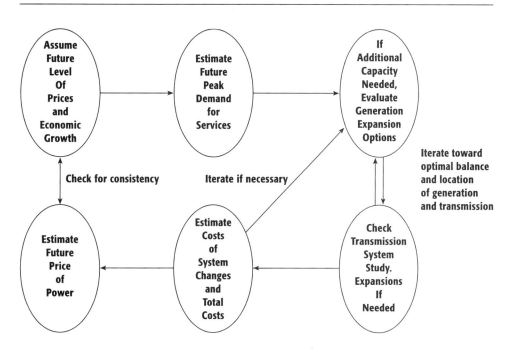

and possibly higher operating costs. Historically, planners have grouped these choices into three types of plants known as base load, intermediate (or cycling) and peaking. (Appendix 3B describes this aspect of utility planning in additional detail.)

Before the 1970s, additional plants were usually the only option that utilities used to address capacity shortfalls. In the 1970s, the cost of new plants drove rates up dramatically, prompting the industry to study whether various types of energy efficiency programs could help to more cheaply supply services. This new planning method, *Integrated Resource Planning* (IRP), became federal policy in the Energy Policy Act of 1992 (EPAct). (Appendix 3B discusses IRP in greater detail.)

When planners have settled on one or two plant and line packages that appear to be reliable, stable, and low-cost, the best alternatives are more carefully evaluated for the expected long-term total costs (bottom middle, Figure 3-1). Using net present value of total costs as the key measure—including the costs of repaying money borrowed to build the plant and paying shareholder dividends, if any—the best options are searched to determine the lowest-cost option.

Political and regulatory considerations enter here. For example, if, under a particular option, it appears unlikely that a necessary permit for a plant or line may not be obtainable, the option may have to be discarded even if the computer says it is lowest in cost. A 1981 Department of Energy report addressed this point:

The planner seeks to match generating unit characteristics with the power system to meet the characteristics of the power system to meet consumer needs at the lowest possible cost. Additionally, the choice of units must recognize numerous constraints including environmental effects, siting constraints, fuel availability, reliability, the financial viability of the utility and the regulatory process. Thus liability is one of the many constraints that the planner must recognize and, in many cases other considerations may far outweigh the planner's reliability concerns.

Permits and siting are discussed in greater detail in the next section. In the meantime, it is important to recognize that it is extremely difficult to separate analytical and technical issues from "non-technical" issues such as the ability to obtain a permit. "Even technical impediments to power transfers ultimately have non-technical causes," noted a group of experts at the National Regulatory Research Institute; "conversely, many non-technical impediments are almost inseparable from important technical issues."[10]

Traditional utility regulation sets the price of electricity equal to the total cost of providing power, including the costs of adding new plants and servicing utility capital. Once planners have estimated the future costs of the preferred system-expansion plans, these costs can be added to the existing system's forecasted cost and translated into the future prices of power (bottom left, Figure 3-1). This brings the planning exercise full circle, and planners can check whether their original assumptions were valid. If not, the load forecast must be adjusted and the entire cycle in Figure 3-1 must be checked to make sure the expansion plan is still best.

Mathematically, this problem of solving for an optimal expansion plan is so large that it must be solved in pieces, each piece analyzed using one or more computer models or simulators. Each partial solution is plugged into the next stage of the process, and the process continues iteratively until a solution is found by trial and error. Figure 3-2 shows a typical sequencing of the various types of computer models a utility might use.[11]

PERMITS, SITING, AND REGIONAL PLANNING

No discussion of utility planning would be complete without reference to the numerous approvals that are required to construct or change power generation and transmission facilities. Wholly apart from the states' present ability to set retail prices and conditions of service, about two-thirds of the states and various agencies of the federal and local governments exercise substantial authority over the ability to build a power facility. Typically, large power plants need as many as several dozen permits from various agencies such as air quality regulators, the Army Corps of Engineers (if waterways will be impacted), land-use permits, and so on.[12]

State permission to build power facilities originally stemmed from the state's desire to control overbuilding at the expense of the ratepayer as well as excessive disruption of public rights of way.[13] In addition, because some state statutes allow the state to exercise eminent domain in acquiring land for power facilities, it was deemed appropriate to verify that a proposed power system addition was of substantial benefit.[14] Because these "certificates of convenience and necessity" (permits to build) were issued by the same

Figure 3-2. Traditional Sequences of Computer Models Used for Electric Utility System Planning

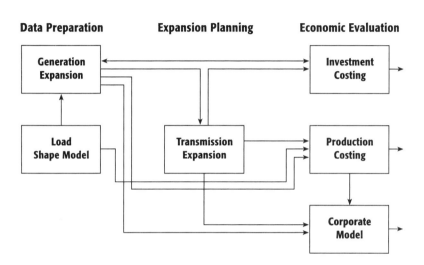

A typical sequence of computer models used for traditional long-term utility system planning. The load shape model embodies a ten-year projection of prices and loads. Note that there is no "feedback" from annual costs (upper right) to load shape, as shown in Figure 3-1. This suggests that this figure dates from the 1970s, before utilities were careful to check for consistency between forecasted cost, rates, and loads (See Appendix 3A).

regulatory body that regularly reviewed utilities' rates, plans, and finances, these approval activities were often viewed as part and parcel of the economic regulatory process. If a utility felt that its own planning process was sound, there was little doubt that regulators would fail to permit an addition.[15]

In the late 1960s and early 1970s, passage of the nation's first major environmental statutes, the growth of the environmental movement, and the establishment of the Environmental Protection Agency heightened awareness of the environmental impacts of energy facilities. In a report typical of this era, the New England River Basins Commission wrote in 1972:[16]

> It now appears that the improvement of environmental quality will become a social objective of unprecedented urgency in New England in the 1970's. As one consequence, new restrictions—and more stringent administration of existing restrictions—on the activities of energy procedures will be sought. At the same time, consumption of electrical energy is projected to rise at an even greater rate than that of population growth, and this increasing demand will create greater pressures for the construction of new facilities for generating and transmitting electricity.

It is imperative that state governments actively participate in the process of resolving conflict between environment and electric reliability. The paramount role of the states in environmental protection is well established and expressly recognized in such landmark federal legislation as the Water Quality Act of 1965 and the Clean Air Act of 1967. Reliability in New England will involve substantial interstate transmission of electric power through a regional interconnection system; as such, many aspects of reliability may be subject to federal regulation. The states, however, also may make vital decisions on reliability, including the policy decision of whether a state will allow construction of new plants for exporting power to other states, and, if so, under what conditions.

At the time this was written, only five states had power facility siting approval processes distinct from public service commission approvals of new plants for economic reasons. By 1984, the situation had changed dramatically. Thirty-three states had facility siting laws that applied to all power generation facilities and transmission (though typically not distribution) lines.[17] About half of these siting laws placed new requirements on public service commissions, requiring them to consider environmental, aesthetic, and other factors outside the traditional realm of power system economics.[18] The other half of the states created new administrative bodies (loosely referred to as "siting councils") that issued siting permits pursuant to specific economic, environmental, and other criteria.[19]

A number of studies have examined the impacts that these expanded approval processes had on regional and national utility planning. Some of these studies are particular to power plants of one kind or another or power plants in general.[20] (In particular, substantial research has examined the source of extremely common construction permit delays in U.S. nuclear plants and compared the regulatory process for nuclear power plants in the United States to similar processes elsewhere in the world.)[21] Other studies of power plant delays are less common, although permitting delays has become an issue of perennial concern to independent power producers as well as traditional utilities.

Transmission lines require fewer permits than power plants, but an environmental impact statement is required if federal lands will be crossed or federal funds are used to finance the line.[22] The primary environmental impacts of transmission lines involve construction, land use, and more recently concerns about the potential adverse effects of the electro-magnetic fields emanating from large power lines.[23] Litigation that attempts to alter the route of a line away from highly populated, environmentally sensitive, or visually attractive areas has become common. When opposition is severe and well-organized, the delays can be enormous and often fatal to the project.[24]

Ironically, a traditional source of difficulty in the siting of new power lines is the regional nature of their benefits. Whereas the public in the area of a new power plant commonly perceive that they will obtain all or most of the output of the plant—not to mention the local property tax benefits and additional employment opportunities created by the plant—transmission lines are (correctly) perceived to exist primarily to strengthen regional reliability, enable cost-reducing exchanges between utilities, and other less-localized benefits.[25] A 1983 study of electric utility regulatory reform by the National Governor's Association neatly summarizes this conflict:

Siting Conflict: Similar problems crop up in relation to siting decisions for power plants and transmission lines. State regulators are often constrained to look only at their own state's needs and concerns in making siting decisions. Yet, generation facilities may be sited most efficiently when the region as a whole is taken into account. The same is true for transmission facilities, yet a state may be reluctant to allow transmission lines to traverse it's territory in cases where little or no electricity will be provided to customers in that state. Such siting conflicts may have an adverse impact on the timely planning, construction and use of power plants and transmissions lines needed to serve a multi-state area.

These difficulties prompted the U.S. Department of Energy to observe:[26]

> The major limitation in planning transmission network configuration is the difficulty of obtaining rights-of-way. Traditionally the rights-of-way were selected on the basis of connecting the power source to the load area by the most economical route. Heightened environmental awareness has increased the difficulty of obtaining rights-of-way, and often the utility must resort to double or triple circuits on a common right-of-way to accommodate new load growth. Such network practices greatly increase the exposures of the transmissions system to...failures.

In spite of these difficulties, the two most complete examinations of transmission line siting both concluded that the actual magnitude of the problem is not large. The National Electric Reliability Council (NERC) reports annually on the status of transmission line additions and identifies those lines that are significantly delayed. The 1989 Federal Energy Regulatory Commission's Transmission Task Force found that, over the previous seven years, a total of 100 circuit miles of high-voltage line was significantly delayed, while 29,000 miles were added with little or no delay across the nation. The Task Force concluded that, "While the projects cited by NERC are of significant concern, the magnitude of the construction problems seem small in comparison to the overall success the industry has experienced in building needed transmission lines."[27] Similar conclusions were reached in an even more systematic 1990 study by the Consumer Energy Council of America.[28] However, at least one other major study (Kelly, 1987) concluded that siting impediments remained serious, numerous, and growing.

In short, the present industry structure and its planning processes are impacted greatly by the need for government construction approvals. With the exception of several highly visible controversies, the impacts thus far have been modest. The more important question is whether siting will become easier or more difficult under restructuring. Many observers of the permitting process have noted the close relationship between the fact that electric utilities are generally considered by the public to be local entities partially devoted to public service, and that this perception has facilitated public approvals. Because restructuring is certain to make utilities less local and possibly viewed as less devoted to public service, this issue is discussed in Chapter 8.

INTERDEPENDENCE IN LONG-TERM PLANNING

In almost all cases, significant additions to one utility's system affect the operating costs and technical features of several neighboring systems. As a result, long-term planning exhibits the same extremely high degree of interdependence as short-term operations. One utility cannot perform a long-term, least-cost planning exercise without knowing the present and future conditions of neighboring systems, including forecasted loads and plant additions. This is true even if each utility maintains roughly enough capacity to match its own system's peak demand, which has been the norm up until now.

Long-run or planning interdependence is the long-term version of the short-term inter-dependence discussed in the previous chapter. In the short run, system operators have a limited ability to correct for all of the costs one system suddenly imposes on another due to an unpredictable outage.[29] In the long run, however, if planners share enough information about each others' system plans, many of the costs imposed on other systems can be spotted in advance—even some that only arise in emergencies. This allows planners to reduce the problem via design changes or changes in the rules under which future system operators work, or possibly to agree on monetary compensation.[30]

In order to explore systems' impacts on each other, planners need to know the approximate location and electrical characteristics of every major planned unit in their area. They do not need to know *everything* about each plant, but they do need to know many of the plant's major attributes, such as size, type, location, and some of its electrical characteristics. Similarly, those who build power plants need to know quite a lot about the existing and planned transmission system in order to decide where best to locate new plants and how best to design them. All this information is used by planners to simulate network behavior to ensure that the system can be dispatched securely after it is built.[31]

Why do utilities put up with what sounds like rather elaborate requirements for sharing information and continual joint studies involving areas well beyond each utility's own system? The answer is that the economic benefits are compelling. Eighty years of successively larger studies of coordinated long-term planning have resulted in numerous and substantial benefits.[32] Depending on the size of the utility involved, long-term coordination can be expected to reduce costs on the order of 10 to 60% over planning and operation on a single-system basis. For example, rural Alaska is one of the few places in the United States where the systems that serve rural villages are not interconnected. Because of the more expensive costs of fuel these isolated systems cost at least twice as much as integrated systems in urban, interconnected parts of the same state.

Why does long-term coordinated planning reduce costs? A report by the Federal Energy Regulatory Commission nicely summarizes the advantages as follows:[33]

> As a load on a utility grows, its reserve margin will decline unless additions are made to its generating resources. Ideally, a utility would schedule new generating units so that its declining reserve would reach the minimum acceptable level at the time just before each new unit goes into service, causing a step to increase in the reserve margin. The step increase results in temporary "excess capacity." If a utility plans perfectly, it will always have some excess capacity,

whose average amount will be proportional to the size of the generating units added. The amount and cost of the excess capacity can be reduced by choosing smaller generating units, added at more frequent intervals, so that there is less fluctuation in the reserve margin. Smaller units, however, have higher busbar power cost, so that a trade-off is involved in selecting the optimum size of the new generating unit. Because the addition of a large unit of best economy represents a smaller percentage reserve increase for a large system than for a small system, a large system tends to not experience decenniums from the "lumpiness" of capacity additions as great as will a small system. However, all systems, large or small, can reduce their excess capacity cost by coordinating capacity construction with other systems, including arrangements to buy or sell capacity. In effect, such coordination allows utilities to add capacity on the basis of a single larger system, reducing their average excess capacity and the cost of carrying that capacity.

Although it may be possible, without prior planning and arrangements, for a utility to sell some of its excess capacity, or to purchase temporary capacity if its construction programs fall behind schedule, such opportunities depend on varying situations of other utilities. Further, the price at which such power is sold or purchased may not be advantageous. With coordinated construction, each utility is assured of a market, at an agreed-upon price, for the excess capacity of a new unit of economic size. It also allows its installed reserve margin to decline below the required margin, prior to the in-service date for a new unit, because it can purchase sufficient firm capacity to make up its deficiency, again at an agreed-upon price.

In summary, for the individual system, construction coordination and the associated agreements for capacity sales and purchases permit a more precise matching of electric supply to load, minimizing cost due to temporary excess capacity deficiencies. Various contractual and coordination agreements are used to achieve such savings. For example, a coordination arrangement may specify that a member of the system will purchase, on a pro rata basis, capacity from a pool unit that is in excess of the owners-utility's needs. Or the coordinating systems may agree to joint ownership of new facilities based on the relative requirements of the parties. This allows individual utilities to add smaller increments of capacity, matched to their load growths, while retaining the scale of benefits of large units.

Greater flexibility in siting is another benefit of coordinated construction. With joint ownership arrangements, a utility need not have all of its generating capacity within its own service area. Such flexibility has the potential for reducing environmental impacts and for reducing construction delays due to litigation. This latter factor can, of course, be translated directly in reserve requirements occasioned by uncertainties in unit installation dates.

The advantages of coordinated planning again raise questions concerning industry structure. In unregulated industries, competitors ought not to benefit from planning jointly. However, because one of the functions of firms is to coordinate investment decisions, it is logical to ask whether the benefits of coordinated planning suggest that perhaps the entities that are planning together would be best in a single firm. Then, rather than having to use complex multi-company processes, planning as an integrated unit would auto-matically occur within a single firm.

Short of combining into one company, planning coordination can be strengthened by a contract or compact, and/or by utilities agreeing to jointly own or operate system addi-tions. Both of these interim measures are used, though not nearly as widely as the potential benefits merit, according to a number of studies. These issues of industry structure are extremely important, and are closely wedded to the forces driving indus-try restructuring. These issues are dealt with in next chapter.

PLANNING INSTITUTIONS AND PROCEDURES

If one utility fails to build enough capacity to serve its customers, system controllers will someday have trouble balancing supply and demand in that utility's entire area—includ-ing parts of other utilities' systems that may have built sufficient capacity for their own systems. When it comes to pass, this shortage of adequacy on the part of one utility will translate into less reliability (more outages) and/or higher costs (caused, for example, by more expensive power purchased under emergency conditions from other utilities).

Generally, several things have prevented this from happening in the past. First, most state-regulated utilities are required by law to serve all customers' demands in their area at levels of reliability that are set by regulators.[34] Hence, regulated utilities violate the law if they willfully underbuild their systems. Removing this "obligation to serve" is one of the most controversial aspects of electric restructuring.

In addition, one of the most important functions of the nine regional reliability coun-cils (discussed in the previous chapter) is to make a regional "code of conduct" that requires each utility to build enough capacity to serve its system, and to do this in a way that does not harm the operation of nearby systems.

Unlike its operating rules, which can be quite detailed, NERC planning criteria can be relatively short and simple. The first three criteria for one of the reliability councils, the Western Systems Coordinating Council, for example, state:

1. Each system should provide sufficient transmission capacity within its system to serve its load and meet its transmission obligation to others without unduly relying on or without imposing an undue degradation of reliability on any other system, unless pursuant to prior agreement with the system(s) so affected.

2. Each system should provide sufficient transmission capacity, by ownership or agree-ment, for scheduled power transfers between its system and any other system. In transferring such power there should be no undue degradation of reliability on any system not a party to the transfer.

3. Each system should conduct the necessary studies to demonstrate that its power transfers, in the absence of loop flow, will not unduly rely on or impose an undue degradation of reliability on parallel transmission capacity of any other system, except pursuant to prior agreement with affected system(s).

Although these principles are simple, implementing them is time-consuming, costly, and sometimes contentious. Utility planning staffs constantly exchange volumes of data about their present and projected systems, and engage in computer modeling exercises involving dozens of utilities over very large areas. Because many of these conditions can be subject to interpretation, studies and arguments can stretch on for many years. Moreover, membership in reliability councils is voluntary,[35] and members have no power to sanction members or nonmembers who violate the rules.[36]

This means that the regional councils largely limit their activities to meeting regularly to exchange planning data and run joint, large-scale simulations to check on the total reliability picture for the grid. They continue to modify their regional operating and planning criteria, usually by a vote of their members. If problems arise and talking about them does not produce a congenial solution, the members ultimately settle their differences before regulators or the courts.

A few power pools handle the long-range planning function as well as short-term interchange and coordination. The New England Power Pool (NEPOOL), for example, uses this process to coordinate the expansion plans of its members:[37]

1. Pool members submit planning data and their individual expansion plans to the pool each year.

2. If the pool finds that one expansion plan will adversely impact another system's future reliability or substantially increase its costs, the pool member whose expansion will be harmful agrees to change its plan. Unlike the urging of a reliability council, this is a binding contractual commitment between all pool members.

3. The pool prepares its own pool-wide expansion plan and compares this "integrated" plan to its individual members' plans.

4. The pool can then designate some number of member-proposed planned generation and transmission facilities, which it finds to be elements of its pool-wide plan, as "pool planned" generators or transmission lines. If this designation is made, all members of the pool may obtain ownership shares or use the pool-planned facilities.

Although this is far from an ironclad regional planning process, it is intended to evolve planning away from many individual expansion plans (NEPOOL has over 90 member utilities) toward a much smaller number of larger pool-planned units in which all members participate. In this way, the pool serves as a vehicle for constructing a smaller number of larger units, rather than many small ones, thus achieving economies of scale in generation plants and lines as well as the other benefits of coordinated planning mentioned above.

The Federal Energy Regulatory Commission has taken a major step toward opening up the regional generation and transmission planning process by requiring all utilties under its jurisdiction to file extensive information on their present and planned transmission system, hour-by-hour system operating costs, transmission planning criteria, and other information.[38] This requirement marks an important point of transition away from the traditional industry, and is included in the discussion of long-term planning in the restructured industry (Chapter 8).

TRADITIONAL LONG-TERM PLANNING: SUMMARY

The long-range activities that keep the power system reliable revolve around several key points. First, the obligation to serve has compelled systems to plan for adequacy, as a clear-cut legal obligation, and to minimize costs subject to this requirement. Extreme events and outages will always occur—indeed, utilities expect and plan for them—but the over-all process has been set up to put reliability (security and adequacy) first, minimum cost second, and to treat everything else (e.g., permits) as constraints on the ideal solution.

As with the short run, reliability and cost responsibilities are mirrored in different industry organizational features. Regional reliability councils eschew any interest in, or respon-sibility for, ensuring that each utility has found the lowest-expected-cost expansion plan; they assiduously limit their studies to ensuring that the total network is adequate and thus reliable. Each utility must couple the regional studies to its own lowest-cost plan that meets regional criteria. This may mean going back to the same utilities with whom reliability studies were conducted to discuss the joint ownership of a plant or line, or any of various other economic solutions that will be examined below. Explicitly or implic-itly, these least-cost plans are subject to the approvals of state regulators. Finally, federal regulators approve the rates for interstate transmission and "promote" interconnection, reliability, and regional planning, but as yet they have little authority over the creation or siting of new facilities.[39]

In reality, the divide has never been as rigid as the responsibilities of the different insti-tutions suggest. First, there are economic tradeoffs between the provision of reliability via added transmission capacity and added generators. Any study of reliability options requires the incorporation of costs in order to make it realistic.[40] In addition, state reg-ulators are viewed as responsible for reliability as well as rate regulation, and regional planning groups understand that every decision they make has potentially enormous financial implications for their members, and must pass muster with utility manage-ments and regulators.

Another important point is the coordinated and, thus far, voluntary nature of the mul-tiutility planning picture. Many states have statewide long-term planning procedures that compel utilities within their jurisdictions to plan together with binding outcomes.[41] However, the natural size of most bulk power markets is much larger than any one state, and regional bulk power planning is now essentially voluntary.

In most cases, this has yielded perfectly acceptable (if not excellent) outcomes, but there are also many stories about how member utilities in these planning bodies reached-stalemates that prevented a system addition that had clear economic benefits to the region. In one instance that I witnessed, three utilities spent more than two years nego-tiating the addition of a major new transmission line, only to see discussions fail and no line built in spite of a mutual acknowledgement that the line would provide substantial regional benefits. Interestingly, the industry has sometimes chosen to downplay these "nontechnical" aspects of planning; for example, Stoll's (1989) otherwise excellent textbook on utility least-cost planning contains only a perfunctory mention of siting and licensing.[42]

More commonly, utility planners have incorporated permitting considerations under the general rubric of "uncertainty." In a 1987 review of utility planning procedures, two consultants concluded:[43]

> One striking feature of our discussions with planners was the degree to which the emerging complexity of planning was called out as the major challenge confronting utilities planners and regulators. There is general recognition that the planning environment has changed dramatically, and that the old methods and tools are not completely sufficient for the new environment. Planners cited three causes of increases planning complexity:
>
> • Increased uncertainty,
>
> • More options for supplying energy services,
>
> • Multiple and frequently conflicting objects.

The substantial increase in uncertainty that now permeates the utility planning environment is a consequence of several factors. Among these are increases in the volume of fuel prices, construction cost and the lead times, competition between fuels and between suppliers, and high uncertainty over changes in regulation.

Utility planning procedures now incorporate the many uncertainties mentioned here using scenario planning and even more sophisticated probabilistic planning tools. For example, one utility planning model simulates thousands of possible combinations of fuel prices, permitting delays, construction costs, and other variables, and estimates the range of possible utility costs 20 years in the future.[44]

Table 3.1 completes the grid that was begun in the previous chapter. Hopefully, the table is a helpful means for distinguishing between short-term and long-term issues and procedures in the industry. The next chapter, explores the link between these traditional technical and economic issues and the number and types of firms in the industry.

ANALOGIES REDUX:
UTILITY PLANNING VERSUS PLANNING IN COMPETITIVE INDUSTRIES

Just as the analogies between utilities and other industries generally were examined in Chapter 1 and from the operations standpoint in Chapter 2, it is useful to compare electric utility planning to long-term planning by unregulated firms. Table 3.2 compares several aspects of utility planning to a typical unregulated heavy industry.

The first category is time horizon and lead time. It is sometimes said that electric power planning horizons are unusually long because electric plants and lines last 40 years or more. Many other industrial facilities do not last quite as long, but, in general, many

Table 3.1. Activities and Procedures that Maintain the Grid

Economic Time Frame	Utility Industry Time Frame	Technical/Economic Objective for Reliable System Operation	Key Present Institutions and Activities
Short Run (immediate to fiscal quarter)	Immediate (seconds to hours) ("Operations Control")	"Security:" Area controllers preserve system *stability* by providing *secure dispatch*	Control areas operated by one utility or a pool for all utilities in area
	Intermediate (days, months) ("Operations Planning")	Emergency and restorative actions by system operators	Voluntary regional reliability councils that establish rules allowing ECC to control all generation and transmission in each control area
		Unit commitment and scheduling	Reliability council exchanges of data and joint studies
Long Run (years)	1–20 years	"Adequacy:" Studying alternatives and acting to ensure that adequate generation and transmission capacity is available	Mandated statewide and utility-specific planning procedures ("Integrated Resource Plans")
			Reliability council "voluntary" planning criteria
			Reliability council exchanges of data and joint studies

heavy industries have very long lead times and long-lived plants. For example, automakers take several years to design, test, and build new model autos, and their production plants are multibillion-dollar, long-term investments.[45]

Similarly, many other industries must undertake long-term commitments on the basis of expectations concerning the cost of, demand for, and price of future products (third row of the table). The first major difference comes in the obligation of suppliers to supply all demand in their area—the obligation to serve all demand within one's service area (fourth and fifth rows of table). This is totally unlike unregulated firms, that can choose to expand or not expand, and are free to turn away new business if profit considerations so dictate.

This relates to another important difference: allocation of the costs of overinvestment. In competetive industries, overinvestment leading to excess capacity harms the profits of the firm, and usually drives prices down everywhere in the market. This can harm other suppliers' profitability, but generally benefits consumers. In regulated utilities, regulators decide how the costs of overinvestment will be allocated between the company and rate-payers. In general, neither of these groups benefits, and prices go up rather than down.

The final three rows in the table refer to the planning processes themselves. First, electric generation and transmission improvements (as well as DSM, load management, and pricing strategies) must all be considered iteratively together. Electric power systems have extraordinarily interdependent parts; in principle, unregulated firms must plan their distribution systems in coordination with their production systems. Similarly, electric power facilities have some of the largest permitting requirements of any industry (and often have their very own state siting legislation and administrative body), but they are not alone in this regard.

The largest difference in planning procedures is—once again—the interdependence of production costs and operations. In some respects, utility long-term planning and investment decisions are the antithesis of long-term behavior in competetive markets. The last thing competitors in most markets want to do is to exchange volumes of data on the costs of their present system, and to cooperatively plan additions to anyone's facilities so as to ensure that no other system is harmed. Competitors want to do just the opposite: keep their long-term plans a secret as long as possible, undercut their opponents, and impose the greatest possible competitive harm allowed under the law on rivals.

The normal incentives to attack competitors were explicitly removed in the utility industry by regulation and the granting of exclusive sales territories. As long as one utility could not legally grow by stealing another utility's customers, no utility had the incentive to compete with or harm another system. Utilities did compete indirectly to lure customers into changing their physical location, or choosing a new location, but this is obviously a more limited form of competition.[46]

Under an industry structure in which all generators compete, it is unlikely that planning can occur in the same fashion. Indeed, if generation is entirely competitive and entirely divorced from transmission, planning methods will change quite significantly. The potential

Table 3.2. Traditional Utility Planning Versus Planning in Unregulated Industries

	Traditional Regulated Utility	Typical Unregulated Industry
Multi-year planning horizon	Yes	Yes
Long-lived capital investments	Yes	Sometimes
Long-ranged projections of future prices and demand	Yes	Yes, though products often less homogeneous
Peak demand in future must be met	Yes	No
Penalty if future undersupply	Set by regulator	Lost sales
Risk from overinvestment	Set by regulator	Lower future profits and prices
Must plan production and delivery infrastructure interactively	Yes, Strongly no inventory or storage	Yes, weakly; inventories and storage possible but not free
Government permits affect planning	Strongly	Varies
Long-run cost interdependence with others in marketplace	Yes	No
Regulation approval can trigger eminent domain	Sometimes	Rarely

for conflict between useful regional coordination and the creation of sound competition between generators is one of the towering uncertainties inherent in electric power restructuring and is a subject discussed in Chapters 7 and 8.

INTERDEPENDENCE, INTEGRATION, AND INDUSTRY STRUCTURE

The interdependence of utility system costs in the short and long term brings the question of industry structure and ownership into focus. In the short run (i.e., when operators are dispatching power plants), does it make for higher or lower costs if one company (as opposed to several smaller companies) owns the entire system under control? If so, then utilities have economies or diseconomies of scale, and the question becomes, "what horizontal and vertical industry structure best serves the consumer?"

The same question can be asked about long-term industry operation. If industry long-term planning processes allow all utilities in an area to plan and build systems that are just as inexpensive as they would be if more or fewer utilities were involved, then multilateral planning procedures and contracts are just as good as horizontal integration—perhaps better. However, if it turns out that these procedures are far from perfect, and that utilities do a demonstrably better job of reducing costs when they own a larger part of their regional system, then perhaps larger systems are better able to reduce long-run costs—even if these larger entities compete with each other.

It it thus time to focus squarely on the organization and governance of the industry—first under traditional regulation and then for the industry in transition.

Appendix 3A

Predicting Future Power Rates and Loads

It is an accepted fact that, for almost every good or service, demand goes down when price goes up, and vice versa. For many years, as electric power rates were dropping, which caused demand to increase, utilities did not pay much attention to the impact of price on long-term use. Indeed, in many early "load forecasts," it was simply assumed that as population increased and the economy grew larger, electric power demand would grow in proportion, irrespective of prices.[47]

Prior to the 1970s, this forecasting approach worked fairly well. For example, Figure 3A-1 provides a 1964 picture of the relatively modest differences between actual loads and predicted loads. It is almost always true that the farther out a forecast goes, the less accurate it becomes. Figure 3A-1 shows that this held true in the 1960s, but even the errors over the longer time horizons were small.

In the 1970s, electricity prices shot up rapidly in many parts of the country. As these increases repeated themselves over a series of years, forecasters noticed that high electric rates were indeed causing power demand to drop. Following a period of analysis and debate, it became universally accepted that, like most goods, electric power demand drops when prices rise. In the case of electricity, however, researchers found an important additional wrinkle: The reaction to an increase in price is not immediate; instead, it stretched out over about five years. Indeed planners mistakenly overlooked the sensitivity of demand to price because not much reaction took place immediately following a price hike.[48]

Researchers found that price response is delayed for some quite understandable reasons. Electricity is used in lamps, motors, appliances, and other equipment that generally lasts more than a year. In order to use less electricity, in response to a price increase, consumers either must use the same equipment fewer hours or buy equipment that is more energy-efficient.

Figure 3A-1. Utility Load Predictions and Price Movements

During the period of price decline and stability, utilities could predict future demand with relative accuracy, often without factoring the price of power into account (Panel A, left). During the 1970s, when prices spiked up (later to drop again), utility forecasts substantially overpredicted future demand (Panel B, right).*

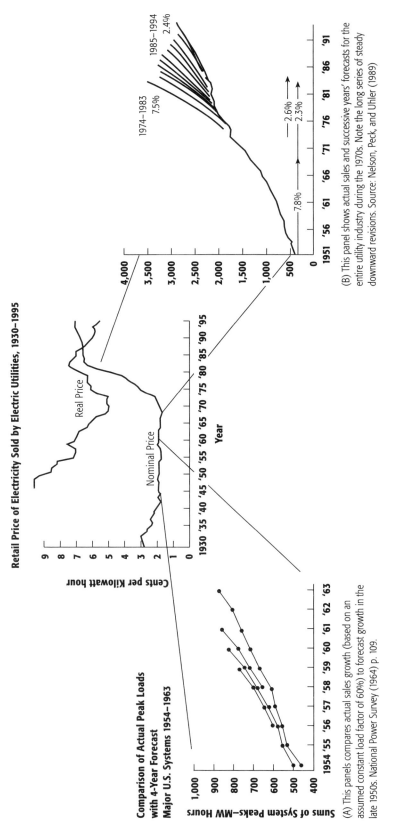

Retail Price of Electricity Sold by Electric Utilities, 1930–1995

Comparison of Actual Peak Loads with 4-Year Forecast
Major U.S. Systems 1954–1963

(A) This panels compares actual sales growth (based on an assumed constant load factor of 60%) to forecast growth in the late 1950s. National Power Survey (1964) p. 109.

(B) This panel shows actual sales and successive years' forecasts for the entire utility industry during the 1970s. Note the long series of steady downward revisions. Source: Nelson, Peck, and Uhler (1989)

*Average retail price of electricity sold to all ultimate consumers. Source: Edison Electric Institute (1978); Energy Information Administration (1995).

Consumers do some of both. The portion of the response attributable to using equipment fewer hours is relatively small, but generally occurs within a year. The much larger effect occurs when higher prices cause a factory or residence to replace its equipment with more efficient versions. This usually occurs over a matter of years following a price hike, as the old equipment wears out or the realization that more efficient equipment will pay for itself sinks in.[49] The income of the household and the general health of the economy continue to play a role as well.[50]

The price increases of the 1970s occurred in many successive years (see Figure 3A-1). Because the effects of one year's increase are felt as far as five years into the future, these increases had a large cumulative effect. Worse still, the effect was easy to miss at first because so little reaction occurred immediately following the first increase.

The result was that, throughout the 1970s utilities were forced each year to readjust their load forecasts downward because they had overpredicted long-term demand. Due to the accumulation of impacts, the corrections utilities made in each successive year's forecast never captured the full price response, and downward adjustments were needed year after year after year. The result was the figure shown on the right side of Figure 3A-1. Due to its resemblance to a hand-held fan, this figure became known as the "NERC fan."[51]

The circumstances that produced the NERC fan prompted extensive changes in utility forecasting techniques. Now, forecasting models rather accurately capture the impacts of price changes far into the future, and most utilities are careful to check for consistency between their assumptions of price increases and the results of their studies of future system costs.[52]

Appendix 3B

Types of Power Plant Additions and Integrated Resource Planning

Planners initially study and predict the future time profile, or load shape, of power use. Most power users have fairly consistent daily and weekly schedules, over the course of which power demand goes up or down quite substantially. This allows planners to develop typical "load shapes" for days, weeks, and years.

The nature of the load shape and the rate at which it changes are strong factors in determining the type of power plant addition that is best for a power system. The left panel in Figure 3B-1 shows a typical daily load shape with two periods of very high peak use.

The shaded areas is the period of time each day that demand is well above the steady, lower level, which is only a few hours. Hence, the power plants needed to serve load on this day will be on only a few hours.

In the right panel of Figure 3B-1, the load shape is much smoother. The shaded area shows that the power plants required to serve load will be needed during many more hours of the day.

In both of these situations, planners must ensure that power capacity equals the single highest use, plus a safety margin. The lowest-cost solution to serving a load shape such as the one in Figure 3B-1 may be not to build a plant large enough to serve this narrow peak, but rather some other alternative that preserves reliability and energy balance. Three possibilities are: (1) arranging to buy power from a nearby system during peak hours; (2) charging more for power during these periods, hopefully causing some to conserve or shift their use to some other time (known as "time of day rates" or "real-time pricing;" or (3) arranging with some very large users to turn them off when demand gets high so that balance is preserved ("interruptible service").

These non-construction options are often cheaper than adding a plant. The idea of "least-cost planning" or "integrated resource planning" (IRP) was developed in the 1970s and early 1980s to urge utilities to look at all options for reliably expanding their systems, and eventually became stated federal policy in the Energy Policy Act of 1992.[53] However, the policy merely requires state regulators to *consider* IRP plans, and there is no formal review or sanction process. As a result, the level of implementation and formality in IRP processes varies considerably from state to state.[54]

Non-construction options notwithstanding, additional power plant capacity may well be a part of a least-cost planning solution. Broadly speaking, the cheapest way to add additional generation to the system for the load shape in the left panel of Figure 3B-1 is to build a plant that does not add much to system cost when it is only operated a few hours a day. This kind of plant is known as a *peaking plant* or a *peaker*.

For the shaded portions of load in this panel, the best generation addition is a plant that can run much but not all of the time and can adjust its output smoothly over time. This is known as an *intermediate* or *cycling* plant. Finally, notice that the unshaded part of this panel is an amount of load that does not vary one bit 24 hours a day. The cheapest way to service this load is with a plant that operates all the time at the lowest average cost and is extremely reliable. The initial cost of this plant can be high because it will be used constantly for many years; it is worth it to have a reliable unit that saves money in the long run. Units of this type are called "base load."

Integrated Resource Planning, Regulation, and Least-cost Incentives

IRP represented a philosphical, operational, and economic departure for the utility industry. Philosophically, the industry had grown up encouraging power use, and was accustomed to viewing its responsibility for adequacy as being paramount. Operationally, the industry had little experience with technological or pricing approaches that *reduced* peak demand, and many did not think that such approaches would work reliably or be cost-effective. Finally, utilities that were regulated in the traditional fashion had no direct profit incentive to *reduce* sales.[55]

IRP changed the philosophy of utility regulation by recognizing that consumers do not demand electricity for its own sake, but rather for reliable and specific amounts of heat, light, and *other energy services*. Under IRP, the obligation to serve was not an obligation to reliably serve all electric power demands without considering how those demands arose; instead, utilities were to serve all energy service needs of customers as cheaply and reliably as possible.[56]

This new obligation, which was made a part of the federal government's IRP mandate for all state regulators and state-regulated utilities was consistent with utilities' profit incentives.[57] The methods regulators used to set prices and profits were adjusted so that utilities could make as much money as they would have absent the use of alternatives that reduced plant additions.[58] However, these profit adjustments never spread beyond a handful of states, and are becoming obsolete with the onset of restructuring.[59]

Operationally, utilities have made significant progress in learning how to cost-effectively reduce and shape demand. Utility load management programs operate sophisticated computers that control thousands of loads in the service area on a momentary basis. Other utilities have paid for the installation of lighting or other building modifications that reduce power demand without reducing comfort or reliability at a cost well below the cost of additional power.[60]

With respect to the mechanics of the planning process, IRP alters the process shown in Figure 3-1 in obvious ways. In addition to studying plant and line additions, the utility studies a variety of demand-side options. Each such option changes future load shapes in a predictable fashion, changes sales and revenues (and possibly prices), and costs a predictble amount of money. Each of these changes is added to their respective part of the Figure 3-1 process, and the result is a plan that combines new generators, new lines, and a set of demand-side measures that, in total, comprises a system's integrated resource plan. Figure 3B-2 shows the schematic for a simplified IRP process; a comparison with Figure 3-1 illustrates the critical process changes that have been introduced by IRP.

Figure 3B-1. Load Profile and Generating Plant Additions

To serve the shaded portions of the future loads shown in (A), the lowest-cost system addition is a peaking plant. To serve the shaded portions of (B) in a future year, the lowest-cost plant is a cycling or intermediate plant. In both instances, traditional regulated utility planning is updated to reflect "integrated resource planning" which considers options other than adding new power plant capacity to serve the shaded loads (e.g., interruptible rates or energy efficiency programs).

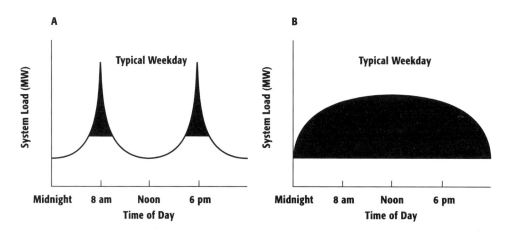

Figure 3B-2. Alternate Version

Integrated Resource Planning Process

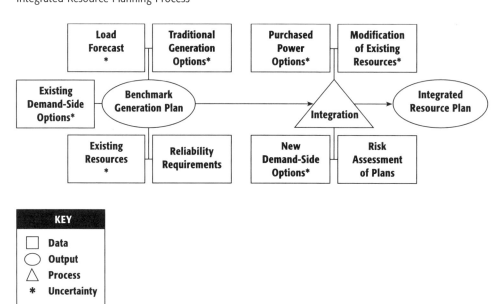

Source: Edison Electric Institute Power Systems Planning and Operations School December, 1995

Chapter 3

NOTES

1 The official definitions of reliability promulgated by the North American Electric Reliability Council (NERC) are: Reliability: "The degree of performance of the element of the bulk electric system that results in electricity being delivered to customers within accepted standards and in the amount desired. Reliability may be measured by the frequency, duration, and magnitude of adverse effects on the electric supply. Electric system reliability can be addressed by considering two basic and functional aspects of the electric system—Adequency and Security. Adequacy—The responsibility of the electric system to supply the aggregate electrical demand and energy requirements of the customers at all times, taking into account scheduled and reasonably expected unscheduled outages of system elements. Security—The ability of the electric system to withstand sudden disturbances such as electric short circuits or unanticipated loss of system elements."

(NERC 1996 Glossary of Terms). With respect to the security/adequacy distinction, a 1979 U.S. Department of Energy report observes that,

Reliability involves the security of the interconnected transmission network and the avoidance of uncontrolled cascading tripouts which may result in widespread outrageous. Adequacy refers to having sufficient generating capacity to be able at all times to meet the aggregate electric peak loads of all the customers and supply all their electricity energy requirements. This distinction is an increasing one: since much confusion exists between these two areas. NERC uses reliability to label the technical electrical engineering problem of protecting against damage from circuit transients and outages. Capacity sufficient to insure supply in the face of scheduled and unscheduled shutdowns of generating facilities is labeled adequacy. In utility planning practice these terms are used more loosely. A "relality" criteria of a loss-of-load probably of one day in ten years may be used to establish reserve margins to assure adequate system capacity. In NERC's lexicon, this is an adequacy issue, not a reliability one, but utility engineers often refer to it under the heading reliability.

2 See DOE (1981), p. 29–33 for an excellent discussion of reliability criteria used in system planning.

3 See Fox-Penner (1990) and Hyman (1992) for nontechnical introductions to electric power concepts and measures.

4 The term coincident peak demand is sometimes used to describe demands that have been aggregated through interconnection. Recognition of the importance of peak coincidence is first attributed to Insull (Platt, 1991), but one of the most important economic explanations was made by Lewis (1949, p. 52). See the discussions in Bonbright, Danielsen, and Kamerschen (1986, p. 496) and Stoll (1989, p. 236).

5 Additional plants and lines are not the only option. See Appendix 3A in this chapter.

6 In the short run, utility controllers try to select at each moment, *from among the existing set of plants*, the combination that meets energy balance *and* provides for system reliability, voltage support, and so on. Mathematically, this is a constrained optimization: find the set of plants in each time period that lowers the total cost of supplying power in that time period. The calculations and data necessary to do this are generally quite straightforward—so much so that the data is now given to a computer and the computer automatically performs economic dispatch in real time.

In the long run, power planners ask exactly this same question, except that now peak demand is rising over time, and they are free to experiment with additions of plants and lines to see which combinations yield lowest-cost dispatches. Mathematically, this is a much larger and more complex problem. Instead of finding the lowest-cost combination given a set of plants, the planner is allowed to test any number of new plant and line options. Since each option will change the system over the entire life of the new facility, the lowest-cost dispatch for every minute of the system operation over the next twenty years or so will be changed.

7 For additional descriptions of the planning process, see Rustbakke (1983) p. 278–279, Federal Power Commission (1967), and Department of Energy (1981) ch. 7. For an excellent but extremely technical reference text, see Stoll (1989).

8 See Rustbakke (1983) p. 283–284 for a good overview of transmission planning and its relationship to generation planning.

9 FERC (1981) p. 22.

10 Kelly (1987) p. 197.

11 In Figure 3-2, the "load shape model" contains future projections of prices and peak demands.

12 A guide prepared to advise utilities on siting issues by the industry's trade association lists 25 government permits required of a typical major fossil fuel power plant. Only one of these permits comes from the state utility regulatory agency (Edison Electric Institute, 1970, p. 7).

13 See Platt (1991) ch. 3.

14 For example, the Illinois Public Utilities Act (Ill. Rev. Stat. Ch. 111-2/3, 1985, Sec 8-406) holds that "[N]o public utility shall begin the construction of any new plant, equipment, or property...unless and until it shall have obtained from the Commission a certificate that public convenience and necessity require such construction." The statute further holds that, when considering such a certification, the Commission must evaluate whether the proposed addition is the lowest-cost feasible alternative, construction will not harm the utility's financial integrity, and other conditions.

15 When I participated in an objection to the issuance of a new plant permit in Illinois in the late 1970s, old-timers at the Commission told me it was the first time in the history of the state that anyone (including the Commission itself) objected to a new facility.

16 New England River Basin Commission (1972) p. 2.

17 Most siting laws exempt distribution lines and small power generators, but the exempt size classes vary greatly. Connecticut's law governs all generators and lines 69 kV. or larger; Maine's law governs transmission systems 125 kV or larger, and New Hampshire's siting law controls plants larger than 50 MW and lines above 100 kV. [Connecticut Public Act 575, 1971; Maine Statutes ch. 476, 1971; N.H. Stats. Ch. 357, 1971].

18 For example, in the Illinois Public Utilities Statute, the Commission cannot issue a certificate for construction unless the plant conforms with the approved statewide plan. In turn, the plan is required to incorporate all cost-effective measures for the use of energy conservation and renewable resources. (Ill. Rev. Stat. ch. 111-2/3, 1985).

19 See Southern States Energy Board (1978, 1985), and National Governor's Association (1986). Approximately 18 states consider or require long-term plans in association with new transmission line approval.

20 See, for example, Federal Energy Administration (1975), Electric Power Research Institute (1983), Edison Electric Institute (1970), Southern States Energy Board (1978, 1985), and other references *infra*.

21 For three prominent examples, see Kemeny (1980), Price (1990), and Rhodes (1993).

22 At present, virtually all major lines built by rural co-ops or municipal utilities use federal funds, and therefore must complete an EIS. In the Western U.S. and Alaska, federal land holdings are so extensive that most long-distance lines traverse federal property at some point, irrespective of their ownership.

As an example of the latter, the Sierra Pacific Power Company has proposed a 164-mile long, 345 kV line from Alturas, California to Reno Nevada. Fifty-six percent of the land traversed is under the jurisdiction of the Department of the Interior, and a Draft EIS has been prepared according to the relevant standards set forth in the Federal Land Policy and Management Act of 1976 (43 U.S.C. § 1701 et seq.). The draft is in two volumes, 878 and 319 pages respectively, including extensive maps and construction plans and discussions of "positive and negative impacts." (Reid and Droffer, 1995, p. 12).

23 For a recent, concise summary, see Lock and Stein (1996), Section 81-20.16(6).

24 One heavily-studied case is the Potomac Electric Power Company's 500 kV loop around Washington, D.C. Proposed in 1972 for large and apparently uncontested economy and reliability reasons, line approval was delayed until at least 1990 by vigorous opposition from local groups and authorities in Maryland. See NRRI (1987) and CECARF (1990).

25 Kelly (1987), Lee and Ellert (1991) p. 198, and Stalon (1991) p. 113. Interestingly, the same phenomenon acting on an intrastate level contributed to the enactment of state siting laws in the 1970s. State facility siting laws typically pre-empt all other zoning and land-use laws, making it impossible for every locality affected by a plant or line to require its own approval. See NERBC (1972), p. 23f.

26 National Governors Association (1983) p. 11

27 FERC (1989) p. 39.

28 See Consumer Energy Council of America Research Foundation (1990), p. 24.

29 Utilities routinely charge each other for service provided under emergency (unplanned) circumstances, and the cost for this power is usually considerably higher than the usual price. However, this approach only provides compensation for that portion of the episode in which the grid continues to function. If an incident in one utility's system causes outages in other utilities, no compensation is typically demanded or paid for the costs of the outage to other utilities or their customers.

These compensation conventions are undergoing scrutiny in connection with restructuring, and some changes are likely. This is taken up in the discussion of ancillary services in Chapter 8.

30 Utilities have found a number of ways of correcting or compensating each other for the cross-system impacts of growth. In some cases, the solution is to alter the system addition, especially the transmission system. Even though flows cannot be directed down specific paths, the total amount and direction of flow on one line can be remote-controlled, and sometimes this helps. Other problems can be solved by utilities agreeing to operate the system a certain way. Finally, if it turns out that one facility affects several systems, it often makes sense for utilities to share the ownership of that facility so that they can divide up its costs and benefits in a special joint ownership agreement.

31 See, Haase (1996), Rustbakke (1983), and Wood and Wollenberg (1992).

32 These studies are cataloged and discussed in Chapter 5.

33 FERC (1981) p. 20–21.

34 Every state public utility regulatory statute contains a requirement that regulated utilities serve all customers. For example, the Illinois Public Utilities Act (Ill. Rev. Stat. 111 2/3, Sec. 8-101) states that "Every public utility shall, upon reasonable notice, furnish to all persons who may apply therefore and be reasonably entitled thereto, suitable facilities and service, without discrimination and without delay." In the regulatory literature, the shorthand term "obligation to serve" is used to describe public utility obligations of this nature. See the extensive legal history of the obligation to serve in chapter six of Nichols (1928); Bonbright, Danielsen and Kamerschen (1988) p. 10; and Kahn (1970) v. I p. 5–7.

35 Until recently, essentially every significant utility routinely joined their reliability council. Recently, Puget Power and Light, a West Coast utility, made history by withdrawing from the Western Systems Coordinating Council over a dispute with other members. Puget's withdrawal suggests that the traditional system is coming under extreme pressure arising from industry restructuring.

36 For a discussion of this point, see FERC (1981) p. 16.

37 NEPOOL Agreement, Sept. 1, 1971, as amended. Also see FERC (1981) p. 68.

38 New Reporting Requirements Implementing Section 213(b) of the Federal Power Act and Expanding Regulatory Responsibilities Under the Energy Policy Act of 1992, and Conforming Other Changes to Form No. FERC 714, 62 FERC 61,297 (1993).

39 A longstanding exception is the creation of transmission lines linked to federally-licensed hydroelectric facilities. Apart from these, it was not until the Energy Policy Act of 1992 that the Federal Energy Regulatory Commission was given authority to compel wholesale transmission service. However, this authority is not equivalent to the authority to approve the siting or construction of a specific line; it merely requires a utility to find a way to provide service over new or existing facilities. Moreover, an additional provision provides that a utiltity need not furnish service if additional facilities are required to render service and it has failed to obtain siting and construction approvals for such facilities following good-faith efforts. Hence, this federal authority does not pre-empt the state or local siting processes described above. [Federal Power Act, 16 U.S.C. § 824, Section 212(h); also see Policy Statement Regarding Good Faith Requests for Transmission Services and Responses by Transmitting Utilities Under Sections 211(a) and 213(a) of the Federal Power Act, as amended and added by the Energy Policy Act of 1992. 64 FERC P. 61,065].

40 "Essentially, there is a tradeoff between adding generating reserve capacity at $200 to $300 per kilowatt for peaking units versus installing additional transmission facilities (say, 500 kV at $300,000/mile) to achieve an equivalent reliability benefit to the system. In general, transmission to some distance can be added much more cheaply than generation, although...transmission only provides access to reserve capacity; it cannot substitute for it." (FERC, 1981, p. 22)

41 For example, the Integrated Resource Management Process in Massachusetts (Massachusetts Department of Public Utilities Docket 89-239). In Illinois, utilities are required to participate in biannual comprehensive statewide energy plans (Ill. Rev. State 111 2/3, ch. 8-402).

42 The neglect is certainly not universal, as exemplified by several very fine works from the late 1980s (National Governor's Association, 1986; Kelly, 1987; CECA/RF, 1990).

43 Hayes and Scheer (1987) p. 6.

44 For additional discussion and examples of advanced planning models in use today, see Caramanis, Schweppe, and Tabors (1982) and Fox-Penner and Spinney (1992).

45 See, for example, Womack, Jones, and Roos (1990) and Dertouszos, Lester, and Solow (1989).

46 For an excellent short discussion of what they call "competition under prevailing institutional arrangements," see Joskow and Schmalensee (1983) p. 20–23. In this section, Joskow and Schmalensee discuss competition to attract new or relocate large industrial or commercial customers; competition to expand the borders of franchise areas ("fringe competition"), competition for franchises themselves, and general competitive comparisons between different utilities ("yardstick competition"). Joskow and Schmalensee discount all of these forms of competition as significant; other authors disagree. Some important countervailing arguments can be found in Weiss (1975) and Carlton (1976). Franchise competition is discussed in more detail in Chapter 4.

47 See the discussion of this point in Fisher, et al (1992). Although most utilities altered their load forecasting equations to incorporate price effects by the late 1970s, many continued to argue that price effects were extremely small. (See, for example, the testimony of Commonwealth Edison in Illinois Commerce Commission Docket 80-0706, 1981). A 1980 report of Stanford University's Energy Modeling Forum found that several prominent utilities, including the Tennessee Valley Authority, used forecasting models with price sensitivity that is only about one-fifth of the price effects consistently used in today's industry (EMF, 1980, p. 20).

48 The 1970s spawned a large literature on the price responsiveness of power demand. Some of the many important works include Edmonds (1978), Bohi (1981), Bohi and Zimmerman (1984), and Fisher, et al (1992).

49 This observation is originally attributed to one of the earliest economic studies of electricity forecasting, Houthakker (1951), further codified by Fisher and Kaysen (1962). As a result of these and later studies, the dominant view of electric power use is that energy-using capital stocks are adjusted over the long term to price increases.

Recent works have examined the stock replacement decision in somewhat greater detail (Hassett and Metcalf, 1991, and Jaffee and Stavins, 1991). Briefly, this research finds that homes or businesses rationally decide not to replace their equipment with more efficient models unless they are convinced that the price increase will last long enough to make the more energy-efficient purchase worthwhile.

50 Today, a typical equation used to predict power demand looks something like this:

Power demands in a future year T in average weather (megawatts) =

A* B*(GNP in year T) + C* (Price increase four years ago)
$\qquad$ + D* (Price increase three years ago)
$\qquad$ + E* (last year's price increase)
$\qquad$ + F* (prices this year)

The equation may also contain terms for expected population growth, changes in household size, expected weather shifts, adjustments for unusual events or new industries entering or leaving the area, and many other terms. Past year's price effects can also be incorporated using a variety of mathematical approaches, including using only one "lagged price."

51 Nelson, Peck, and Uhler (1989). A precursor article, Peck and Weyant (1985), is also valuable.

52 For discussions and examples of modern techniques that account for price effects in system planning, see, Fox-Penner (1988) and Fisher, et al (1992) and the references therein.

53 EPAct (P.L.102-486; 42 U.S.C. § 13201) Section 111 holds that "Each electric utility shall employ integrated resource planning. All plans or filings before a state regulatory authority to meet the requirements of this paragraph must be updated on a regular basis, must provide the opportunity for public participation and comment, and contain a requirement that the plan be implemented."

54 See Edison Electric Institute (1994) for a recent state-by-state review.

55 There are two disincentives to doing DSM. The first is that less investment in power plants causes a lower rate base, yielding lower profits for the utility. The second is that total sales of electricity are lower than they would otherwise be, so the utility has a reduced opportunity to amortize the costs of its existing system from sales. See Moskowitz (1989) for a seminal discussion, as well as Nadel, Reid, and Wolcott (1992) and Comnes, et al (1995).

56 See Fox-Penner (1994).

57 EPAct (P.L.102-486; 42 U.S.C. § 13201) Section 111 holds that "The rates allowed to be charged by a state regulated electric utility shall be such that the utility's investment in and expenditures for energy conservation, energy efficiency resources, and other demand-side management measures are at least as profitable, giving appropriate consideration to income lost from reduced sales due to investments in and expenditures for conservation and efficiency, as its investments in and expenditures for the construction of new generation, transmission and distribution equipment."

58 Several schemes were developed by regulators to remove these disincentives. To address rate base concerns, some regulatory agencies allowed utilities to treat the cost of a DSM program as a capital investment that could be placed in rate base. Other regulators told utilities that, if their DSM programs met performance targets, the utility's rate of profit on its existing rate base would be increased slightly, raising allowed profits.

Neither of these approaches have proven popular with many utilities. Placing DSM assets in rate base is not popular because tax and corporate accountants don't like rate base to consist of assets that are neither tangible nor easily identified. Increasing the percentage return on rate base based on performance does not have this problem, but it is difficult to specify in advance a level of measurable DSM performance that regulators and utilities both think is fair for the purposes of granting rewards.

As to the disincentives caused by lowering the level of sales, regulators have developed formulas that "decouple" allowed profits from the level of sales. These formulas can be complicated to administer, and are only used in some states, but they do a reasonably good job of removing sales drop disincentives and ordinary circumstances. However, when unusual conditions cause sales to be much higher or lower than the levels used to calibrate these formulas, regulators or utilities no longer believe the formulas, and the approach can become contentious.

59 For discussions of the limited success of state programs to counter DSM disincentives, see Chamberlin, Brown, and Reid (1992) and the July, 1996 issue of the *Electricity Journal*.

60 There is widespread agreement that most DSM programs have, in the aggregate, lowered the total cost of electricity service to the nation. However, the precise magnitude of the savings is in dispute, along with the cost-effectiveness of a number of DSM programs. To sample this extensive and sometimes complex debate, see Eto, et al (1996).

Chapter 4

Determinants of Electric Industry Structure

The formal economic concepts of industries, firms, and products are central to the next stage of the restructuring discussion. Products—the outputs of firms—are considered to be within a single market if they are good substitutes for one another. The context in which substitutability is evaluated is critical. In some contexts, all brands of automobiles may be good substitutes; in others, the relevant group of substitutes may be only four-door compacts. A market is an economic arena in which sufficiently substitutable products compete for buyers.

A firm is an organization that produces a set of products, or equivalently, competes in a set of markets for those products. There is an expansion of scale when a firm makes more of one product; when it chooses to produce additional, significantly different products, this is an expansion of scope.

At the boundary of a firm, it must buy goods and services to bring them inside the firm (or sell them to make money). Within a firm, transactions are controlled by management according to a vast array of the incentive, control, and monitoring schemes that are customarily used to guide the business activities of firms.

The boundaries of firms and industries can be viewed as having two dimensions that economists call *horizontal* and *vertical*. Vertical boundaries refer to the portion of the chain of production activities, from raw material to finished product, the firm controls by ownership.[1] A company is completely *vertically integrated* if it owns everything from the land on which all of its raw materials are produced to the retail store in which its product is sold. No firm satisfies this description—all producers buy something from someone—but a farm that sells its own produce in an adjacent retail store is highly vertically integrated.

Horizontal structure refers to the number and size of firms supplying products at any discrete point in a supply chain. There may be dozens of small firms selling in a given product market (e.g., restaurants in a small community), or there may be only a handful of firms, as in the manufacturing of commercial aircraft. The concepts of monopoly (one seller in a product market), duopoly (two sellers), and oligopoly (a small group of sellers) are shorthand descriptors for important horizontal structures.

No theory completely explains the economic determinants of firm size and scope and industry structure. However, for the electric power industry, transactions cost economics has proven to be the clearest, most useful organizing principle.[2]

TRANSACTIONS COSTS AND VERTICAL BOUNDARIES

In unregulated markets, responses to market forces and antitrust laws are the primary determinants of the horizontal and vertical boundaries of the firm. No one would dream of legislating the degree of vertical integration or the number and type of products a firm must offer, much less requiring that a firm serve every customer in a given geographic market.

According to a well-accepted principle of economics, unregulated firms decide the level of vertical integration by asking, "Is the overall production chain cheaper if several parts of the chain are owned by one company, or should the elements of the chain remain separate and sell competitively to other parts of the chain?"[3] If one stage of production yields something that cannot be easily bought or sold, or two adjacent stages of production must be very tightly coordinated, deciding the price and terms of the transaction are outweighed by the advantages of creating a competitive "intermediate" market. On the other hand, if the intermediate product lends itself readily to competitive trade, and the well-known tendency of competition to force costs down makes the two stages operate more cheaply than a single firm, there is no reason to integrate.

Procuring goods via competitive markets incurs costs as well as benefits. Competition means that the buyer must take the time to search for and evaluate all product offerings, haggle over the price and terms of the sale, worry about whether delivery will occur as promised, and pursue remedies against the seller if it does not. Both buyer and seller need to agree on an explicit or implicit contract; they must carefully account for payments and deliveries, and they may need to involve lawyers, accountants, and other experts. These costs are commonly called *transactions costs*, i.e., the costs of using market transactions in place of the control that firms exercise within their boundaries.[4]

Internalizing transactions costs within a firm through vertical integration lowers costs because managers have relatively greater latitude controlling the workers and other resources of the firm. Managers can reorder production plans without having to rewrite contracts or haggle over a new set of prices for internal exchanges. Indeed, as law professor Ronald Coase explained 60 years ago, the ability to reduce transactions costs is the entire reason why firms are created.[5]

Short of running experiments, how can a company determine whether it is cheaper to buy an input or integrate further upstream and produce it within the firm? Economists studying vertical integration (VI) have found that several aspects of vertical production chains are associated with integration. Where these attributes (discussed below) are in the production technologies and organizations that make up the chain, integration rather than competition is often the cheaper alternative.[6]

The first attribute associated with VI is *specialized assets*, i.e., assets that cannot be moved to other geographic areas or used for other purposes. Both conditions are important in determining whether an asset is economically specialized. Airplanes are specialized assets; their sole purpose is to fly people or freight, and their parts are not usable for other purposes. However, they are obviously quite mobile, and can move to any geographic market in the world. Conversely, a shopping mall is not very portable, and taking it apart to sell off its pieces or move it is not very economical, either. However, a building of this nature is extremely versatile; if it does not work as one kind of store, there are many other potential uses for the space.

The more specialized the assets used in a production chain, the more likely it is that VI will be present. The reasons are fairly intuitive. If a special production asset is not used for the one thing it is good for, it is an asset that is costing money and producing nothing. The chances of this happening are smaller if the machine is part of a production chain commanded by management, instead of a stand-alone machine that must go out and hunt for customer after customer.[7]

Another aspect of asset specificity is the possibility of so-called holdup and bargaining costs. In production chains containing a series of specialized assets, each link in the chain can hold the other links hostage if it is not happy with the arrangements. Because specialized assets by definition cannot be easily moved or quickly replaced, there is an incentive to try to renegotiate contracts to "get a better deal." Even before the deal is struck, there is typically much more bargaining involved as each link in the chain tries to obtain the highest percentage of the value of the entire chain's output—the ultimate retail product.

The effects of uncertainty, limits on firms' ability to share information, and the inability to guarantee a steady supply of inputs create a final group of transactions-related VI determinants. If a firm requires a steady supply of an input to keep its production processes running efficiently, it often prefers to own the supplier rather than depend on a contract for supply. A firm becomes more efficient when it has more complete information about its inputs or its downstream distributors. If a firm treats these suppliers as competitors, they may not be willing to share information or coordinate planning as extensively as they would under a joint ownership arrangement.[8]

Another production aspect associated with VI is the need for momentary coordination in order to maintain production. A foundry, for example, cannot economically keep a molten kettle of steel hot while it searches for customers to sell it to, haggles over price, and possibly fails to find a taker. It is more efficient to choose boundaries for the firm at points in the production chain where inputs and outputs need not be closely coordinated with the next link up or down.

When competitive transactions are involved, it is desirable for buyer and seller to make the terms of the exchange explicit—in other words, to form a contract that sets forth all the possible changes that might occur, and how these changes affect the price paid and the terms of service. In economic jargon, this is referred to as a *complete* contract, i.e., one that sets out every possible contingency that might occur, and who will pay what under each contingency.[9] Vertical integration eliminates this problem by eliminating the contracting process and internalizing the planning for contingencies.

An inability to write satisfying contracts is also associated with extremely complex problems in which a computation of the correct value or price for a transaction is difficult due to the sheer volume of information required. In this case, the costs of gathering and processing the information needed to determine price may add significant transactions costs. When senior management judgment is also required in order to make decisions and evaluate calculations, the amount of attention management must devote to these computations may divert it from more important tasks. When the amount of information is large, and the range of uncertainties inherent in various aspects of the decision is large, the problem of grappling with competitive input purchasing may simply dwarf management's ability to comprehend and act in the best interests of the firm.[10]

Finally, VI is associated with transactions and processes that have large externalities, or "spillover effects," that are eliminated by integration. Suppose, for example, that in one area all of the good locations for broadcasting radio signals were on one hill, and that one radio tower was interfering with all other signals coming from this hill. If a single owner owned all of the radio signals emanating from the hill, he or she would have the incentive to solve the problem in a manner that served the best interests of all of the senders and recipients of radio signals.[11]

All of these costs—bargaining, renegotiating contracts, searching for information, coping with availability and other forms of uncertainty, and sometimes contractual litigation and disruptions—may be reduced by backward or forward integration. There are dozens of instances in which the transactions costs associated with using competitive markets exceed the benefits that competition imparts.

Some aspects of transactions costs—e.g., the costs of monitoring complex, time-sensitive contracts—are related to the costs and capabilities of information technologies. As vertical structure is ultimately determined by the tradeoff between the benefits of competition and transactions costs, declines in the costs of infotech may affect the costs of controlling resources within an organization less than they affect the costs of monitoring external contracts, thereby ultimately reducing the level of VI in many industries.[12] Although evidence of this is thus far weak, in the electric power industry, it is clear that technology is enabling modes of interaction that were unthinkable 20 years ago, and information and communications technologies will play an enormous and essential role in restructuring.

THE DETERMINANTS OF HORIZONTAL INDUSTRY STRUCTURE

Horizontal structure can refer to the number of firms making one product or to the range of products made by a single firm. The former is the traditional realm of horizontal structure and the theory of competition itself. The technologies and costs of production and distribution may lead to an atomistic market that is full of hundreds of small firms or to a highly concentrated market of a few large competitors. A "natural monopoly" is a unique form of horizontal structure in which a market can be served most efficiently not by a small or large number of competitors, but rather by a single firm.[13]

Acquisitions and mergers have an impact on horizontal structure. Antitrust laws and other restrictions on "horizontal combinations" ensure that all markets for private goods in the United States are reasonably competitive and that no firm is able to exercise monopoly power over consumers. Merger policy in the electric utility industry raises a number of issues that are explored in in Chapter 9. For the current discussuon, it is sufficient to note that horizontal expansion is set by the firm based on its success in the marketplace, except perhaps when expansion occurs via merger.

The electric utility industry was originally considered a very special form of horizontal structure: a natural monopoly. In a natural monopoly, all demand in a given area can be supplied most cheaply by a single firm. As seen in Chapter 1, this contention has been disputed for years. A number of economic studies of utility system costs have been conducted, all attempting to determine the extent of scale economies of power production. Research concerning the effects of the power industry's scale economies on horizontal structure is discussed below.

Expansions of scope (i.e., making additional *different* products) and vertical integration decisions are similar questions of boundary-setting. Both hinge on the managers of the firm determining whether it is more profitable to own and operate a larger or more diversified firm. If adding another product to the firm's product line increases profits or reduces average costs, the firm experiences *economies of scope*.

Unregulated firms are free to choose their level of vertical integration and their expansions of scope. The firm's success in the marketplace, which is not under the unilateral control of the firm, decides its share of sales and profits. Using vertical integration, economies of scope, and anything else management can think of that improves service or lowers cost, a firm vies against its competitors to gain share. If all goes according to theory, the resulting number of firms and their corresponding share will constitute a competitive equilibrium in which several competitive firms of varying sizes but roughly equal quality and cost serve all customers.

THE ECONOMIC STRUCTURE OF THE ELECTRIC UTILITY INDUSTRY

An industry is a collection of firms that produce similar or identical products—i.e., the group of firms that sell into the same product market(s). The electric utility industry is generally regarded as the aggregation of firms that generate, transmit, and/or distribute power. The structure of the industry—the number and type of firms, their degree of vertical integration, and the range of products they make—can be viewed as the result of four major determinants (Figure 4-1). The first two factors (top of figure) are the traditional internal and external cost characteristics that shape horizontal and vertical structure in most markets. These forces are present in most markets, but they depend on the past, present, and future state of the technologies used by the industry and its suppliers, as well as by changes in the pattern of demand.

In the utility industry, the other two main determinants are the industry's legal and regulatory restrictions and the role and effectiveness of the various voluntary activities of the industry, such as the reliability councils. Although these legal and institutional influences are present in all industries, they have played a far larger role in the power industry than in most other sectors.

The rest of this section briefly reviews the literature on each determinant's contribution to the structure of the power industry. It is very useful to examine each influence in turn, but bear in mind that what is known about each is influenced by other kinds of determinants—particularly in this industry. For example, economists have tried to use statistical analysis to determine whether VI is less costly than *deintegration* (i.e., selling power from generator to transmitter and then reselling it to the distribution system).

Figure 4-1. Determinants of Electric Utility Industry Structure

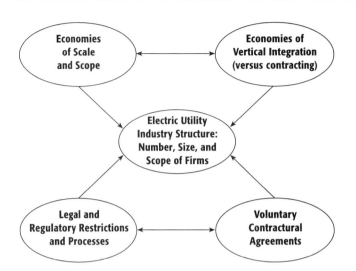

This is a highly simplified model of the restructured industry, so this is effectively a crude measurement of the benefits of restructuring itself. However, the measurements available thus far come mostly from heavily regulated or nonprofit firms that may not behave like deintegrated firms in future, less-regulated structures.

Another important example is the influence of scale economies (horizontal) on vertical integration economies, and vice versa. Each of the three stages of production can possess scale economies, and it is sometimes difficult to separate these from vertical economies. It is for this reason that Professors Joskow and Schmalensee admonish that "[T]he close links between transmission and generation make it potentially misleading to analyze either in isolation."[14] Although economists and policy makers have shown tremendous interest in whether utilities have economies of scale, they have until recently paid little attention to vertical economies or the relationship between the two.

Ultimately, none of the four major determinants in Figure 4-1 is independent of the others. Every sort of determinant exerts direct and indirect influences on the others, which suggests that a more accurate (but messier) rendition of the Figure would have lines from every circle to every other circle.

The relationships between the four determinants has occupied political economists and students of utility regulation for decades. Much of the debate over the origin of utility regulation and the industry's structure can be seen as a debate over which of the determinants in Figure 4-1 came first, or which caused the other determinants to become important. For example, a number of scholars argue that the natural monopoly cost attributes were the result of regulation, while others argue the opposite.[15]

The purpose of this text is decidedly not to resolve these disputes. For the purpose of studying the future of the industry, it is sufficient for now to agree that the four determinants in Figure 4-1 do exist and will continue to play a role regardless of how they came into being. Moreover, there is no disputing that all will remain related in complex ways in every future industry scenario. As a result, an inability to determine what caused what in the past must not cause future relations between these determinants to be ignored, however they may come to pass. Of all the tall orders that this (or any other) study of restructuring face, this is the tallest.

ECONOMIES OF SCALE: A SURVEY BY STAGE

Most large utilities in the United States are vertically integrated, most small ones are not, and the most reliable sources of cost data include the costs of all three stages of production in the integrated firms. Integrated utilities' costs do not show in which stage(s) the natural monopoly effect is occurring, and some economists argue that the statistical tests for scale efforts in integrated firms are unreliable.[16] However, these estimates are valuable in at least one unique respect. The reported costs of integrated firms include not only the sum of the costs of the three stages, but also any costs required to connect the stages, as well as to manage and finance the entire firm. By comparing the costs of integrated firms with the costs of similar, deintegrated firms, the savings achieved through vertical integration itself may be roughly inferred—not simply the cost characteristics of the individual stages.

The two most widely cited studies of so-called "firm level" economies of scale were conducted by economists Christensen and Greene (1976) and Huettner and Landon (1978). The former finds that economies of scale are exhausted at about 3,800 MW of size, while the latter estimates that declining average costs cease between 1,600 and 3,000 MW. These results were and are often cited as reasons to reject natural monopoly-style regulation.[17] However, the natural monopoly model of utility regulation does not require that costs decline forever as utilities grow ever larger. Rather, it is sufficient that the firm have declining average costs over a range that includes the sizes of most areas regulators have chosen to make into a single franchise. (Had regulation envisioned a single natural monopoly of national size, the law would have to provide for a single franchise for a national power company.)

A number of more recent studies have re-examined electric utility scale effects—in some cases using explicit tests for subadditivity (see Box 1-2) rather than declining marginal costs. Nelson and Wohar (1983), Atkinson and Halvorsen (1984), and Joskow and Rose (1985) all found that scale effects are not exhausted in most integrated utilities. In an extremely careful study, Henderson (1985) found modest scale effects for generation and very substantial scale effects for transmission and distribution. In a recent review of these findings in the *Yale Journal of Regulation*, Professors Douglas Gegax and Ken Nowotny conclude:[18]

> In short, a large body of evidence indicates that the electric utility industry has not exhausted economies of scale. Regulators cannot justify a policy to encourage entry into the electric utility industry solely on the basis that economies of scale in generation have been exhausted.

One recent study of utility costs casts a shadow over this rather surprising conclusion and another generally supports it. Economist Keith Gilsdorf tests a sample of integrated electric utilities and concludes that there is "no evidence of subadditivity for vertically integrated utilities."[19] However, George Washington University Professor John Kwoka finds essentially the opposite, concluding that overall subadditivity exists, particularly for utilities whose systems contain most of their own generation, transmission, and distribution.[20] The economics of generation are now shifting toward new decentralized or "distributed" technologies. These studies predict future reductions in *plant* economies, but the timing is uncertain and these findings should not be mistaken for *firmwide* economies of generation.[21]

Scale Economies in Transmission and Distribution. Starting with the less controversial segments, it is widely accepted that the transmission stage of the industry has declining average costs (and therefore subadditive costs)[22] over the range of outputs observed today.[23] This result is suggested by the engineering economics of power transmission. The vast majority of the costs of moving power are the carrying charges associated with line construction. However, the capacity of a power line increases as the square of the line's voltage, while the cost of building a line increases linearly with voltage.[24] For example, a 200 kilovolt (kV) line will cost twice as much to build as a 100 kV line, but will carry four times as much power. Accordingly, the average cost of transporting a unit of power declines with larger and larger lines—precisely the traditional natural monopoly condition.[25] In its 1994 Transmission Policy Statement, the FERC concluded that "it appears that transmission will remain a natural monopoly for the near future.[26]

Most observers also believe that the distribution stage of the industry is a natural monopoly, although agreement is not universal. Recent reports by the U.S. Office of Technology Assessment and the Federal Energy Regulatory Commission,[27] as well as many other studies,[28] voice this conclusion. Steve Henderson, an economist recently at the Federal Energy Regulatory Commission, carefully examined scale effects in transmission and distribution and concluded that "the average cost of transporting electricity is about 17.7% higher than the long-run marginal cost," precisely equivalent to the old test of natural monopoly as well as a clear finding of subadditivity.[29] Henderson argues that his findings apply to virtually every size of utility, from the smallest to the largest observed systems.[30]

In a minority dissenting view, economists Walter Primeaux (1975) and Leonard Weiss (1975) challenged the natural monopoly notion, the former by examining cities in which, by historic accident, two electric utilities competed with each other for sales. In a critique of Professor Primeaux's work, Professors Joskow and Schmalensee argue that "too much attention has been paid to his results, and we conclude that they do not cast appreciable doubt on the proposition that distribution is a natural monopoly."[31] Most members of the industry, observers, and regulators agree.

Scale Effects in The Generating Sector. Most studies of generation alone find that economies of scale *in individual units* grew until the mid or late 1970s and then flattened out or even declined. *Units* are the technological building blocks of which plants are made— usually a combination of boilers, combustors, and turbines. A *plant* is generally a single location that usually contains one or more identical or non-identical generating units.[32] Most plants are multi-unit.

Strictly speaking, generation economies of scale can be examined by drawing a box around a single plant and measuring the costs of all power emanating from that site. For the sole purpose of establishing whether scale economies exist, it does not matter why average costs decline as the amount of generating capacity at the site increases. However, it is useful to distinguish between plants getting cheaper because units are getting larger, versus plants getting cheaper because more units (possibly smaller in size) are built inside the same site. To do this, there must be a distinction between economies of scale in units and plants.

Over a period of more than 50 years, as electric power units grew larger, their average costs declined. This prompted utilities to add increasingly large plants to their systems. Figure 4-2 displays U.S. Department of Energy (DOE) data on the average size of new and total installed generating plants since 1938. As shown in the leftmost set of points on the figure, the average size of *new* generating plants grew steadily from the late 1930s through the 1970s.[33] The same is true for units during this period (not shown on the Figure), which increased from an average size of 22 MW in 1938 to perhaps ten times this amount by 1980.[34] Throughout the period, generation scale economies were driven more by increased unit size, which roughly grew by a factor of ten, rather than by increases in the number of units in each plant. Successively larger units, although slightly less reliable,[35] were more fuel-efficient[36] as well as less costly to construct per unit of output.[37]

As late as the mid 1970s, this trend was expected to continue. In 1970, the White House Office of Science and Technology Policy reported that, "in the next twenty years, new capacity will come from 250 huge power plants in the range of 2,000 to 3,000 megawatts each."[38] The then-young nuclear power industry was predicting units as large as 3,000 MW each.[39]

To put it bluntly, this never happened. After decades of bigger-is-cheaper, technological change started making smaller units just as efficient as the largest ones. Units of all sizes continued to improve, but the advantages of larger units in greater operating efficiencies and construction cost diminished or were reversed. Figure 4-3, for example, shows that manufacturers' average costs for gas turbines (the most popular kind of unit in use today) are virtually identical for all sizes of plants above 100 MW.[40]

The industry responded to the flattening of unit economies in predictable ways. In the middle panel of Figure 4-2, the line of solid black boxes shows that average new plant sizes in the 1980s were far below the levels seen in the 1965–79 period, though still above 1960 levels and exhibiting a slight upward trend. It appears possible that the 1965–79 period, during which new power plants grew to sizes far beyond past experience, was an aberration, and that the equilibrium trend is closer to the 1980–84 data. In any event, the data suggest that the average size of new plants is increasing, but on a new trend line below the large sizes of the early 1970s.

The middle panel of Figure 4-2 also displays data on the average size of new *units*. Recalling that units increased in size steadily prior to 1980—in some periods rather dramatically—it is clear that this is where the trend has been reversed. The size of the average units added during the 1980s (shown on the line with cross marks in the middle panel) trends down from the 200 MW range to the 100 MW range. Interestingly, there

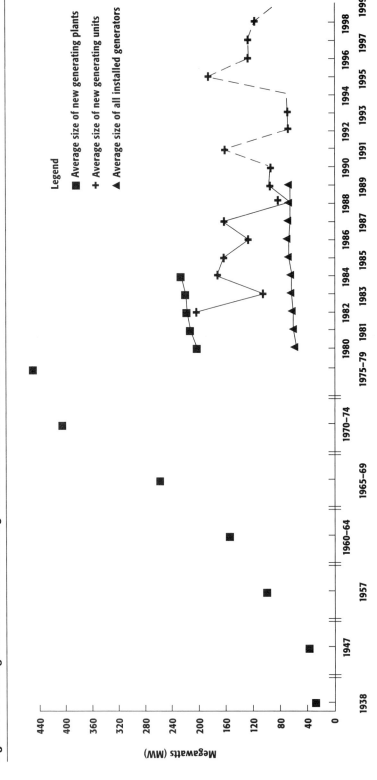

Figure 4-2. Average Sizes of New Generating Plants and Units, 1938–1999

Legend
■ Average size of new generating plants
+ Average size of new generating units
▲ Average size of all installed generators

Source: Messing, Fresma, and Morell (1979) p. 11
Joskow and Schmalensee (1983), Table 5.3
U.S.D.O.E. Electric Power Annual (various years)
U.S.D.O.E. Inventory of Power Plants (various years).

Figure 4-3. Manufacturer's Cost of Natural Gas-Fired Turbine-Generators, 1991

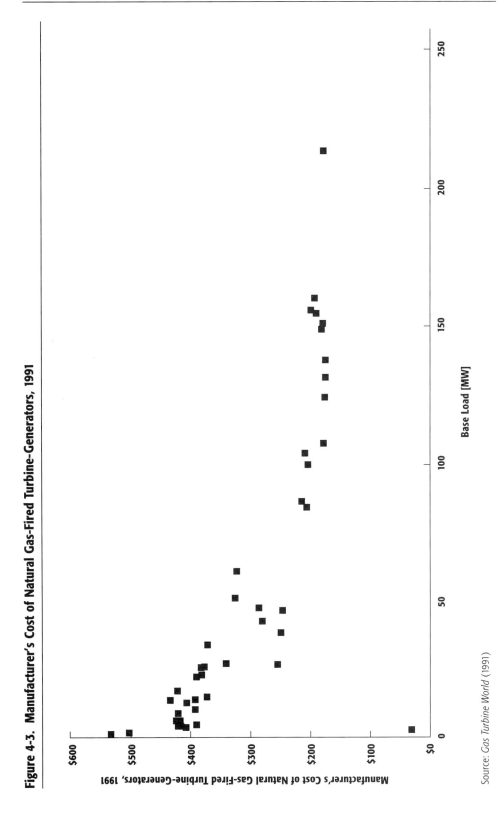

Source: *Gas Turbine World* (1991)

appears to be no change in this range of units sizes through 1999, as demonstrated in the rightmost dashed line in the Figure. This line illustrates the industry's projections of new unit sizes through 1999, as reported to DOE. At least for the foreseeable future, average unit sizes appear to be holding steady within a range utilities apparently find optimal under present circumstances.[41]

A variety of factors may account for the inflection point in unit sizes. Although recent plant data is harder to come by, it is generally the case that most plants contain more than one unit, and that the trend has been toward a larger number of somewhat smaller units at each plant site.[42] Hence, the facts reflected in Figures 4-2 and 4-3 are not sufficient to prove that average plant size has flattened out or is getting smaller. However, these data indicate that the combination of technological change and economic conditions that made progressively larger units the additions of choice has ended in the United States.

The overall conclusion concerning economies of scale in generation is somewhat more complicated than the simple view that generation economies are a thing of the past. Trends in unit size choices and costs certainly suggest a flattening of unit economies. However, it is likely that average costs for *plants* still decline over the size ranges most utilities operate within. Economies of scale in generation now appear to come from things like placing more units at one plant site, rather than from higher boiler temperatures and pressures, as in the past.[43]

More importantly, scale economies in electric utilities almost certainly continue to exist, but are no longer due to ever-larger individual units. Instead, these economies come from the cost advantages of owning *and operating* many large (but not overlarge) and diverse units and interconnecting them with transmission. More than one stage of the industry must be examined, and operating as well as financing advantages must be included to gauge optimal firm scale. Indeed, some interpret the present wave of mergers of utilities into extremely large firms as evidence that even today's large utilities do not realize all economies of scale.

As the industry separates into firms that specialize in owning and operating large numbers of power generators, economies of scale may arise from the ability to diversify geographically, mass marketing, and operation and maintenance economies. These are much the same scale advantages that many unregulated national brands or franchises enjoy, and they may contribute to industry concentration in a deregulated generation sector. On the other hand, because modern scale economics arise largely out of the integration of generation and tansmission, and restructuring is likely to sever much of this linkage, the net influence on scale economies is uncertain.

VERTICAL INTEGRATION AND ELECTRIC PRODUCTION TECHNOLOGY.

Many of the aspects of electric technology that were examined in Chapters 2 and 3 suggest that this industry is an extremely likely candidate for vertical integration. Indeed, the industry is an almost textbook case, as it matches virtually every attribute closely associated with VI.

To begin with, electric utility assets are quite specialized and geographically immobile. Power plants, power lines, transformers, and other utility hardware are generally made for only one purpose. They are also expensive to disassemble and move, although this is not unheard of. Once it is in place, the momentary coordination needed to keep the system running is precisely the kind of situation that lends itself to post-contract strategic behavior— e.g., if one major element of the power system refuses to cooperate, the entire system can be brought to an immediate halt. Moreover, this kind of strategic behavior hurts much more than the just the links of the specific vertical partners of the "bad actor." A single major power outage can impose costs on hundreds of other power plant operators and millions of customers—an enormous potential for externalities if ever there was one. And on a more positive note, a decision by one transmitter or generator to add capacity could reduce costs for many other generators or transmitters—a positive externality, in economic terms.

Finally, the high degree of interdependence in power systems can lead to information overload and difficulties in writing contracts that can specify who is responsible for what during unusual or emergency conditions. Because power systems operate moment-to-moment, there is little time for before-the-fact contract renegotiation or discussion.

All of these conditions suggest that the managers of electric power firms would find it cheaper to vertically integrate than to operate separate generation, transmission, and distribution companies. In their extensive study of electric utility deregulation, professors Joskow and Schmalensee devoted considerable attention to the transactions costs aspects of vertical deintegration. After examining the attributes of the industry, J&S concluded:[44]

> Investments in electric power supply facilities have many of the characteristics that economics professor Williamson identifies as leading to internal organization (vertical integration) or complex, long-term contracts. Where segments of the electric power systems must be linked by long-term contract rather than internal organization…long-term contracts will emerge. There is no guarantee that such contracts will lead to more efficient outcomes than internal organization.

Economists studying the cost structure of the electric utility industry have generally found this to be the case. By comparing the costs of a variety of vertically integrated and unintegrated power firms, and controlling for differences in fuel and other exogenous factors, economists find that it has been substantially cheaper for one firm to own and operate generation, transmission, and distribution facilities that are roughly as large as the sales it makes to its service area.

Remarkably, not much empirical work on this subject was done by economists prior to the 1980s, apart from incidental work in the context of investigations of scale economies. As interest in generation competition increased, and economic techniques improved to the point where several stages of a vertical chain could be analyzed at once, economists turned their attention to careful measurements of VI.

Economist Steve Henderson (1985) performed statistical tests on electric utility cost data and concluded that generation, transmission, and distribution costs were subadditive. Kaserman and Mayo (1991) estimated that vertical integration reduced aggregate utility costs by about 12%, and George Washington University economist John Kwoka (1996) estimated that VI produced savings of 13–14% (Kwoka, 1996). The sole empirical dissenting view came from economist Keith Gilsdorf (1995), who originally argued (Gilsdorf, 1978) that vertical integration was present, but later concluded on the basis of statistical tests that vertical subadditivity could not be proven. Gilsdorf nonetheless concluded that "Although this result is consistent with pro-competition policies, it does not necessarily support complete industry divestiture, since economies of scope between stages may exist in the absence of [statistical evidence of] subadditivity."

Kwoka finds that the savings are greatest for utilities whose generation, transmission, and distribution capacities are approximately equal to their sales. This points out an important and previously-neglected relationship between scope and scale economies within present-day utilities. Integrated firms that have approximately equal amounts of generation, transmission, and distribution capacity appear to be most efficient in the traditional structure. According to Kwoka, at sizes of about 12 million MWh/year in sales—roughly the size of the large investor-owned utilities (IOUs) in the United States—these economies diminish.

Research into horizontal and vertical economies based on today's utilities provides an important first look at the question of optimal market structure. However, the research can shed only a little light on a future that relies on more complete markets and different forms of contracts between buyers and sellers. Because the ability to write different types of contracts in the utility industry has been largely a question of regulation, it is appropriate to now examine the twin institutional factors at the bottom of Figure 4-1.

LEGAL, REGULATORY, AND CONTRACTUAL DETERMINANTS

In Figure 4-1, the two determinants of industry structure yet to be examined are legal and regulatory strictures and voluntary agreements and associations. In the utility industry, these are so closely associated that it is easiest to discuss the two in an integrated fashion. Moreover, a wide range of cultural and political forces have played a particularly strong role in the utility industry. As is proper, these "non-economic" forces have impacted the industry's laws and institutions quite visibly throughout the industry's history.

There are four major types of utility firms in the United States. Investor-owned utilities are corporations with publicly-traded stock, much like other large private U.S. companies. Two other types are: (1) non-profit corporations owned by their customers (rural electric cooperatives or co-ops), or (2) instrumentalities of state and municipal government units, including public utility or irrigation districts. Finally, the federal government owns a handful of utilities that sell power, generally at wholesale, from federally-owned generating plants, prompting the name "Power Marketing Administrations" (PMAs).

Each of these types of entities is subject to an extensive but different set of rules, ranging from the type of assets they may own to accounting and rate-making practices to environmental statutes. Some governmental rules apply equally to all types of entities; many do not. The differences stem from the fact that each type of utility is established under different organic statutes at the federal or state level. Table 4.1 illustrates some important differences in the laws and regulations governing utility entities today.

The Franchise. The single most *common* regulatory feature of the retail electric indus-try has been the existence of exclusive retail franchises. Every state has some provision in its public utility laws that divides the state into service or franchise areas.[45] These areas define territories in which one utility—be it a rural co-op, IOU, or irrigation dis-trict—has exclusive rights to sell all retail power.[46]

The ostensible economic rationale for exclusive retail franchises relates directly to the idea of natural monopoly: If one firm can serve a given area most cheaply, but one firm will also hold monopoly power, then the government picks one firm and controls its prices and services to keep costs at the low level that only a natural monopoly can achieve.[47] Of course, this idea requires that power distribution be a natural monopoly. In the abstract, the distributor need not get the power from vertically integrated generators and transmission lines.

State laws that award franchises usually require that each franchised distributor have sufficient capacity on hand to serve all customer needs, now and in the future. This requirement is not rooted in the economics of natural monopoly; instead, it is a conse-quence of combining exclusivity with technical practices that are necessary to keep regional grids adequate and secure. As discussed in Chapters 2 and 3, this "adequacy" requirement is also referred to as the *obligation to serve.*

The requirement that franchised retail distributors provide for all demand is not the same as a requirement that distributors *own* all of the generation and transmission they need. Indeed, public utility laws have generally not mandated VI, although PUHCA and other aspects of utility regulation can be read as encouraging it. In any case, an absence of complete VI in the industry ensures that some distribution-only utilities have to buy generation and transmission from others.

Economists and policy makers have argued periodically that utilities should be regu-lated not via prices and plans, but rather by regularly having to compete for an exclusive franchise. The most famous call in modern times is Demsetz (1968), though there are others who have sounded this theme consistently for many years.

Prior to the 1920s, the awarding of franchises, often for short periods or non-exclusively to promote competition, was the primary means of controlling the industry. For a variety of reasons, some practical and some recognized in theory, this means of control proved unsatisfactory.[48] However, the idea that nearby franchises offer a "yardstick" by which other public or private franchises can be measured survives to this day.[49]

There has been a steady stream of franchise-related competition and litigation during at least the past 35 years.[50] Nevertheless, for reasons of historical experience and theoret-ical uncertainty, state regulators and municipalities have been reluctant to rely on fran-chise competition alone for retail electric services, and electric utility franchise changes have been a relatively small area of activity over time.[51] Of more importance, one key-stone issue in utility restructuring is whether to abandon the retail franchise concept entirely in favor of a choice of retail suppliers. This idea is examined at in length in Chapters 9 and 10.[52]

Table 4.1. Summary Legal and Regulatory Features Influencing Utility Industry Structure

	Integrated IOU	Municipal Public Utility(a)	Rural Electric Co-op(b)	State Power Authority ©	Bonneville Power Admin.	Other PMAs	Tennessee Valley Authority
Exclusive retail franchise	Yes	Yes	Yes	No	No	No	No? (d)
State retail rate regulation	Yes	Yes 13 states (e)	Yes 19 states (e)	N.A.	N.A.	N.A.	N.A.
Enforcement of adequacy	Franchise and state utility statutes	Implied or state franchise	Limited (f)	No obligation	No obligation	No obligation	No obligation (f)
Enforcement of IRP or LCP	State statutes	No explicit enforcement	REA rules (g)	No obligation	Federal Statute (g)	No obligation	Federal statute (g)
Subject to state siting authority	Yes	Usually	Usually	Varies	No	No	No
Bulk power sales regulated by FERC	Yes	No (p)	Yes (h)	No (p)	No, except limited FERC review (m)	No, except limited FERC review (m)	No
Owns most of its own transmission	Yes	Seldom	Seldom	Sometimes	Yes	Yes, except SEPA	Yes
FERC auth. to compel wheeling via order (n)	Yes	Yes	Yes	Yes	Yes	Yes	Yes (l)
FERC auth. to compel "open access"	Yes	Yes (j)	Yes (j)	No (j)	No (j)	No (j)	No (j)
Exempt from taxes	No	Yes (k) or Partial	Yes (k) or Partial	Yes	Yes	Yes	Yes

Continued on next page

Table 4.1. Summary Legal and Regulatory Features Influencing Utility Industry Structure *(continued)*

Notes:

a. Includes Public Utility Districts and joint-action agencies that sell power at wholesale owned or controlled by distribution munis, and may own or control generation and transmission resources.

b. Includes "G&T" cooperatives that own or control generation and transmission resources and sell power at wholesale to distribution co-ops.

c. Usually sell only at wholesale to public distribution utilities, but occasionally sell directly to large retail customers.

d. 16 U.S.C. § 831(i).

e. See NARUC (1993).

f. Co-ops must consider extending service to all who want it, but need not do so if the costs are burdensome. 7 CFR Ch. XVII 1710.103.

g. See note 41 *supra*, the Northwest Power Planning Act (94 Stat 2697), and 16 U.S.C. § 831m-1, respectively.

h. Unless presently a borrower from the U.S. Rural Utilities Service (formerly the Rural Electrification Administration).

i. TVA can suspend and appeal a FERC wheeling order (Federal Power Act Section 212).

j. The recent FERC open-access order does not apply to munis, but section 211 of the Energy Policy Act of 1992 gives FERC authority to compel transmission under the procedures and conditions set forth in the Act. FERC Order 888, Docket RM95-8-000, April 24, 1996, p. 160–164.

k. Many municipal utilities and co-ops pay "payments in lieu of taxes" to state and local governments or they pay taxes at reduced rates.

l. FERC's order may not violate TVA's restriction on power sales outside its permitted sales area as set by statute.

m. The FERC may only approve, disapprove, or remand rates for agency reconsideration; it may not modify rates proposed.

n. If the owner of transmission facilities used for the sale of power at wholesale.

o. Unless inconsistent with BPA's "other statutory mandates" or would impair BPA's ability to assure adequate and reliable service to loads in the Pacific Northwest.

p. 16 U.S.C. § 824 (f)

IOU Regulation. In every state except Nebraska, *retail* power sales by IOUs are regulated by state utility commissions.[53] State utility regulation extends to virtually every aspect of utility operations, but regulators' most important functions are to set retail rates and conditions of service and to approve or modify the construction plans of utilities and their attendant costs. Rate making, planning, and cost recovery approval exist independent of whether the retail utility purchases its power on the bulk power market or generates it within a vertically integrated system. Box 4-1 briefly reviews state utility regulation.

Investor-owned utilities are also subject to the Public Utility Holding Company Act of 1935 (PUHCA). This law prohibits utilities from forming holding companies unless they are registered with regulators at the Securities and Exchange Commission, owning non-utility businesses, or operating an "unintegrated" (noncontiguous) system. In addition, utilities not exempt from the law must obtain approval to issue securities, acquire assets, or engage in transactions with affiliates, and are subject to other restrictions as well. All of these conditions can be traced to a period of corporate abuses during the 1920's (see Appendix 4A).

PUHCA can be viewed as imposing a number of important limits on utilities. First, limits and approvals are needed to issue utility securities. Second, PUHCA limits utility deintegration and geographic diversification, even for utilities whose activities are strictly within the power business. Finally, PUHCA discourages utilities from diversifying into non-utility businesses, such as real estate.

One particularly important aspect of PUHCA's controls concerns "affiliate transactions." Whenever one firm owns two businesses (one unregulated and one subject to cost-of-service regulation and an exclusive franchise [see Box 4-1]), and the unregulated business can sell a product or service to the regulated one, the firm faces an incentive to increase its profits by inflating the price of the unregulated product.[54] This kind of transaction, which harms economic efficiency and is unfair to the ratepayers of the regulated business, is loosely termed *self-dealing.*[55] PUHCA prohibits any transaction between two "affiliates" within a holding company unless these transactions meet the standards prescribed by the Securities and Exchange Commission and are performed "economically and efficiently" at prices that are fair to both parties.[56] Many state public service commissions have similar restrictions, but if the contracts between affiliates are in interstate commerce, the states cannot regulate them outright.[57]

Many provisions of the law apply only to the 11 electric and three gas utility holding companies that chose to register rather than dissolve.[58] However, the law also acts on all other utilities by keeping them from taking actions that would cause them to become subject to the Act. For example, it is largely due to PUHCA that utilities in one state do not buy utilities in other states unless their service areas intersect (thus meeting the "integrated system" test). It is also one of the reasons that utilities have been cautious about diversifying into other lines of business.[59]

To make matters more complex, PUHCA contains a number of exemptions that allow "exempt" utilities and holding companies to engage in the practices that registered holding companies may not engage in. Approximately 152 holding companies (gas, electric, telephone, and combined) are exempt. The chief reason for exemption among electrics is that the utilities are mainly intrastate or are undiversified operators of a single contiguous system.[60]

Differences in the restrictions faced by exempt and registered holding companies has produced increasing complaints that they are being treated unfairly by this aging legislation.[61] Moreover, they point out that the Act serves little purpose if most of the utilities in the nation either do not fall under its jurisdiction or are specifically exempt. Those that favor retaining the law (or merely retaining some of the law's protections in other forms) argue that it exercises an important influence over the structure of the IOU portion of the industry by preventing the formation of more holding companies or other prohibited activities.[62] For its part, the Securities and Exchange Commission has recommended three options for altering PUHCA, one of which is repeal and another of which is simplification but not complete elimination.[63] It is extremely likely that some or all of PUHCA's provisions will be changed as part of utility restructuring.

With the exception of federal regulation (discussed below), state utility regulation; state statutes on franchises, planning, and facility siting; and the strictures of PUHCA constitute the most important legal and regulatory determinants of the boundaries and behavior of investor-owned firms. Public and federal utilities operate under somewhat different sets of rules.

Federal Regulation of Interstate Sales. The marketplace in which utilities buy and sell generation is a wholesale or *bulk power* market, also referred to as "sales for resale." Rates and conditions of service for most sales in this market are regulated by the FERC under the Federal Power Act (FPA) irrespective of whether the buyer or seller is itself state-regulated.[64] The FPA created the Federal Power Commission (FPC), which held the original federal regulatory authority. In 1978, the National Energy Act changed the FPC into the Federal Energy Regulatory Commission (FERC). In industry structure terms, bulk power regulation creates a regulated market in which (mostly voluntary) contracts can be made, with rates and conditions set or reviewed by the FERC. (The main types of transactions/contracts are discussed below.)

In 1927, Naragansett Power of Rhode Island sold power to the Attleboro, Massachusetts Steam and Electric Company for retail distribution. The Rhode Island public utility commission attempted to set a rate for the sale, but the utilities objected, arguing that the sale was clearly in interstate commerce. The Supreme Court agreed, and hence the sale was not subject to the jurisdiction of either Rhode Island or Massachusetts regulators.[65]

The "Attleboro gap" in states' authority to regulate interstate sales led to calls for *federal* regulation of bulk power sales. Campaigning for President in 1932, Franklin D. Roosevelt described utility holding companies as a "monstrosity," and that "the methods used in building up these holding companies were wholly contrary to every sound public policy." Roosevelt concluded "that power has grown into interstate business of vast proportions and requires the strict regulation and control of the Federal Government."[66] Following his landslide victory, Roosevelt made good on his vow by amending the Federal Water Power Act of 1920 to create the Federal Power Act of 1935 (FPA).

Box 4-1. State Utility Regulation in One Easy Lesson

State regulatory commissions have the power to set utility rates and terms of service, but they must set prices so that the utilities are able to collect sales revenues that are sufficient for them to recover all of their costs.[1] In this context, the cost of providing electric power includes revenues large enough to pay stockholders and bondholders a fair return on their investments.[2] This rate setting approach suggests that regulators must set revenues equal to costs plus investor return. Regulation that sets rates in this manner is sometimes called *cost-of-service rate making*.[3]

Three primary commission activities stem from this basic responsibility. First, the Commission exercises oversight over all aspects of the IOU's costs. This oversight occurs in the form of pre-specification of expenditures, requirements that utilities use competitive bids, ex-post audits of utility purchases, and ex-ante or ex-post reviews of spending on particular construction plants. (The latter became particularly contentious in the early 1980s, when many state regulators ruled for the first time that utilities' costs of constructing nuclear plants were excessive or "imprudent." They did not allow utilities to pass on a portion of their construction costs by disallowing the inclusion of these costs in rate calculations.) It is for this reason that exacting accounting rules are prescribed which utilities must follow in their bookkeeping and their reports to regulatory agencies. In contrast, least-cost planning and IRP statutes are merely an extension of the longstanding regulatory duty to ensure lowest costs.

The second main activity is choosing the portion of the allowed rate that provides utility investors their fair return on investment (ROI). Whereas costs are predictable in advance and measurable very precisely after the fact, utility commissions must rely on financial theory, market conditions, and the advice of experts in deciding the amount of profits that investors are entitled to earn from utility revenues.[4]

Choosing the rate of return on invested capital, auditing and permitting other costs, and determining the future level of sales is roughly sufficient to determine the average price a utility must charge to collect its required level of revenue. The third and final major activity of retail regulators is to decide the rate structure, or the differences in prices between all of the different services offered by the utility. Generally, one or two different retail services (tariffs) are offered to each class of customers; the common classes are residential, commercial, industrial, and government agency customers. For any given level of average revenues required, there are a number of theories of how rates ought to be set.[5] The present and future structure of retail rates as examined in more detail in Chapter 10.

Most state regulatory agencies are commissions of appointed or elected officials who follow administrative law procedures. Formal (and often separate) proceedings are held on utility planning and forecasting processes, cost levels, and many other subjects; these ultimately contribute to rate making proceedings ("rate cases") or regulatory approvals for new plants or other outlays. Regulatory proceedings folllow administrative law rules of procedure. The decisions of most state regulatory agencies can be appealed to the state and sometimes federal courts, but only on a relatively narrow set of issues.[6]

The foregoing is the barest description of what has become a complex and elaborate undertaking. Many excellent textbooks explain the theory and practice of state utility regulation, including Troxel (1947), Clemens (1950), Garfield and Lovejoy (1964), Phillips (1993), Kahn (1971), the Illinois Governor's Office of Consumer Services (1978), and Bonbright, Danielsen and Kamerschen (1988).

1   Rates are set in conjunction with forecasts of sales in order to compute revenues. (See Doherty (1977), Chernick (1983), and Fox-Penner (1988).) The exact nature of the service offered at the price is also specified to avoid the utility obeying the regulator's price, but offering a less valuable product to customers at the allowed rate. A tariff is a combination of the rate and a description of services offered.

2   This requirement is important from two perspectives. First, if private investors provide capital to a regulated utility, and regulators cause the firm to lose its capital, this can be interpreted as an unconstitutional governmental taking of property without compensation. Second, if investors in utilities repeatedly lose money, they will cease to lend to the industry, or will require very high interest and dividend payments to compensate themselves for their risk. See Clemens (1950), Kahn (1971), and especially Bonbright, Danielsen, and Kamerschen (1988, ch. 14).

3   Although cost of service is well-accepted as the fundamental basis of rates for regulated electric utilities, other theories of rate making have played a role in state commission proceedings. These include basing price on the value of service, considerations of fairness or equity between customer groups, social and political forces, and policy considerations such as the encouragement of energy efficiency. These considerations often play a role in the decisions of public service commissions. For additional discussion, see among many other works on the subject, Bonbright (1961), Bonbright, Danielsen, and Kamerschen (1988, chs.6-10), Kahn (1971), Posner (1974), Peltzman (1971, 1976, 1981), Joskow (1973), Moorhouse (1986), Noll (1985), Shephard and Gies (1974), Crew (1980) and Zajac (1978, 1985).

4   See Kolbe and Read (1984), Abel (1984), Copeland (1978), and Kahn (1971).

5   Ramsey pricing is named after economist Frank Ramsey, who is credited with formalizing the insight that profits are maximized when prices follow a pattern inverse to the buyer's sensitivity to rising prices. Among many important references, see Baumol and Bradford (1970), Brown and Sibley (1986), Olson (1974), Kahn (1978), Caywood (1972), Cicchetti (1975), Mann (1977), Shephard (1966), Vickrey (1955), and Wilder (1983).

6   Kahn (1971), Posner (1974), MacAvoy (1970), Navarro (1982), Noll (1985), Owen and Braeutigam (1978), Mitnik (1980), and Barkovich (1978) discuss state regulatory processes in additional detail.

Attleboro was sufficient to settle the federal government's authority to set rates for sales between utilities, and the Federal Power Act established the Federal Power Commission (FPC) (now renamed the Federal Energy Regulatory Commission, or FERC).[67] As discussed in the previous chapter, the FERC is charged with setting rates and terms for such sales, as well as issuing licenses allowing production at hydro sites and other matters.

The FERC's authority over the interstate *transmission* of electric power for third parties, or wheeling, (as distinct from power sales)[68] has been the subject of almost continuous disagreement. There has never been much question that transmission *rates* for interstate transactions were under Commission jurisdiction; the controversy has centered on whether all transmission is interstate and whether transmission owners can be required to transmit power for others.

For many years, litigation ensued over whether the FPA gave the FPC jurisdiction over "intrastate" transmission.[69] If a distribution-only utility purchased power from a generator within the same state, was this transaction in interstate commerce? Cognizant of the physics of the transmission system, in which any large-scale flow of power affects the entire transmission grid in a region, one school of thought argued that it was an absurdity to assert that a large "in-state" transaction did not have an impact on interstate power sales. Indeed, because loop flow is the inevitable byproduct of all bulk power transactions, a major flow within a state is sure to have impacts on all of the major lines in the area, some of which almost always cross state lines.

This physics-based view was eventually endorsed by the Supreme Court.[70] Since then, federal authority over transmission rates has not been questioned, and attention has centered on the authority to compel wheeling, and the methods of setting transmission prices, whether compelled or voluntary.

Over a long series of court cases and amendments to the Federal Power Act, the Commission's authority to require transmission service for third parties has slowly been strengthened. The original Act gave the FERC (nee FPC) broad powers to order interconnection between most types of utilities,[71] and even more extensive authority in wartime. (It also has the duty to "promote and encourage interconnection and coordination" and to divide the nation into regions for this purpose.)[72] However, the Act did not give the Commission the authority to order the construction or siting of new transmission lines or any action that would compel the addition of more generating capacity, nor did it allow it to *require* utilities to transmit power for others, or "wheeling."[73] Indeed, Congress apparently considered and rejected provisions that would have given the Commission authority to mandate transmission service.[74]

For many years, distribution-only utilities complained that the owners of transmission lines (chiefly, but not exclusively, the IOUs) were uncooperative in allowing the use of transmission lines for wheeling power to municipalities.[75] The owners of transmission replied, in turn, that wheeling often prompts inefficiencies in the operation of the power system (these claims are examined in Chapter 6). Although periodic complaints were filed before the Commission and the courts, transmission for others continued to be regarded as wholly within the discretion of utilities.

A turning point occurred in the 1970s. First, the Supreme Court ruled that outright utility refusals to provide transmission could be viewed as violations of the antitrust rules, irrespective of regulators' control over transmission rates and terms.[76] Second, Amendments to the Atomic Energy Act required nuclear plant licensees to provide transmission access to the plants.

In 1978, the federal government sought to establish a class of independent, unregulated power generators. Congress recognized that these generators would be transacting on the wholesale power market and would therefore be subject to FERC jurisdiction, absent a specific exemption. Moreover, these generators could engage in only two possible transactions: a sale to the utility in whose territory they were located or a sale to a distant utility, which required the local utility to wheel power over its lines.[77] Rather than give the FERC authority to require wheeling in this limited instance, Congress elected to require the local utility to purchase power from these generators or to wheel it to someone who would.[78]

The Commission's first authority to compel wheeling came in 1992 with the passage of the Energy Policy Act (EPAct).[79] This legislation gave the Commission clear authority to compel transmission across the lines of most utilities, but only after a hearing and decision that established that a number of conditions were met. Briefly, before issuing an Order, the FERC must ensure that (1) all affected parties are notified and heard, (2) reliability is not "unreasonably impaired, "(3) the interests of the transmitting utility's own customers are not unduly disadvantaged, and; (4) existing contracts are honored. FERC's ability to determine the prices utilities could charge for compelled wheeling was preserved as well. In short, EPAct gave the Commission compulsory wheeling authority for the first time in 60 years, though only upon request, and only upon a limited set of conditions and procedures.[80]

The FERC has stepped from this conditioned, case-by-case approach to a blanket requirement that all jurisdictional utilities offer transmission to all qualified requestors on terms and conditions that are comparable to those transmission-owning utilities provide themselves.[81] This new FERC policy is the cornerstone of the industry's new structure and is discussed at length in Chapters 6 through 8.

Municipal, County, and State Utilities. Municipal utilities (and their close cousin, Public Utility Districts)[82] are the second-oldest form of utility organization. Their rates are regulated by state commissions in about 13 states; in the rest, retail rates, construction and purchasing plans, and most other activities are determined by the government-owner.[83] Most munis purchase some or all of their power on the bulk power market, meaning that they are distribution-only utilities, and many do not own any transmission or generating plants.[84]

All muni bulk power sales for resale and purchases are subject to FERC rate approval. FERC's interconnection authority extends to municipal systems. Under the 1992 EPAct amendments to the Federal Power Act, FERC was explicitly given authority to order wheeling on the transmission lines of the relatively small number of munis that have them. However, the recent FERC "open access" rule does not apply to munis; those seeking wheeling over muni lines must get it voluntarily or request a FERC hearing and order.[85]

Rural Electric Co-ops. Rural electric cooperatives are not-for-profit entities that are owned by their members and are governed by the Rural Utility Service (RUS, formerly the Rural Electrification Administration), a branch of the U.S. Department of Agriculture. The original type of co-op was distribution-only, and purchased all power from federal PMAs or other sources. Recently, regional generation-and-transmission ("G&T") cooperatives have been formed by groups of distribution co-ops. Whether distributor or G&T, co-ops can usually set their own retail rates,[86] are not governed by state utility least-cost planning statutes,[87] and are able to borrow funds from federally-backed lending sources available only to co-ops.[88]

When a group of distributor co-ops form a G&T, they typically sign an "all requirements" contract with the G&T which designates the G&T as their sole source of power. G&Ts combine facilities they build and own with purchases from the bulk power market to supply their distributor co-ops. Rate making jurisdiction over wholesale G&T sales and rural co-ops is exceptional in several ways. The FERC disclaims jurisdiction over G&T

sales to distributors, making these one of the few bulk power sales outside FERC's purview. In place of FERC, states may regulate G&T distributor or distributor retail rates as long as they are not pre-empted by RUS rules.[89]

One important difference between IOUs and all other utilities is that IOUs are subject to taxation. Essentially all of the non-IOU utilities are exempt from federal income tax, and many are also exempt from some or all state and local taxes.[90] Moreover, most munis and some PUDs can issue tax-exempt bonds to finance utility plants. (Rural electric co-ops do not issue tax-exempt bonds, but they are able to borrow at below-market rates due to federal financing.)

The impact that differences in tax policies have on various types of utility entities has been a subject of perennial dispute.[91] A number of analysts have attempted to quantify the "subsidy" that tax exemption has provided to public utilities; others have responded with arguments as to why IOUs have different sorts of subsidies.[92]

The PMAs and TVA. Federal electric utilities have their own unique legal and administrative structure. First, most federal hydroelectric generation facilities are under the control and jurisdiction of either the Army Corps of Engineers or the Water and Power Resource Service (formerly the U.S. Bureau of Reclamation), which is a unit of the Department of the Interior. These agencies control the operation of the plants, and must pay all of their operating and maintenance costs from their federally appropriated accounts. The decision as to whether to build additional facilities officially rests with the Army Corps or the Service, but they must must receive a federal appropriation for each specific project. Importantly, the responsibilities of the Corps and Service are not founded soley on the provision of power, but rather include the maintenance of navigable waterways, flood control, and the provision of irrigation water. As a result, analyses of the need for new facilities by these agencies incorporate considerations that are very different from capacity expansion studies by traditional utilities.

Authority to market the power from these facilities resides with the Secretary of Energy, who "shall transmit and dispose of such power and energy...at the lowest possible rates ...having regard to the recovery of the cost of producing and transmitting such electric energy, including the amortization of the capital investment allocated to power...." The Department of Energy has delegated this rate making authority to four Power Marketing Administrations, or PMAs, which are able to set their own rates (subject to the cost-of-service principles just described) for wholesale sales to distributors. In addition, DOE has the authority to "construct or acquire, by purchase or other agreement," all transmission facilities required to deliver its power "on fair and reasonable terms and conditions." The four PMAs—Bonneville, Southwestern, Southeastern, and Western—all own their own transmission except Southeastern, which leases capacity on other utilities' lines.[93]

One of the most important requirements in the FPA pertaining to federal electric power is known as the preference clause. Section 825 of the Act simply states that "preference in the sale of such [federally generated] power shall be given to public bodies and cooperatives." Because the combined demand of the nation's public and cooperative utilities far exceeds generation from federal dams, this clause has the effect of denying IOUs any firm supplies of federal power. IOUs are often able to purchase such power, but only under conditions of temporary surplus, and when the power cannot effectively be sold to preference customers.

The PMAs began as all-wholesale entities selling all of their output to munis, co-ops, and PUDs. They later gained authority to sell directly to certain industrial customers. In all cases, they are not legally subject to FERC rate making jurisdiction, either for power sales or transmission transactions, nor are they subject to state rate-making or siting jurisdiction. All undertakings by the PMA are subject to the National Environmental Policy Act (NEPA), which requires an environmental impact statement and review. This requirement is not unique to PMAs; many G&T projects by co-ops and IOUs require a federal license or use some federal funding, thus triggering NEPA requirements.

The Tennessee Valley Authority (TVA), the final member of the federal utility realm, differs from the four PMAs in a number of respects. First, TVA's charter includes economic development of the Tennessee Valley region. The agency owns its own generating facilities and is authorized to plan facilities and set rates without the approval of any other executive agency, and it is authorized to borrow money for construction on its own accord. The agency effectively has the exclusive right to serve all distribution utilities in its area, but it cannot sell its power outside its area. Like co-ops and munis, the TVA can be ordered to wheel on a case-by-case basis, but is not subject to the recent open-access rule.

TRANSACTIONS AND CONTRACTS WITHIN THE REGULATORY FRAMEWORK

Within the legal and regulatory framework just described, IOUs and public utilities enter into several types of agreements and contracts. Requirements agreements, which may be "full" or "partial," are intended to be substitutes for owning one's own capacity. The seller is required to supply all of the buyer's net demand, not a fixed number of megawatts. If a distributor is responsible for maintaining adequacy under state law or reliability council rules, a requirements purchase effectively transfers this responsibility to the seller.[94] For example, many distribution-only co-ops sign "all requirements agreements" with G&T co-ops to which they are connected. The G&T then has the responsibility to buy or build all capacity needed by the distributor co-op.

At least until recently, most requirements sales were made on a delivered basis—i.e., no separate transmission was required for the buyer to receive the power. In order for this to occur, buyer and seller had to be physically interconnected, and the system had to be operated in a manner that credited the buying utility with the capacity when the reliability councils or system operators tallied up resources.

The terms "firm" and "non-firm" are sometimes applied to bulk power sales. A firm sale simply means that the capacity sold under the agreement cannot be withdrawn by the seller without a long period of advance notice—usually five years or more. As the name implies, non-firm power is surplus power that is sold on an as-available basis. Non-firm power is often traded by utilities with almost no notice—a few hours or less—if one utility happens to have some extra and another needs extra, and there are no transmission limits that hamper the trade.

Coordination transactions are a kind of non-firm transaction that, in Professor Joskow's words,

> emerged to encompass the short-term purchases and sales of electricity engaged in by interconnected integrated utilities to make possible the economical use of generating plants owned by proximate utilities and to ensure reliability. That is, utilities traditionally owned sufficient generating capacity to meet their loads, and relied on short-term coordination transactions to ensure economical and reliable joint operation of these facilities.

Joskow further observes,

> This category of wholesale transaction has expanded in recent years to encompass virtually all voluntary bilateral wholesale contracts that do not involve an open-ended obligation by the seller to provide for the requirements of the purchasing utility. [Joskow (1989) p. 131]

Multilateral Agreements and Power Pools. The contracts described thus far are bilateral, involving only a buyer and seller, but there are also important multilateral associations and contracts in the industry, an example of which is a power pooling agreement.

Just as a full-requirements purchase can be seen as a substitute for owning one's own generation, and a bilateral non-firm purchase can be seen as a one-time opportunity to trade higher- for lower-cost generation, pools may be viewed as a contract that tries to achieve many of the same benefits that a single firm (consisting of all pool members) would achieve.[95] Of course, the question of whether a contract can achieve the same or better net benefits as a single owner of all of the assets contractually committed to the pool is a prototypical question for transactions costs economics as a determinant of industry structure, and is the main topic of discussion in the next chapter.

The wholesale marketplace for electricity has always been significant, being the only source of power for thousands of distribution utilities. In spite of the high degree of vertical integration in the portion of the industry owned by IOUs, almost 40% of all power sold at retail was purchased rather than generated "in-house" as of 1995. This wholesale trade splits roughly into one-third for pooling transactions, one-third firm or requirements sales, and one-third economy interchange.[96] Wholesale trade has continued to increase, reaching 1,927 billion kWh in 1994—more than double 1985 total transactions.[97]

LEGAL AND REGULATORY FEATURES: SUMMARY.

The legal and regulatory landscape of the electric power industry is a rich but sometimes bewildering collection of public and private firms governed by a complex set of state and federal rules.

The most striking difference suggested by the summary in Table 4.1 is the inconsistent coverage of state and federal regulation. In one state, muni's may face full-scale state rate regulation; next-door, an identical utility may not. Similarly, FERC's authority to require transmission systems to provide "open access" applies only to IOUs. Restructuring will create dramatic changes in this patchwork, but this does not necessarily imply reduced complexity. Ironically, restructuring may cause the state-federal rulebook to become even more varied and complicated—whether or not retail price regulation becomes widespread.

Appendix 4A

Holding Companies and the Public Utility Holding Company Act of 1935

At the turn of the century, America was dotted with approximately 5,000 tiny electric utilities, almost all of which were isolated (non-interconnected) plants serving small areas within existing cities.[98] Between 1900 and 1929, entrepreneurs began buying up small utilities to form larger and larger systems.

To construct plants and lines, utility owners had to raise substantial capital from the securities markets of the early 1900's, which were largely unregulated. Entrepreneurs soon recognized that relatively little capital could be used to control utilities if most of the plant was purchased by issuing non-voting debt and only a small portion with equity. Holding companies were a means of extending this method of leveraging small amounts of capital into ever-larger control across many firms. To create a holding company, a majority of the already-meager equity in one highly-leveraged company is purchased by another firm, the majority of whose stock is acquired by another company, and so on. Proceeding in this fashion, one holding company, the Standard Gas and Electric Company was able to control assets worth $1.2 billion with an controlling investment of only $23,000.[99]

Through this leveraging, companies of unprecedented scope were created. In the five years between 1910 and 1915, holding company revenues doubled from $20 million to $40 million as ten new holding companies were formed; between 1915 and 1920 revenues more than quintupled, to over $250 million; and by 1925 the number of holding companies exceeded 50 and revenues exceeded $450 million.[100] In one banner year, 1926, there were more than 1,000 mergers of small systems into the large holding companies. By 1930, 90% of the nation's electricity systems were controlled by 19 holding companies.[101]

As investigators would later document, the largest financial magnates of the "roaring twenties," including J.P. Morgan, Samuel Insull, and Cyrus Eaton, used legitimate as well as illegitimate financial practices to build these companies. Holding company managers inflated the value of their assets, manipulated their accounting procedures, engaged in "self-dealing" and other fraudulent affiliate transactions, and engaged in predatory investment and pricing behavior in order to build their empires.[102]

Coupled with extraordinary leveraging, these practices made the holding companies highly vulnerable to the stock market crash of 1929.[103] The subsequent outrage of investors in these systems, many of whom had lost much or all of their investments—along with widespread political antipathy toward "big business"—led to passage of the Public Utility Holding Company Act of 1935 (PUHCA).

Summary of the Legislation. PUHCA required all holding companies either to register under the Act or to obtain an exemption. Exemptions from federal regulation are granted for systems that are: (1) "predominantly" within one state; (2) "predominantly" operating utility companies that operate in a group of contiguous states; (3) not in the public utility business, or are in the business only outside the United States; or (4) only temporary arrangements.[104]

For companies that are not exempt from the law, the following provisions apply:

Integrated, Contiguous, Utility-Only Systems. Registered holding companies must operate an integrated system, "i.e., a group of related operating properties within a confined geographic region susceptible to local management."[105] Non-utility businesses can be operated only if they are "reasonably incidental, or economically necessary or appropriate" to the operation of the utility.[106]

SEC Approval of Securities and Asset Sales. The Securities and Exchange Commission set standards for the issuance of holding company securities, and must grant prior approval before a holding company can acquire assets or securities.[107]

Affiliation with Other Holding Companies. SEC approval is required before any utility can become an affiliate of another public utility.[108]

Interaffiliate Transactions. Sales of electricity or other goods and services between affiliates within a registered holding company system are subject to strict standards that are designed to prevent "self-dealing" and other abuses.[109]

Coordination with State Regulation. PUHCA directs the SEC to condition its approval of securities transactions on utility adherence to state commission findings and orders.[110]

Economic Benefits of Holding Companies

Due to the well-chronicled history of financial abuses associated with the holding company era, many continue to view holding companies as anti-competitive attempts to monopolize an industry at the expense of innocent investors. However, wholly distinct from the negative financial histories of these organizations, holding companies were an important means of bringing economies of scope and scale to the industry. Historian Thomas Hughes writes, "Contrary to popular opinion, the origins and development of several leading electric utility holding companies are to be found rooted more deeply in technology and management history than in finance."[111] Clemens (1950) describes a number of economic advantages provided by holding companies, including operating economies, superior management, improved access to capital, risk diversification, and the evasion of inefficient local rules. Clemens writes:

> The fact that many of these promoters were first-rate operating men has often been lost sight of midst the general criticism to which they were subjected. S.Z. Mitchell, first head of the Electric Bond and Share Company, was recognized as one of the industry's best executives. The much abused Samuel Insull and even the flamboyant Henry L. Doherty were likewise men whose abilities extended beyond financial manipulations. These men are not to be confused with the Hopsons and the Foshays, who were hardly more than stockjobbers. Even these latter, however, had the brains to hire the best management obtainable. Service and management companies pooled the experience of all companies and supplied each with counsel and advice not obtainable otherwise.

These benefits had been acknowledged years earlier by the Federal Trade Commission, which noted that holding companies had been the main means by which isolated utilities became interconnected and gained access to larger units, better engineering, centralized purchasing, high-voltage transmission, and better financial management. The FTC found substantial evidence that holding companies improved service and reduced costs and rates, precisely as the then-current theory of natural monopoly for utility services predicted.[112]

It thus appears that many legitimate economic determinants that guide businesses toward the most efficient size and shape contributed to the formation of the early holding companies. From this, it follows that the 1930s' reaction to the abuses associated with holding companies, while certainly based on real abuses, has led to binding limitations on industry structure.[113] Absent PUHCA, it is certain that the electric utility industry would look very, very different than it does today.

Chapter 4

NOTES

1 Similarly, the terms upstream and downstream are used to describe activities occurring in firms or markets nearer to or farther from the raw materials stage, respectively.

2 Bonbright, Danielsen, and Kamerschen (1988, p. 57–66) provide an elegant overview of transactions cost economics. In addition, the authors suggest two alternative paradigms for explaining public utility regulation, which they label the public and private interest paradigms, respectively. The first term is shorthand for the traditional view that utilities are regulated because they are natural monopolies. Evidence in favor of this view is discussed in Chapter 2 and the remainder of this chapter. The "private interest" paradigm asserts that regulation exists to protect the monopoly rents earned by utilities, which utilities share with their regulators. This view was discussed in Chapter 2.

Bonbright, *et al* see transactions cost economics as a third means of explaining the existence of utility regulation. The authors conclude that no one of the three paradigms offers a complete explanation of the existence of regulation. In contrast, transactions costs considerations are used as the basis of this chapter's analysis because present industrial organization theory recognizes them as the primary determinant of the boundaries of all firms, unregulated or regulated, except where the latter's boundaries are affected by regulation. This approach is therefore ideal for examining what might happen to the boundaries of firms, their behavior, and therefore market structure and behavior when regulations that have strongly affected transactions costs between and within firms are changed. This is essentially the approach used in Joskow and Schmalensee (1983) for wholesale competition.

3 The observation is attributed to Coase (1937); later seminal works include Stigler (1971), Alchian and Demsetz (1972), and Williamson (1975).

4 It is important to note that the relevant transactions costs are the *difference* in these costs incurred as a result of using competition in place of common ownership and internal coordination. Coordinating production takes specification, negotiation, accounting, and other services; differences between competition and coordination costs are strictly questions of cost magnitude and the ability to reduce them under alternative arrangements. See Goldberg (1989, p. 22) and Niehans (1987).

5 See Coase (1937, 1988), Alchian and Demsetz (1972), and Williamson (1979).

6 This discussion emphasizes the transactions costs theories of vertical integration. Economists recognize a number of other vertical integration determinants not related to the reduction of transactions costs. Firms may integrate to avoid taxation, regulation, or other government restrictions; they may also integrate forward or backward to attempt to gain monopoly power. See Blair and Kaserman (1985), Carlton and Perloff (1990, ch. 16), and Perry (1989) for excellent general discussions.

Although these non-transaction-cost theories are extremely important in many industries, they are not emphasized in the traditional electric utility industry because the latter is extensively regulated at all levels. Vertical foreclosure issues have occasionally surfaced in electric utility merger proceedings, and distribution-only utilities have raised vertical foreclosure issues in antitrust and regulatory proceedings. (See Joskow, 1985). As the industry becomes more unregulated, these issues may become increasingly relevant.

7 See Klein, Crawford, and Alchian (1978), Masten, Meehan, and Snyder (1988), Perry (1989), and Williamson (1989).

8 See Perry (1989, Section 4), as well as Arrow (1975), Carlton (1979), Crocker (1983), and Riordan and Sappington (1987).

9 For additional discussion, see Macaulay (1963), Goldberg (1980), and Hart and Holmstrom (1986).

10 The economic literature refers to this phenomenon as *information impactedness* or *bounded rationality*. See Williamson (1975, ch. 3).

11 Vertical integration to internalize externalities is discussed in Carlton and Perloff (1990) and Perry (1989) as well as many other references. Much of the theory stems from Coase's (1960) landmark work, which is expanded in Cowan (1990).

12 Malone and Rockart (1991) make this claim. However, in an unpublished conversation Professor David Levine of the Haas School of Business notes that infotech improvements also can enable expansions and improvements *within* firms, so that their impact on VI is not unambiguous. Perhaps this is why Perry (1989) finds no long-term trend toward or away from VI in American industry as a whole. Also see Cash, et al (1994) and Clemons (1991).

13 See Box 1-1, the next several sections, Kahn (1971, VII, Ch. 4), and Sharkey (1982).

14 Joskow and Schmalensee (1983) p. 41.

15 See Chapter 2, Stigler (1962, 1971), Rosner (1974), Moorhouse (1986). For a less formal view, see Hyman (1988) p. 66.

16 Among others, Joskow and Schmalensee (1983) argue that statistical studies of firm-level scale economics based on the data reported to regulatory agencies cannot be used to test for the existence of natural monopoly. In brief, the reasons they cite are (1) differences in the spacial and temporal aspects of utility demands; (2) differences in utility franchise area production possibilities, such as indigenous hydro sites; (3) learning economies are captured by third-party firms that share their experience with other utilities; (4) pooling and coordination serve to level the costs of participating utilities; and other items.

17 Huettner and Landon state, "Clearly the above findings question the natural monopoly status of this industry and raise serious issues as to the appropriateness of current public policies towards it and the generating sector in particular." (1978, p. 907).

18 Gegax and Nowotny (1993), p. 71.

19 Gilsdorf (1995) P. 137.

20 Kwoka (1996).

21 See the discussion of distributed generation in Chapter 14.

22 Single-product firms with true declining average costs (the old version of natural monopoly) automatically qualify as natural monopolies under the more accurate test of cost subadditivity.

23 Federal Power Commission (1964) V.I.P. 153;

24 Olson (1978) p. 44.

25 The only other significant cost element in transmission is the power lost as heat in the transmission line, and this also declines as voltage increases. (Fox-Penner, 1990c). The remaining costs of transmission are estimated to be about 1% of the cost of the transmission line itself (Olson, 1978, p. 50).

26 Transmission Pricing Policy Statement (RM93-19-000) October 26, 1994 p. 16

27 Report of the Transmission Task Force Report (FERC, 1989. P. 67) and the U.S. Office of Technology Assessment (1989).

28 The economic literature includes Zardkoohi (1986). Hjelmfelt (1979). Joskow and Schmalensee (1983). Weiss (1975). And Olson (1970). See also the U.S. Department of Energy (1979). FERC (1989, p. 67), and Federal Power Commission (1964) p. 28ff.

29 Henderson (1985) p. 88.

30 Ibid, p. 90.

31 Joskow and Schmalensee (1983) pp. 61-62. Joskow repeats this conclusion six years later in Joskow (1989).

32 Discussions of plant and unit economics are in Joskow and Schmalensee (1983), pp. 48–58; Cowing and Smith (1978); Landon and Huettner (1978); Hirst (1979); French and Haddad (1981), Fisher, Paik, and Schriver (1986); and Perl (1982).

33 An average new unit size of almost 500 MW by 1980 is reported in Table 5.3 of Joskow and Schmalensee (1983), but is shown in Table 2 as a *plant* rather than a unit size. It is difficult to believe that the Table 5.3 numbers are units, as this would produce a large discontinuity between the Messing, et al. Data prior to 1960 and the Department of Energy data beginning in 1980, both shown in Figure 2.

34 Messing, Friesma, and Morell (1979), Olsen (1970), Hughes (1971), Joskow and Schmalensee (1983, Table 5.3), and others chronicle the steady increase in unit size and operating pressure between the 1930s and the 1970s.

35 Haddad and French (1981) state, "It is generally accepted that small units are more reliable than larger ones." Representative reliability data are in Joskow and Schmalensee (1983). p. 48.

36 Messing, Friesma, and Morell (1979). P. 13.

37 Joskow and Rose (1985).

38 Statement of David S. Schwartz before the Subcommittee on Antitrust and Monopoly, Judiciary Committee, United States Senate, S. Res. 334, Part 2, June 9, 1970.

39 Hughes (1971) p. 70.

40 Even with this information, however, the manufacturers of utility plants are still not unanimous in their opinions concerning optimal future generator sizes. In the Department of Energy's Advanced Turbine systems program, which is helping to build the next generation of fossil and renewable power generators, manufacturers have chosen to build plants as small as 5 MW and as large as 400 MW.

41 The middle panel line marked with diamonds in Figure 4-2 displays the average size of *all* units in operation (new as well as existing). The data show that, in spite of the trend toward relatively smaller new units, the average size of all operating units is still growing slightly. This is because there are many old, very small units on the system which are reaching retirement. As these units drop off, the remaining units in operation have a larger average size. The data indicate that this trend will probably continue at least through the 1990s.

42 Messing, Friesma, and Morrell (1979, p. 17) discuss this trend in detail, as do Joskow and Schmalensee (83, p. 52). Both sets of authors argue that there are cost savings in licensing, permitting, environmental compliance, fuel purchasing, and other factors that have driven utilities to make increasing use of multiple-unit sites.

43 However, even this conclusion comes with a caveat or two. For example, the Japanese reported the construction of a 2,800 MW 8-unit gas plant. Using language much like that of utility technical reports in the 1950s, the announcement claimed that technological breakthroughs enable these plants to operate at higher temperatures than ever before, enabling higher efficiencies than ever before to be acheived. *Electric Utility Week*, July 26, 1993. We cannot rule out further techinological change that makes larger, multi-unit plants increasingly cost effective.

44 Joskow and Schmalensee (1983) p. 126.

45 Franchise laws are sometimes called territorial allocation schemes. For some examples, see Oregon Revised Statutes ch. 756–758; Utah Sections 54-4-25 UCA), and Montana State Statutes, Section 69-5-101 *et seq.* Although it is extremely old, the most extensive discussion and legal history of the franchise concept is Nichols (1928).

46 There are a few areas within states where competing franchises have been allowed, or where franchises are non-exclusive for reasons other than competition. Economist Walter Primeaux (1975a, 1975b, 1986) argues that franchise exclusivity is purely a protectionist device, and that areas with dual franchises have been well-served by competition. Because Primeaux's arguments can be interpreted as asserting that the distribution stage of the industry is *not* a natural monopoly, it is fair to call this a minority view. Joskow and Schmalensee (1983, p. 61) and Joskow (1989, n. 7) critique Primeaux's methods and conclusion. In the latter, Joskow concludes that, "While in many states electric utility franchises are technically non-exclusive, economic and regulatory barriers to the creation of directly competing distributions systems give most incumbents a de facto exclusive franchise."

47 The history of the utility industry suggests that the marriage of exclusive franchises and state regulation was driven by factors beyond the economics of natural monopoly. For example, the granting of multiple franchises in the early days of the industry led to confusion, corruption, and constant disruption of roads and other utility rights of way in American cities—nuisances citizens traded for a single regulated franchisee (see Platt, 1991, ch.3 and Tollefson, 1996). Some political economists argue that the combination of exclusive franchises and retail rate regulation was devised by the industry as a means of guaranteeing itself a source of monopoly rents. See the discussion in chapter 1 *supra* and Kahn (1971) v. II, p. 117.

48 For extensive discussions, see Phillps (1969), p. 86ff and Clemens (1950) ch. 4.

49 President Roosevelt and many other public power advocates of the 1920s and 1930s espoused this idea. In 1932, President Roosevelt agreed to create the Bonneville Power Administration "to provide a yardstick to prevent extortion against the public and encourage wider use of electric power." (Tollefson, 1986, p. 110). Most public utility texts discuss yardstick competition; for a diverse example, see Frankena and Owen (1994) p. 124–125, Kahn (1971) p. 104, Hellman (1972) Vennard (1968), and the references in the prior note. A related idea holds that franchises compete for customers who are starting new businesses or relocating, and choose locations based on the availability of reliable, low-cost power.

50 See Vince and Fogel (1995), Fairman (1995), and Ridley (1995) for recent discussions of the issue. Joskow (1986, p. 206) notes, "what passes for franchise competition nearly always involves municipalities that are trying to take over the distribution function from an IOU or that are considering offers by one or more investor-owned utilities to take over the distribution facilities owned...by a muni." An indication of the level of franchise transfer can be obtained from the dates of establishments of munis reported in Public Power's Annual Statistical Directory. In 10 year intervals, new retail muni formation has been:

1990–1994   2

1980–90   33

1970–80   21

1960–70   38

(Public Power V. 54 no. 1, Jan. 1996. These data exclude joint action, wholesale, or marketing utilities, which generally do not hold franchises). Similar results are reported in Coopers and Lybrand (1993).

As to the mechanics and difficulty of establishing a new municipal franchise, see Clemens (1950) p. 89, "The Fight Against Municipal Purchase," and the testimony of William G. Moss in Case No. 88-00043, United States Bankruptcy Court, State of New Hampshire describing a study of municipalizations and attempted municipalization in the U.S. during the past 20 years, and Coopers and Lybrand (1993).

A related idea holds that franchises compete for customers who are starting new businesses or are relocating, and choose locations based on the availability of reliable, low-cost power. It has been observed that industries whose production processes use large amounts of power, notably the aluminum industry, behave this way. For the vast majority of power customers, however, the effect is usually dominated by other considerations. See, for example, Barkenbus (1987), Carlton (1979), and Calzonetti (1987).

51 Based on their dates of establishment as reported in *Public Power* magazine, January 1996, approximately 38, 21, and 33 new municipal utilities were formed in the 1960s, 1970s, and 1980s, respectively. One noteworthy, highly litigated example of municipalization involving the town of Massena, N.Y., in the 1980s is described from the utility industry's viewpoint in Edison Electric Institute (1985).

52 Professor Joskow (1986) provides a modern critique of the concept of franchise regulation in electricity. Among other things, Joskow notes that only about 6% of a utility's costs are variable costs for the distribution system. Since variable costs are the part of a utility cost structure competing distributors can most readily minimize, Joskow concludes the scope for franchise-competition-induced efficiency gains is minor.

Ironically, Joskow's argument is the mirror image of early arguments for franchise regulation. In the early days of the industry in the Pacific Northwest, hydroelectricity was extremely cheap and distribution reportedly accounted for over 80% of the cost of delivered power. This gave rise to one early argument for franchise competition (Tollefson, 1996, p. 108).

A more fundamental critique concerns the decisional mechanics of periodic franchisee selection. The selection of a franchisee will be much like the selection of winning municipal supplier bids, but franchises typically do not specify price and run for very long periods (twenty to 100 years). Many observers believe that one-time selection by political officials, coupled with the profit making leeway afforded by a very long contract to supply an essential service with no price control, constitutes an invitation for trouble. Drawing in part on observed municipal experience with cable TV franchises, the best-known exposition of this case against franchise competition in the academic literature is Williamson (1976).

53 Nebraska has no investor-owned utilities; all its utilities are publicly- owned. In Nebraska (and most other states as well) the latter are self-regulated by the governmental entity that owns them, so no public utility commission is needed.

54 Under COS rate making, a regulated utility charges its captive customers for 100% of its out-of-pocket costs, as reflected in their accounting records. When the regulated firm pays an affiliated company an above-market price for a good it could obtain elsewhere, the regulated utility appears to have higher costs, and therefore is entitled to higher rates, with little loss of sales since its customers are captive. Meanwhile, the unregulated affiliate who receives the inflated price makes a handsome profit on the transaction. See Fox-Penner (1990e) for additional information on the conditions of cost and demand required for this mechanism to hold.

55 See Phillips (1969, p. 560), Kahn (1971, v. I, p. 20) and Fox-Penner (1990e) and the references therein.

56 PUHCA Section 13 (b). Important limits on this authority were placed in *Ohio Power v. FERC,* 954F. 2d779.

57 See the comments of the National Association of Regulatory Utility Commissioners and the testimony of the Ohio Consumer Counsel.

58 The registered electric holding companies and their headquarters are Allegheny Power System (New York, NY), American Electric Power (Columbus, OH), Central and South West (Dallas, TX), Cinergy (Cincinnati, OH), Eastern Utilities Associates (Boston, MA), Entergy Corporation (New Orleans, LA), General Public Utilities (Parsippany, N.J.), New England Electric System (Westboro, MA), Northeast Utilities (Berlin, CT), The Southern Company (Atlanta, GA), and UNITIL Corporation (Exeter, NH).

59 One important exemption has evolved since 1978 to allow utilities to own independent power plants. This exemption began in the Public Utility Regulatory Policies Act of 1978 (PURPA) and originally applied to cogeneration and renewable energy generators given federal sanction under this law, known as "Qualifying Facilities." (16 U.S.C. § 2601, Sec. 210(e)). Congress expanded the exemption to any facility qualifying as an "Exempt Wholesale Generator" in the Energy Policy Act of 1992 via registration with the FERC under its EWG rules (FERC Order 550, Feb. 10, 1993; Order 550A, April 20, 1993).

60 These exemptions are found in Sections 3(a)(1) and 3(a)(2) of the Act, respectively. See *Electricity Journal* (1990) for a useful statistical overview of holding companies as of that date. Also See U.S. Securities and Exchange Commission (1995).

61 See, for example, William T. Baker, Jr. "The Case for Repeal of the Public Utility Holding Company Act of 1935," speech before the DOE-NARUC Second National Electricity Forum, April 20–21, 1995, Providence, RI. Also see report of the Committee on Banking, Housing and Urban Affairs, U.S. Senate, to accompany S. 1317, 104th Congress Report 104-365, Sept. 9, 1996.

62 For expositions of this view, see Comments of the Coalition for PUHCA before the United States Securities and Exchange Commission, File No. S7-32-94, February 7, 1995; Comments of the National Association of Regulatory Utility Commissioners (NARUC) in the same docket, and the testimony of Larry A. Frierman, Esq., Office of the Ohio Consumer Counsel, before the Senate Banking, Housing, and Urban Affairs Committee, June 6, 1996, and Cooper (1994). For one state commissioner's vociferous views, see "Life After PUHCA Reform: Not A Pretty Picture, Says Russell", *Energy Daily* v. 19, N. 54, Wed. March 20, 1991.

63 See the U.S. Securities and Exchange Commission (1995) and the testimony of S.E.C. Chairman, Arthur Levitt, before the subcommittees on Telecommunication and Finance and Energy and Power, Committee on Commerce, U.S. House of Representatives, Aug. 4, 1995.

64 The Federal Power Act created the Federal Power Commission which held the original federal regulatory authority. In 1978, the National Energy Act changed the FPC into the Federal Energy Regulatory Commision (FERC).

65 *Rhode Island Public Util. Comm. v. Attleboro Steam and Electric Co.,* 273 U.S. 83 (1927). See Clemens (1950) p. 425.

66 Tollefson (1996) p. 110.

67 In the 1980s, the Supreme Court issued a series of decisions that reversed the Attleboro precedent. Briefly, the Court found that states could regulate wholesale power transactions if the transaction was primarily intrastate and had far more impact within a state than in the region. See *Arkansas Electric Cooperative Corp. v. Arkansas Public Service Commission,* 461 U.S. 375 (1983) and Nowack and Taylor (1996), Section 2.043.

68 In engineering terms, the distinction between the provision of transmission services and generation services is subtle and sometimes *de minimis.* In legal terms, transportation of electricity for a third party (wheeling) is considered a different service than the sale of power itself. Also see Norton and Richardson (1996) Section 82.01.

69 In *F.P.C. v. Southern California Edison Co.,* 376 U.S. 205 (1964), the Supreme Court held that FERC jurisdiction attaches to any intrastate sale occurring on a system in which "some of the energy on the system originated out of state." (Norton and Richardson, 1996, p. 82–10.) The matter was largely put to rest in *F.P.C. v. Florida Power and Light,* 404 U.S. 453 (1972), when the Supreme Court found that the operation of the bulk power grid was such that every major sale affected the system in a manner that affects interstate commerce, and therefore triggers federal jurisdiction.

70 Ibid.

71 FERC has authority to order interconnections between and among IOUs, rural electric cooperatives, and municipal utilities. However, it cannot order interconnections with federal power agencies and the Tennessee Valley Authority (FPA, Section 202). See Clemens (1950) p. 434 and Plum (1938).

72 16 U.S.C. § 824a.

73 See Clemens (1950, p. 424).

74 Justice Douglas' opinion in *Otter Tail Power v. U.S.,* 410 U.S. 366 (1973) [*"Otter Tail"*} contains an extensive history of legislative attempts to turn the transmission system into a common carrier, starting with the intentions of some of the original framers of the Federal Power Act. The latter is also mentioned in Munson (1985, p. 83).

75 See Hellman (1972) and the discussion in *Otter Tail,* n. 38 *supra,* as well as *Gulf States Utilities Co. v. Federal Power Commission,* 411 U.S. 747 (1973). For extensive legal discussions, see Fairman and Scott (1977), Norton and Early (1984), Reiter (1983) and Green (1990).

76 Around 1970, a utility seeking transmission from a the privately-owned Otter Tail Electric Company in Minnesota sued Otter Tail under the antitrust laws for denying transmission service to the distributor. The Supreme Court ultimately ruled in favor of the distributor, holding that Otter Tail's status as a regulated IOU did not exempt it from anti-competitive behavior under the antitrust laws. (*Otter Tail v. U.S.,* 410 US 566 (1973). This suit led investor-owned utilities (many public utilities are exempt from antitrust statutes) to expand their offering of transmission service, though not nearly as much as distribution-only utilities and later independent power generators wanted. Antitrust issues in the restructured industry are discussed further in Chapters 7 and 8 below. For discussions of transmission-related antitrust issues in the traditional industry, see Kahn (1971), Joskow and Schmalensee (1983, p. 15), Breyer (1982) and especially Weiss (1975) and Joskow (1985).

77 Constructing a transmission line out of the local service area to a utility willing to buy or wheel power was generally considered infeasible, although some independent producers within small distances of buyers have built dedicated lines.

78 See Fox-Penner (1993).

79 Energy Policy Act of 1992, amending the Federal Power Act, Sections 211 and 212.

80 As an indication of the degree of administrative complexity involved, as of May, 1995, approximately five transmission orders had been issued by the FERC. Most proceedings required more than a year, though one reportedly required only four months. Interestingly, requests have come from every segment of the industry, and request service over the lines of every segment. For example, one case involves an IOU requesting service from another IOU, while another pits one municipal agency against another. This very limited sample stands in contrast to the commonly-held (and probably accurate prior to EPAct) view that transmission litigation was primarily transmission-dependent publics versus transmission-owning IOUs.

81 Order No. 888, Docket No. RM95-8-000, April 24, 1996.

82 Public utility districts, which are sanctioned by state laws in Washington and Oregon, irrigation districts, county power authorities, and other local government agencies sometimes supply electricity in their area. These agencies are quite similar to munis, but generally extend beyond urban areas to encompass one or more counties, including rural areas. See Tollefson (1996) and Vennard (1968).

83 NARUC (1993, Part B, Table 1). Munis may also be subejct to state PSC accounting rules, as in Ohio (Nowak and Taylor, 1996). Most states' energy facility siting rules do not exempt municipally-owned facilities, though many munis do not own generation or transmission plant—only distribution lines, which typically do not fall under siting laws.

84 American Public Power Association (1996) and Energy Information Administration (1995d).

85 To attempt to reduce any resulting imbalance in the availability of transmission service, FERC Order 888 allows IOUs subject to its jurisdiction to require reciprocal service from non-jurisdictional utilities as a condition of providing service to owners or non-jurisdictional transmission. However, where the provision of reciprocal service would jeopardize the tax-exempt status of the facilities of a state or local utility, the latter is exempt from reciprocity pending issuance of new I.R.S. guidlines.

86 Cooperative retail rates are regulated by state commissions in about 19 states. (NARUC, 1993, Part B, Table 1).

87 However, the Rural Utility Service has promulgated rules that make least-cost planning a condition of receiving federally-backed financing. See 7 CFR CH. XVII, Subpart H.

88 Rural electric coops may apply to the Rural Utility Service for a variety of low-interest loans. One program provides loans from the Federal Financing Bank at interest rates between 5% and 2% per year; another program provides credit guarantees for loans from private lenders. See 7 CFR CH. XVII Section 1700.20.

89 *See Arkansas Electric Coop Corp v. Arkansas Public Service Commission* 461 U.S. 375.396 (1983) and *Wabash Valley Power Association v. Rural Electric Administration.* 903 F. 2N445 (7th Cir. 1990).

90 Some public utilities make "payments in lieu of taxes" (PILT) to state and local governments.

91 For example, Venard (1968) p. 40–50, points out that IOUs paid taxes equal to about 5% of the value of gross plant in 1968, while the comparable average for all public utilities was about 0.75%.

92 Pace (1972), Kiefer (1982), Joskow and Schmalensee (1983), p. 16–19.

93 Kaufman and Dulchinloss (1986).

94 However, "partial requirements" service is also possible, which simply means that the buyer is entitled to effectively own a fixed amount of capacity that is less than the buyer's total needs.

95 Professor Joskow makes this point in his comments in the Technical Conference concerning Independent System Operators and [REFORM?] of Power Pools, docket RM95-8-000, Jan. 24, 1996, p. 4–5.

96 FERC (1989) p. 18. Also see Kelly (1987) p. 26ff.

97 EIA, 1995b, p. 4.

98 Federal Power Commission (1964) p. 18.

99 Bonbright and Means (1932, p. 1 16), quoted in Kahn (1971, v. II p. 73).

100 Clemens (1950 p. 490).

101 (EIA, 1993c, p. 13). Clemens (1950, p. 499) notes that around 1930 the nineteen largest holding companies earned about 77% of industry and six large independent companies earned an additional 11%. About 12% was earned by the remaining 130 or so holding companies and all systems owned by municipalities or independent.

102 In 1928, the Federal Trade Commission began a seven-year, 101-volume study of the practices of holding companies; the Interstate Commerce Committee of the U.S. House of Representatives conducted a similar study between 1933 and 1935. These reports produced extensive documentation of (Securities and Exchange Commission, 1995, p. 3):

...nineteen general categories of abuses: issuance of securities to the public that were based on unsound asset values or on paper profits from intercompany transactions; extension of holding company ownership to disparate, nonintegrated operating utilities throughout the country without regard to economic efficiency or coordination of management; mismanagement and exploitation of operating subsidiaries of holding companies through excessive service charges, excessive common stock dividends, upstream loans and an excessive proportion of senior securities; and the use of the holding company to evade state regulation.

Tollefson (1996, p.93) recounts the sworn testimony of a J.P. Morgan partner familiar with Morgan's predatory tactics used to acquire additional holding company properties.

103 Fifty-three holding companies with $1.6 billion in assets (in then- current dollars) went bankrupt between 1929 and 1936, more than a third of the 151 holding companies studies by the FTC (including the legendary Samuel Insull's Chicago-based utility empire). Twenty-three more companies were in arrears for payment on an additional $1 billion of preferred stock as of 1938. (U.S. Securities and Exchange Commission (1995, p.4).

104 PUHCA Section 3 (a).

105 PUHCA Section 2(a)29.

106 PUHCA Section 11 (b)1.

107 PUHCA Section 7.

108 PUCHCA Section 9 (a) 2.

109 PUHCA Section 12.

110 PUHCA Sections 8, 18, 19, and 33.

111 Hughes (1983, p. 393).

112 Federal Trade Commission (1937) Part 72A. Also see Hughes (1983) p. 393 and Clemens (1950).

113 One example might be the experience of the Sacramento Municipal Utility District, which sought the services of another utility to help it operate its troubled Rancho Seco nuclear plant. No other utility would enter into a contract for management of the plant out of a fear that such a contract would create a holding company subject to SEC jurisdiction. "PUHCA Blamed for SMUD Troubles in Finding Rancho Seco Operator." *Electric Utility Week*, Sept. 8, 1988.

Chapter 5

Industry Structure: Past, Present, and Future

Tables 5.1 and 5.2 show the present pattern of ownership of generation and transmission plants in the industry by type of owner. There are presently about 220 IOUs,[1] over 900 co-ops, about 2,000 public systems, and ten federal utilities.[2] There are five main federal utilities: BPA, TVA, SWAPA, WAPA and SEPA. The remainder are much smaller and/or function via one of these five. On average, IOUs are 40 times as large as the publics: the average IOU sells about 9 million MWh, the average public and co-op sell 0.20 and 0.24 million MWh, respectively. However, among the publics there are a large number of substantial systems—more than 50 publics selling more than 2 million MWh a year, for example. The average federal PMA is about half the size of the average IOU, or about 4.6 million MWh of annual sales.

In addition to the tremendous variation in average size among the nation's utilities, there is a corresponding and equally large variation in the degree of vertical integration. Almost all IOUs are vertically integrated—although restructuring is widely expected to alter this—and the vast majority of other entities are either distribution only (munis, co-ops) or wholesale power sellers only (the PMAs, G&T co-ops, state power authorities). IOUs produce about 75% of the power they sell, and much of the remaining 25% consists of short-term coordination sales—temporary exchanges from surplus generation available for a few hours or days. Conversely, public systems purchase all but about 17% of their retail sales. As shown in Table 5.1, in 1994, IOUs owned about 77% of all generation; co-ops and publics owned 3.6% and 1%, respectively. The remainder was owned by independent generators that are exempt from rate regulation—so called non-utility generators or independent power providers (IPPs).

As shown in Table 5.2, transmission capacity is also predominantly owned by IOUs, which own 530,000 miles of transmission lines, or about 80% of the industry total. Public power agencies own much of the rest (14%), while the federal utilities own very little. At the distribution level, the principal assets to own are transformers and low-voltage lines, and distributors' ownership of these assets is roughly proportional to the number of ultimate customers and customer density.

These statistics can be stylized into a portrait of the 1994 utility industry that is shown in Figure 5-1A. About four-fifths of the industry is integrated and the remaining one-fifth is a combination of generators selling at wholesale to a slew of mostly very small munis and co-ops. In this picture, the wholesale power market serves two rather different functions. For the integrated IOUs, it is a means of providing reliability and exchanging low-cost supplies. For the munis and co-ops, it is their main source of supply.

TABLE 5.1. Ownership Profile of the U.S. Electric Utility Industry, 1995

Ownership Category	Number of Firms (c)	Capacity 1991 (MW) (f)	Percent of Total Industry Capacity
Investor-Owned (a)			
Integrated	198		
Wholesale Only	19		
Transmission Only	3		
Total	220	558,508	68.8%
Rural Electric Co-ops (b)			
Total	922	31,925	3.9%
Public, Non-Federal			
Municipal Utilities	1,818		
Public Utility Districts	75		
Irrigation Districts	9		
State and Mutual	75		
Total	1,977	87,268	10.7%
Federally-Owned			
Total	10	66,989	8.2%
Non-Utility Generators (e)			
Total	44	66,633	8.3%
INDUSTRY TOTAL		811,323	100% (d)

Notes:

(a) Does not include service or holding companies, which are generally non-operating.

(b) Includes co-ops that are exclusively generation and transmission ("G&T"), as well as utilities that are integrated or are distribution only.

(c) Source: Electrical World Directory, McGraw Hill, 1995.

(d) Totals may not add due to rounding.

(e) EIA Electric Power Annual, 1994, viii

(f) Energy Information Administration, forms EIA-860, 861 Public Power Annual Statistical Issue, Jan. 1996.

TABLE 5.1A. Profile of U.S. Electric Generating Capacity, 1993

Installed at Year End

Fuel Type	Number of Units	Capacity (b) (MW)	Average Unit Size
Coal	1,223	324,790	266
Petroleum	3,360	76,947	23
Gas	2,170	142,573	66
Water (a)	3,531	18,378	5
Nuclear	109	107,849	989
Renewable (c)	3,466	74,154	21
Total	13,859	744,691	54

Planned Additions

Fuel Type	Number of Units	Capacity (b) (MW)	Average Unit Size
Coal	16	6,919	432
Petroleum	84	4,571	54
Gas	278	28,516	103
Water (a)	NA	NA	NA
Nuclear	0	0	NA
Renewable (c)	NA	NA	NA
Total	378	40,006	106

Notes:
(a) Includes pumped storage, large and small hydroelectric
(b) Nameplate capacity
(c) Other than hydroelectricity

Source: Energy Information Administration (1994) Tables 1–4

TABLE 5.2. Ownership and Voltage of Overhead Transmission Lines, 1994

	Circuit Miles 1994	Percent of Total
Investor-Owned Utilities	530,478	80%
Federal Power Marketing Administrations	32,995	3%
Rural Electric Co-ops*	33,652	3%
State and Local Public Power Agencies	95,816	14%

Distribution of Transmission Lines by Voltage Rating, 1994

Nominal Voltage	1994p
Total	659,289
22,000–30,000	82,717
31,000–40,000	96,276
41,000–50,000	34,430
51,000–70,000	111,465
71,000–131,000	95,313
132,000–143,000	68,502
144,000–188,000	24,934
189,000–253,000	67,896
254,000–400,000	47,483
401,000–600,000	26,396
601,000–800,000	3,876

Source: Edison Electric Institute Statistical Yearbook (1994)

Hearings of the Subcommittee on Energy and Water Development, House Committee on Appropriations, 104th Congress 1996, p. 903

1994 Statistical Report Rural Electric Borrowers. Rural Utilities Service 201-1

* Not identified as circuit-miles

Box 5-1. The U.S. Generation Portfolio—Today and Tomorrow

Table 5.1A shows the number, size, and type of electrical generating plants in the United States at the end of 1993 and projected additions through 2003. The Table shows that the fuel and size profile of generating units is changing quite slowly, notwithstanding the changes the industry is undergoing. The largest fraction of existing generating plants (43%) are coal-fired, with an average size of about 300 MW; the next largest group are natural gas-fired plants, which are only 70 MW on average. The remaining large groups of existing plants are oil and renewable (principally hydroelectric) units averaging 25 MW each. A group of about 110 nuclear plants are of by far the largest average size—approximately 1000 MW/plant.

According to the projections U.S. utilities have filed with the Department of Energy, future plant additions will be dominated by natural gas plants, which will be about 100 MW each in size, representing 62% of all planned additions. No nuclear additions are planned, but there will be significant coal-fired additions, and coal will still be the dominant electricity source in 2003. Relatively small renewable energy plants will continue slowly to gain shares, with larger gains expected in the middle decades of the twenty-first century.[1]

To some extent, the projected choices of unit size and fuel type are unique to the United States. However, there are several broad generation trends all over the world. Large coal, nuclear, and (generally smaller) oil-fired plants dominate the world's power supply (and will for decades to come), and natural gas-fired power is on the rise worldwide in mid-sized plants (100-400 MW).[2] Plans to build large new nuclear plants are limited largely to Japan, China, and some other parts of Asia;[3] renewable energy is making its largest inroads in developing countries that have low-cost renewable resources and little electric power infrastructure.[4] There are many, many 1,000 MW or larger plants on the drawing boards, in addition to plants in every lesser size class.

---

1 Many considerations enter into the selection of unit sizes, and one must be cautious about drawing conclusions from these simple figures alone. However, it is noteworthy that the world's largest and most advanced power system still has literally hundreds of comparatively very small plants in it. Even with a flattening out of economies of scale, most of these plants are much smaller than the units planned for more competitive and technologically advanced future. In Chapter 4, a body of research is reviewed which shows that most small U.S. utilities have never adequately achieved economies of scale.

2 International Energy Agency (1996) and Economic and Social Commission for the Asia and the Pacific (1993).

3 Energy Information Administration (1995) p. 49ff. Also see "Asia Delivers an Electric Shock," *The Economist*, October 28, 1995.

4 See, among many other works that come to this conclusion, "The Battle For World Power," *The Economist*, October 7, 1995; Johansson, et al (1993) ch. 1; and Energy Information Administration (1995).

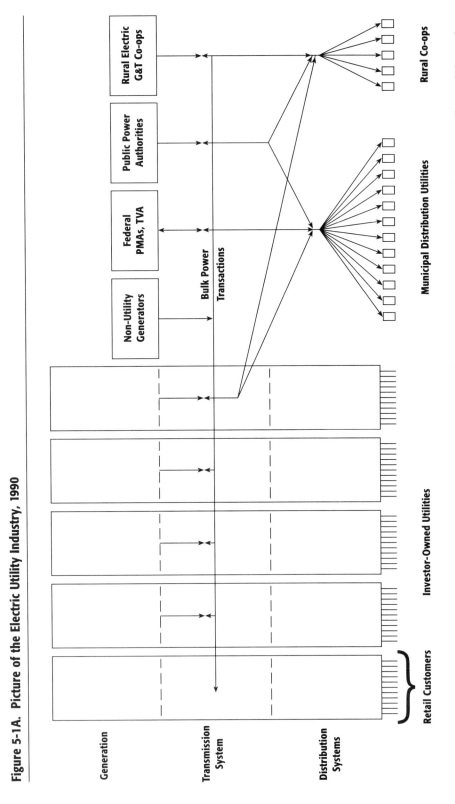

Figure 5-1A. Picture of the Electric Utility Industry, 1990

About 75% of the industry is vertically integrated. IOUs own approximately as much generator and transmission capacity as they distribute to their customers. The remainder consists of wholesale sellers (NUGs, PMAs, Public Power Authorities) who sell on the bulk power market to several thousand (mostly small) muni and co-op distributors.

Figure 5-1B. Electric Utility Industry Structure with Wholesale or Retail Competition

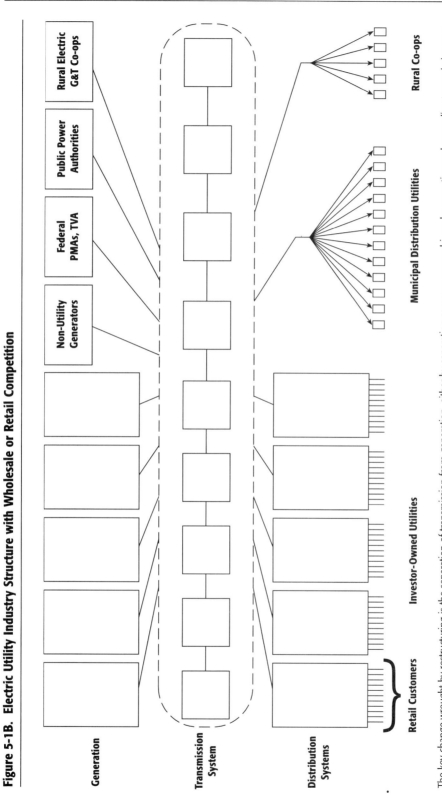

The key change wrought by restructuring is the separation of transmission from generation, either by separating common ownership or by operating and expanding transmission as a non-discriminatory "common carrier" for all generators. Under wholesale only competition, generators sell to distribution utilities, who then re-sell all power in their area at retail. Under "direct access" or "customer choice," the distribution systems also become non-discriminatory common carriers, and retail buyers shop and purchase directly from generation companies.

The ownership entities shown in Table 5.1 are subject to the patchwork of laws and regulations shown in Table 4.1. This ownership and regulatory organization is far more complicated than those of other utility industries and the power industries of other nations. Indeed, until recently, most other electric utility sectors in the rest of the world—developed or developing—were state-owned monololpolies.[3] "In diversity and complexity," wrote the 1964 National Power Survey, "the U.S. electric utility industry is unique among the power industries of the world."

One important point to remember is that Figure 5-1 is a national composite picture of the industry—it obscures enormous and important regional differences. "[E]ven in its present advanced form," write industry experts Thomas Lee and Frederich Ellert, "the power transmission network of the United States is primarily a regional network."[4] Many national studies of the power industry, including the National Power Surveys, have been organized and pursued region-by-region, rather than via national aggregates and averages.[5]

Different regions of the country have very different geographical features (which, for example, make transmission lines easier or harder to site), differential access to and prices for generating fuels, different degrees of pooling, and different attitudes toward public versus private power. In addition, the patchwork of regulation and jurisdictions has a definite regional slant. As an example, the Northwest Power Planning Act strongly influences regulatory policies across the entire Pacific Northwest region. For many years, the electric grid in Texas was not interconnected to the rest of the United States, and therefore escaped much of FERC's jurisdiction. In the rural Midwest, a utility recently reported it had never had a single problem getting siting approval for a transmission line; in the crowded Northeast, another utility mockingly complained that "we can't even add an extension cord to our system without years of hearings."

Viewing the North American power industry as a group of regional markets is increasingly becoming accepted as a most useful geographic construct.[6] However, the boundaries of the regions are not necessarily the same for all applications, and all of the regions are interconnected to varying degrees.[7] As increased competition causes markets to become more and more regional in nature, which is roughly the natural size for many bulk power markets, regional boundaries will certainly become increasingly important, and state boundaries will be less so.

EVOLUTION OF THE INDUSTRY STRUCTURE, 1890–1940

Until the 1920s, most electric utility systems were smaller than the states in which they were located, and many were not interconnected to their neighbors. In 1920, there were about 5,800 distinct companies, almost all serving part of a large city or one or two small towns at most. The wholesale transactions in this market were overwhelmingly sales from one local company to another nearby distributor.

The small utilities of this era were financially unstable and often unreliable. Franchises were not yet commonly exclusive, and many private companies used aggressive political tactics to obtain franchises, especially in the hard-fought large cities.[8] Cities' experience with utility-related corruption and unreliable service led to one movement that favored the public takeover of utilities and a second that argued in favor of private ownership with regulation by statewide administrative bodies.

Figure 5-2. Average and Maximum Sizes of Electric Utility Steam Plant Boilers Ordered Each Year, 1946–64

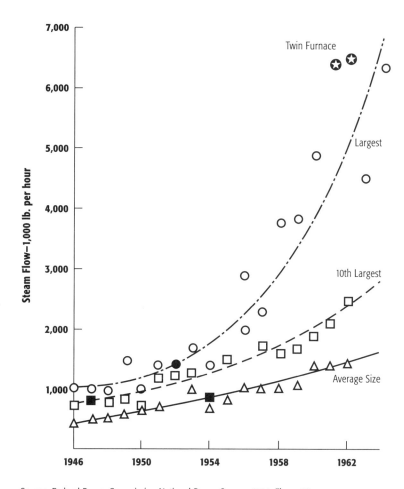

Source: Federal Power Commission National Power Survey, 1964, Figure 39

The debate between the proponents of public ownership versus private ownership and regulation was ideological, politicized, and hard-fought.[9] Although it is often said that the forces favoring state regulation "won," the battle actually yielded no surrender.[10] Between 1907 and 1920, roughly 30 states established state regulatory commissions that were authorized to control the retail rates and investments of in-state utilities (most remaining states followed later). At the same time, however, private and municipally-owned systems grew at about the same pace, each numbering about 2,100 by the mid-1920s.[11]

In the 1920s and early 1930s, two major changes began to occur. First, transmission and control technologies enabled wider and larger-scale interconnections, allowing utilities to start to realize the many economies of coordination discussed in Chapters 3 and 4.[12] Utilities increasingly either stretched across more than one state or were trading power

127

with other systems across state lines. The maximum voltage of transmission lines (and, correspondingly, the maximum economic transport distance) zoomed from a few thousand volts in 1900 to over 250,000 volts in 1925, and comparable advances were made in generator and system control.[13] (Maximum transmission voltage in the industry remained at 345 kV from the late 1930s until almost 1960, when transmission voltages entered a second stage of rapid growth.)[14]

The second major change in the 1920s and 1930s was the emergence of a belief that the nation's hydroelectric sites were an important source of regional economic development and belonged in the public domain.[15] This belief grew in part out of the use of federal hydro facilities that proved important for national defense industries in WWI, as well as out of a belief that national development of hydro sites would bring a variety of development benefits to the rural areas that contained the sites. Finally, the establishment of rural and municipal utilities, many of which needed more power than they could supply, were an ideal market for federal power.

The third major development of the period was the growth of IOU holding companies—enormous pyramidal financial companies that owned and controlled hundreds and eventually thousands of successively smaller IOUs. In the years between 1915 and 1930, holding companies (discussed in Chapter 4) changed the IOU portion of the industry from an atomistic collection of 2,100 firms to one in which 19 holding companies controlled 90% of all IOU systems and operated in as many as 24 states at once.[16]

The enactment of the FPA and PUHCA ended the utility industry's first era and began its second. Other New Deal laws produced additional change. In 1933, Roosevelt created the Tennessee Valley Authority, followed by the Rural Electrification Administration in 1936, and the Bonneville Power Administration in 1937. Federally owned or financed plants and lines grew from about 1% in 1935 to almost 15% today. Federal transmission lines also began to interconnect with those of IOUs, co-ops, and state power authorities.

These changes yielded an ownership structure in the early 1950s like that shown in the first column of Table 5.3. The number of IOUs stood at 523, reflecting the breakup of many of the 151 holding companies and the rest of the IOU population. Interestingly, the number of municipal systems stood at about 2,300—virtually unchanged in over 30 years. Following the establishment of REA, the number of rural electric co-ops zoomed to almost 1,000 by the 1950s; the percentage of rural customers that were electrified rose from 11% in 1935 to 98% by 1966.[17]

The 1950s opened the third era in the utility industry. Some analysts call the decades of the 1950s and 1960s the industry's golden era, as every segment of the industry expanded with few problems. Almost everywhere, the cost of producing power declined as a result of larger generating units and more large-scale interconnection and coordination—although power costs were never homogeneous in all parts of the country.[18]

Table 5.3. Evolution of Utility Industry Structure, 1952–1991

	1952	1956	1960	1964	1968	1972	1976	1980	1984	1987	1991
Number of Utilities											
IOUs	523	424	383	366	256	274	249	240	221	210	203
Munis	2,337	1,884	1,873	1,819	1,790	1,769	1,755	1,753	1,814	1,812	1,807
Co-ops	963	930	925	935	932	923	933	936	936	937	928
Other	0	85	89	94	93	80	96	144	159	101	170
Federal	7	9	10	24	26	40	37	41	38	37	37
Total	3,830	3,332	3,280	3,238	3,097	3,086	3,070	3,114	3,168	3,097	3,145

Note: 1952 Federal includes state-owned utilities

Generating Capacity

	1952	1956	1960	1964	1968	1972	1976	1980	1984	1987	1991
Kilowatts in Thousands											
IOUs	64,349	91,145	128,450	167,704	220,766	314,353	415,504	477,083	514,863	552,705	573,023
Munis	6,010	8,325	11,499	15,199	19,429	23,049	30,602	34,958	36,717	39,378	40,424
Co-ops	532	795	1,390	2,017	3,434	6,704	9,947	15,422	24,738	26,359	26,453
Other	1,648	2,096	4,313	9,022	12,473	14,239	23,402	27,509	32,841	34,858	34,496
Federal	9,678	18,336	22,350	28,343	34,956	40,408	51,707	59,083	63,304	64,666	65,561
Total	82,217	120,697	168,002	222,285	291,058	398,753	531,162	614,055	672,463	717,966	739,957
Percentage of Total Capacity											
IOUs	78.27	75.52	76.46	75.45	75.85	78.83	78.23	77.69	76.56	76.98	77.44
Munis	7.31	6.90	6.84	6.84	6.68	5.78	5.76	5.69	5.46	5.48	5.46
Co-ops	0.65	0.66	0.83	0.91	1.18	1.68	1.87	2.51	3.68	3.67	3.57
Other	2.00	1.74	2.57	4.06	4.29	3.57	4.41	4.48	4.88	4.86	4.66
Federal	11.77	15.19	13.30	12.75	12.01	10.13	9.73	9.62	9.41	9.01	8.86
Average Size of System—MWs											
IOUs	123.04	214.96	335.38	458.21	862.37	1,147.27	1,668.69	1,987.85	2,329.70	2,631.93	2,822.77
Munis	2.57	4.42	6.14	8.36	10.85	13.03	17.44	19.94	20.24	21.73	22.37
Co-ops	0.55	0.85	1.50	2.16	3.68	7.26	10.66	16.48	26.43	28.13	28.51
Other	NA	24.66	48.46	95.98	134.12	177.99	243.77	191.03	206.55	345.13	202.92
Federal	1,382.57	2,037.33	2,235.00	1,180.96	1,344.46	1,010.20	1,397.49	1,441.05	1,665.89	1,747.73	1,771.92

In terms of the scale of generating units and the importance of interconnection, this period represented the expanded realization of cost fundamentals that had been recognized as early as the 1890s. Generating plants became larger and cheaper, and proportionately fewer small units were added. Figure 5-2 shows that the largest utility boilers ordered grew in capacity almost sevenfold between 1906 and 1964; maximum plant size grew from a few dozen megawatts to more than 600 in a single plant. During the same period, the *average* unit industrywide doubled in size and became 50% more efficient.[19] Figure 5-3 shows that the growth in high-voltage transmission was even more dramatic, tripling during the 1950s and tripling again during the 1960s. By contrast, during the next 20 years, high voltage transmission less than doubled.

Figure 5-3. Total Miles of AC Transmission Lines (230 KV and above) Total Industry

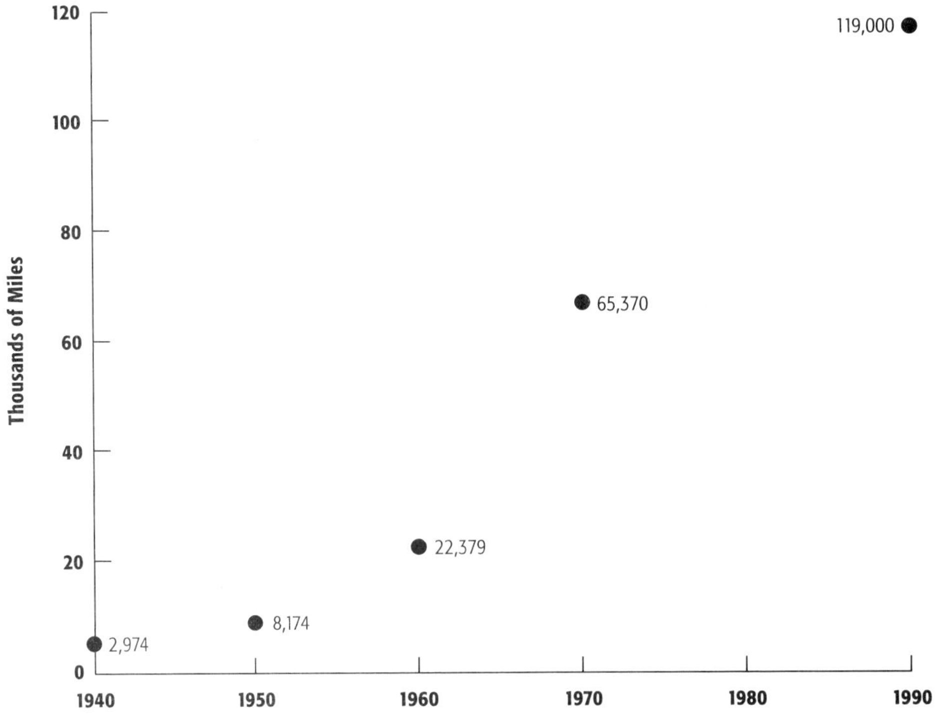

Source: FPC, 1970, PI-13-4. 1990 Estimate based on EEI (1994)

More interestingly, the main structural change during this period involved the horizontal re-massing of the IOU portion of the industry into fewer and larger systems. Table 5.3 shows that between 1952 and 1976 the number of munis declined until 1968, flattened, and then rose a little. The number of co-ops remained constant throughout the period. The average size of these systems grew substantially, but from a very small initial size.

Modest growth of public power utilities was dwarfed by IOU horizontal integration. PUHCA constraints notwithstanding, the number of IOU's dropped from 523 in 1952 to fewer than half of that (276) in 1976. During the same period, increased per-customer demand and mergers expanded the average IOU by a factor in excess of 20. This growth is attributable to a number of conditions, including the economies of operating larger firms, the tendencies of cost-of-service regulation, and a political climate that was favorable to the IOUs. Leonard Hyman concludes his review of this era by saying[20]:

> "For electric utilities, the postwar period was one of reorganization out of the holding companies, mergers for some smaller utilities, minimal need for rate relief, declining costs and prices, motivation to add to the rate base, satisfied investors, and acceptable (although unspectacular) returns for owners."

The industry's fourth era began in the 1970s, when rising energy prices, nuclear plant cost overruns, the flattening of unit economies, a slowdown in demand, and widespread IOU financial problems put the industry on a path toward today's restructuring. (This is discussed in Chapter 2.) The fourth era's key structural watersheds were the birth of independent generators and the first steps toward widespread transmission access. Meanwhile, IOUs continued to become larger and fewer. As Table 5.3 shows, between 1972 and 1991, the total number of munis was little-changed and their average size grew by about 30%. In contrast, the average IOU grew 70%, partly due to roughly 70 IOU mergers.

COORDINATION AND INTEGRATION RESEARCH, 1950–1980

Embodying widely-held views of the national interest, the Federal Power Act created federal power regulation "for the purpose of assuring an abundant supply of electric energy throughout the United States with the greatest possible economy and with regard to the proper utilization and conservation of natural resources."[21] In the decades that followed, policy makers, industry leaders, and economists have watched the evolution of the industry structure (Table 5.3) and the patchwork of regulation and questioned, "Is this a rational industry structure? Should the scope of regulation be relaxed or increased and/or should ownership or integration changes be fostered?"

The FPC was nominally responsible for promoting an industry that provided power at the lowest possible cost. However, its authority was limited to encouraging interconnection and setting rates for wholesale sales. It could not compel interconnection, it could not compel the enlargement of transmission or generation facilities, and it had very little influence over federal and public power as a whole. Studies of industry structure during this period therefore focused less on open-ended inquiries into optimal industry structure and more on narrower questions pertaining to optimal coordination, cooperation, and regulation of the existing sets of players.

In the context of these inquiries, the word "coordination" plays an interesting dual role. In the short-run, coordination usually meant more interconnection between utilities and more sales of power, ideally in regional pools that employed economic dispatch. Because power pools and coordination exchanges all use markets, this kind of coordination can be seen as an increased use of the marketplace—even if the sellers were not vying for each others' customers and sold at regulated prices. More importantly, the short-run objective of "maximum coordination" in this context is identical to the objective achieved in competitive markets—i.e., all users of (bulk) power in each area should be able to obtain it at lowest cost.

In the long-run, coordination generally meant transmission and distribution planning and expansion using the lowest-cost collection of lines and plants. Because unit and line economies of scale were present until the 1970s, this was often read simply as aggregating the loads of smaller utilities that individually did not need large additions and seeing whether the joint or partial ownership of plants by small (or small plus large) utilities would be cheaper than each utility planning for its own growth.

In both time frames, inquiries of this nature can be seen as questions about the transactions costs framework and firm boundaries discussed in Chapter 4. These studies ask whether the formulation and adherence to complex contracts between many small utilities (e.g., to jointly build a large plant) or between small and large utilities (e.g., to form a pool) worked better or worse than more horizontal or vertical integration.

A 1979 Department of Energy study summed up the spectrum of views on the overall issue:

There is a body of opinion that believes that the major benefits that are economic for society to achieve from physical interconnection, coordinated planning and development, and integrated operations are now being captured, and that future potentials to the extent that they are advantageous and attractive, will be captured in a timely fashion through the natural evolutionary extension of existing bulk power supply systems. This opinion holds that most if not all of the advantages of diversity, economic dispatch, reserve sharing, economics of scale, and other benefits of system coordination have already been achieved and that what remains unutilized is not significant. Included in this body of opinion are some who seriously question the wisdom of altering a well-established, time-tested existing system which has worked well.

There is another body of opinion that is convinced that there are substantial unrealized opportunities for net beneficial gains in economy, reliability, and energy conservation through greater joint action than is now taking place or than is likely to materialize without outside inducement of some sort. That opinion holds that if there is no increased federal intervention in the proceedings, it is too much to expect that a satisfactory national bulk power supply system will evolve as a result of a multitude of independent "corporate" decisions made on a disjointed, utility-by-utility basis. It believes that satisfactory evolution is particularly unlikely in view of the fact that those decisions are driven to different narrow-based objectives, different management capabilities and aspirations, different planning assumptions, and different policies involved in the construction and operation of bulk power supply facilities. In short, that opinion suggests that if no affirmative steps are taken by government intervention, an effective national bulk power supply system will never come, or if it does come, substantial net benefits will be lost in the interim.

The studies that addressed this broad controversy during the past 30 years can be placed in two overlapping categories. One group of studies examines interconnection and industry structure issues from a more national perspective, including the possibility of transferring power from region to region. This group includes a number of studies of large interregional transmission lines that can be used either to ship large amounts of power from one region in surplus to another in deficit (often due to the time of year),[22] or to substitute transmission for coal shipments.[23] In its grandest form, this group includes a series of periodic proposals to create a "national power grid" of very high-voltage, high-capacity lines that could be overlaid on the existing transmission system, much like the interstate highway system is built atop the rest of the nation's roadways.[24]

The second group of studies focused primarily on optimizing the benefits of interconnected operations and existing ownership-integration structures within the various regions of the United States. Included in this group are studies of power pooling and joint planning by region or within a state.[25]

National Studies of Interconnection and Coordination. The 1964 National Power Survey was a bold initial effort at quantifying the potential national benefits of stronger intra- and interregional transmission ties. Estimating the benefits of seasonal diversity, greater reliability with less generation, and the construction of larger units via joint planning—but not estimating the costs of achieving these benefits—the Survey found substantial potential national savings could go unrealized due to suboptimal coordination. The 1964 Survey estimated that retail power costs could decline 27% by 1980, but roughly half of this decline might not be realized due to inadequate pooling and planning.

It drew attention to smaller systems, which it felt were far from achieving integration and coordination economies, noting several factors (Federal Power Commission (1964) p. 275):

> There are few physical or economic obstacles in the way of small system growth and improvement, but legal and psychological barriers must be considered. The principal legal barriers are [that]…"[T]he legal authority of local public agencies to enter into interconnection and pooling agreements, leasing arrangements and joint ventures, including the creation of jointly owned entities to construct new facilities, is limited in various respects which sometimes prevent entirely, and sometimes substantially affect the nature of, proposed pooling arrangements.

> Psychological barriers also stand in the way of full coordination between the small municipal and cooperative systems and the investor-owned systems. These barriers are the fruit of decades of intra-industry animosities, and they impair many opportunities for small system improvement. Because of their distrust

the municipalities and cooperatives are often hesitant to sacrifice any of their autonomy by purchasing power from investor-owned systems or engaging in joint projects with them regardless of the prospect of large economies.

As this was written, the industry entered a new era of coordinating organizations. Most of what we know as power pools today were formed between 1963 and 1970. In addition, the industry voluntarily created the regional reliability councils discussed in Chapter 3. These developments prompted increased coordination, interconnection, and planning, but researchers continued to question whether these activities were sufficient to realize all achievable gains.

Several influential works cast doubt on the question. Economist William R. Hughes (1969) compared the industry's actual costs in 1964 to the costs of an industry organized into 20 or 30 large, regional, "optimally coordinated" firms and concluded that the potential for savings from further coordination was on the order of 4 to 10%.[26] Shortly thereafter Harvard Professor (and now Supreme Court Justice) Steven Breyer and economics Professor Paul MacAvoy examined the pooling that had occurred to date, as well as a number of industry studies of interconnection benefits, and concluded, "that coordination is seriously inadequate. In fact, the problem was still as troublesome in 1972 as it seemed to the Federal Power Commission in 1964."[27]

Breyer and MacAvoy spent some time trying to understand why utilities failed to pool when the cost savings were so apparent. They concluded that "several obstacles that grew out of the regulatory process...militated against combining small units into larger ones." However, they stopped short of blaming the problem entirely on regulation, instead noting that competition between regulated utilities also played a significant role.

Breyer and MacAvoy's study was followed by a number of economic studies of the pooling phenomenon. In general, these studies found that pooling provided large theoretical and somewhat smaller actual benefits. Lack of participation in pools was due to transactions costs, inability to agree on the allocation of benefits (which is really another transaction cost), and discouragement (or at least a lack of incentives) from state regulators.[28] These observations agree with the views of many pool participants, who often found agreements hard to reach. Moreover, small utilities participating in pools often felt that pools provided disproportionately small benefits to members that had no generation and/or were small in size.[29]

Between 1975 and 1981, a spate of federal studies capped off this area of inquiry. One DOE study asked whether present levels of reliability were adequate and whether the combination of NERC self-regulation and state PUC oversight was creating an efficient system. This study was promoted by Congressional concerns that the reliability councils were not causing sufficient investment in reliability, nor proper tradeoffs between reliability and cost.[30] The study concluded that many aspects of reliability planning and regulation were deficient, but it did not recommend any significant shift in responsibilities or procedures.

In response to interest from several U.S. senators, the Department of Energy also conducted a mammoth study of the costs and benefits of establishing a national power grid.[31] This study was remarkable not so much for its main conclusion—i.e., that a national grid was not economical—as for its large second volume of work papers. To this day, these commissioned papers remain among the most detailed analyses of several utility industry topics, including jointly-owned generation projects and "governmental impediments" to greater system efficiency.[32] Recommendations called for substantially greater FERC authority to order interconnection, wheeling, and pooling as well as regional regulatory compacts to plan and regulate wholesale power markets on a regional basis.[33]

Another section of this DOE study examined the potential for cost savings by increased interconnection and pooling within each major U.S. region, assuming no changes in regulation or ownership of the system. The study found that benefits exceeded costs only in the Southeastern portion of the United States.[34] However, in this region, the cumulative benefits, net of the costs of increased interconnection and sales, were over $10 billion (undiscounted) over ten years, or about 3.5% of regional costs. This study was echoed by several studies that were conducted by state agencies and utility organizations which found that most of the gains of pooling were already being achieved, with significant exceptions attributable to a host of idiosyncracies.[35]

In 1981, the Federal Energy Regulatory Commission examined the same question; this time with an eye toward determining the extent to which existing regulations helped or harmed optimal pooling and planning. FERC's specific findings included the following:[36]

- The largest unrealized gains continued to be in larger units for groups of smaller utilities, the flattening of unit economies of scale not withstanding. However, the "effective inclusion of such systems in coordination agreements has been one of the more difficult coordination problems, and continues to be a major issue."

- Tight pools, which involve regional economic dispatch and (ideally) joint planning, were most effective but rare, due in part to the transactions costs of forming and maintaining a large, multi-party organization. Additional and/or perceived disincentives to pooling included state regulatory concerns about the loss of control of the system, fear of exposure to allegations of anti-competitive behavior, and possible competition or animosity between prospective pool members.

- Although interconnection and bilateral exchange continued to increase, there were actully fewer formal pools in 1981 than in 1970. Regional reliability councils played a useful role, as did multi-lateral alternatives to pools that were not so cumbersome to negotiate.[37]

- Unrealized gains from greater coordination were "perhaps on the order of 1 to 2 percent" of total costs, though not distributed evenly.

The FERC concluded:

> The Nation's 3,600 separate electric power enterprises are operated by a great diversity of agencies, some investor-owned, others owned by cities, states, counties, public utility districts, and cooperatives, as well as by the Federal Government. Together, they provide this country with a system of power supply which at the retail level is generally responsive to local needs and local control. However, the large number of separate systems, coupled with rivalries and controversies between segments of the industry, has frequently resulted in economically meaningless boundaries for utility system planning and operation which undoubtedly cost the power consumers of this country millions of dollars every year in wasted opportunities for cost reduction. These boundaries can be transcended without losing the benefits of the existing pluralistic institutional structure if all segments of the industry, and all the individual systems within each segment, would realize that their ideological differences are no bar to working together in establishing stronger regional and interregional power pools. To do so would strengthen all and diminish none.

Summing up this collection of studies is difficult, as regional disparities and other points of controversy have often made it impossible to reach a consensus. The Department of Energy observed in 1979 that:

> None of the studies conducted over the last 15 years has been accepted uncritically. For the most part, the conclusions are not unambiguously decisive. What seems to be in greatest doubt is whether a combination of benefits, some of them unevaluated, can convert otherwise marginal interconnection concepts into clearly advantageous or acceptable options. Also in doubt is whether most of the economical advantages from physical interconnection, coordinated planning and development, and integrated operations that might emerge in the future will be captured in the normal course of events by the evolutionary extension of existing bulk power supply facilities and joint action arrangements, or whether to capture those advantages will require increased government involvement or altered institutional frameworks.

This otherwise accurate statement obscures several important and recurring results. First, it is almost a tautology that under the existing set of institutions and rules, most of the net benefits of coordinated planning and operation are being achieved. This merely says that under present incentives and rules, utility managers of all types are doing all that for which the perceived benefit exceeds costs. If an apparent benefit of greater coordination is going unrealized, it is likely the the benefit is outweighed by an unobserved tansactional, political, or institutional barrier.

But studies that were willing to simulate or study the impact of relaxing the present patchwork of regulations, incentives, and institutions found consistent and significant benefits, generally within regions. Time and again, studies of optimal regional structure have found that greater integration of rival public and private systems, increased transmission access, and regional rather than state regulation of both short- and long-term functions could provide net cost reductions of several percent or more.

PARTIAL DEINTEGRATION AND COMPETITION, 1978–96

By the end of the 1970s, it was becoming clear that little could be done to change utility industry structure without changes to impediments identified in the studies above. At the same time, policy makers were not anxious to significantly disrupt or dismember an industry that was pluralistic, politically sensitive, and had a virtually unmatched record of price reductions and public service.

Against this background, opportunities for important incremental progress arose. By the mid-1970s, a significant group of observers argued that generation should be deregulated.[38] At the same time, studies of the energy efficiency of conventional power plants found that generators that used the heat that most power plants eject into the air conserved substantial amounts of energy.[39] Combining these two points of view, the Carter Administration and the 96th Congress decided it was appropriate to begin experimenting with deregulated, deintegrated generation for power plants that would contribute energy conservation or fuel diversity benefits.[40]

The Public Utility Regulatory Policies Act of 1978 (PURPA), along with associated rules by the FERC, implemented limited generation competition.[41] PURPA required independent generators either to use renewable fuels or to reduce their energy waste. Renewable facilities were limited in size, and utilities could not own majority interests in these plants. Generators meeting these criteria were called "Qualifying Facilities" or QFs. Congress required the nearest utility to a QF to buy all of the power it offered, with the price to be set at the utility's "avoided costs."

These changes paved the way for the growth of unintegrated generators that stood as competitors to utility-owned, integrated plants. Integrated IOUs as well as unintegrated public distribution utilities had to choose between building and owning new plants or executing contracts with unregulated generators. For the first time in its existence, the electric power industry and its regulators faced the sort of "build-versus-buy" determination that unregulated businesses regularly confront.

As utilities and regulators began to implement PURPA, numerous problems arose. The areas of friction between utilities and QFs included access to transmission, the price paid for QF power and backup services from the utility, the amount of QF power purchased by utilities, and various aspects of the complex contracts between QFs and utilities. Appendix 5B describes these problems in greater detail.[42] State regulators, who often found themselves refereeing disputes between utilities and QFs, frequently found it extremely hard to compare the cost of power to alternative sources.

Typically, utilities were required to buy power from QFs under "take or pay" contracts (i.e., contracts that left them no choice but to pay for and use QF power each month). As the number of these contracts grew, utilities became concerned about the impacts of non-utility purchases on their overall financial costs and risks. Briefly, utilities argued that long-term contracts with QFs reduced their financial and operating flexibility and increased the risk of insolvency.

These claims were never demonstrated empirically, but they have an unassailable theorectical foundation. (See Appendix 5B.) Because changes in risks implied by replacing integration with contracts are often difficult to assess, they are frequently neglected aspect of utility restructuring.

One of the more apparent risks, an unavoidable byproduct of reliance on long-term contracts at fixed prices, was the chance that estimated future prices would not reflect future costs. Many QF contracts fixed a price for power for the next 10 to 20 years, which meant that everyone had to guess the future prices of boiler fuels that co-generators would burn, as well as other elements that make up the cost of power. If contract prices turned out to be too high, utilities would be paying prices that would be higher than what their avoided costs actually turned out to be; if they were too low, the reverse would occur. Hence, utilities bore the risk of above-market fuel price projections and co-generators the reverse.

As it happens, many regulators guessed high, and some utilities have long-term contracts under which they are paying prices that are much higher than the cost of power from other sources. It is estimated that the present discounted value of that portion of utility obligations to QFs is approximately $40 to $50 billion in total.[43] Niagara-Mohawk Power Company, a New York utility, estimates that its QF contracts have an above market cost of $3.3 billion—roughly $60 per customer per year in added retail rates.[44] The comparatively high prices that some utilities are paying for QF power under approved long-term contracts are forcing some utility rates up to the point where utility customers are complaining. The problem is particularly severe in areas where nearby utilities did not buy similar QF power and/or have much lower rates.[45]

In response to some of these concerns, the 1992 Energy Policy Act (EPAct) removed most of the restrictions on the type of generators that could sell deregulated wholesale power, creating a new entity known as "Exempt Wholesale Generator" (EWG).[46] Unlike QFs, EWGs could be of any size or fuel type and they did *not* have a guaranteed right of sale to the nearest utility at avoided cost. However, they were unregulated by FERC and exempt from PUHCA. EPAct did not allow sales from EWGs to ultimate users. Accordingly an EWG's mission was to find an integrated or distribution utility buyer and to find transmission to that buyer.

EPAct contained an even more important change in transmission policy. For the first time since electric regulation began, federal regulators had authority to compel wheeling across a utility's transmission line. Wheeling terms, conditions, and rates could be ordered by the FERC, but only on a case-by-case basis, and only following rather extensive notice, appeal, and hearing procedures.[47] These changes, which expanded transmission opportunities for non-utility generators and public power, actually expanded demands for even less restrictive and cumbersome transmission access.[48]

Whatever their other effects, the many problems of PURPA and EPAct implementation did not stop non-utility generators from becoming the single most important source of new power supply in the United States and a large industry in its own right. Deintegrated generation has grown steadily from its pre-PURPA levels of 18,000 MW to levels of about 68,000 MW in 1994—an increase of over 50,000 MW. Figure 5-4 charts the trend in the growth of utility-owned (integrated) and non-utility generation since 1987. During this period, a steadily larger share of all new plants added to the system have been non-utility. In 1994, fully 78% of all net generator additions were built by independents. In short, the industry's answer to the question of whether to build or buy is largely "buy."[49]

Many reasons account for the steady increase in non-utility purchases. First, and in spite of the present concerns about above-market QF contracts, there is a widespread perception that the rivalry introduced by NUGs caused the cost of new generators to decline significantly. In a somewhat circular chain of cause and effect, competitive, highly cost-conscious private plant developers are credited with forcing utilities to pare down the costs of their units. They are also credited with forcing manufacturers of plants to find cheaper ways to build smaller plants, further eroding large-unit economies of scale and thereby strengthening the case for independent power.[50]

Figure 5-4. Growth in Utility and Non-utility Capacity Additions Since 1987

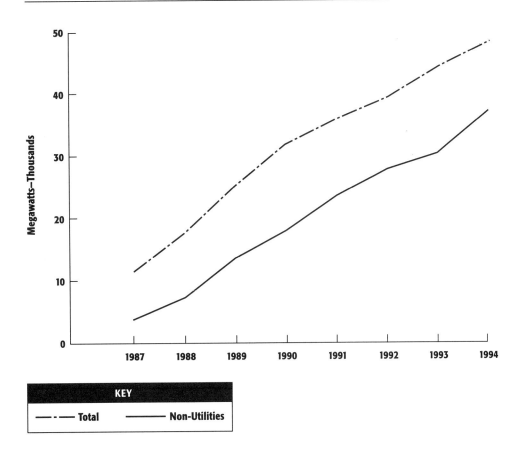

Source: Energy Information Administration; Edison Electric Institute

Non-utility purchases also grew because utilities quickly adopted an "if you can't beat 'em, join 'em" attitude and started unregulated subsidiaries to invest in QFs and EWGs.[51] These well-capitalized "utility affiliates" quickly became significant players in the fledgling NUG industry, accounting for about one-seventh of "non-utility" operating capacity by 1989 and about half the new NUG capacity on the drawing boards at that time.[52] Since this time, utility involvement has remained strong, with utilities or former utilities strongly involved in many of today's leading NUG firms.[53]

Finally, despite the constant stream of headaches and problems, contracting between NUGs and utilities has worked out better than some had predicted. Although they do not like overpaying for NUG power, utilities have sometimes enjoyed the flexibility to cancel or accelerate NUG contracts if any of the many uncertainties surrounding their plans changes. They also appreciate not having to bear the risks of construction cost overruns or the failure of plants for engineering reasons. Finally, and with a number of specific exceptions, the ability of most utilities to operate their systems reliably and to use economic dispatch has not been diminished by contracts with NUGs as much as some had feared.

If, for these and other reasons, utilities are effectively deintegrating all new generation, then isn't restructuring already happening? There is no question that new generation is separating itself, but 90% of the industry's total IOU capacity remains integrated, and opinions are divided as to whether vertical divestiture will be extensive or rapid. Meantime, the co-existence of vertical integration, state and federal regulation, and deregulated generating plants has produced success; nevertheless there is lingering, almost universal dissatisfaction. If limited generation competition and transmission access has shown such promise, why should deintegration and transmission access be less than complete?

ELECTRIC UTILITY RESTRUCTURING: THE EPOCHAL CHANGES

The restructuring of the electric utility industry, which is now underway, consists of creating a universal wholesale (and perhaps retail) power marketplace in which vertical integration will be effectively eliminated. All distribution systems, whether integrated or not, will have to buy from the marketplace; all generators, whether integrated or not, will have to compete to sell their power—and will get paid only if their offers are among the winners selected. Large integrated utilities will be forced to compete to sell to their own downstream units, not unlike a firm that has told its in-house parts business that it will now have to compete to sell to the assembly division. Meanwhile, the 2,000-odd distribution systems will shift from being the odd men out in an industry of huge integrated players to a small version of the downstream norm.

The second part of this epochal change consists of a movement to allow individual retail customers to contract directly with the generation source of their choice. Many see the idea of eliminating exclusive franchises and letting individual retail customers choose their generation supplier as an irrefutably logical extension of a universal competitive wholesale marketplace. It also touches a deep political nerve in a nation in which monopolies (regulated or not) are never popular and individualism and competition are cherished ideals.

Under a retail choice model, the distribution utility has a franchise requirement to *distribute* all power over its one set of lines, but not to purchase it or ensure adequacy, universal service, and so on. The distribution utility becomes a transporter only, much like the transmission system will be in the universal-wholesale market.

The rest of this volume is devoted to examining the benefits, implications, and tradeoffs inherent in these new structures. Before proceeding, however, a few notes are in order.

1. **None of these changes spell anything approaching complete deregulation.** Even if exclusive retail franchises, retail rate regulation, and the regulation of power generators are eliminated, there will still be only one transmission and distribution system, and the rates and terms of T&D services will remain regulated. At a minimum, this means that competition and regulation in various industry segments of the new industry will have to be carefully joined, and that transmission and distribution owners must be regulated and incentivized intelligently and effectively.

2. **Restructuring is revolutionary, but it is not as different as it seems from 30 years of proposed changes (such as increased power pooling).** Since the 1960s, study after study has called for utilities in a region to voluntarily form into "tight" power pools in which all plants in a region are economically dispatched. In pools like this, secure dispatching is maintained and pool generators generally have access to most members' transmission.

   If a competitive wholesale market works according to competitive market theory, and a tight pool works as it should, the resulting benefits and costs to every generator and every distributor in the region ought to be roughly similar, particularly in the aggregate. In other words, the universal wholesale market model is a more competition-oriented way of establishing region-wide, least-cost operation of the grid in the short run than policy makers have repeatedly proposed.[54] The biggest change is that *all* of the generators in the market will be (or will act as if they are) unregulated competitors.

3. **In the long run, changes wrought by retail restructuring are revolutionary—and uncertain.** The phenomena of customer choice and competition in electric utility must adapt to the uniquely interdependent nature of the electric power system (Chapters 2 and 3). Establishing protocols to ensure that the system will remain adequate and reasonably stable in the long run requires a tremendous departure from an industry structure that has been built around an *intentional absence of competition.*

   Will competitive generators and a regulated transmission grid be able to plan together without jeopardizing robust competition between generators? Will the number of generation firms decline to the point where generation becomes an oligopolistic industry that bargains with the grid? Will states continue to control the siting of generation and transmission, and will this frustrate regional plans?

Restructuring will put enormous new pressure on public power systems to adapt to the new market structure. Because these systems are already largely wholesale buyers, they stand to gain from a more robust wholesale market and better transmission access. Moreover, some publics can borrow money more cheaply than regulated IOUs or non-utility generators, and some of their power is the cheapest in the nation. What do they stand to gain from restructuring? And if they do not alter their structure, will this create an even more diverse regulatory patchwork than what currently exists?

4. **Will the transactions costs of competition outweigh the benefits?** Restructuring can be viewed in terms of the economic tradeoffs between vertical and horizontal integration and the more extensive use of contracts awarded on the basis of competition. This reflects a widely-held view that the cost-reducing benefits of competition between generators outweighs the benefits of vertical integration that presumably caused the large IOUs and publics to integrate in the first place.[55]

This may well be true, but the meager quantitative research on hand hardly makes the case for vertical deintegration a "slam dunk." Uniform transmission access and unleashed generators of all stripes will probably create more effective competition, but there may be a significant loss of vertical coordination benefits—depending on how good a market system is developed. When retail choice is added, transactions costs increase substantially. So, hopefully, will the variety of products and services, the consumer benefits of choice, and reductions in prices and costs.[56]

To examine these issues in more detail, we bid farewell to the electric industry of the twentieth century and enter the industry of the twenty-first.

Appendix 5A

Contracting Risks and the Cost of Financing Power Industry Capital

When investing in any business, investors evaluate the probabilities that they will get larger or smaller earnings in various future, highly uncertain scenarios. A long-term contract that places risks on the business which it cannot pass off or diversify may increase its chances of insolvency or low-earnings. Whether the contract improves or diminishes total risk can only be judged by careful evaluation of the exact terms of the contract and the best available relevant forecasts.

Power purchase contracts place a number of risks on the buyer of power (what if the plant is not ready when it is needed? What if environmental laws require changes of fuel types?) and the seller (what if the buyer refuses to pay?).

Financial analysts who examined contracts between QFs and utilities in the past ten years found that these documents created significant risks for both buyer and seller. For example, sellers typically bore all risks of construction cost overruns, the unavailability of boiler fuel, and could be liable to pay a penalty if the plant was late or was operated unreliably. Conversely, utilities bore many mirror-image risks from these contracts.

The new risks borne by utilities led some analysts to suggest that signing too many purchase contracts would result in a much riskier utility, in which investors should demand a higher expected return. This led to the view than NUG purchase contracts would cause a utility's cost of borrowing money to increase. Perhaps because so many other utility risks were also in flux at the time, statistical confirmation of the proposition that NUG purchases raised the cost of capital could not be found.

At the same time, NUG developers were arguing that it was *they* who were being given the lion's share of the risks inherent in the power transaction. Among observers, this raised the question of the transactions costs of vertical deintegration in another form. Did creating long-term contracts between a NUG and a utility cause *both* entities to become riskier than they would be if they were integrated and jointly owned? If so, a single owner could be a lower-cost solution, particularly if weight was placed on scenarios in which one party or the other could not honor its end of the contract, and an uncertain court or regulatory resolution was required.

In short, an increase in the aggregate risks faced by the industry, (increasing financing costs), should serve as warning sign that perhaps the new market structure may be lacking. (For more discussion, see Energy Information Administration (1994); the August/September, 1993 *Electricity Journal*; National Independent Energy Producers (1995); Kahn (1991); and Brown, Lewis, and Ryngaert (1994).)

Appendix 5B

Contracts Between QFs and Utilities

Sections 210 and 211 of the Public Utility Regulatory Policies Act of 1978 (PURPA) required utilities to purchase all power offered by "Qualifying Facilities" (QFs) at the utility's own "avoided cost." Although purchases at avoided cost were supposedly neutral toward utilities, most of the latter believed that the net costs of QF purchases far outweighed their benefits.[57] First, utilities and QFs had to design and build physical interconnections, much like interconnections between utilities. Because the utility was forced to purchase from the QF, it understandably had little incentive to spend time and money on interconnection studies—the same kind that utilities jointly conduct on a voluntary basis when there are mutual benefits.[58]

In addition, the rules that limited QFs to certain fuels and technologies meant that QFs were much more limited to base load facilities. (See Appendix 3B). Because QFs were not owned by utilities, the utilities had very limited rights to alter the output from QFs.[59] (Unless a utility's expansion or integrated resource plan happened to call for new base load plant where the QF chose to locate, a QF represented a suboptimal long-term addition.[60] In the short term, because QF output was not controllable, the QF could not be dispatched in merit order, which caused production costs to be greater than otherwise.)

If a utility's plans called for building another power plant, the QF could be paid the construction and financing costs that would be avoided if the utility could rely on the QF for firm power equivalent (from system planning, reliability, and plant configuration standpoints) to the plant it would have built. Under these conditions, a utility and QF could agree on a contract that paid the cost of the avoided plant, as well as the energy produced by the plant. In exchange, the QF typically agreed to continue to supply power for a long period of time under terms and conditions that gave the utility confidence that the QF purchase obviated the need for the new plant.

Due to the high degree of technical interdependence in power systems, these contracts were extremely technical and complex.[61] It was not unusual for them to run into the hundreds of pages, with technical sub-agreements filled with engineering formula. For example, if a utility, as a member of a reliability council or a power pool, was required to follow various operating procedures at all of its plants, the QF in turn had to agree to follow these procedures or it could jeopardize the utility's membership.[62]

Because the prices utilities paid for plant-displacing long-term contracts were the only ones that enabled new QF construction, and a utility would sign contracts for at most a number of megawatts equal to the size of the plant it would otherwise build, these deals became highly-sought after prizes. Utilities whose filings with regulators showed a new plant on the horizon were often flooded with requests to negotiate contracts with many disparate developers. Utilities were consumed trying to discuss, negotiate, and evaluate contract offers; developers were equally unhappy not knowing the outcome or even the timetable for the arduous, expensive, and unpredictable life-or-death negotiation process.[63]

The rules implementing PURPA made state regulators responsible for determining avoided costs and thus payments between utilities and QFs.[64] This forecasting task was difficult and controversial, and soon led to state regulators voluntarily or involuntarily refereeing innumerable disputes between QF developers and utilities over various aspects of PURPA implementation. These dockets overloaded and/or exasperated many already-strained public service commissions.[65]

Regulators' experience in these proceedings illustrate much of the discussion about vertical-integration-versus-contracts at the beginning of Chapter 4. First, it is just plain difficult to evaluate the differences between two proposed generating plants, neither of which has been fully sited nor designed, and no one knows how well or for how long the plants will be operated. More fundamentally, a contractual entitlement to the output of a plant is much different than owning the plant itself.[66]

The difficulties of comparing long-term power purchase with ownership were illustrated in a state regulatory proceeding in Maryland which compared two units that Baltimore Gas and Electric proposed to build to a co-generation plant proposed by Cogen Technologies, Inc. In a process that took more than two years, the Maryland Commission's examination of the two expansion alternatives identified no less than thirty significant differences between the two options, including:

- Date of plant completion and lifetime of the plant or contract;

- Environmental compliance expenditures were different because different laws applied to utility versus non-utility units;

- The costs of delay or construction cost overruns would have greater impacts on power rates under the utility-build option;

- Variable operating costs were different;

- The utility plant was more dispatchable (controllable);

- The non-utility plant was subject to fewer taxes, affecting its bid price;

- The plants had different levels of liability and other insurance;

and the list went on. Despite an unusually careful comparison, the Commission was understandably unable to place a value on all of the differences, and ultimately had to decide between the two options on the basis of more than the estimated net total cost.[67]

Finally, many QFs found good locations for their plants in areas where utilities either did not need power or had low avoided costs. PURPA gave the local utility the option to wheel power to another buyer; consequently, requests for wheeling increased greatly after PURPA. Because no one could compel wheeling, and policies concerning the price of wheeling were unsettled, wheeling negotiations were long, contentious, and usually fruitless.[68]

Some of the disadvantages of frequently simultaneous open-ended negotiations were reduced when regulators hit upon the idea of requiring utilities to hold competitive procurements to solicit bids for QF power when it was determined that a new plant was needed. This allowed utilities to clearly state their technical and contractual requirements for developers in advance and to solicit bids from developers who agreed to adhere to them. Regulators could then compare the cost of the plant bid to the estimated cost of the utility's own proposed construction project.[69]

Chapter 5

NOTES

1 The main three groups of retailers (IOUs, munis, co-ops) have about the same share of total retail customers that they do of generation (EIA, 1994 Figure 3).

2 FPC (1964, p. 15).

3 Hunt and Shuttleworth (1996) p. 15–19 note that only two other nations had regulated private utilities and discuss the main structural alternatives. See also Gilbert and Kahn (1996).

4 Lee and Ellert (1996).

5 See FERC (1989) p. 18ff for a brief snapshot of regional attributes and differences.

6 A number of studies from the 1970s that call for regional regulation or power market share reviewed below. Also see the discussion in Chapter 1 (e.g., Cohen, 1977), as well as Stuntz (1992), Stalon (1992), Curtis (1992), and Massey, et al (1992).

7 The regional institutions that have developed in the industry to date have differing boundaries. For example, the boundaries of the reliability councils do not correspond to the boundaries of most Regional Transmission Groups or power pools.

8 "The City Councils of that day were notoriously corrupt," writes Hyman. "Franchises were granted on receipt of payoff and franchises were not renewed when the franchisee fell out of political favor." (Hyman, 1981, p. 68). See esp. Ch. 3 of Platt (1991) and the discussion of franchise competition in the previous chapter.

9 See Ramsey (1937), Platt (1991); ch. 3, Munson (1983) chs. 5, 6; Tollefson (1996), Funigiello (1973), Vennard (1968), Gruening (1964), and McDonald (1962), to cite just a few of the many useful perspectives on this long-standing controversy.

10 Looking at the growth of IOU power versus public power, for example, Munson asks "How did IOUs achieve this remarkable victory?" (Munson, 1984) p. 106.

11 Federal Power Commission (1970) Table 2.1. Also see Vennard (1968) p. 30.

12 For an excellent, lengthy historical review of utility interconnection activities, see the U.S. Department of Energy (1979) Also see Hughes (1983) and Tollefson (1996).

13 See Federal Power Commission (1964), v. I p. 30.

14 See Federal Power Commission (1971), ch. 13.

15 For additional discussions relating to this paragraph, see Fungiello (1973), Tollefson (1996), Munson (1983), and Vennard (1968). Munson's chapter 9 contains expressions of these sentiments during the 1980s as well.

16 Energy Information Administration (1993), p. 7; Securities and Exchange Commission (1995) Part I.

17 Vennard (1968, p.p. 259, 265). This reference contains the IOU's view of REA; thus may be compared to REA's own history (USDOA, 1982).

18 The 1964 National Power survey noted that rates in New England were 29% above the national average. (FPC, 1964 p. 230).

19 FPC, 1964, p. 67

20 Hyman (1992) p. 93

21 Federal Power Act Section 202 (a), 16 U.S.C. § 824 (a).

22 These exchanges are often called seasonal exchanges or seasonal diversity. Because different regions of the United States use different fuels and experience different seasonal cycles, weather-driven power plant usage varies widely across the U.S. Air conditioning loads are high in Florida in the summer, for example, with somewhat less power used in the winter. In Maine, the pattern is reversed. As another example, in periods of heavy rainfall the Pacific Northwest hydroelectric plants often have substantial surplus power. One of the largest interregional transmission lines runs from the Pacific Northwest south into power-hungry Los Angelas and the Southwest, thus demonstrating the feasibility of large scale interregional transfers where surpluses are fairly common and the cost of the surplus power is low enough to justify the added cost of long-distance transmission.

Seasonal diversity has been a controversial topic for many years. Most of the "national" studies considered seasonal diversity as one of the main reasons to create a national grid or more interregional ties. However, except in certain cases such as above, most major studies in past decades found that seasonal diversities of load and price disparities were not consistent enough to warrant large-scale expansion of interregional ties. Federal Power Commission (1971), p. I-18-20, Edison Electric Institute (1972) and U.S. Department of Energy (1979) v. II. ch. 7. (Two exceptions are the Federal Power Commission (1964), which estimated a 6% savings in national capacity from seasonal diversity, but admitted its numbers were very rough, and which did not estimate the costs of achieving diversity exchanges, and the U.S. Department of the Interior (1968), which was not widely accepted by the industry.) (U.S. Department of Energy, 1980, p. 6)

Other studies of expanded national transmission have considered seasonal diversity as one of several benefits; these studies also generally do not find strong seasonal benefits.

As real and expected real fuel prices have declined since these studies were conducted in the 1970s, and transmission lines have become more costly and more difficult to site, one would not expect an increase in unrealized seasonal exchanges. However, one of the benefits cited during the comparatively recent utility merger between Utah Power and Light and Pacific Power and Light was seasonal exchange benefits.

23 This is the "coal by wire" idea, in which generators located near coal mines send power over wires, rather than sending coal trains to distant power plants. Although some studies find that coal-by-wire is economically justified, and there is a significant amount of this concept embedded in the U.S. system today, unrealized opportunities do not appear large. See Congressional Research Service (1976), U.S. Department of Energy (1979) ch. 6, and Caizonetti, *et al* (1989).

24 The idea of a national power grid captured the imagination of some as early as the 1920s, but the idea was not considered remotely practical until the 1960s. In 1977, four Senators introduced legislation to create a nationally-owned, coast-to-coast power grid. This grid was intended to avail the nation of all of the potential benefits of multi-regional interconnection: increased reliability, seasonal exchange (see note immediately above), and energy security. See U.S. Department of Energy (1980) and (1979) v. II, ch. 2.

25 These studies are similar in nature to those utilities and regional councils themselves do—recall that modern G and T planning *requires* that the entire regional network be studied to determine the effects of a new plant or line. The difference between the studies examined in this Chapter and the thousands conducted routinely in the industry is that the latter generally accept as given any contractual, marketplace, or regulatory impediments and model various alternatives. Industry structure studies do just the opposite: they seek to determine the scope for lowering costs by altering regulations or other industry features.

26 Hughes (1971). See the discussion of this work in Kahn (1971) v. II. p. 74 and Breyer and MacAvoy (1974) p. 96.

27 Breyer and MacAvoy cite studies by Mabuce, Wilks, and Boxerman (1971) and Casazza and Hoffman (1969), both of which find that the optimal size for an integrated utility was 40–50,000 MW and 15–30,000 MW, respectively. These studies were characteristic of studies by industry engineering-economic experts at the time, based on computer models and calculations of hypothetical systems. As noted in the text, these studies are not designed to analyze the nature or cost of impediments to greater coordination, but rather simulate hypothetical systems that are free of most or all impediments and transactions costs. They are very useful for demonstrating the theoretical optimum, but they cannot determine how much closer to optimum one can get via regulatory or other changes.

28 Joskow and Schmalensee (1983) p. 83 also review these points in useful detail. See, among other works, Cramer and Tschirhart (1983), Herriott (1985), Mulligan (1985), Gegax and Tschirhart (1984), and Caputo and Mulligan (1985).

29 These issues are discussed at length in ch. 16 of the 1964 National Power Survey, and in FERC (1981) ch. 6. They are discussed to a much more limited extent in other National Power Surveys and in the otherwise encyclopedic U.S. D.O.E. (1979). To see an address admonishing small public systems to overcome their reluctance to participate in pools, see Chase (1970).

30 The specific Congressional mandate for the study was section 209 of the Public Utility Regulatory Policies Act of 1978. According to D.O.E. (U.S. D.O.E., 1981 p. 2), the study was the result of a compromise between those in Congress who wanted direct federal regulation of reliability and those who felt the present system was adequate.

31 Also, in 1975, the newly-established Federal Energy Administration commissioned a study of a proposal that examined federally-chartered regional wholesale power administrations regulated by a new type of regional regulator. The report saw regional grids and regulators as a solution to resolving conflicts between the existing system and regional optimality (D.O.E. 1979, 1980a).

32 The authors of this section of the study found considerable incentives favoring joint utility projects that (in their opinion) possibly outweighed a number of disincentives to joint action. The incentives included reduction in load forecasting and financial risk; the disincentives included utility credit quality from weaker co-participants, inability to agree on cost allocations, state regulators' unwillingness to allow participation of funding for out-of-state facilities, and inability to gain transmission access. (U.S. D.O.E., 1979, ch. 11.)

33 U.S. D.O.E. (1979) ch. 12.

34 U.S. D.O.E. (1979) ch. 5.

35 Similar conclusions were reached in a study of the MOKAN pool in the Missouri-Kansas area and in various other studies of central dispatch (e.g., Jensen and Scott (1988). Two important caveats apply to these studies. First, they generally allow for little or no change in regulation or ownership, so there is little change in the incentives or disincentives to cooperate. Second, these studies often examine only short-run production (fuel and other variable) costs, so they do not reflect the benefits of additional coordination in planning or construction.

36 FERC (1981) p. 166–167

37 For example, joint ownership of a generating unit is a simpler contract to agree on than a long term pooling agreement.

38 See Chapter 1 *supra.*

39 See Fox-Penner (1990s) and references therein.

40 Ibid.

41 For this paragraph, see 18 CFR 292 or Spiewak (1987).

42 US. Department of Energy (1986) and Spiewak, ed. (1987). The frequency of complaints led Congress to hold hearings on PURPA implementation problems in 1986 [Committee on Energy and Natural Resources, 99th Cong. 2nd Sess. 99-820, June 3, 5, 1986, and Subcommittee on Energy Conservation and Power, House Committee on Energy and Commerce, 99th Cong., 2nd Sess 146, June 11. 1986.] These hearings led to a series of FERC regional conferences devoted to comments on the problems of implementing PURPA *[Cogeneration; Small Power Production–Notice of Public Conference and Request for Comments, FERC State Par 35.011, Docket No. RM87-12-000).*

43 "Consultant Says Stranded Cost in $163 Billion–$73 Billion for Generation." *Electricity Journal*, March, 1995, p. 5. Many other analysts have put forward widely varying estimates of these costs, but many do not separate PURPA contracts from other "stranded costs." Of those that do, estimates appear to be in this range.

44 Eric Durr, "Cogeneration Plants Drive Up NiMo Rates, Officials Say." Watertown, N.Y. Daily Times, May 13, 1996; Toni Mack, "Don't Shoot the Power Company," *Forbes*, April, 1996. As noted in the text, PURPA was intentionally designed to leave utilities neutral to QF power, while co-generators were free to make as much profit as possible as long as their sales price was at the utility's avoided cost. Where avoided costs were set generously, some co-generation companies reportedly made handsome profits. In May, 1996 *Forbes* magazine profiled a number of non-utility plant developers who became multimillionaires from their investments in QFs. In the profile, utility consultant Kent Knutson says, "These entrepreneurs are making 6 to 8 cents per kilowatt-hour when the market dictates two to three cents. The [buying utility's] customer gets nothing." Forbes' reporter replies that "Knutson's right, of course, but you can't blame business people for taking advantage of dumb laws cooked up by politicians who don't really know what they are doing."

As a caution, however, note that estimates of the difference between presently observed power prices and the avoided costs in long-term QF contracts cannot be characterized as the magnitude of avoided cost mis-forecasting on this basis alone. The correct comparison is between QF contract prices (which can be observed) and the cost of what utilities would have built had they not purchased. Some observe that the latter may have included power plants whose costs per unit of output would also now be above market–perhaps further above than QFs. Also see Knutson (1996).

45 At least one utility, Niagara Mohawk, has threatened to declare bankruptcy in order to free itself of its contractual obligations to QFs. Citing high rates from QFs as one (though not the only) reason, one municipality in Niagara's service area formed a municipal utility and is purchasing power at much lower rates from the New York State Power Authority.

46 EPAct Section 711, implemented by FERC in *Filing Requirements and Ministerial Procedures for Persons Seeking Exempt Wholesale Generator Status,* III. F.E.R.C. Stats. and Regs. Para 30,964 (1993). Another change from this period was the FERC's experimentation with more flexible rates for integrated, regulated bulk power sources, known as "market-based rates." For an excellent overview, see Tennenbaum and Henderson (1991).

47 Sections 211 and 213 of the Federal Power Act, as amended by EPACT, implemented in *Policy Statement Regarding Good Faith Requests for Transmission Services and Responses by Transmitting Utilities Under Sections 211(a) and 213(1) of the Federal Power Act, as Amended and Added by the Energy Policy Act of 1992,* III FERC Stats and Regs. Para 30,975 (1993).

48 Penn (1994) reviews the changes in transmission policies mandated by EPAct and the FERC's implementation as of 1994.

49 This trend is also reflected in wholesale power sales statistics. Wheeling has almost quadrupled, from 67,000 GWh in 1975 (4.5% of IOU generation) to 228,000 (10.3% of generation), in sixteen years. (Frankena and Owen, 1994, p. 48). Coordination sales—the type utilities exchange in pools or bilateral short-term transactions—are dropping as total wholesale power sales increase. (EIA, 1995b). For additional general background, see EIA (1993b).

50 In 1992, my colleagues and I examined a large sample of power plants built by non-utilities and utilities. We chose plants that were as similar as possible, all gas combined cycle power plants, and we attempted to control for differences in the costs of labor and materials in the different parts of the country where these plants were constructed.

Construction Cost by Region

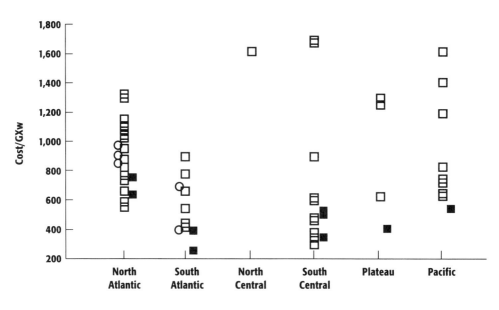

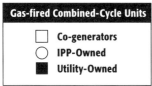

Source: FERC Form 1 Organization & Independent Power Plant Firm (COGEN) Handy-Whitman Index of Public Utility Construction Costs, Bulletin 133

We expected to find that a comparison of this nature was difficult, and that no two power plants were the same. Indeed, we ultimately reached the conclusion that this method wasn't too useful for projecting the hypothetical cost of a plant, except as a crude average. But what we were surprised to find is that there was no way to distinguish between the costs of utility and non-utility plants. The latter did not appear to be systematically cheaper than utility plants–we simply couldn't tell the difference. In checking these results with industry colleagues, we generally got the same reply: utilities had adopted the leaner construction methods of the non-utility developers, and were extremely averse to cost overruns.

51 Utilities could not own a majority share of a QF without violating PURPA and the Holding Company Act, but EPAct repealed both of these limitations.

52 National Independent Energy Producers (1989) ch. III.

53 According to McGraw-Hill (1996), 14 of the top 20 IPP firms in the world are former utilities or utility affiliates, including the four largest.

54 The similarity in estimated industry-wide cost savings from extensive regional pooling called for by the FERC in 1981 (and other studies of this period) are strikingly similar to those estimates to accrue from FERC's open access restructuring via competition in Order 888. In the former, FERC and others estimate that 1–5% of the costs of electric power could be reduced by uniform regional pools; in the Order 888 Environmental Impact Statement, FERC estimated the cost savings from its expansion of wholesale competition to be $5 billion a year, about 1.7% of the nation's annual power bill. (Order 888, p. 693).

55 Whether these benefits can be achieved without complete divestiture remains the subject of much debate, and will be discussed in Part II below. So far, however, we are pursuing a hybrid strategy: we are setting rules for the industry that ostensibly create fair and robust competition between generators, but we aren't requiring vertical divestiture. The FERC's epochal restructuring "MegaNOPR," RM95-8-000, culminating in Order 888 on April 24, 1996, explicitly eschews vertical divestiture. As of this writing, most federal and state utility restructuring bills do not mandate divestiture, though some encourage it. See Chapter 7 below.

56 Earlier in this chapter we noted that much research has identified historic-political and transactions cost impediments to an efficient industry. The size of these impediments bears greatly on what we can expect to gain from industry restructuring today. Suppose there is a differential pattern of unrealized benefits that averages 1–2% of costs, as the FERC estimated in both 1981 and 1996. Utilities in various regions aren't realizing these cost savings due to some combination of mistrust and other "non-economic" factors, as well as the costs of entering into and participating in complex agreements.

Ideally, restructuring at the wholesale level is designed to remove many of these impediments and realize the attendant cost savings. If restructuring is entirely successful, it will get them all. However, the restructured market will have new transactions costs of its own. Wholesale buyers and sellers will have to spend what it takes to participate in the marketplace; we'll see in the next two chapters that these costs are non-trivial. Moreover, there is probably a loss of vertical integration economies widely observed when restructuring forces integrated utilities to de-integrate, whether defacto or dejure. If the benefits of VI are near those estimated by Kwoka, the potential losses for the fourth-fifths of the industry could be as large or larger than the benefits of increased wholesale competition, depending on how effectively the latter addresses the full complement of impediments to optimality in both time frames.

Professors Baumol, Joshow and Kahn (1995) make this point in a recent Edison Electric Institute Monograph:

> There is a danger that introducing more competition into electric generation will produce inefficiency if it results in loss or dilution of the economies of vertical integration between generation and transmission. We have already expressed the view that operating and cost complementarities between these two activities probably explains the virtual universality heretofore of their integration within single corporate entities. The entry into generation of non-integrated competitors clearly requires the development of alternative mechanisms to coordinate these complementary

functions. This need would of course become intensified if, as some parties advocate, all utility-owned generating facilities were divested from the transmission and distribution companies, in order to avoid all possibility of the owners of the latter facilities discriminating in the access they provide to competing generators —a proposal on which we take no position in the paper…

In a vertically integrated utility transmission is not 'priced' separately. The system is planned internally in a way that takes account of all relevant short-run marginal costs, transmission constraints, losses, network externalities, and cost complementarities (economies of scope).

In a de-integrated environment one must place prices on all of these services and find some way to internalize externalities and cost complementarities. If we can successfully unbundle the transmission functions, identifying all of the monopoly grid services now being provided, assign clear property rights to the facilities that supply them, measuring what is supplied from whom to whom and properly pricing each, we can permit more thorough going competition in generation without concern about its causing inefficiencies. If we are unable to do so, it is entirely possible that the efficiency benefits flowing from competitive generation could be entirely dissipated or indeed more than entirely offset by the deteriorated coordination between that sector and transmission. Insuring against these possible consequences is an extremely important responsibility of policy makers, and it can not sensibly be postponed. This should be a key topic for policy makers as they contemplate industry restructuring, the role of competition, and how the industry will be organize to promote efficient competition.

57 See, for example, "Cogenerated Power Irritates Utilities," *Wall Street Journal*, October 3, 1985, p. 6. In simple terms, utility profits are positively related to its total value of its assets approved by regulators, or its "rate base." If a utility builds and owns a power plant to supply its own needs, the value of its assets increases.

There are a number of cases in which a utility's financial condition is not improved by building a plant. First, added power may not be needed or cheaper power may be purchasable from another utility or a non-utility plant. In either case, regulators may not allow the plant to be included in the calculation of profits. If so, any money the utility spent on the plant comes out of its accumulated past profits and diminishes earnings. Many public service commissions now subject all proposed utility plants to strong scrutiny, either as part of an IRP proceeding, prior to approving construction, or in an after-the-fact dockets commonly called prudence reviews. In other situations, adding assets may make a utility less sound financially, diminishing its value to investors. In 1986, Metropolitan Edison's CEO told a newspaperman that 'shopping around is infinitely less risky and costly than building plants, with their financial, legal, and safety problems.' (Andrew Cassel, "Met Ed Flexes Muscle After Three Mile Island," *Chicago Tribune*, June 29, 1986, p. 8B.) (For additional discussions of this view, see Sawhill and Silverman (1983, 1985) and Gilbert (1989)).

Still others viewed these possible positives as more subtle financial negatives. Because purchases from a non-utility did not add to rate base and increased the possibility that any of a utility's other existing or planned plants would be viewed as surplus, and therefore subject to a prudence disallowance.

58 PURPA requires the QF to pay for the full cost of the interconnection, but the utility must still study, design, and approve the construction of the facility. Regarding interconnections, see U.S. Department of Energy (1986) p. 8.43 and Spiewak (1987) p. 24.

59 By the late 1980s, many non-utility generators offered partial or complete "dispatchability," subject to complex contract provisions. See Kahn, et al (1990).

60 In practice, a utility seldom can locate a power plant at precisely the best possible location from the standpoint of cost and reliability for its whole system. Instead, the best of the apparently feasible sites is selected, and the system is modified to accommodate a plant at this location. Sites selected by QF developers were sometimes better than sites available to utilities and sometimes less useful. In any case, this particular aspect of the difference between QF and utility-owned units is amenable to analysis.

61 Small facilities such as residential solar-electric panels or windmills (100 kW or less) were required to be offered simple, standardized contracts (FERC Rules in 18 CFR 292). These small facilities were too small in size and number to affect the economics or operation of utility systems, so it was not difficult for most utilities to comply with this requirement. See Hamilton and Bros (1985).

62 For the advice of one authority on QF contract negotiations during this period, see Goodwin (1989s, 1989s, 1990).

63 See the comments in FERC Docket RM87-12-000, Note 35 *supra*, and *Regulations Governing Bidding Programs,* FERC Docket RM88-5-000, March 16, 1988, n. 28.

64 See Burns, et al (1982), American Gas Association (1987),. Past issues of two now-defunct magazines, the *Cogeneration and Small Power Monthly* (pub. Scott Spiewak) and the *Avoided Cost Quarterly* (pub. McGraw Hill) also chronicled avoided cost determinations.

65 Under PURPA, most of the day-to-day job of implementing the law was made the responsibility of state public service commissions (a controversial approach that narrowly won Supreme Court approval in *FERC v. Mississippi*, 456 U.S. 742.) Though FERC was to be a "court of appeal" for state PURPA proceedings, it rarely entertained challenges to state implementation. The basic policy was articulated in *American Ref-Fuel*, 47 FERC Par 61,161; the most famous exception was its *Order on Petitions for Enforcement Action Pursuant to Section 210(h) of PURPA*, EL95-16- 000, February 23, 1995.

66 For two broad collections of views on the subject, see Synergic Resources Corp. (1989) and the August/September, 1993 special issue of the *Electricity Journal*. Also see National Independent Energy Producers (1990), ch.10 and Joskow and Schmalensee (1983).

67 *Re: Baltimore Gas and Electric Company*, Case No. 8241, Phase II Order No. 69938, 83 MD. P.S.C.113, May 21, 1992.

68 Regulators have heard consistent expressions of concern over the availability of transmission for QFs since PURPA was enacted. See Chapter 7 following and U.S. Department of Energy (1986), ELCON (1987), Rosenthal (1988), Public Utilities Reports, Inc. (1990), National Independent Energy Producers (1990), Fox-Penner (1993).

69 Regarding utility competitive procurements for power and their relationship to PURPA and IRP, see the U.S. Department of Energy (1987); National Independent Energy Producers (1987, 1990); the FERC's Notice of Proposed Rulemaking, *Regulations Concerning Bidding Programs*, RM88-5-000, March 16, 1987; Synergic Resources Corp. (1989); Rosenthal (1989); Kahn, *et al* (1989); Fox-Penner, O'Rourke, and Spinney (1990); Dworzak and Long (1990); and the author's testimony before the Pennsylvania Public Utility Commission in Docket No. P-890366 (containing a capsule history of QF bidding).

Part II

COMPETITIVE ELECTRICITY MARKETS

Chapter 6

Building-Blocks for the New Power Marketplace

The technology of electric power systems (Chapter 3) makes it difficult to view generation and transmission as separate transactions or products. If all major generating plants and transmission lines are jointly controlled and optimized to keep the grid stable and secure and to use the system in the lowest-cost manner, where does generation stop and transmission begin?

When most capacity was integrated, and both generation and transmission were regulated, the division of generation and transmission into separate kinds of products was difficult. In some contexts, as in allocating assets for rate making purposes, it was sufficient to simply label each physical part of the system as G or T. In other contexts, such as economic analyses of the potential for monopoly power, some economists concluded that what utilities sold at wholesale was a bundle of transmission and generation called "delivered bulk power."[1]

Similarly, in some contexts, restructuring can be seen as the mechanical separation of generation and transmission assets, perhaps into separate companies. But because operation of the system will continue to mean that the system controller balances total generation and controls the transmission system, deciding what constitutes "generation" and "transmission" in this context becomes much more difficult. If the system controller decides that a "NUG must be turned off suddenly (e.g., to preserve system security following a transmission system outage), is the cost of this outage a cost of generation or transmission? Indeed, as will be seen, the separation of G and T into separately "priceable," legally enforceable products requires rather elaborate institutions and procedures that, when applied, collectively have the effect of creating a number of complete G and T products with difficult-to-define costs.[2] This chapter examines some new and altered transmission system ideas and institutes that undergird new utility markets. In the following chapter, these building blocks are used to construct new competitive power markets.

WHEELING AND TRANSMISSION: THE TRADITIONAL VIEW

In an industry in which most systems are integrated, designed, and operated to deliver power from a nearby set of plants to a specific service area, transmission by itself was not too common. System designers and operators thought more about how the system would work (in both the short and long run) to satisfy its own franchise customers ("native load").

When transmission alone was needed in this traditional setting, it was often in the form of wheeling—i.e., the transmission of power from a plant outside a utility's service area to the border of that area. Recall, for example, that federal law gives preference for low-cost federally generated hydroelectricity to public distribution systems. This power must be wheeled from large hydro sites around the United States to thousands of public power distributors that are typically many systems away from the federal hydro plant.[3]

Philosophically, a wheeling transaction was not seen as a sanctioned responsibility of utilities; it was, instead, something that might be accommodated as a discretionary matter. This view stems logically from the practice that each franchise utility, public or private, designs and obtains approval for its operations. This model emphasizes the utility's responsibilities to its franchise area, with capacity built almost entirely to be adequate for native load. Hence, requests for wheeling, not built-in to the utility's plans, were understandably viewed as affecting what would otherwise be the least-cost status quo.

Also in this traditional setting, regulators tied rates for transmission to cost-of-service principles. For natural monopolies, or more properly, firms with subadditive costs, the most common pricing standard used by regulators is a regulated unit price equal to average cost.[4] The problem is that measuring the cost of a transmission transaction, particularly according to the deviation-from-status-quo construct, is a complex task.

The cost impact that wheeling has on a native-load-based system is comprised of components, some of which may cause substantial cost increases.[5] These components map nicely into the two main time frames. In the short run, a wheeling transaction may require the utility doing the wheeling to adjust its generator dispatch and transmission line use. If its previous dispatch was lower in cost, its new production costs are higher.[6] These adjustments also cause transmission lines throughout the system to change. When lines become more heavily loaded, power flowing on the lines loses a larger share to heat, impacting deliveries to the utility's native load as well as wheeling deliveries.[7]

If the wheeling is non-firm (i.e., interruptible at the wheeling utility's discretion), wheeling rates follow the common convention that the wheeler need not pay for the capital costs associated with transmission.[8] However, if the wheeling transaction is firm, industry conventions call for allocating a pro-rata portion of the transmission system's capital costs to the wheeler. Measuring the capacity of a transmission system is much harder than measuring the size of a power plant, but a common approximation is to use the aggregate capacity of the utility's plants.[9] Allocation of capital costs in this fashion squares with the underlying idea of maintaining responsibility for native system adequacy as well as the traditional distinction between the product notions of energy and capacity.

If projected or requested wheeling transactions are large or numerous, they may affect present or projected dispatches to the point where the system is so far from cost-optimal or unreliable[10] that new facilities are necessary. In this case, some of the cost of building the new facility is attributable to the need for wheeling. This *incremental capital cost* of a transmission addition is quite measurable; the issue becomes controversial when these incremental costs must be allocated between the various users of the new facility.[11]

Finally, the element of wheeling cost which is referred to as *opportunity costs* has both a short- and long-term element. Opportunity costs refer to savings from altered operation or construction that a utility *could have been able to realize* if a temporary or unforseen cost-reducing power trade came along, but could not do because a wheeling transaction was occupying the transmission capacity necessary to complete the deal. These opportunity costs can be measured only by estimating the benefit of transactions that never happened—a difficult calculation to make, even for power experts.[12]

THE NEW PARADIGM: TRANSMISSION SHOULD FACILITATE COMPETITIVE GENERATION

Much of the world is now shifting toward a new paradigm for transmission costing. The basic premise is that the public interest is best served by a transmission system that treats all users equally, and charges prices for transmission services that allow for effective competition between generation sources. In this new model, there is no status quo to use as the baseline for computing changes in least-cost dispatch and there is no native load for which the transmission system is primarily responsible. In other words, the purpose of transmission is to create a functional, reliable marketplace in which generators can compete to sell downstream (rather than to provide benefits to an integrated, franchised distribution company).

To achieve the objectives of the new paradigm, transmission pricing and access policies must adhere to several principles.[13] First, the terms and conditions of access should be non-discriminatory so that a substantial set of existing generators and new entrants have a fair chance to compete. This includes offering a wide enough range of products and services as to allow the marketplace some flexibility and dynamism, but not so much variation in individual transmission arrangements that equivalent treatment cannot be ensured.

The principle that the transmission system should not exercise market power over generators may be viewed as part of the nondiscrimination principle. However, the idea is just a bit broader because it encompasses the idea of not exercising monopoly power over all generators equally, to the detriment of consumers, as would occur if the transmission system charged monopolistic prices. This is why transmission rates have been and will remain regulated for some time.[14]

The second principle is one of the basic requirements for economic efficiency in the transmission pricing context. Economic efficiency requires that all goods should trade at prices that come as close as possible to the marginal total cost of providing the good. This condition holds for all goods unless they are public goods, such as clean air or national defense, which do not have well-defined marginal costs. Externalities must be factored in to the measurement of marginal cost, but the presence of externalities does not alter the condition.[15]

Because transmission is a subadditive cost industry, a third basic condition is required. In these industries, marginal cost is often below average cost, which means that charging users marginal cost does not earn the transmission system enough revenue to pay all of its costs. In one recent situation, five Spanish economists concluded that charging short-run marginal costs in the transmission system they studied recovered less than 30% of the total cost of the system. According to these economists, similar results have been observed in many other systems around the world.[16]

This is an age-old problem in economic regulation, and many theoretical and practical solutions have been suggested or applied. The traditional practice of setting wheeling rates equal to *average* rather than marginal costs is one of the easiest ways to address this third condition. Although it automatically violates the second condition (i.e., price equals marginal cost), it is easy to administer and generally considered a reasonable compromise on all sides. Other approaches to providing the revenue shortfall when rates are set to marginal costs include two-part pricing: (1) all users have an access fee and pay marginal costs per unit; (2) the revenue shortfall is raised through taxes or fees on other activities.[17]

This third condition is also a bridge between the short- and long-run time frames. Revenue recovery is essential to keep the existing transmission system solvent. It is a bigger problem when adding a system expansion for which the marginal cost for one user may be quite small but the total construction cost is large. The difficulties of ensuring that prices raise sufficient revenue to cover system expansion are discussed with other long-run issues in Chapter 9.

In addition to these broad principles, pricing approaches should be legally enforceable, simple to understand and administer, and predictable enough to give affected parties the ability to plan and adjust. In addition, political considerations usually dictate that a sense of fairness prevail, and that any strongly-held public interest objectives associated with the industry be provided for in one fashion or another. These important and pragmatic considerations should be kept in mind when specific bulk power marketplace organizations are examined in the following chapter.[18]

TRADITIONAL WHEELING: PRICING AND PROBLEMS

The traditional methods of pricing a transmission transaction were developed by joining the cost-impact-on-native-load philosophy with traditional utility rate-making.[19] This approach has been rapidly supplanted by the new paradigm. Nevertheless, it is useful to review the problems that arose when the traditional approach confronted the emergence of competitive generation.

Typical traditional wheeling agreements were classified either as firm or non-firm.[20] As in bulk power sales, firm wheeling cannot be curtailed except in times of system emergencies; non-firm wheeling can be interrupted at the whim of the provider. Wheeling agreements also specify the point of origin—usually a specific generator or a utility system's border—and points of receipt, such as the border of another utility's system or one or more substation portals to a distribution system.[21]

Firm wheelers pay a pro-rata share of the transmission system's capital or carrying costs. Because a firm transaction obligates the wheeling system to ensure that transmission capacity is always available, the length of the transaction is carefully specified, and canceling the transaction requires a substantial notice period so that both the wheeler and the system can make alternative arrangements.[22] (In many cases, rates do not depend on the total distance over which power is wheeled; these rates are called "postage stamp" due to their similarity with distance-insensitive postal rates.) Several more elaborate methods of allocating fixed costs factor in the type of transaction and the amount of system impacted—e.g., *Megawatt-Mile* and *Impacted Megawatt-Mile* pricing.[23]

In firm transactions, the FERC allowed average or incremental capital costs and estimated line losses to be recovered, but has generally resisted allowing recovery of other more difficult-to-demonstrate opportunity costs.[24] Acknowledging that these costs exist, the FERC has been concerned that an open-ended policy of allowing opportunity costs could allow a transmitting utility to benefit from the exercise of market power. If, for example, a regulator constrained a monopolist to charge prices that were lower than monopoly levels, and then asked the monopolist to measure the opportunity cost of regulation, the answer would be the difference between regulated and monopoly prices. If the monopolist could add this differential back into the regulated price, the result would be a price right back up at monopoly levels.

Rather than pursue an approach that could be construed along these lines, the FERC established an opportunity cost pricing policy that capped wheeling rates at the pro-rata share of the embedded (already incurred) costs of the existing system *or* at the pro-rata costs of all incremental facilities and redispatch costs needed to provide service, but not both.[25] (This approach quickly became known as "or" pricing.) The FERC reasoned that this approach ensured that a utility could not be harmed by accommodating a wheeling transaction because a utility could always decide to add the incremental capacity, which by definition was of sufficient size and cost to allow the wheeling to take place without adverse system impacts. Because the utility was always free to charge the wheeler its pro-rata share of these upgrades, the utility's native load could not be harmed by wheeling at such rates.[26]

Utilities and NUGs divided bitterly over this aspect of federal transmission policies. Transmission-owning utilities generally felt that "or" pricing naively assumed that utilities could add transmission capacity without incurring costs and risks that would be far larger than the compensation they would receive from pro-rata cost sharing for the upgrade. Due to the difficulties of getting siting approvals, the uncertain length of the construction period, and escalations in construction costs, it is commonly (and usually correctly) assumed that adding transmission capacity costs more per unit than existing capacity. Furthermore, transmission cannot be added in anything other than rather large increments—usually much larger than a single generator's wheeling transaction.[27] Hence, any upgrade a utility may have to build would leave it with the carrying costs of surplus transmission capacity. This capacity could be subject to a prudence write off and/or could encourage yet more wheeling. Epitomizing this view, former Federal Power Commission Chair Joseph Swidler said in a 1991 speech,[28]

> Utilities do not build transmission capacity that they know they will need. They build only what is required for safety, adequacy and reliability of service, including emergency reserves, plus an increment for foreseeable growth in loads. Any part of the capacity appropriated for other uses is likely to be at the expense of one or more of the uses for which the capacity was installed, unless the purpose of the appropriation is to make possible a transfer of loads of the host utilities' own customers to an Independent Power Producer (IPP). Here the use of the transmission capacity would not change, but the utility would lose a customer, and perhaps its solvency. The answer of the IPPs is for the host to build more transmission capacity. They show little awareness of the obstacles, delays and cost, not to mention the uncertainties, of building new transmission lines under

present conditions. Some companies have given up on siting lines after a decade of trying. The IPPs are willing to pay the modest fee for the use of existing capacity, but their willingness to shoulder the cost and risk of building new capacity to replace the capacity they have appropriated, is doubtful at best. The utilities' customers seem likely to pick up the cost for the replacement capacity.

On the other hand, IPPs and others observed that the FERC's obligation to facilitate an efficient, competitive bulk power marketplace was broader than an obligation to preserve lowest costs to native load customers. If firm wheeling capacity was prohibitively expensive or was arbitrarily kept by the utility that owned it or was given to favorites, an open competitive generation market could hardly get started.[29]

This argument was complicated by the fact that many IPPs seeking wheeling were effectively competing with generation owned by the same utilities that also owned transmission systems and provided wheeling. This added another potential cost to a utility's willingness to wheel, namely the costs of idling its own generating plants while it purchased from NUGs. If NUG purchases rendered its capacity surplus, the utility faced a possible financial penalty from state regulators.[30] This issue, part of the problem known as "stranded costs," is discussed in greater depth in Chapter 6.

Although antitrust rulings make outright refusals to provide wheeling illegal, utilities viewed the FERC's transmission pricing policies as providing nothing in the way of an affirmative incentive to provide third-party transmission. NUGs pointed out that, irrespective of the FERC's pricing policies, utilities could be *required* to wheel by the FERC only upon a case-specific finding of fact.[31] Hence, if utilities took their time negotiating with wheeling requesters and performing all of the necessary rate making calculations, NUGs would be forced to either endure long negotiation periods or prosecute expensive cases before the FERC in the hopes obtaining wheeling orders. The FERC ultimately found these prospects unappetizing, burdensome, and ineffective for NUGs seeking wheeling.[32]

CONTRACT PATHS AND LOOP FLOWS

Other problems with the traditional approach did not arise directly from the growth of non-utility generation, but were greatly exacerbated by it. The first such problem was *contract path pricing*.

When the points of origin and receipt span more than one company's transmission system, it is often necessary to determine whose system is being used, and therefore who is entitled to payment. For example, in Figure 6-1, a hypothetical generator is wheeling from utility A to utility F in a highly simplified utility system. As discussed in Chapter 2, power flows everywhere in the network between the generator and system F. Purely for purposes of paying for wheeling, the utility industry developed a rather arbitrary notion of a contract path. The path specifies which systems all parties agree to assume that the power flows over for payment purposes. In Figure 6-1, there are several possible paths from generator G to system F (for example, one such path travels through nodes 1-4-8-10-15-19).

Figure 6-1. A Simple, Hypothical Power Network

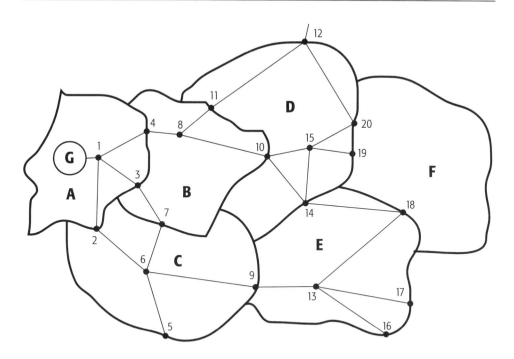

This single transmission grid has one generator labeled G, twenty delivery nodes, and six utility service areas (A–F).

To complete a wheeling transaction, the wheeler and each of the utilities along at least one contract path had to study the wheeling transaction and determine that it did not have too large a negative impact on reliability or the costs of service for native loads in each intervening system. Most utilities with transmission systems have studied their systems to the point that they can quickly tell how much power can flow across the borders between their systems and other systems (sometimes referred to as the *interfaces)* during any particular time period. Therefore, the process of checking to see whether transmission capacity is available for a particular transaction usually involves utilities along a contract path checking to see that the capacity across the interfaces have suffic-ient capacity to accommodate the transaction.[33]

Of course, it is a fiction to assume that all of the power wheeled (to take the example in Figure 6-1) from generator G to utility F would go across the agreed-upon contract path. If the contract path in Figure 6-1 was through utility B and D, it is very likely that some or even much of the power will flow through the transmission systems in areas C and E as well (in addition to flows in other utility areas not shown in the picture), but because the contract path is through B and D, the wheeler only has to pay these two systems. Figure 6-2 (published in a 1989 Federal Energy Regulatory Commission Report) shows an example of the loop flow resulting from a power transfer from Ontario, Canada (OH on the figure) to the New York Power Pool (NYPP). The contract path for this 1,000 MW flow is along lines that go directly from Ontario across the St. Lawrence Seaway to New

York. However, as shown on the figure, only about half the power flows along this path. The rest flows through the three lower loop paths on the Figure, affecting utility systems as far away as Kentucky and West Virginia.[34]

The problems with this case-by-case, contract path approach are readily apparent. But why does this practice continue to exist more than 50 years after its inception? "Loop flows" were not as serious a problem prior to the 1970s as they became during that decade because utilities usually negotiated with each other to keep the problem within acceptable bounds.[35] However, when third-party wheeling requests became more common, utilities became less interested in overlooking loop flow and saw less mutual benefit in finding cooperative, negotiated solutions.

Figure 6-2. Loop Flow Example

Interconnected System Response for Ontario Hydro to New York Power Pool 1,000 MW Schedule

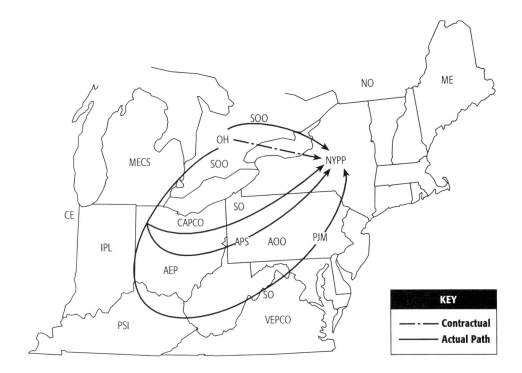

Source: "ECAR/MAAC Interregional Power Transfer Analysis." ECAR/MAAC Coordinating Group, June 1985.

Loop flow causes the 1,000 MW scheduled to flow along the contract path (dashed line) from Ontario Hydro to actually flow on the paths shown (solid lines). Half the scheduled flow occurs on the paths outside the contract path.

Figure 6-2B. Trends in Technology and Cost of Electric Energy.

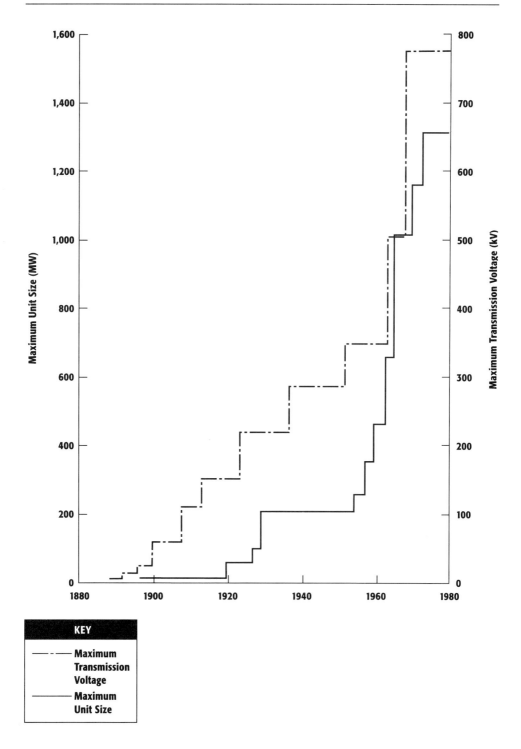

Figure 6-2B.1. Trends in Technology and Cost of Electric Energy.

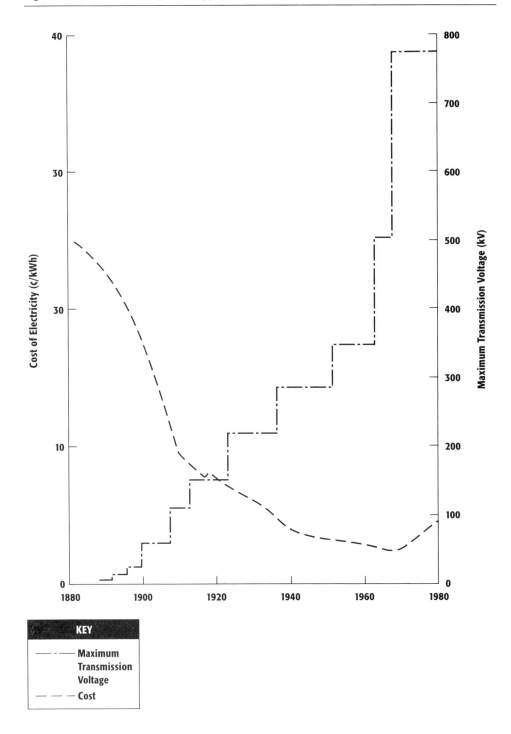

A final problem with the traditional approach was that it was extraordinarily cumbersome and potentially very expensive for the wheeler, particularly with regard to firm wheeling.[36] In the first place, there was the difficulty and expense of having to negotiate with every utility in between generator and load, checking the capacity of their interfaces and so on. Secondly, application of a pricing policy that required the wheeler to pay a pro-rata share of capital costs based on the ratio of the wheeling transaction to total system load resulted in disproportionately large payments when wheeling across more than one system. This phenomenon became known in the industry as transmission rate *pancaking*.[37]

RTGS AND NEW PRICING POLICIES: STEPS TOWARD OPEN ACCESS

Between the passage of the Energy Policy Act of 1992 and Spring 1996, the Federal Energy Regulatory Commission began adopting policies that laid the groundwork for open access. The first change arose out of a legislative compromise that almost became part of EPAct, but was reached too late to include in the final legislation. This compromise called for the formation of voluntary "Regional Transmission Groups" (RTGs) composed of utilities that owned transmission, utilities that did not, independent generators, and other interested parties in a region. In the never-enacted compromise, these groups were created to develop regional access and pricing policies and to arbitrate specific disagreements between RTG members.

The FERC quickly seized upon the idea and issued a policy statement concerning RTGs.[38] In it, the FERC explained the objectives it hoped RTGs would reach:

> The commission believes that RTGs can be alternative vehicles for attaining the same goals inherent in the new section 211: promoting competition in generation, improving efficiency in both short-term and long-term trading in bulk markets, and reducing the cost of electricity to customers. RTGs can provide mechanisms for encouraging negotiated agreements and resolving transmission issues without resorting to the procedures under the sections 211 and 213 of the FPA. As such, RTGs should reduce the need for potentially time-consuming and expensive litigation before the Commission. To that end, the Commission is announcing the a general policy of encouraging the development of RTGs, and providing guidance regarding the basic components that should be included in RTG agreements filed with the Commission.

The Commission went on to outline elements that RTGs are required to contain in their bylaws: broad-based membership including users and owners of transmission, fair voting and dispute resolution procedures, coordination with adjacent RTGs, and consultation with state regulators in the RTG's region and contiguous geographic areas. Two final requirements were particularly important and controversial. First, an RTG had to require member utilities to expand their transmission facilities in order to accommodate all wheeling requests. Second, the RTG had to develop methods of coordinated regional transmission planning: The Commission asked that requests for RTG approvals provide "as much detail as possible" concerning the proposed regional planning protocols.[39]

Although the Commission did not condition RTG approval on specific RTG pricing proposals, observers noted that the FERC clearly hoped that RTGs would help resolve the bitter disputes over transmission access and pricing.[40] As it happens, a number of RTGs were formed relatively soon after the policy statement and went on to obtain Commission approval.[41] Although the Groups emphasized the Commission's requirements, including planning procedures, they did little to resolve pricing issues.[42]

Following a series of individual cases that set pricing precedents, the Commission elected to clarify its transmission pricing policies in a 1994 policy statement.[43] This policy statement did not fundamentally alter the Commission's "or" pricing policy, but it did encourage substantially greater latitude in designing tariffs consistent within "or" limits. In particular, the policy encouraged utilities to depart from contract path and postage stamp pricing, and to set forth "principles" that transmission pricing proposals must adhere to—principles quite similar to those just discussed above.[44] Significantly, the Commission repeatedly recognized that advanced pricing proposals had to balance simplicity, transparency, and administrative burdens with elegance, complexity, and the ability to send more accurate pricing signals.[45]

THE FERC ADOPTS OPEN-ACCESS TRANSMISSION

Following the years of disputes and proceedings chronicled above, the FERC convened an inquiry into transmission access and pricing, wholesale competition, and stranded costs. This inquiry led the FERC to issue Order 888 (See Box 6-1), a final, deeply historic shift toward the view that the transmission system should facilitate generation competition.[46] "The future is here," FERC Chair Moler said when issuing the Order, "and the future is competition. It is a global trend, and in North America, we are at the forefront. There is no turning back."

Most importantly, Order 888 required every transmission-owning utility over which FERC had sufficient jurisdiction to create an open access policy that gave other users of its transmission comparable access to its transmission system at precisely the same terms, conditions, and rates as the transmission system charged its own generators, if it was integrated and owned generators. The Order did not force integrated utilities to place their G and T assets in different companies; they merely had to run them as if they were different systems, and to charge prices and require terms and conditions of the utility's own generators as if they were independent. To ensure this "comparability", an accompanying order from the FERC (Order No. 889) required utilities to post the availability of transmissions capacity on electronic bulletin boards that are accessible to anyone with a computer and modem.[47]

Although FERC did not have the authority to compel utilities over which it does not have jurisdiction (See Table 4-1) (to file similar open access policies, it allowed the utilities under its jurisdiction to require reciprocity from non-jurisdictional systems. The FERC's hope was that non-jurisdictional systems would be sufficiently interested in gaining open access to FERC-regulated systems that they would adopt their own similar policies.[48]

As the embodiment of a historic shift from the notion of integrated utilities with a franchise responsibility to an industry in which transmission's role is to promote fair and efficient generation competition, Order 888 is unquestionably one of the most important changes ever to occur in electric utility regulation. However, the Order itself stopped short of describing the mechanics of a specific pricing and access policy that would achieve comparability. Instead, utilities were required to propose their own versions.[49]

Fortunately, the short-run elements of transmission costing schemes, which are more suited to open access than traditional wheeling have been discussed by academics and industry experts for a number of years. (Long-run pricing is more difficult, and will be discussed in Chapter 8). Most of these approaches stem from an approach known as electricity spot pricing.

TIME-VARYING ELECTRICITY PRICES AND SPOT PRICES

Locational spot pricing begins with the constantly-changing cost conditions on each point of an operating electric power network and translates them into a price signal to all users and generators. As discussed in Chapter 2, a single system controller performing economic dispatch of all units in the area continuously selects from among immediately available units to create the lowest system-wide production costs. At any moment when the control area is balanced, the system controller can tell the added cost of supplying one more or one less unit of power by looking at the variable costs of the marginal plant. This is the short-run marginal or incremental cost of generation for the whole system, often referred to in the power business as system lambda. Figure 6-3 shows an actual system lambda for one utility over the course of a week. Note that it spikes up regularly one or more times a day and varies over a wide range of costs within a typical day.

Spot pricing begins with system lambda and then adds several cost components. First, as system demand approaches the total generating capacity of the system, it rapidly becomes possible that the next unit of generating capacity simply does not exist. To prevent a breach of this limit, electric prices must increase enough to reduce total demand below the limits of the system. In other words, spot prices must shoot up when demand comes close to system capacity by enough to force demand below capacity, and price during these periods exceeds the marginal cost of producing power in the last-to-be-used generator in the system. The difference between system marginal generating cost (i.e., lambda) and spot prices under these conditions is sometimes referred to as the scarcity rent or capacity premium.[50]

The second addition to system lambda in spot and real-time prices is a geographic component. The same type of computer program that analyzes all available plants and comes up with the lowest cost mix can also analyze each separate part of the system and determine the impact of transmission constraints on these parts. Because of these transmission constraints and the need to keep the grid up and running, the generating plants in some parts of the system will be a little more costly than average. In these cases, there are several geographically distinct system lambdas, each leading to a geographic spot price.

Box 6-1. The FERC's Order 888

On April 24, 1996, pursuant to its authority under the Federal Power Act, the FERC issued a rule requiring every utility over which it has transmission jurisdiction to file an open access transmission tariff.[1] Under this tariff, a jurisdictional utility must provide transmission service to all requesters under terms and conditions that are comparable to those that the utility provides itself for transmission of its own generation to its own customers. Indeed, each transmitting utility that also generates must begin to act as if its generators are purchasing transmission services no differently than if they were unintegrated, unaffiliated generators.[2]

As noted in Chapter 5, the FERC does not have authority under the Federal Power Act to order or approve transmission directly to retail customers (direct access, or customer choice). All of the provisions of Order 888 and the required tariff apply only to wholesale transmission transactions. In this context, the buyers or customers for transmission services are either distribution systems that purchase generation from NUGs, federal PMAs, or other utilities, or generators attempting to sell to distributors.[3]

The tariff filing was required to offer, at a minimum, traditional "point to point" transmission service as well as network service. The rates for these services must be just and reasonable, non-discriminatory, and consistent with the present state of the FERC's transmission pricing policies. Each tariff is required to have a number of features, including:

• Utilities may reserve existing transmission capacity to provide for the growth in native load and the loads of others for whom it has existing contractual obligations, including those for whom it is now doing long-term wheeling. However, after the reservation period, this capacity must be offered for sale during the period prior to its need.

• When curtailing firm service for reliability reasons, the utility may not discriminate between its own transmission and that performed for others.

• In addition to bilateral reciprocity (see the discussion in the main text), transmission service must be offered to all members of a power pool or RTG if it is offered to one such member.

• Bilateral transmission rights may be resold ("reassigned") by the buyer to others. Because network service is more case-specific, the FERC did not allow network service reassignment.

In addition to the tariff, the FERC also required jurisdictional utilities to establish electronic bulletin boards on which they will be required to post the availability of transmission capacity on an essentially "real-time" basis. Although many details remain to be worked out, these information systems are intended to dramatically reduce the time required to compute available transmission capacity and to negotiate a transaction under the new tariff.

The Commission recognized that the widespread provision of transmission under open access tariffs might cause transmitting utilities to incur costs that it had not generally recognized in many tariffs up to this time. The Commission labeled the activities that gave rise to these costs *ancillary services* and identified six such services that utilities had to provide, but could charge for on a case-by-case basis under specific pricing principles.

The FERC also recognized that widespread open access transmission could lead to greater stranded generating costs. The Commission found that, for most contracts or agreements, existing customers who cause generating assets to become stranded (no longer saleable due to their high cost) must pay their pro-rata share of the costs they are abandoning. In other words, distribution systems were free to shop for cheaper competitive generation than they were currently using, but they could not escape paying the cost of capacity they had committed to use.[4]

In a separate request for comments which was issued concurrently with the Order, the Commission asked for comments on a proposal that would replace both point-to-point and network service with a "capacity reservation" tariff patterned after natural gas pipeline tariffs. The Commission also required power pools to file open access tariffs for all pool transactions (much like those for individual utilities), and required that bilateral coordination agreements also conform to the rule.

Viewed in the context of the decades of controversy that preceded it, Order 888 can be seen as a compromise that acknowledged important points on both sides of the debate, but nonetheless firmly established the new paradigm. For NUGs and other buyers or sellers of bulk power, the FERC adopted an open system that includes far greater guarantees of transparency, rapid service, and non-discrimination. For utilities, the Commission continued to recognize the importance of allowing capacity dedicated to native load, recovery of transmission cost impacts beyond incremental losses, and stranded cost recovery.

---

1 See the discussion on reciprocity in the main text. The FERC estimates that its jurisdiction extends to approximately 163 IOUs. The rule also encourages utilities to form regional groups and file group tariffs applying on a regional basis (among other things, this could reduce pancaking).

2 The rule contains "standards of conduct" applying to utilities that own both generation and transmission designed to prevent affiliated generators from obtaining "preferential treatment."

3 In an associated and somewhat controversial finding, the FERC asserted that the Federal Power Act gave it jurisdiction over all rates for transmission of power, including any transmission rate a distribution utility developed to transmit bulk power across its system to a retail customer. This meant that the FERC admitted to lacking jurisdiction to order "retail wheeling," but it retained the right to set prices for such transaction if they came to exist by other means.

4 The stranded cost provisions of the rule are complex and are discussed in Chapter 16 *infra*.

Figure 6-3. Actual System Lambda ($/MWh)

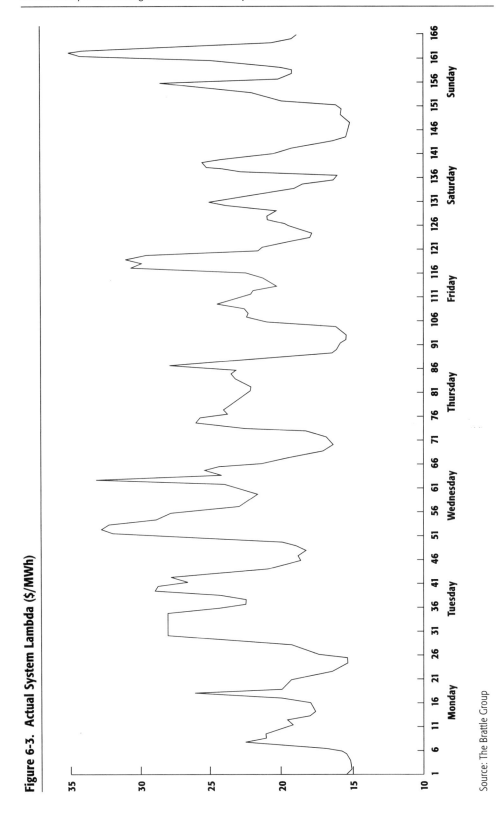

Source: The Brattle Group

Figure 6-4. Professor Hogan's Congestion Pricing Framework

Bilateral Transaction Between IPP and Yellow LDC Increases Congestion

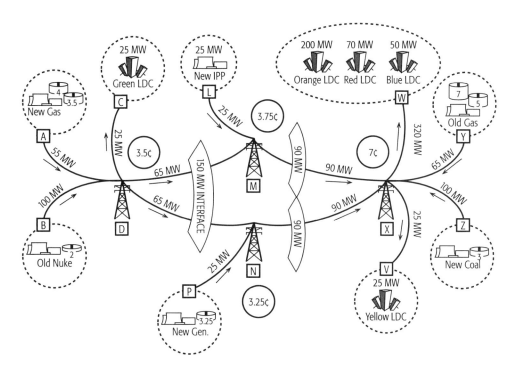

Constrained Transmission Interfaces cause spot pricing to differ at different locations and times. In this example (which is only a momentary snapshot in time), the two fully loaded 90 MW interfaces prevent generators with 3.5¢ marginal cost to supply node X. As a result, node X must be served with plants whose marginal cost are 7¢/kWh. "Spot prices" at M,N, and X at the time of this snapshot are thus 3.75¢, 3.25¢, and 7¢/kWh, respectively.

Source: Hogan (1995).

As an example, Figure 6-4 shows a system that is centrally dispatched for lowest possible cost, but has very different generating costs at different points in the system. At points D, M, and N, which are not constrained with respect to each other, generating costs are around 3.5 cents per kWh. However, at point X, marginal costs are 7 cents/kWh—the marginal cost of the least expensive unit that can supply point X from among the plants at V, Z, and Y. Were it not for the lines leading to X from the lower-cost plants in the lefthand side of the system, these cheaper plants would send power to point X. Hence, the presence of an internal transmission constraint has caused the spot price at point X to be 7 cents, while it is about 3.5 cents elsewhere in the system, in spite of the use of economic dispatch. As a result, spot prices are time- and location-sensitive due to the marginal cost of adding more power in, or removing power from, a specific point in a grid at a specific time.

Because the amount of power demanded by electric customers is constantly shifting, and power plants and lines unpredictably go off-line, and controllers are constantly having to adjust plant output (including reactive power) to maintain voltage, marginal costs at every point in a power grid change rapidly throughout the day. To make things workable, spot prices are usually set once or twice each hour, a period during which demands and the availability of plants usually do not change too much. Costs associated with short-term balancing of the system are recovered, for example, as part of anciallary service charges, discussed in Chapter 7.[51]

Spot prices are an ideal conceptual basis for implementing the new transmission paradigm because they do not distinguish between integrated utility generators and NUGs, or make any other distinction based on the ownership or regulatory status of the generating unit or transmission system. As three of the pioneers of spot pricing put it, "the formula is the same regardless of who owns and controls the generators, transmission/distribution system, and customers."[52] However, in order to work as a fair price signal, spot prices must be computed accurately (or accurately reflect the interaction of many independent buyers and sellers). They must be available to, and trusted by, all market participants. To ensure accuracy, the system operator or some other entity must have extensive information about line loading and generator availabilities and costs, and must be able to immediately make calculations and share information.

Another useful feature of spot prices is that they resolve the product distinction between generation and transmission in a unique way. In a spot pricing world, transmission exists only to bring power at one location and spot price to another location and spot price simultaneously. The value of one unit of transmission between two nodes is simply the difference in spot prices at these two locations. This difference in locational spot prices is often referred to as a "congestion cost."

This sounds quite simple, but there are a number of subtleties to keep in mind. First, if there are no transmission constraints between two points in a network, the principles of economic dispatch call for them to be at the same spot price. Otherwise, the system controller is failing to increase generation from the low-cost units at the low-cost node and thereby lower system costs. However, the controller must also ensure that no other costs are imposed on the network (such as increased losses).

Assuming this is not the case, something must be constraining controllers from equalizing marginal generating costs. The cause of the constraint may be that the transmission lines are *temporarily* overloaded and will soon be back to normal. This temporary constraint may be due purely to transmission congestion, or may reflect secure dispatch constraints, or both. If the spot price is different over a long period of time in the same direction, the system is probably chronically overloaded and may benefit from expansion. Indeed, it is the savings the system could realize from expanding access to the lower-cost node that might justify the cost of the upgrade. Accordingly, depending on the nature of the constraint, spot price differentials may be signaling a long-run price or a short-run price; the difference requires additional study.

Time and location-varying marginal costs have always existed on interconnected power systems. These costs simply have not been displayed to most users of electricity, nor used as the basis for buying and selling power other than between utility systems. For some time now, anyone (with appropriate permission) could have connected a readout to a large node on the power system and read off spot prices at that point. (In theory, this could be done at every outlet in a home, much as real-time stock market prices are available almost anywhere today.) The economic advantages of making time-varying electricity prices more widely available are discussed in Chapter 10.

Finally, remember that spot prices are often short-run marginal costs. As noted above, charging all users these prices does not recover anywhere near enough revenue to pay the carrying costs of the generation and transmission system. This means that something more than spot prices based on marginal costs will be needed to govern the expansion of capacity—the subject of Chapter 8. First, however, the mechanics of an open-access wholesale market are examined in more detail.

FUTURES CONTRACTS FOR POWER

An electricity futures contract is an agreement to sell a specific quantity of power at a specific future time and place for a price that has been determined well in advance of the delivery date. In effect, a futures contract allows risk-takers to offer guaranteed future power prices to risk-averse buyers, even if it means possibly guessing wrong and over-paying (relative to the ultimate future spot price) at the time of delivery.

Regulated utilities have long traded with each other at locational spot prices, and, in theory, could have traded forward contracts that resembled futures contract (sometimes they did).[53] For a variety of reasons, however, interest in a futures market for power was limited until the competitive generation market grew significantly.

The recent surge in interest prompted the New York Mercantile Exchange to develop and offer two electricity futures contracts that began trading in the middle of 1996. These futures contracts represent agreements to buy or sell 736 megawatt-hours delivered over a future one-month period at two particular locations—the California-Oregon border and the switchyard at the Palo Verde nuclear power plant. (These locations were chosen because they are central nodes in the power grid in the Western United States.)

Some observers argue that the nonstorable nature of electricity hampers the establishment of an effective market for power futures, while others believe that the innovation is a logical element of a restructured industry.[54] Futures will be briefly discussed again in Chapter 8.

Chapter 6

NOTES

1 In the FERC proceeding examining the merger of Utah Power and Light and Pacific Power and Light, economists appearing on behalf of applicants argued that transmission was not a distinct product. The FERC rejected the argument, stating that "transmission is a separate product market from the bulk power market since it can be sold separately and…cannot be substituted [for bulk power]." 45 F.E.R.C. 61,095 at 61,284.

2 Kirby, Hirst, and Vancoevering (1995) and Hirst and Kirby (1996) elaborate on the points made in this paragraph.

3 Another common situation involves power plants that have several owners, possibly as part of a pool or association whose purpose is to enable several smaller utilities to achieve scale economies by investing together in a single large plant. If one utility owns all of the lines to the plant, the remaining joint owners need wheeling over these lines to their system borders. See FERC (1981) p. 35.

4 See Chapter 9 for additional discussion.

5 For general discussions of measuring the costs of transmission transactions, see Shuttleworth (1996), Graves (1995), Metague (1990), and Fox- Penner (1990). Professor William Hogan (1993) notes that congestion effects (see below for definition) can double the price of power; elsewhere he gives an example in which one additional megawatt of transmission in one part of a system causes system controllers to adjust seven times as much power elsewhere on the system. Alvardo (1996) notes that the additional losses due to a wheeling transaction can be 20% or more of the power transported.

6 In some cases, a wheeling transaction happens to allow a utility to change its dispatch in a manner that reduces its production costs. Similarly, some lines become less heavily loaded due to the wheeling, reducing losses. In these instances, the short-run cost impacts of a wheeling transaction may be *negative* [See Alvardo (1996) p. 25 for an example]. However, this is almost never true in the long run under traditional pricing, and new modes of transmission pricing automatically incorporate these effects (see the discussion of spot pricing *infra*.)

7 Methods of computing losses can be found in many power systems engineering books, such as Wood and Wollenberg (1995) and Stoll (1989); important technical papers; include Alvardo (1978) and Ilic, *et al* (1996).

8 The FERC encapsulated its policies on non-firm pricing as follows: "For non-firm transmission service, the Commissioner has permitted rates, to reflect, in addition to the variable cost of providing the service, with the provisions that pricing must reflect the characteristics of the service provided." (RM93-19-000, Oct. 26, 1994, mimeo p. 4).

9 Professor William Hogan (1996a, p. 38) writes, "the problem of establishing a fixed capacity for a transmission interface is well understood to be difficult or impossible." The rule of thumb is based on the traditional situation wherein a utility has built a transmission system designed to deliver its generation to load.

10 See Edison Electric Institute (1988).

11 See FERC (1989) p. 59f, Metague (1990), and Hughes and Felak (1996) p. 38–39 for discussions of the costs of transmission system expansions in this context and the discussion *infra* on this topic.

12 The FERC's definition of opportunity costs makes these points quite clearly:

Opportunity costs are not the costs associated with the actual facilities used to provide transmission services. Rather, they are the costs a utility incurs when, as a result of providing transmission service to a third party instead of keeping the transmission for its own use, the utility has to forgo opportunities to reduce its generation cost. For example, if the utility has limited excess transmission capacity to the third party instead of using the transmission for its purpose, it may be unable to engage in economy energy sales or economy energy purchases that would otherwise be made in order to reduce rates to the utility's native load customers (the costumers for whom it was a statutory, franchise, or contractual obligation to provide service).

(*Pennsylvania Electric Co.,* 60 F.E.R.C. Para 61,034 at 61,129 & n. 1 (1992), *aff'd sub nom. Pennsylvania Electric Co. v. F.E.R.C.,* 11 F.3d 307 (D.C. Cir. 1993). Baumol and Sidak (1995), Tye (1991), Tye and Lapuerta (1996), and the FERC's (1989) Transmission Task Force Report also discuss various aspects of opportunity costs.

13 For additional discussion, see Baumol and Sidak (1996), Hughes and Felak (1996), Shuttleworth (1996), Hogan (1995), and FERC (1989).

14 The various possible forms of transmission system market power and their impacts are discussed in FERC (1989) p. 73ff.

15 Most economics and regulatory theory textbooks describe these conditions, e.g. Baumol and Blinder (1991) chapter 34 or Rosen (1988) chapter 7. For a specific example, see Kahn (1971) v. I p. 69.

16 Perez-Arriaga, I.J., *et al* (1996).

17 Among hundreds of important works on the subject, two good general discussions may be found in Bonbright, Danielsen, and Kamerschen (1988) ch. 17 and Kahn (1971) chs. 3, 4. An excellent discussion of this in the transmission system context is in FERC (1989) p. 96ff.

18 The importance of keeping transmission pricing transmission was recognized in the FERC'S recent landmark restructuring rule (order 888) as well as their Transmission Pricing Policy Statement [copy cites] as well as in Hughes and Felak (1996.) Also See Bonbright, Danielsen, and Kamerschen (1988) ch. 17 for a discussion they call "Limitation of Marginal Cost as a Basis for Optimal Pricing."

19 Under the Federal Power Act, virtually all wheeling transactions are under the jurisdiction of the FERC (See Chapter Five). See FERC (1984) and Hughes and Felak (1996) for additional background and examples of traditional wheeling tariffs.

20 In many instances, several grades of priority were established. These priority levels set forth the conditions under which the wheeling utility can unilaterally interrupt the transaction.

21 Impartao (1995) lists the attributes of a traditional wheeling transaction as: Amount, location, and path(s); recallability (firm v. interruptible), curtail ability, duration/term, pricing, obligations to take, flexibility, capacity increase obligations, and ancillary service obligations (see Chapter 8). Imparto notes that interruptible service can be interrupted for any reason at the discretion of the provider, whereas many firm transactions can be temporarily curtailed under emergency conditions.

22 Typical contracts surveyed in FERC (1984) call for one to five years' advance notice of cancellation.

23 Hughes and Felak (1996) describe cost allocation alternatives for the capacity portion of wheeling rates.

24 See Edison Electric Institute (1991) for a review of "indirect costs" the FERC has allowed wheeling utilities to recover in approved tariffs and contracts.

25 The FERC's policy was first articulated in Northeast Utilities, Op. No. 364-A, 58 F.E.R.C. Par. 61,070 at 61,203, *aff'd in pertinent part*, 993 F. 2nd 937 (1st Cir. 1993). This policy explicitly intended to balance an objective of protecting native load customers from undue hardship with the need to promote a competitive generation market and to protect against the use of transmission for monopoly power (The case in which this finding was made was a merger of two large integrated utilities that owned a major portion of the transmission system in New England. Hence, the F.E.R.C. faced a statutory responsibility to ensure that the merger did not enable or promote market power in generation or transmission.

The "or" policy was applied in a number of subsequent merger proceedings and was restated and updated in *Pennsylvania Electric Co.,* 60 F.E.R.C. Para 61,034 at 61,129 & n. 1 (1992), *aff'd sub nom. Pennsylvania Electric Co. v. F.E.R.C.,* 11 F.3d 307 (D.C. Cir. 1993). Following a broad inquiry into transmission pricing principles that started in late 1994, the Commission issued a Transmission Policy Statement (*F.E.R.C. Statutes and Regulations Preambles* Par. 31,005. This statement affirmed "or" pricing, though it also opened the door to transmission rates that used more realistic calculations of the actual flows and impacts of transmission transactions rather than contract paths. The Commission stated that "greater pricing flexibility is appropriate in light of the significant competitive changes occurring in wholesale generation markets, and in light of our expanded authority under the Energy Policy Act of 1992." (*Id* at par. 31,136).

During this period, the Commission also issued a Policy Statement Regarding Regional Transmission Groups, 64 F.E.R.C. Par.61, 138. In the Commission's original view, one of the roles of these groups was to work out new transmission pricing approaches. According to Maize (1994), this effort was unsuccessful, and RTGs evolved much more into long-run planning entities.

26 Non-firm transactions often last only a few hours and by definition can be interrupted if they impose too large a cost burden on the wheeling system. Under these conditions, non-firm wheeling should never impose a cost penalty on a utility larger than the revenue it receives (otherwise it should interrupt the transaction). As a result, variable costs recovered in non-firm wheeling rates were often limited to added line losses.

An even easier approach, also approved by the FERC, was to allow a series of short-term buy-sell transactions. Applying this approach to a non-firm transaction between generator G and System F in Figure 7-1, generator G would sell to utility A, utility A would sell to utility B, utility B would sell to D and D would sell to F. Each utility that bought and resold the power would be allowed by regulators to add a small percentage to the price it paid before reselling to the next. Additional, somewhat flexible arrangements have been allowed for economy interchanges between utilities. See Frankena and Owen (1996), p. 59.

27 See note 12 *supra.*

28 Remarks of J.P. Swidler before the 1991 Meeting of the North American Reliability Council, quoted in "Questions for the Record from Commissioner Trabandt," FERC Docket PF91-1-000, mimeo p. 10–11.

29 See National Independent Energy Producers (1995) for the view of this argument.

30 Many state regulatory policies require that utilities collect revenue only for power plants that are "used and useful." If a power plant is sitting idle because a utility has voluntarily or involuntarily purchased power from a NUG, regulators may decide that the power plant is no longer useful. The utility will then have to pay the plant's carrying charges out of its shareholders' equity. For the NUGs view of the situation, see National Independent Energy Providence (1995).

31 Under Section 211 of the Federal Power Act, as amended by Energy Policy Act of 1992, the FERC could order wheeling, but only following a determination that reliability will not be impaired, the public interest is served, and that "the rates, terms, and conditions" of the wheeling promote economic efficiency. Costs must be allocated to the wheeling transaction as much as is feasible. In addition, any utility ordered to transmit power may apply to cease such transmission if the capacity used for the transmission was originally surplus and is now needed for native load.

32 The FERC first discussed utility reluctance to provide wheeling in FERC (1989), p. 85, citing Penn (1986). Six years later, in its Notice of Proposed Rulemaking that led to its historic open access decision, the FERC devoted extensive discussion to this subject, concluding that its powers under Sections 211 and 212 of the Federal Power Act were not sufficient to promote competition among generators, and that utilities were continuing to provide inadequate transmission access. (Promoting Wholesale Competition Through Open Access Non-Discriminatory Transmission Services by Public Utilities; Recovery of Stranded Costs by Public Utilities and Transmitting Utilities; 60 FR 17662, April 7, 1995, Section III.D.)

33 Metague (1990), Hogan (1995, 1996), and Alvardo (1996) all give extensive examples of computations of transmission costs across crowded interfaces.

34 The impacts of inadvertent flows on crowded transmission interfaces can be even longer. Hogan (1995a) gives real-world examples where additional flows on part of a system requires seven times as much capacity to be curtailed in order to preserve system security.

35 See the discussion in Casazza (1993, 1996), FERC (1989) p. 62ff and in Stalon (1991) p. 115, 139. NERC planning studies occasionally require refereeing by the FERC also helped to mitigate some problems. Stalon (1991) marvels at the fact that traditional utility relations kept loop flow litigation to a minimum—he could recall only one regulatory dispute in his tenure as commissioner but he then concluded that this was unlikely to last under deregulation: "The failure to create a system wherein transmission-owning utilities compensate one another for parallel flows created by short-term transmission transactions will heavily tax the gentleman's agreement under which the current system operates, probably more heavily than the system can support. Anecdotal evidence suggests that the system has already been unduly taxed by profit-seeking IOU's. If the system shows signs of fraying under the limited trading of the last decade, when it has operated largely as an open system with little ability on the part of the profit-seeking players to influence parallel flows, one need only imagine the potential for inefficiencies that appear as competition in transmission services is legitimized and phase shifters make possible many profitable beggar-thy-neighbor actions."

36 An early review of the full range of impediments was Kelly (1987).

37 Pancaking can be illustrated with this example. Suppose that a wheeler successfully negotiates a contract path from A to B to D to F. Suppose that A, B, and D each have peak system loads of 1,000 MW, and that generator G wants to wheel 100 MW. Under the typical convention of paying a pro-rata share of each system's capacity that is used, the wheeler pays one tenth of system A's carrying costs (100 MW divided by 1,000 MW system capacity). The same payment (computed the same way) is paid to systems B and D, so the total payments per dollar of system costs are:

(100/1000) x [A's carrying costs] + (100/1000) x [B's carrying costs] + (100/1000) x [D's carrying costs]

Now suppose that the wheeler could treat systems A, B, and D as one system with a peak load of 3,000 Megawatts (the sum of the three systems alone) and carrying costs also equal to the sum of the three systems. In this case, the wheeler would pay:

(100/3000) x [Sum of carrying costs of systems A, B, D]

The first approach results in a payment three times as large as the second, in spite of the fact that the basic principle of paying for one's share of system use is identical. Pancaking is discussed in Kelly (1990), p. 46.

38 Policy Statement, RM93-3-000, July 30, 1993.

39 Ibid, p. 11ff.

40 The Commission also hoped RTGs would address protocols for resolving inadvertent and loop flow problems, standardize regional grid operating procedures, and deal with other "technical issues." Ibid, p. 7.

41 Approximately six RTGs formed in most major regions of the U.S. For an example of FERC's review and approval, see Order Accepting Compliance Filing, Southwest Regional Transmission Association, 73 F.E.R.C. Par.61,147, Oct. 31, 1995.

42 For the comments of one observer on the failure of RTGs to address pricing, see K. Maize, "FERC Looks to RTGs on Transmission Pricing," *The Electricity Journal*, June 1994, p. 14–16.

43 Policy Statement, RM93-19-000, October 26, 1994.

44 Ibid, p. 13. The following box aligns the text discussion to the FERC's articulated principles.

Transmission Pricing Principles That Promote Efficiency and Competitive Generation...

...As discussed in this text	...As expressed by the FERC
• Non-discriminatory equal access; costs and prices treating all transactions and users equally	• Non-discrimination and comparability as expressed in Order 888
• Prices reflecting marginal cost	• Prices that promote the efficient use of the transmission system
• Prices that recover cost of transmission system	• Prices that permit the recovery by transmission of all "legitimate, verifiable, and economic costs"

45 Ibid, p. 8, p. 21, p. 22.

46 Promoting Wholesale Competition Through Open Access Non-Discriminatory Transmission Services by Public Utilities; Recovery of Stranded Costs by Public Utilities and Transmitting Utilities, Order 888, RM95-8-000, RM94-7-001, RM95-9-000 and RM96-11-000, April 24, 1996.

47 Order 889, April 24,1996, Docket No. RM95-9-00, modifying 18 CFR Part 37.

48 Firms may integrate backwards by acquiring the resources that are pre-requisites to its stage in the process of getting goods to markets. Thus, a company generating electricity could backward integrate by owning coal mines and railroads. Forward integration is the opposite direction toward the final disposition of the product in their market. It is too early to tell whether this strategy will result in universal participation by all transmitting utilities in open-access regimes that meet FERC's conditions. A number of non-jurisdictional utilities are preparing their open access policies, and many state restructuring proceedings are attempting to ensure that all transmitting utilities do so.

49 However, it did ask for comments on a proposal that would eliminate point-to-point wheeling entirely in favor of transmission tariffs that reserved capacity on the entire transmission network. (RM96-11-000, April 24, 1996).

50 See Bohn, Schweppe, and Tabors 1991, Schweppe, et al (1988), and Vickrey (1971) for three important and useful discussions of spot pricing.

51 Strictly speaking, demands fluctuate constantly during the spot pricing period, and also generally have an upward or downward trend. This introduces substantial challenges in the definition of ancillary services discussed in Chapter 8. The optimal period for setting spot prices has been the subject of several challenging theoretical treatments, including Ilic, et al (1996), McGuire (1996), and Alvarado (1996).

52 Bohn, R., F. Schweppe, and R. Tabors (1989), p. 274.

53 See Chapter 9 of Hunt and Shuttleworth (1996) for a description of various sorts of utility conttacts.

54 The positive view is epitomized by Blackmon (1986) and Krapels and Stagliano (1996); concerns have been expressed to me by consulting economists Phillip K. Verleger, Jr. (unpublished conversation, 1996) and John Treat of Booz, Allen, and Hamilton (American Enterprise Institute Spring Policy Forum, May 17, 1995.) Also see Daniel Southerland, "Watt's Next for Area Electric Utilities? Power Futures." *Washington Post* 4/2/96 p. E1, and J. Javetski, "Trading in Electricity Futures Advances Deregulation Process. *Electrical World*, May 1996 p. 66.

Chapter 7

Competitive Power Markets: The Short Run

With its new requirement for comparability, the FERC rejected the PURPA-era model of independent generators selling to integrated utilities in favor of a model in which all generators compete to sell to all distributors. If retail competition is allowed, generators can sell directly to customers. But nothing in the advent of competitive generation has changed the fact that the system controllers must continue to monitor and sometimes control all generators and lines, adjusting the area's dispatch to meet load and security needs. When generators and distributors strike a deal, someone somewhere needs to ensure that the system can accommodate the transaction.

There are two basic models of how competition and the need to preserve grid integrity can be combined. These two approaches, which have become known as the *poolco* and *bilateral* models are polar extremes. In real world restructuring, the two concepts are usually integrated into a hybrid or parallel model, employing both concepts. To understand the strenghts and weaknesses of the two models, a brief detour into economic theory is necessary.

COMPETITIVE MARKETS AND ECONOMIC EFFICIENCY

The oft-acknowledged efficiency advantages of competitive markets arise out of the fact that buyers and sellers voluntarily find each other and execute trades of mutual benefit. The self-correcting, self-optimizing nature of these markets is intuitively understood. To understand power market debates, the actual mechanics of competition must first be examined more closely. These mechanics can be illustrated with a simplified market for a single good in which it is assumed that:

- The good is homogeneous enough that any variation in the quality or other attributes of the product can be readily reflected in the price of the good;

- Transactions costs are small. This is a concise way of saying that information is freely available about offers and purchases, so it is very easy to compare offers to buy and sell and to reach agreements to trade. Furthermore, buyers and sellers can freely and cheaply resell goods to each other;

- There are a large enough number of buyers and sellers, relative to the size of the market, that no seller or buyer has the ability to influence price by witholding output. New sellers and buyers can freely enter and exit;

- Buyers ("demanders") have the almost universal trait that the higher the price, the less of a good they want to buy. This is known as the "law of demand" and is graphically depicted in the form of a downwward sloping "demand curve" (Box 7-1); and

- Individually and as a group, sellers have increasing marginal costs—i.e., the industry as a whole is not a natural monopoly. Because it has already been assumed that there are many sellers, they can be arranged in increasing order of their marginal costs to create the common picture of a supply curve. (Box 7-1.)

In the economics literature, these common assumptions are given the shorthand *homogeneity, perfect information, low transactions costs, atomistic buyers and sellers, free entry,* and *traditional supply and demand functions.* Like many assumptions in economics, these are never expected to apply perfectly. The test of a robust theory is whether it works when the assumptions are moderately—rather than perfectly—accurate.

Suppose there is an open-air market for this good, with the supply and demand curves magnified so highly that individual sellers' and buyers' portions of the supply and demand curves can be seen. In Box 7-1 (Panel 2), widget sellers *A, B, C,* etc. have been stacked up in increasing order of their marginal costs—i.e., the cost of them selling one more widget. Widget seller A is willing to sell for any price higher than its cost; the same holds true for B, C, and so on. A can offer the lowest price, but once A sells all of his widgets, the next lowest price that could be available is *B's* offer price, which is just above *B's* cost.

The demanders in the marketplace display roughly the reverse behavior. The demand curve shows the amount each demander is willing to pay and the quantity that he or she is willing to buy at that price. Demander *A* is willing to pay the highest price for its desired quantity of widgets; *B* is willing to pay a bit less for its desired supply, and so on. (The maximum price a demander is willing to pay is called his or her reservation price).

Suppose now that all of these buyers and sellers are unleashed in a marketplace and they can bargain at will. Seller *A* can hawk her wares to almost every buyer because she can make a profit selling at the price every demander from *A* to *H* is willing to pay. There is far more demand than *A* can supply at her price, so she sells out quickly. Suppose we assume that the sale was made to buyer *D* at *D's* reservation price. Similarly, B can profitably sell to everyone from *A* to *K,* and sells out quickly; suppose the deal is made with *A* at *A's* reservation price. This process continues until H attempts to sell to all remaining buyers and none are left with reservation prices above cost.

After this first round of contracting, the prices in the marketplace are not at all the same. Some low-cost sellers have found customers with high reservation prices, such as *B* selling to *A.* These sellers are making very high profits. Other low-cost sellers who happened to have sold to low-priced customers are not making nearly as much as they could. It is even possible that there are some buyers who are willing to pay more than the prices other buyers are paying but cannot find a seller.

When sellers and buyers finish this first round and look at each others' arrangements, they quickly conclude that improvements are possible. Seller *A* sees that, in spite of the fact that her costs are lower, she has gotten a lower price for her product than *B,* whose

Box 7-1. Panel 1. A Simplified Competitive Trading Market

The typical graph of supply and demand in a competitive market shows a smooth, downward-sloping demand curve and an upward sloping supply curve. Supply intersects with demand at the equilibrium on market-clearing price.

If the supply and demand curve is magnified to the point where individual buyers and sellers could be identified, the two curves might look more like panel 2.

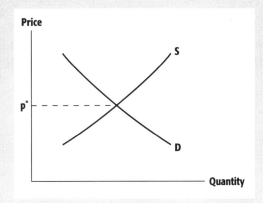

Panel 2

This figure shows that seller A could sell profitably to almost any buyer at the buyer's reservation price. Seller E, however, can profitably sell only to buyers a–h, whose reservation prices exceed E's marginal cost.

In this market, if all sellers randomly contract with buyers according only to the rule that the price paid be between the buyer's reservation price and the seller's marginal cost, no one equilibrium price will be observed initially in the market place. However, assuming that all buyers and sellers see all other transactions, the random scatter of prices will eventually converge on all trades occurring at a single market-clearing price p*. At this price and quantity combination, society makes the best possible use of private resources (economic efficiency).

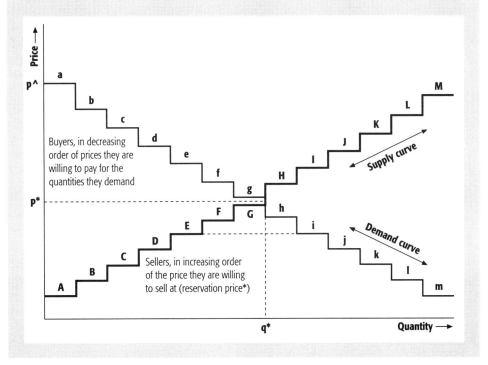

costs were higher. Because this is a free market, she can approach *B's* buyer *(A)* and offer a price below what *A* was offered by *B*. This prompts *A* to change suppliers and sets *B* off on a search for new customers, displacing someone else's sales. Because *B* can match any other seller's price (other than *A,* who has already sold out), she is certain to find a buyer and sell out as well. Meanwhile, buyer C figures out that she can outbid any other buyer, except *A* and *B,* and sets out to make sure all sellers do not sell out at prices that are lower than what she is willing to pay.

It takes some mathematics, but it can be shown that many rapid, successive rounds of recontracting result in the entire marketplace converging on a single "market-clearing" price p*, which every seller sells at and every buyer pays. This price occurs at the point where the marginal seller G can just sell at price p* and buyer *G* will just buy, and total supply quantity q* equals total demand at this price. This is also the price-quantity combination in which the marginal cost of the next unit of supply equals price (marginal revenue).

The converged equilibrium price and quantity resulting from this process maximizes economic efficiency. Sellers make the largest possible total industry profit, yet they continue to face competitive pressure to reduce costs and conserve resources. Buyers also receive the highest aggregate benefit because most buyers are paying a price that is much lower than the highest possible price they are willing to pay.[1] Importantly, society is not wasting resources by giving business to inefficient producers, and it is making sure that all buyers who really place a high value on the good get their value, while those who really do not value the product as much as it costs society to produce it do not get it. Finally, although it is not a condition of economic efficiency, the fact that this marketplace clears a single, observable price confers a sense of fairness to all participants.

Depending on how they are set up, markets for power can approximate this behavior, particularly in the short run. The poolco and bilateral approaches to market organization each attempt to produce an outcome that is something like the idealized marketplace described above, but in very different ways, and with somewhat different strengths and weaknesses.

Simplified versions of the two approaches are pictured in Box 7-2. In a poolco, an entity much like the dispatcher of today's "tight" power pools schedules and dispatches all generators, buys power from them, and sells power from the pool to all customers. In a bilateral model, generators contract directly with power buyers and purchase transmission service from the grid. If direct access is allowed, buyers can be distributors or retail users; if not, buyers are distribution systems only.

THE POOLCO APPROACH

The poolco model traces its lineage to economic dispatch by tight power pools. In these pools, the system operator looks at the full range of units available to the pool every hour and performs economic or merit-order dispatch, subject to security constraints. Economic dispatch means that the operators first choose the plant with the lowest marginal or incremental cost of production and use as much of this plant as possible, then they do the same for the unit with the next lowest cost, and so on until demand is

Box 7-2. Panel 1. Poolco and Bilateral Wholesale Power Markets

In a poolco or "mandatory pooling" market, all generators sell into a pool run by the Independent System Operator (ISO). The ISO also controls the transmission systems of all utilities. There is no difference in this model between contracts and flows: generators' financial trades are all with the pool as are all their physical sales of power. Similarly, all purchase transactions and all power flows are between buyers and the pool. If the market allows wholesale competition only, the buyers are distribution companies (distributors), such as IOU or publicly-owned systems,who then resell the power at retail. If the market allows retail choice, buyers may be individual power customers, power marketers, distributors, or other entities.

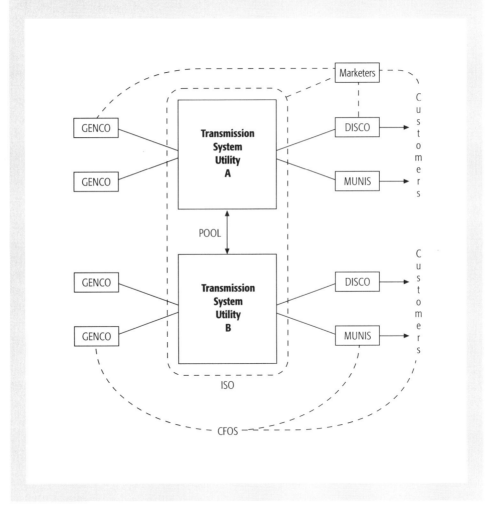

Box 7-2. Panel 2. Bilateral Wholesale Power Market

In a bilateral market, the ISO continues to operate the transmission system and possibly a spot market for surplus power as well. As shown by the *solid* lines, all power continues to flow from generators through transmission to distributors and then to customers.

The dotted lines in the figure show that, in this model, marketers that do not alter power flows can make financial agreements to aggregate, bundle, or sell power from generators to the transmission system or to distributors directly. Generators can also directly contract with discos.

If retail competition was allowed, generators and marketers could also make contracts with customers directly.

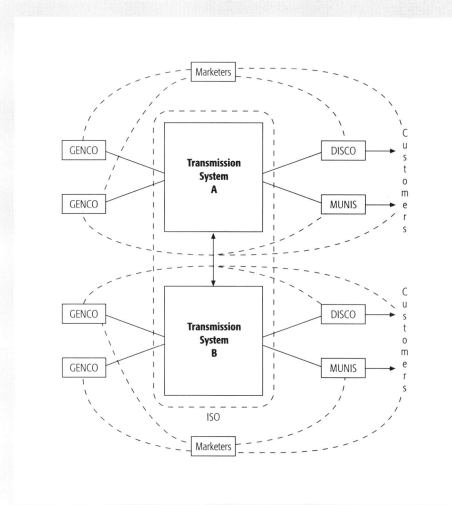

Source: Adapted from *Optimal Operating Arrangments in the Restructured World: Economic Issues*, Raymond S. Hartman and Richard D. Tabors, December 1995.

satisfied. Recalling from Chapter 3 that base load plants are by definition those with low marginal costs, these can be expected to be dispatched full-out all the time, followed by intermediate plants, and then peaking units, with high operating costs and use only during periods of high demand.

The simple dispatch of lowest-to-highest cost plants is adjusted in several ways to maintain grid security. Operators maintain spinning and standby reserves, which requires that some units be ordered to run at lower power or remain on standby at the most feasible points in the system. They also make changes in substation settings, line use, and plants near load centers to provide voltage or reactive power. They may also modify the dispatch order, using less from a cheaper plant than the plant is capable of using, due to congestion on the transmission system. Box 7-3 shows that this process results in a "supply curve" and a market-clearing price something like the ideal market in Box 7-1.

When the pool units were all regulated, all of the costs and accounts of regulated units were subject to regulator scrutiny. Absent special permission, the prices of all transactions (including pool sales) had to be at cost-of-service rates. The marginal cost of each unit was public information, and no deviation from marginal cost was allowed.

The poolco model works in similar ways, except that the units participating in the dispatch exercise are competitors—many or all are unregulated (Box 7-3, Panel 2). Because costs are now confidential information, rather than published, regulated information, the poolco operator cannot use incremental costs to decide who to dispatch first. The solution is to hold an auction, at which each generator bids a price and quantity of power for each hour of, say, the following day. Based on the bids, the system operator runs simulation programs, decides for each hour or half-hour on the combination of bid prices that stack up much like the regulated marginal costs did before, and alters the simple ordering if needed for reliability.

When the poolco operator gets to the point where all estimated demand for the hour is satisfied, the price of the last generator to be dispatched becomes the system-wide pool price, much like system lambda in today's regulated systems. The poolco operators then announce the "winners" of the auction, i.e., all plants they will dispatch. With the exception of out-of-merit-order plants that must be dispatched due to congestion or voltage support, these are all the plants with costs up to and including the last plant dispatched.[2]

The results of the auction set the price that all buyers who want power from the pool have to pay during the hour. This price is very much like the "market-clearing" price p* in Box 7-1: It is a single price that approximates the maximally efficient, highest profit outcome. This makes sense, in part, because the poolco model roughly satisfies the assumptions underlying competitive markets.[3] First, the poolco's rules automatically enforce product homogeneity: The terms of the auction specify exactly the attributes of what the pool buys and sells, how much the attributes can vary, and what quality variations are worth. There is also extensive and common information because the pool is operated as a regulated entity that must post its selling and buying prices and buy and sell without discrimination. There is a single market-clearing price that resembles the

Box 7-3. Panel 1. Traditional Pool-Type Economic Dispatch

Economic dispatch schedules the lowest-incremental-cost-units first and then proceeds in order up to the point where the supply of generation meets demand. The incremental cost of the last unit dispatched is the system lambda or marginal cost, p*. This mimics the market-clearing price p* in Box 7-1.

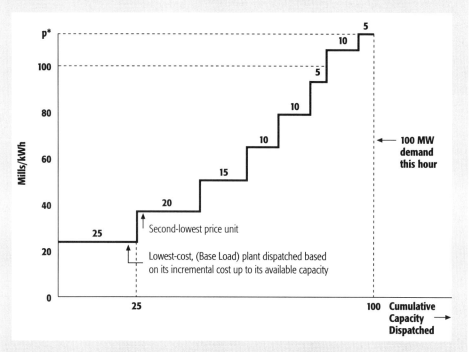

Panel 2. Poolco Competitive Dispatch

If the same units are available as in the traditional dispatch, *and* if each unit bids the same price as its past regulated incremental cost, and if no new transmission constraints or other network needs prompt a change, the dispatch of "winners" would be identical to traditional economic dispatch. However, it is highly likely that many generators may choose not to bid their past marginal cost, as they now face competitive pressures and have no revenue guarantee.

Hourly Bids

Plant	Quantity (MW)	Price (¢/kWh)	Plant	Quantity (MW)	Price (¢/kWh)
Winning Bids			Losing Bids		
A	25	1	G	20	2.3
B	20	1.2	B	5	2.4
C	15	1.4	E	10	2.5
B	10	1.6	I	5	3.0
D	10	2.0			
D	5	2.05			
E	10	2.1			
F	5	2.2			
Total Winning Bids	100				

idealized market above, with maximum aggregate sellers' surplus. The only additional assumption needed to achieve a competitive outcome is an absence of concentration or collusion of buyers and sellers—an assumption that is important both in theory and practice.

The perception of fairness that is conferred by a single market price is another attribute of the poolco model that some policy makers find important. If the set of all buyers includes many small municipal distribution systems (as there are in the United States) as well as some potentially very large buyers, the fact that all buyers get the same hourly price may be an added fairness benefit for which there is no sacrifice of economic efficiency—at least in theory.[4]

From the standpoint of most buyers, volatile prices are an unfortunate counterpart to equal prices. In a poolco model, pool prices are the only way of signaling to generators that more or less power is needed, and to buyers that power is scarce and should be conserved. If price must ration demand every hour, and demand for electricity rises and falls precipitously with little price sensitivity, then prices may really have to rise and fall dramatically to perform their balancing function. The result is that pool prices are equal to everyone, but extremely variable and often quite unpredictable.

The volatility of pool prices depends on dozens of factors ranging from unexpected plant outages, which reduce supply, to highly-localized weather changes, which are unpredictable and can impact demand dramatically. To get a flavor for the volatility of prices in a poolco, first note that during an average day, it is normal for system marginal cost (lambda) to vary by a factor of ten or more. (In retail equivalent, the very rough range might be from about a penny per kWh in the dead of night to ten cents during the busiest part of the day). Differences in the daily or weekly averages of these hourly prices reflects supply and demand conditions on the system due to generator and line outages, seasonal demand increases, and many other unforeseeable events. These averages exhibit just as much volatility as daily prices, and in a less predictable pattern. Figure 7-1 shows the volatility in average weekly prices in the United Kingdom's pool, which is examined more completely below. As of January, 1996, the highest price buyers paid for power in the six years of pool operation was about $1.54 per kWh—an astonishing *50 times* typical average levels.[5]

Few buyers feel comfortable buying in markets with such volatility. Yet in a poolco market, the pool itself offers no way to guarantee a long-term price or supply for electricity. This is a gigantic departure from today's practices, where virtually every power plant is built either by an integrated utility or by a NUG developer who has signed a contract at prices that are not time-varying or pool determined for a period of many years. Attempting to sell all power in hourly increments is akin to telling all homebuilders that they must continue to build homes, but they can no longer sell them using long-term mortgages— they must sell them one day at a time. Mirror-image considerations apply to buyers who face the prospect of paying extremely volatile pool prices with no immediate source of long-term price stability.

Figure 7-1. Contracts for Differences

Allow Bilateral Transactions

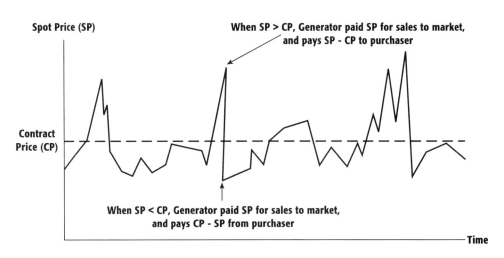

Source: Hogan (1995a)

For goods that are much less a necessity than is electricity, economists have long observed that most buyers prefer some stability with regard to prices and supplies. In poolco or hourly power markets, there is every indication that buyers and sellers will want to create contracts that specify prices and supplies in advance and with a measure of certainty. In a poolco market, these contracts become financial instruments, not power transactions. (These arrangements are examined in more detail two sections hence.)

THE BILATERAL MODEL

In a power marketplace that is organized solely around bilateral trades, buyers and sellers individually contract with each other for power at whatever price, and under whatever terms and conditions they agree upon. One municipal utility may purchase all of its capacity needs for many years with a fixed price contract, while another might decide to play the spot market and purchase its needs only one hour at a time. Like most bilateral agreements, no permission is required from an outside authority to allow the trade, price is not regulated, and performance disputes between buyers and sellers are settled according to the terms of the contract or by recourse to the legal system.

The obvious question raised by this model is, "how can anyone be sure that the transmission system can accommodate the collection of contracts the marketplace comes up with?" In this model, all transactions have to be provided to the system controller, who then analyzes all of the trades in each forthcoming time period, and determines which ones are infeasible due to network constraints. Because the order in which the transactions are analyzed and scheduled can easily determine which transaction takes the system over its limits, and therefore must not be allowed to occur, it is especially important that this be done without bias or discrimination and with some clear rules for prioritization.

As part of evaluating and scheduling bilateral trades, network controllers take most of the same actions to preserve network integrity as they must take in the traditional and poolco models. It is still necessary to make sure that the network has adequate spinning and standby reserves; reactive power must be provided where it is needed for voltage support; and excessive congestion must be mitigated by redispatching units—even if this violates the terms of some bilateral trades. Moreover, in an emergency, the controller must retain rather extensive authority to abrogate contracts and shift power flows around in order to keep the grid up. (Terms of this nature are required in all present transmission agreements and will be required in the future irrespective of whether the marketplace adopts a poolco or bilateral approach.) To do these things with a minimum disruption of contracts, system controllers need to have some uncontracted generation that *they* can adjust to keep the system in balance. This means that the network operator purchases some generation for system purposes from available "swing units."

Unlike the poolco approach, the bilateral model emphasizes neither homogeneity nor a single market-clearing price. Also, neither the operators of the network nor a poolco "clearinghouse" actually buys or sells power. Instead, all trades are between buyers, sellers, and a variety of market facilitators or "middlemen," variously described in modern utility terms as aggregators, brokers, or power marketers.[6] A network authority is required to decide which transactions the network can accommodate, provide and charge for transmission services, and provide the services needed to run the network reliably. (These issues are discussed at the end of this chapter.)

Because bulk power market prices are mostly or entirely deregulated in the bilateral model, buying and selling prices will not be posted or controlled by the government. Indeed, prices may even be confidential because the system controller who evaluates and schedules the contract need not know the price—only how to prioritize the transaction in case it cannot be accommodated. Instead, buyers and sellers will have to shop and discover prices the way they do in all other unregulated markets—i.e., through advertising, market information services, and comparison shopping.

A hypothetical set of outcomes for this bilateral market is shown in Box 7-4. The top portion of the figure lists the hypothetical transactions; the lower shows how these transactions would be placed in the dispatch by the transmission system operator. Note that even if the transmission operator schedules all transactions in order of priority (firmness, duration), the dispatch may leave out generators that are less expensive than those that won contracts. This would not occur in an ideal marketplace, but in many competitive markets a cluster of prices rather than a single price is the norm.

SPOT PRICES AND LONG-TERM POWER CONTRACTS

The poolco model lends itself to spot pricing, as the pool process automatically results in an hourly or half-hourly spot price for the system as a whole. Moreover, the calculations needed to determine locational differences in spot prices in each period are automatically computed as part of the process of determining the system-wide price and providing for reliable operation. These prices are available for posting, allowing all market participants to know where there is congestion in the system.

Box 7-4. Bilateral Contract Dispatch Based on Firmness and Duration Priorities

In a bilateral environment, the dispatcher does not know contract prices; dispatches are based on agreed-upon priority rules that are modified if necessary to preserve grid stability. Note that because since 115 MW of contracts were executed, but demand was only 100, the dispatcher curtails the two lowest-priority contracts.

The top panel shows the same dispatch in a different presentation, namely cumulating capacity dispatched from lowest to highest price. The dispatcher does not know prices and cannot do this. It is clear that a supply curve can be constructed from whatever set of bilateral contracts are selected for dispatch at one time. It may not be the lowest-cost supply curve created (in theory) by poolco or traditional dispatch, but its still a supply curve.

Contracts

Generator (seller)	Distributor (Buyer)	MW	Firm or Non-firm	Duration (hours)	Price (mils/kWh)	Priority
E	c	20	F	18	12	3
A	g	15	N	24	20	4
M	a	10	F	24	16	2
B	f	10	N	4	10	6
D	d	15	F	48	25	1
F	a	25	F	48	14	1
L	k	5	N	16	5	5
C	e	10	F	2+	7	2
G	h	5	N	4	5	6
		115				

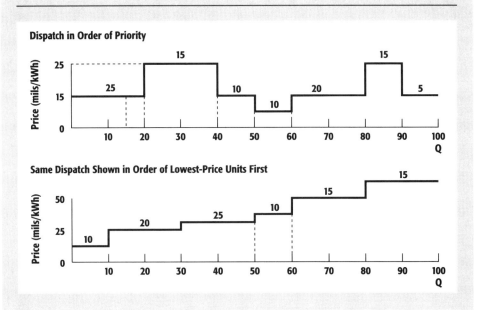

Dispatch in Order of Priority

Same Dispatch Shown in Order of Lowest-Price Units First

Assuming this market and its institutions and regulators are working effectively, the fact that a regulated, non-discriminatory entity computes and posts spot prices provides a basis for a particular approach to long-term contracts. These contracts, known as contracts for differences (CFDs), guarantee a price and quantity for power between a buyer and seller.

Once there is a poolco structure with posted locational spot prices, a CFD is an extremely simple idea (Figure 7-1). A CFD simply provides that the "seller" (actually, the financial risk taker or guarantor) agrees to "sell" to the "buyer" a specified schedule of power at pre-specified prices over whatever time period the two agree to—often several years. This contract guarantees that the buyer gets the contracted quantity and price. However, unlike in the bilateral model, the way this occurs is not via an actual power sale that the pool must schedule. Instead, the buyer continues to buy all power through the pool, including the amount contracted in the CFD. All the "seller" does in the CFD is compute the difference between the price he or she has guaranteed the buyer and the price the buyer paid to the pool and pay the buyer the difference (if positive) or collect it (if the buyer paid less to the pool than the contract price). The name "contract for differences" is thus quite apt—what the buyer and seller have contracted for is really the payment of the difference between the contract price and pool price—a financial amount, not a quantity of power at all.[7]

In the bilateral market, there is no need for CFDs because any sort of long-term quantity or price guarantee can be built into the terms of the contract. There is also no automatic determination of spot prices by the network operator. As in today's marketplace, however, it is extremely common to find small mismatches between the amount of power a distributor is demanding and the amount it has contracted for (as read from meters at delivery points). Because power consumed must equal power supplied at every moment, a bilateral marketplace will have to supply the balancing amount of power instantly—a service performed today by the control area utility. The source of power for settling imbalances will likely be a spot market created by pooling all available surplus generation in the area, including power from the "swing" units that operators must have to ensure stability.

In addition, regional formal or informal spot markets are likely to arise out of precisely the same kinds of incentives that were observed after the first round of contracts in the example in Box 7-2.[8] Suppose, for example, that in one part of the system a distributor had a contract at a very attractive price for more firm power than it needed, while another distributor happened to need more power that day. Just as in today's power pools and economy interchanges—which are essentially simplified spot markets used to supplement existing bilateral or integrated power supplies—these kinds of markets would undoubtedly spring up in a bilateral world. Indeed, because most power trading today is bilateral, and spot transactions remain extremely common as well, it is probably more accurate to say that today's spot markets would continue or expand if bilateral contracting under open access remains the predominant model.[9]

OWNERSHIP AND CONTROL OF TRANSMISSION: TRANSCOS AND ISOS

The poolco and bilateral models just discussed were simplified almost to the point of absurdity in order to illustrate their fundamental character differences. A welter of challenges awaits anyone who tries to implement either of these ideas in real electricity markets.

The first set of issues involves the division of responsibilities over owning, operating, and regulating the physical transmission system and the market institutions (e.g., the pool). Putting aside the need for regulation, the poolco model requires:

1. One or more owners of the existing transmission system and related assets;

2. One transmission system operator, which we have loosely referred to as the system controller or operator;

3. An entity that conducts the hourly auction, announces pool prices, and schedules winners; and

4. An entity that serves as the financial intermediary or "clearinghouse" for all pool purchases and sales.

In a bilateral marketplace, the first two entities are needed, as is the regulator. No clearinghouse is necessary (buyers pay sellers directly in this model), but in place of the hourly auction there still must be a system-operated spot market to handle sudden system needs.

As discussed earlier, one of the keystones of restructuring is the effective or actual deintegration of generation from transmission to promote effective G competition. FERC's Order 888 requirement that all transmission owners must provide comparable service to self-owned and independent generators was meant to implement this idea. The separation of transmission system *ownership* from system *operation* adds another, perhaps even stronger layer of protection to this scheme. The creation of an "Independent System Operator" (ISO) is an idea that has gained almost universal acceptance in the restructuring debate. As Professor Hogan notes, under an ISO, "the easy-to-state but hard-to-enforce principle of comparability would be transformed into an easier to enforce principle of non-discrimination."[10] The role of the independent agent, though considerably different in bilateral and poolco models, is nonetheless seen as extremely important. Utility consultant Frank Graves notes that "after two years of debate, the creation of an ISO may be the only thing poolco and bilateral advocates agree on."[11]

The essence of this idea in its simplest form is that one or more companies own the transmission system in a control area—transcos or gridcos—while a single entity, not controlled by transmission owners (or all other market participants), performs scheduling, dispatching, auctions, and other grid operations. As with so many other facets of restructuring, there are dozens of variations on the concept of separating ownership and control of transmission. Unsurprisingly, these involve different jargon for similar concepts, regional differences, differences in the attitudes of public and private utilities, and all the rest of the fault lines along which utility debates often fracture. Reflecting these differences in scope and character, proposed names for the grid operator include ISO,

Independent Grid Operator (IGO), Independent Tariff Administrator (ITA), and others.[12] Box 7-5 shows the organization and operation of a poolco market, in simplified terms, with a separate transco and ISO.

Two obvious candidates instantly emerge as logical organizations to become ISOs. Power pools that presently facilitate trading or dispatch plants are the kid sisters of full-fledged ISOs, at least in theory.[13] Regional Transmission Groups (RTGs), formed according to FERC's requirements that all transmission market participants be well-represented, are an even more broad-based alternative. At present, many power pools and some control-area utilities are developing proposals to become ISOs of one kind or another. These proposals have proven to be controversial because most pools were formed by the largest utilities in their regions.[14] Critics warn that if the ISOs are to become truly independent the utilities that founded and nurtured these pools for many years may have to give up much of their present influence or control over the pool.[15]

The dramatic character of this issue was illustrated when a group of 18 public utility commissioners from around the United States took the unusual step of issuing a joint declaration calling for strict ISO independence. The declaration said that "FERC, the states, and Congress must insist on the creation of ISOs that have the authority to operate and improve regional transmission systems, and that are truly independent from the owners of the generation resources."[16]

Concerns over the conflicting issues and demands facing ISOs has made ISO governance an active area of discussion. Apart from regulatory issues, discussed below, the main questions include:

- Should the ISO be a for-profit or not-for-profit entity?

- Who may select the voting members of the ISO's Board of Governors? Should the board have composition requirements?

- What voting procedures and rights should board members have? What dispute resolution procedures are needed?

- Should the owners of transmission lines or any others have special rights?

- How should the ISO be organized and administered below the governing board level?

- What is the relationship between the ISO, the actual pool or spot market, and related institutions, such as the reliability councils?

The complexity inherent in this last point is illustrated by the pool that is now operating in the United Kindgdom. Keeping in mind that the transmission system in the United Kindgdom is a fully independent and separate firm, the pool nonetheless is structured as an organization that is separate from the gridco and is composed of representative voting members from all segments of the market. The organization, overseen by an executive committee, has 20 working committees as well as an independent auditor, a contract with a settlement house, and a separate "funds administrator."[17]

Box 7-5. Simplified Poolco-ISO Market Organization and Operation

1. Generators bid to Pool.

2. Pool picks and notifies winners in best dispatch (Pool may be run by ISO).

3. ISO dispatches generators according to winning dispatch over transmission systems of all in the ISO's region.

Transmission owners do not control dispatch or transmission, but receive regulated rates for the use of their system.

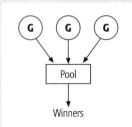

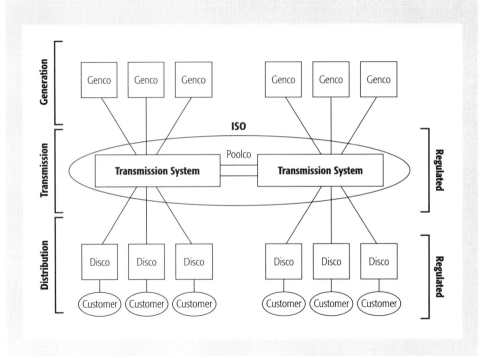

In Order 888, the FERC did not require the establishment of ISOs. However, it stated that it "wished to encourage" voluntary ISO formations, and provided 11 principles it requires in order to approve ISO agreements.[18] In compact form, these ISO principles[19] require:

1. A fair, non-discriminatory governance structure.

2. No financial interests in any power market participants.

3. A single, open-access tariff for the entire ISO area.

4. The ISO has responsibility for system security.

5. The ISO may control system dispatch in its area for effectuating pool or bilateral dispatches.

6. The ISO can identify and resolve transmission constraints.

7. The ISO has incentives to act efficiently.

8. The ISO's transmission and ancillary service pricing (see below) promotes efficiency.

9. The ISO must post transmission availability in real time on electronic bulletin boards.

10. The ISO must coordinate with adjacent control access; and

11. The ISO must have a dispute resolution procedure.

Significantly, the FERC did not impose on ISOs any responsibility for expansion of the transmission system to accommodate all requests; this responsibility remained with the owners of transmission. (This and other long-run issues under restructuring are addressed in Chapter 8.

Box 7-6 is a rather thorough diagrammatic exposition of the transmission ownership and control options developed by a group in the U.S. Pacific Northwest. The main options identified by this group are:

1. A mutual transmission agreement that has been reached and implemented by all transmission-owning utilities in the region;

2. An ISO that operates the system, but does not have authority or responsibility for planning system expansions ("ISO-lite, or ISO-L in the figure);

3. An ISO that operates the system and can build or order utilities to build transmission (ISO-O, for ownership); and

4. An independent transco that owns and operates the transmission system.

The main attributes of these models are shown in the top panel of Box 7-6. The bottom panel addresses additional details and variations on these four models, including such issues as the legal structure of ownership. For example, the ISO-L can be organized as a federal, cooperative, for-profit, or non-profit entity, or it can be formed via an inter-state compact between state governments.

REGULATING ISOS AND TRANSCOS

Separating ownership and operation of the transmission stage of the industry is a clever solution to U.S. interest in competitive generation markets and simultaneous disinter-est in vertical divestiture. Unfortunately, this solution to potential market inefficiencies presents a significant and underappreciated regulatory challenge.

Box 7-6. Characteristics of Transmission Alternatives

The top panel shows the main ownership and regulation alternatives for grid governance. Each of the main attributes of these alternatives is shown on the left. The bottom panel expands the two main alternatives involving ISOs into several more detailed alternatives.

Basic Characteristics	ICA	IGO-Limited
Transmission ownership	Utilities	Utilities
Independence from power mrkts	None	Operational
Need to allocate contract paths	Yes	Some
Upgrades		
Who plans	Utilities	IGO-L
Who budgets and builds	Utilities	Utilities/others
Who has eminent domain		Utilities
Maintenance		
Who sets standards		IGO-L
Who sets budgets		Utilities
Who schedules		IGO-L
Who performs system maintenance		Utilities
Operations		
Who schedules		IGO-L
Who curtails		IGO-L
Who maintains reliability		IGO-L
Who back stops ancillary services		IGO-L
Who has 211 responsibility		Utilities/IGO-L
Who do you ask for service		IGO/Utilities
Who regulates transmission	FERC/PUCs	FERC
Transmission compensation	Utility tariffs	Lump sum

Issues	ICA	Limited IGO–No Transmission Ownership				
		Federal	Coop	Non-profit	Interstate	For profit
Type of Entity	Contract	Federal cooperation	Members of the cooperative	Non-profit corporation	Interstate compact	Corporation
Investment Capital	None	Small capital needs	Small capital needs	Small capital needs	Small capital needs	Small capital needs
Working capital	No change	Reserves & customers	Reserves & members	Reserves & customers	Reserves and customers	Reserves & shareholders
Governance	NRTA	DOE & Congress	Customer board	Customer board	Customer board	Private board
Who regulates	No change	FERC/PUCs	FERC/PUCs	FERC/PUCs	FERC/PUCs	FERC/PUCs
Dispute resolution	NRTA/FERC	NRTA/FERC	NRTA/FERC	NRTA/FERC	NRTA/FERC	NRTA/FERC
Incentive structure	No change	Management performance	Mgt./perf. & dividends	Management performance	Management performance	Mgt./perf. PBR & Profit
Who bears risks	No change	Customers	Customers & members	Customers	Customers	Shareholders and customers
Who files tariffs	Owners	IGO/Owners	IGO/Owners	IGO/Owners	IGO/Owners	IGO/Owners
Who appoints boards?	Customers	President	Customers	Customers	Governors	Shareholders

Box 7-6. Characteristics of Transmission Alternatives

The top panel shows the main ownership and regulation alternatives for grid governance. Each of the main attributes of these alternatives is shown on the left. The bottom panel expands the two main alternatives involving ISOs into several more detailed alternatives.

Basic Characteristics	IGO/Owner	Transco
Transmission ownership	utilities/IGO-O	transco
Independence from power mrkts	complete	complete
Need to allocate contract paths	no	No
Upgrades		
Who plans	IGO-O	
Who budgets and builds	utilities/IGO-O	
Who has eminent domain	utilities/IGO-O	
Maintenance		
Who sets standards	IGO-O	
Who sets budgets	utilities/IGO-O	
Who schedules	IGO-O	
Who performs system maintenance	utilities/IGO-O	
Operations		
Who schedules	IGO-O	
Who curtails	IGO-O	
Who maintains reliability	IGO-O	
Who back stops ancillary services		
Who has 211 responsibility		
Who do you ask for service		
Who regulates transmission	FERC	FERC
Transmission compensation	lump sum	purchase

Issues	IGO Owner/Operator					Transco
	Federal	Coop	Nonprofit	Interstate	Forprofit	
Type of Entity	Federal corporation	Members of the cooperative	Non-profit corporation	Interstate compact	Corporation	Corporation
Investment Capital	Private markets	Private markets	Private markets	State revenues	Private markets	Private markets
Working capital	Reserves and customers	Reserves and members	Reserves & customers	Reserves & customers	Reserves & shareholders	
Governance	DOE and Congress	Customer board	Customer board	Public board	Private board	Private board
Who regulates	FERC/PUCs	FERC/PUCs	FERC/PUCs	FERC/PUCs	FERC/PUCs	FERC/PUCs
Dispute resolution	NRTA/FERC	NRTA/FERC	NRTA/FERC	NRTA/FERC	NRTA/FERC	NRTA/FERC
Incentive structure	Management performance	Mgt./perf. PBR & Profit	Management performance	Management performance	Mgt./perf. PBR & Profit	Mgt./perf. PBR & profit
Who bears risks	Customers	Customers and members	Customers	Customers	Shareholders & customers	Shareholders & customers
Who files tariffs	IGO	IGO	IGO	IGO	IGO	Transco
Who appoints boards?	President	Customers	Customers	Governors	Shareholders	Shareholders

There is no disputing that both the remaining transcos and the ISOs will be monopolists in their respective areas. As has been discussed earlier, transmission itself has the cost attributes of a natural monopoly, not to mention strong siting and expansion issues. An ISO is, by definition, a monopolist in the services that FERC requires it to provide exclusively, as well as with regard to its possible administration of spot markets or pools and other functions.

Regulating these two monopolists for efficient short-run behavior will be an extraordinary challenge. Regulating the transcos may not be too difficult, but only because all the challenges are passed to the ISOS. Specifically, most ISOs that are distinct from gridcos are expected to lease or sign contracts enabling them to use, charge for, and entirely control the use of all of the transcos' assets. In exchange, transcos will have to receive revenues that adequately compensate them for the use of their property. Hence, rate-setting proceedings similar to those employed today will have to occur periodically to set the revenues that transcos are entitled to collect from their ISO lessees.[20]

It is at the ISO stage that regulation faces its unique challenge. The ISO will own essentially no assets, may be a for-profit or not-for-profit entity, and is responsible for adhering to FERC pricing policies as well as providing its contractually-determined payments to the transco. In the first place, it is not clear that the monies collected from transmission rates will equal required revenues. Shortfalls may be collected from regulated access charges levied on all transmsission users. The ISO may be barred from getting additional funds from other activities, but if it is a public entity, taxpayer-provided working capital or a credit authority may be available.

It is also difficult to see how strong performance incentives can be placed on an entity that has no assets and little accountability to anyone other than its board of governors and federal regulators. If the ISO is a nonprofit agency, there is no financial cushion that can be used to signal owners that performance is better or worse than expected. If the ISO is profit-making but has no assets, what determines the appropriate level of profit? This issue affects not only regulators' need to prevent possible anti-competitive behavior, but also affects the simple incentives of the ISO to manage and staff itself efficiently.[21]

Although this issue has received less attention than it deserves, the few policy makers and researchers who have examined ISO regulation have made somewhat discouraging observations. Many note that the separation of ownership and control of assets has often been found to pose large regulatory problems and inefficiencies, and that the problem would be more familiar and easier to address if the ISO and transco were united as well as independent.[22] In this case, the transmission system's allowed profits could be set by conventional rate-making principles and this level of profitability could be used as the point of departure from which the ISO could be rewarded for good performance and punished for bad.

For those who do not find this an option, the remaining choices appear to be (1) public control of the ISO, i.e., constituting the ISO so that it is essentially self-regulated, and (2) adopting some form of incentive regulation for the ISO's profits. The former leaves regulatory oversight FERC as a second fail-safe layer; the latter poses the challenges discussed above, as well as long-run issues that are discussed in Chapter 8.

Finally, just as there is a danger that the ISO will under collect revenues due the transco, there is also a need for regulators to ensure that the ISO does not exercise its authority to drive up the price of electricity above the levels set by competitive generation plus its required transmission revenues. Some observers argue that too much ISO authority creates an excellent means of facilitating collusion, particularly if the ISO's governing board is comprised of a small group of sellers and wholesale, regulated buyers.

Moreover, the establishment of an ISO does not eliminate the phenomenon of transmission congestion. However, just as the FERC was prompted to loosen its transmission pricing policies as part of its move to open access, a firmly independent transmission operator facing a revenue requirement arguably should charge congestion costs in order to send the correct price signals to users. If the ISO can charge prices well above costs for congested transmission links, it will return to transmission owners the sort of handsome profits that those with market power make. To keep price signals at the appropriately high levels but prevent excess profits, regulators will have to forego congestion pricing (and thus economic efficiency) in favor of physical allocation schemes or they must force the ISO to use its profits to expand the system or issue refunds to all customers. In other words, the creation of open-access and ISOs does not eliminate the fundamental problems that are inherent in setting prices for a constrained transmission grid; it merely places them outside the traditional context of incumbent-versus-entrant.

Hence, ISO regulators will have to balance the ISO's need to earn enough revenue to pay transco revenue requirements and maintain appropriate cash or credit reserves against the need to keep prices at efficient levels. This balancing act will require careful attention to ratemaking on both sides of the ISO's ledger, and may require innovative approaches to incentive-based regulation as well. It will also require oversight to ensure that neither the ISO nor transmission owners nor generators exercise market power.

INTERCONNECTED OPERATION ("ANCILLARY") SERVICES

There is no question that isolating all of the costs of transmission and allocating them in a manner that collects adequate total revenue and provides accurate price signals for generators is a precondition for efficient competition.[23] If the basic transmission pricing paradigm (e.g., FERC's "or" pricing) does not capture all of the costs of transmission under open access, the remaining costs must be identified and their allocation determined. If not, the transmission system will end up bearing these costs or it will pass them onto system users, perhaps in hidden or undesirable ways.

The U.S. industry has coined two terms to refer to the services that are required to run the network as part of the provision of open-access transmission: *Interconnected Operations Services* (IOSs) or *Ancillary services*.[24] Although these services together constitute only a small portion of the price of electricity, Kirby and Hirst (1996) estimate that for the U.S. industry their total cost is on the order of $12 billion, which is more than enough cost to have a significant impact on the marketplace.

IOSs represent the frontier of restructuring in two ways. First, defining the things that transmission system operators do in the form of discrete services that can be accurately costed is the *sine qua non* of efficient vertical deintegration. Second, defining and accurately pricing ancillary services is the key to resolving the debate between those who complain that integrated utilities have never been able to recover the full cost of providing wheeling and those who insist that transmission access must be comparable for all, but must also charge each user for all aspects of the costs they incur.

In the wake of open access, IOS becomes the crux of this debate. If all users of the system have equal rights and responsibilities, and services must be provided which benefit all users jointly, who has the responsibility for paying for and providing these common goods? In the comments received by the FERC on its open access rule, this was an extremely contentious point. Whereas in the prior wheeling regime transmission providers felt uncompensated for the costs imposed by wheeling, transmission providers now argued that in an unbundled world they would still not be adequately compensated. In a sure sign of increased competitive pressure, most commentors wanted marketplace participants other than themselves to have responsibility for these services.

Discussions of specific IOSs can be confusing at this stage because the range of services grid operators provide or manage can be disaggregated or decomposed in different ways, depending in part on the overall structure of the market into a poolco, bilateral, or hybrid model. To complicate matters further, different terms have been coined to describe the same services.[25] Many of these differences can be traced to differences in regional system topologies, ownership, and organizational structures (longstanding differences were examined in Chapter 4). Other differences can be ascribed to different standards that have been set by the regional reliability councils.

In Order 888, as part of its establishment of universal comparable access, the FERC devoted considerable discussion to the issue of ancillary services.[26] The FERC agreed with a position taken by NERC and others that a number of ancillary services were integral to the provision of transmission and therefore had to be included as part of open-access transmission service. The Commission noted that other IOSs could be defined or quantified, but did not believe that these services had to be part of open access tariffs. By neither requiring nor forbidding these additional services, the FERC left open the question of whether the costs allowed in open-access tariffs were adequate to compensate transmission owners for all of the costs of providing service.

Table 7.1 displays the IOSs that the FERC requires as part of open access tariffs. The first two columns of the table show the name and description of the service. The third column shows Oak Ridge National Laboratory's very approximate estimate of the cost of each average U.S. transaction. (Remember that because of regional differences this may be a purely arithmetic construct.[27]) The names and definitions adopted by the FERC in large part reflect the recommendations that it received, and are universally acknowledged to be a first attempt that is in need of additional study and revision.[28]

The unbundling of transmission service into various IOSs is greatly influenced by the overall organization of the marketplace. Because the FERC and most utilities today function in a bilateral world, the costs shown in Table 7.1, as well as those in many utility tariffs, are defined by reference to bilateral contracts. A poolco organization can take many of the costs shown in the table and roll them into the hourly spot price, thereby bundling some ancillary services with hourly energy. Beyond this, cutting-edge researchers are finding that restructuring is accelerating progress on new computational and operational methods for power system control and cost allocation which could enable ancillary services to become much more transaction-allocated.[29]

In every production process, there are some costs that cannot be attributed to any specific transaction. In a factory, for example, the cost of maintaining security guards and a set of auditors is largely invariant to specific products and their marginal costs. These common costs, often referred to as "overhead," must be allocated into the prices of all of the products made by the factory.

Because transmission systems are a monopoly resource in their areas, regulators have not been willing to leave the allocation of common-cost ancillary services entirely up to the transmission system operators. In particular, transmission owners who also own generators might be tempted to favor the latter with low allocated common costs. The rightmost columns of Table 7.1 briefly indicate the extent to which the cost is common or allocable to specific causal transactions.

The final major issue concerning IOSs is the system's means of providing them. In some cases, it is obvious that only the system operator can provide the service. In such instances, the ISO is a natural monopolist in that particular IOS and its price must fall within the purview of the ISO regulator. Other services can be purchased by the ISO from generators or users by various procurement schemes, so competition could discipline their prices. Many argue that where possible, outside procurement is best because competition to sell these services to the ISO will lead to lowest costs as well as product innovations and a reduced need to regulate the ISO. However, markets for many ancillary services are very formative, and it is still too early to declare many of them permanent features of the new power marketplace.[30]

With respect to IOSs, the United Kingdom took a simple approach. In the U.K. pool, the grid operator provides and procures all ancillary services each day and computes the total costs of doing so for each *period.* An equal share of these costs for each half-hour's operation is added to that half-hour's market-clearing generation price. The cost of transmission constraints plus an added margin for ancillary services, together called *uplift,* constitutes the difference between the "pool purchase price" and the "pool selling price." This approach is much less "unbundled" than the evolving U.S. marketplace.

THE UNITED KINGDOM AND CALIFORNIA: TWO MAJOR RESTRUCTURING EXAMPLES

The creation of competitive power markets as described in this chapter is now under way in several states of the United States and many nations around the world.[31] One of the best-watched restructurings occurred in the United Kingdom in 1990, when the national government privatized the power system in all of England and Wales and simultaneously established a poolco marketplace. In this market, generators submit bids to the operators of the pool. The pool and the grid operator (the National Grid Company or NGC), which both owns and operates the transmission system, have no financial ties to generators or distributors. The pool announces winners and provisional hourly pool prices at 4 p.m. the day before. In real time, NGC dispatches to actual load, provides IOSs from swing generators, and keeps track of the differences between expected and actual dispatch and the generators from which it has purchased surplus generation.

Table 7.1. Interconnected Operating Services Required in U.S. Open-Access Transmission Tariffs, 1996[1]

Service	Description	Measurement Methods	Approx Cost (mills/kWh)	Allocation and Procurement Notes
Scheduling, System Control, and Dispatch	Costs of conducting poolco auction or collecting and evaluating bilateral contracts; unit or transaction scheduling (unit commitment). Also presumably includes creating, maintaining, and operating the area control center, including the communications and control network, and the accounting services necessary to perform scheduling and dispatch.	Transmission provider or ISO's accounts reflect costs for all of these functions. Total cost of providing these functions does not vary with each specific transactions, but increases with the approximate total volume and size of transactions.	0.2	Cannot be procured externally. Allocated on a per-kW or per-kWh basis or the equivalent.
Regulation and Frequency Response	As system loads fluctuate minute-by-minute, generators must be available to instantly increase or decrease loads to match fluctuations. These reserves are controlled by the ISO.[2]	(1) System production costs, absent load-following reserves, can be compared to actual production costs via simulation to determine variable cost differences; (2) Load-following units may have greater maintenance costs over time and shorter lifetimes; these cost elements must be estimated; *and* (3) The capital and maintenance costs of generator and control equipment specific to load-following can be estimated over time.	0.5	Difficult to allocate to individual transactions. Tradeoff between computational simplicity and transparency and sending correct price signals to system users.[3]

continued on next page

Table 7.1. Interconnected Operating Services Required in U.S. Open-Access Transmission Tariffs, 1996[1] *(continued)*

Service	Description	Measurement Methods	Approx Cost (mills/kWh)	Allocation and Procurement Notes
Operating Reserve: Spinning Reserves	A quantity of generators (not service load) must be kept fully warmed up and ready to take over within seconds in the event of a generator or transmission line failure.	Essentially identical to above.	0.5	May evolve to become transaction or seller-specific allocation, but at present is considered a common cost.
Operating Reserve: Supplementary Reserves	Generation kept on standby so that it can be started rapidly in the event that generators or lines suddenly fail—but not as rapidly as spinning reserves above.	(1) Units kept in reserve may have greater maintenance costs over time and shorter lifetimes; these cost elements must be estimated; (2) The capital and maintenance costs of generator and control equipment specific to load-following can be estimated over time; (3) When outages occur, system production costs absent load-following reserves can be compared to actual production costs via simulation to determine variable cost differences.[4]	1.8	Same as above

continued on next page

Table 7.1. Interconnected Operating Services Required in U.S. Open-Access Transmission Tariffs, 1996[1] (continued)

Service	Description	Measurement Methods	Approx Cost (mills/kWh)	Allocation and Procurement Notes
Energy Imbalance	Over a period of a week or a month, the totalenergy contracted for by a buyer does notperfectly equal the quantity delivered. Periodically, an adjustment must be made to even the accounts of the buyer, seller, and whom ever has provided the residual energy.	Routinely measured as part of the metering and billing of bulk power sellers and buyers.		Need not all be provided by the transmission provider or ISO. Possibly purchased competitively or obtained from a spot market pool. In a poolco model, automatically provided as part of market-clearing process.
Loss Compensation	In the aggregate, the difference between total power generated and total power delivered to all customers.	Losses vary greatly over short time periods and depend on all transactions on the system at any one time. They can be determined ex-post by comparing generation and loads and can also be estimated via computer simulations. Losses occur in the form of additional variable costs of production for generators who increase their output to compensate for losses.	1.3	Same as above. Average losses can be allocated to individual transactions in the presence of significant real-time system information; see text .*supra*

continued on next page

Table 7.1. Interconnected Operating Services Required in U.S. Open-Access Transmission Tariffs, 1996[1] *(continued)*

Service	Description	Measurement Methods	Approx Cost (mills/kWh)	Allocation and Procurement Notes
Voltage Support and Reactive Power	Dispatchers at the control center alter the settings on transformers, transmission lines, and other downstream grid-connected equipment to provide sufficient reactive power in areas where needed. Dispatch of some generating units may also be altered from a base case in which no reactive correction is needed.	(1) Accounting data records the capital and operating costs of equipment used only for voltage support, such as capacitors and reactors. (2) Increases in production costsover a no-compensation status quo can be estimated via computer simulation.[5]	0.4	Same as above

1 Source: Adapted from FERC Order 888, Kirby and Hirst (1996), and Hirst and Kirby (1996).

2 Grid operators provide load-following by selecting (via competitive auction or other means) units to be load followers and then setting these units to automatically adjust their outputs as needed. Typically, roughly one percent of the generators on line at any time are programmed in this manner.

3 Hirst and Kirby (1996) p. 24 suggest that the cumulative impact of real-time covariance with system imbalances might be useful; they have also suggested the use of average deviation from average load, though they admit that both these methods are complex.

4 Additionally, the opportunity cost of keeping standby capacity available, rather than in use, may be estimatable. Under present FERC pricing policies, recovery of this cost component would probably not be allowed.

5 The opportunity costs of devoting generation to VAR support can be similarly estimated.

In the British restructuring, the original buyers were privatized distribution-only utilities known as the Regional Electric Companies (RECs). Each REC purchased all its needs from the pool. If a REC wished to obtain price assurances, it could enter into CFDs with generators. On a phased-in schedule, successively smaller customers were allowed to purchase directly from the pool and to enter into CFDs for price guarantees on their own—a poolco version of direct access.

The British model has worked well in many respects, but it has also illustrated many of the risks of restructuring in general and of poolcos in particular. The experiment has undoubtedly been greatly affected by the fact that the British government privatized its generators into two very large companies, National Power and PowerGen, and, until recently, kept ownership of the nation's commercial nuclear plants. For its first years, the pool had three very large, well-established bidders and only a handful of much smaller independent producers. According to published reports, the large bidders soon learned that they could influence the pool's winning price by controlling the amount and price of capacity they bid. This influence led some to call for bilateral trading outside of the pool so that small producers would have another way of selling; others called for deconcentrating the generation market.

Other experiences in the United Kingdom illustrate the promise and perils of restructuring. The new marketplace prompted a fall in the price of pool power from its original government-set levels to levels approaching short-run marginal generating costs. At times of low demand, pool prices are extremely low, rising by factors of ten or more within a day during periods of high demand. The volatility of pool prices has increased, and average prices rose so high during 1994 that the Regulator imposed a price ceiling (Figure 7-2). Because deregulation also prompted a fall in the price of natural gas, many gas-fired independent plants have been built. Seeking to avoid price volatility and guarantee supply and price, many RECs have purchased interests in these plants, reintegrating an industry that was just deintegrated!

The system established to regulate the pool, transmission company, and the RECs was brand new upon privatization. It has come under increasing criticism for allowing excess profits and anticompetetive behavior. A number of researchers have concluded that problems with regulation and competition have prevented retail prices from being as low as they should be, and that U.K. retail electric prices are higher than those in less competitive systems. (Appendix 7A discusses the U.K. experience in more detail.)

Another bold and enormously contentious effort by the state of California's public service commission recently culminated in the enactment of state legislation to restructure the California industry.[32] Drawing on the U.K. experience, the poolco versus bilateral controversy, and a history of animosity between public and private utilities in the state, California opted for a model that combines an ISO with a poolco for the formerly regulated IOUs and a bilateral marketplace for independent generators and public power. The utilities (and other generators that wish to do so) bid into a *power exchange* much like the U.K. pool, while generators and buyers who prefer to sign long-term contracts do so and submit them directly to the ISO. Using procedures that are creating a great deal of controversy and are not yet finalized, contracts and bids will be combined into a single dispatch.[33]

Figure 7-2.

This figure shows average weekly prices for power purchased from the U.K. electricity pool ("Pool Purchase Price") from the pool's inception in 1990 to December 1995. The point at which voluntary maximum prices were imposed by the Regulator (February, 1994) is marked as the "Offer MMC' decision. ("Offer" stands for the Office of the Electricity Regulatory, and MMC for the Monopolies and Mergers Commssion, both arms of the British government.)

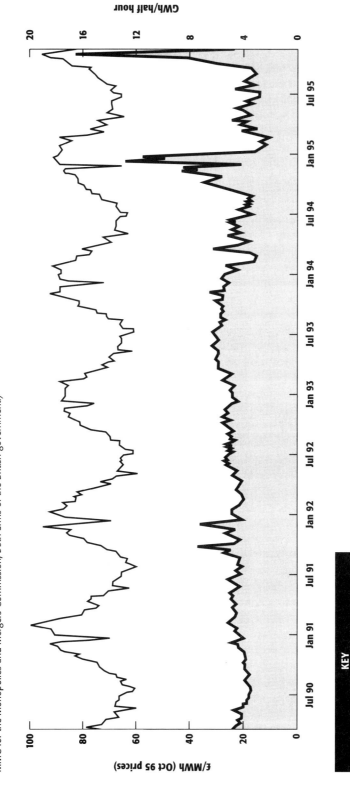

Source: Evan (1996)

209

KEEPING GENERATION COMPETITIVE

The U.K. experience with excessive concentration in generation serves as a stark reminder that deregulated generation is not synonymous with competitive generation. Effective competition requires a reasonable number of independent buyers and sellers of generation. In the United States, at least once before, the owners of deregulated generation found that horizontal concentration was very profitable.

Limiting ourselves to the short-run has important implications for the analysis of market power. First, the kind of market power concerned with here involves the ability to raise short-run prices above competitive levels, but only briefly. (Long-run market power is examined in the next chapter.) The long-run role of high prices and profits as an inducement to attract new competitors—*entry*, to use the formal economic term—cannot necessarily be relied upon to quell prices that have been driven up by market power.

Recall from Chapter 2 that it is an absolute technical necessity in utility grids that energy supplied must equal energy demanded at all times. If demand exceeds supply at the originally prevailing price, price must be raised immediately so fast-responding buyers can reduce their purchases. If this does not work, system controllers will start to curtail loads or, in some cases, protection circuits will disconnect loads by themselves.

This means that if physical rationing is to be avoided, prices rising to "whatever the traffic will bear" is necessary. What might make this market-clearing phenomenon anti-competitive or at least viewed by some as "anti-consumer?" There are at least two interconnected possibilities. First, sellers and/or the system operator might act to prolong the high prices longer than they are needed. But this only makes sense if whoever is doing the prolonging earns a higher profit. This makes the issue of who profits from price rationing an important issue.

In the United Kingdom, where market power with respect to short-term prices has been a problem, these are precisely the two issues that the Regulator has examined. Focusing on the bidding behavior of the existing set of generators, the Regulator found, for example, that these generators were withholding capacity from bidding for one period and then submitting it during the next. The generators made money from this "gaming" because the pool rules said that whenever capacity was insufficient, the pool would pay a premium during the next period to try and induce generators to bid more power. Had the generators kept their availability steady, prices would have been lower.

Similar short-term market problems occur in areas that become isolated by virtue of transmission constraints. If for a few hours each day or week, some generators cannot ship power into a region, then the generators in that region will have to balance supply and demand. If the region is not relatively large and diverse in its native generation supply, prices may spike up to the benefit of the ISO, local generators, or both. Professor David Newberry (1995a) p. 16 describes this phenomenon nicely:

...the root of the problem is that transmission constraints periodically fragment the market into smaller sub-markets, in each of which the competing generators will be few in number, and substantial local market power. Existing pool rules insulate the consumers from the full impact of this local market power, but at some cost. If the PSP [pool selling price] were regionally determined on the basis of the cost of supplying that region at the constrained-on price, then consumers might choose either to contract for the services of constrained-on plant, or might reduce demand in periods of high local prices, and the local market might be made contestable to entry and hence subject to competitive pressure. This solution raises other non-trivial problems, for regional pools have fewer competing generators, and give more opportunities for market power, unless constrained by trading between pools. In that case, charges for transmission need careful regulation to avoid the exercise of local market power by the transmission company. The design of a system of regional trading and pricing is under active scrutiny in Australia and the United States and awaits resolution.

There are no simple fixes for this kind of short-term phenomenon. Some observers argue that bilateral markets are less prone to such problems and that short-term bidding is used mostly for balancing. Both theory and the British experience serve as reminders that regulatory authorities will have to watch electric spot markets with vigilance and care.

POOLCO AND BILATERAL MODELS: CONVERGENCE AND DISCORD

There is now a widespread acknowledgment that there are strengths and weaknesses in both the poolco and the bilateral models.[34] For many, it is easier to conceptualize (and therefore have more confidence in) competitive outcomes in poolco markets because they are arguably easier for regulators to monitor to ensure that competition is working properly. In addition, many feel that this model provides the greatest assurance that all users—large or small, integrated or independent—will have access to power at the same price as the larger, more skillful, and possibly more powerful market participants.[35] Conversely, and perhaps based on the U.K. experience, others argue that poolcos *facilitate* collusion and market power.[36] Table 7.2 shows two authors' summary of the key aspects of these two models.

Bilateral model advocates emphasize that most competitive markets are not picture-perfect. They do not treat every buyer or seller to the identical outcome. In practice, a variety of aggregators and brokers can be expected to diversify product offerings and lubricate bilateral competition.[37] Researchers in California, simulating the behavior of a poolco-type dispatcher, have and demonstrated that the selection of the lowest-cost set of units in a poolco auction may be extremely complex or even indeterminate in some instances. If there are more than one least-cost solutions, the ISO has a basis for favoritism and generators have a basis for litigation if they are not selected.[38]

Another area of debate concerns the ability of the two market structures to promote product innovations.[39] Bilateralists note that poolco markets require that all trading must be in the form of one essentially regulated and homogeneous product. Every other

Table 7.2. Comparison of the Poolco and Bilateral Paradigms

Key Aspects

Model (in it purest form)	Generation Competition	Independent System Operation (ISO)	Transmission Ownership	Transmission Constraint Impacts	Contracts for Power
Poolco	Multiple sellers competing to serve load with pool acting as central buyer of and seller of all power	ISO has overall control of generation as well as transmission to keep supply-demand balance and maintains system security	Multiple owners have no control of transmission operations	Price paid by buyers reflects expenses associated with relieving constraints; determined ex post	Cannot accommodate any contracts that involve physical power flows because all flows are to/from pool; all contracts can be purely of a financial nature
Bilateral	Multiple sellers and multiple buyers who can mutually determine price, terms and conditions of each transaction	ISO has overall control of generation as well as transmission to keep supply-demand balance and maintains system security	Same as above	Various proposals to relieve constraints	Allows the undertaking of any contracts involving physical flows agreed upon by seller and buyer

Source: Gross and Balu (1996)

212

product in the power marketplace must be purchased in the form of a financial trans-action (such as a CFD) or fashioned from pool purchases. This, bilateralists argue, stifles the creative juices that markets thrive on. Following airline deregulation, for example, the industry developed many new products and innovations that had not been foreseen under regulation: hub-and-spoke scheduling, no-frills carriers, paperless tickets, frequent flyer clubs, and so on. Had airline deregulators enforced the types of airlines, flights, or scheduling permitted after deregulation, these innovations might never have developed.

Poolco advocates respond to these concerns by pointing out that there is always a trade-off between the value sellers and buyers derive from standardizing products and the diversity produced by competition. Many industries voluntarily adopt standards or prac-tices to facilitate customer purchases—from the familiar underwriters laboratory seal to the Energy Star label awarded by the Environmental Protection Agency. Many industries rely on very standardized building-blocks to fashion highly-customized products. New homes, for example, may be highly-individualized, but virtually every board, nail, pipe, and fixture is likely to meet one standard or another. In short, the right level of stan-dard-setting and homogeneity can promote competition, not stifle it.[40]

Other points raised by bilateralists are the extensive disruptions and transactions costs that are involved in establishing a poolco model in the United States. It is true that the United Kingdom designed its poolco from scratch without much fear of stockholder lit-igation or approvals from state and federal regulators. The establishment of a poolco in California has proven to be a task of almost gargantuan proportions. But poolco propo-nents have also noted that there is ample U.S. precedent for the model (stemming from traditional tight pools),[41] and others have noted that the establishment of an ISO is more difficult in a bilateral world.[42]

Bilateralists, in turn, respond by noting that not enough is known about the future evo-lution of the industry to impose a rigid procurement framework on it, nor have the skills to wisely regulate this framework been developed. They stress that the essence of com-petition is the decentralization of marketplace authority—the antithesis of a central authority conducting hourly auctions.[43] They further stress that communication and computer technology is at the point where bilateral traders can literally trade power in real time, with the system operator's role limited to maintaining a computer program that automatically determines whether a trade can be scheduled, the necessary ancillary services required for the trade, and so on. Were such a system to evolve, bilateralists argue, regulation would have a much smaller role and the scope for new services on the supply and demand side would expand market efficiency greatly.[44]

Another point of disagreement is the difficulty and delay that would be involved in estab-lishing a poolco market. As the present regime is largely bilateral, it is arguably true that establishing poolcos would require substantial changes in industry practice and regula-tion. The negotiation and approval of a lease between the transco and the ISO could take substantial effort.

This argument is certainly supported by California's experience as it tries to establish a statewide poolco. That effort, now in its third year, has resulted in unprecedented controversy requiring state legislative intervention for resolution. Simply establishing the computer software required to select bids and operate the pool will cost a quarter of a billion dollars.[45]

Because there is some truth to all of these points, policy makers are searching for ways to combine the diversity of bilateral contracting with the comforting transparency of poolcos. The richness of the debate over market institutions and governance, product homogeneity, and the importance of fairness and trust emphasizes that this is an extraordinarily difficult matter. The extent to which any of the proposed solutions work may depend on subtle questions of transactions costs (some in new areas of law and economics) as well as the effectiveness of competition and regulation in areas where there is little experience with either.[46]

It is far too early to determine the course of future events, but three directions are evident.

1. If tensions over the equitable allocation of the costs of transmission escalate, and regulators find it difficult to referee the bilateral market, the momentum will shift toward poolcos.

2. Conversely, if restructuring continues to advance in communications and control methods for power networks, and this enables valuable new bilateral trading regimes. the marketplace will likely evolve in this direction.

3. If neither direction gains favor, the industry is likely to continue down a complex hybrid path.

Appendix 7A

Electric Restructuring in the United Kingdom

Prior to 1990, a single, state-owned utility owned and operated the electric system of England and Wales. In March, 1991, Parliament divided and privatized the industry, setting up competition in some segments of the industry and regulation in others. Figure 7A-1 shows the basic structure of the industry following privatization.

MAJOR ELEMENTS OF THE NEW INDUSTRY:

Competitive Generators. Of the nation's approximately 70 generating plants, approximately 35 were privatized within a single, investor-owned generating company ("National Power"). The remaining non-nuclear plants were privatized in a second generating company ("PowerGen"). The nation's nuclear plants were kept in public ownership until recently, when they were privatized as well.

Figure 7A-1. Changes in the Structure of Electricity Supply in England and Wales- 1990

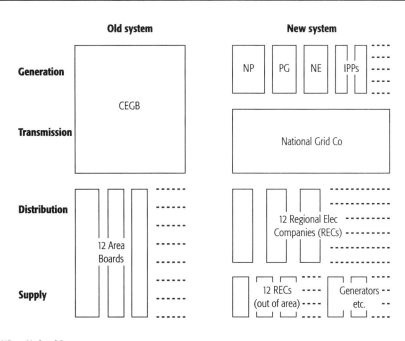

Notes: NP = National Power
     PG = PowerGen
     NE = Nuclear Electric (state-owned)
     IPPs = Independent Power Project (all part-owned by RECs)

Source: Mackerron (1995)

In addition to these three large incumbents, independent power producers were allowed to enter the market, subject to the technical rules of connection to the grid and siting procedures. Prices and profits of all generators were determined by their payments from the pool.

The National Grid Company. The nation's transmission system (National Grid Company, NGC) was spun off into a corporation that was owned collectively by the regional distribution companies. Recently, this was privatized into an investor-owned company as well.

The Pool. The pool is a mutual selling and buying club that conducts auctions for electricity supply for each half-hour of the following day. All physical purchases of power occur through the pool; CFDs, futures contracts, and other financial agreements do not involve the pool.

The Regional Electricity Companies (RECs). The distribution utilities were privatized into 12 regulated, franchised regional IOUs.

Regulation. In place of a U.S.-style regulatory commission, Parliament chose a single individual known as the Regulator with a staff of approximately 100 professional assistants.[47] The Regulator was charged with ensuring that the pool mechanism created effective competition between generators, constraining prices to competitive levels, monitoring the operation of the pool and the transmission company to ensure nondiscrimination and efficient transmission policies; and regulating NGC and the twelve RECs.

Regulation of the RECs and NGC did not employ cost-of-service regulation as in the United States, but rather "price-cap regulation." This regulatory approach sets a maximum average price for electricity but does not control the profits of the REC; if costs can be reduced without raising prices above the cap, all such added profits become the REC's. The maximum price is allowed to rise with the rate of inflation ("Retail Price Index", or RPI), less a discount for productivity improvement known as X, giving this form of regulation the common name "RPI minus X."

In addition to his ability to reset prices under the price cap mechanism, the Regulator enforces the terms of a license, similar to the U.S. concept of a franchise, for NGC and the RECs. The license includes such provisions as nondiscrimination, an obligation to serve customers who remain with the franchise, and the maintenance of quality standards.

Retail Choice and Franchise Customers. Upon privatization, retail customers with peak demands of 1 MW or more were allowed to purchase from a competitive supplier other than their REC and/or enter into contracts-for-difference as price guarantees. As of 1994, customers with demands larger than 100 kW were allowed to shop, and by 1998 RECs are required to allow retail competition for all customers.[48] As of 1995, approximately one-half the largest customer group and 42% of the 100kW+ group were no longer purchasing from their franchise REC.

Pool Prices, Market Power, and Gaming. All generators submit the availability of their plants and their bids to the pool. The pool selects and announces the winners and the pool's purchase price for each half-hour. NGC purchases interconnected operating or "ancillary" services from generators, such as reactive power and load balancing. The cost of these services, known as "uplift," is added each half hour to the pool purchase price for buyers. The pool also adds a capacity component to the sales price based on a formula that tracks the balance of supply and demand.

The existence of three very large bidders and only a handful of much smaller independent plants has raised questions as to whether the market for generation is truly competitive. The U.K. regulator and researchers have found that the large bidders were able to influence the winning pool price by withholding capacity or otherwise altering their bid strategies. For example, one provision of pool rules provides that plants the gridco must dispatch due to transmission constraints are paid their bid price irrespective of the winning pool price. When the owners of these "constrained on" plants discovered this, they began bidding unusually high prices.[49]

The questionable degree of competition and the evolution of market forces has prompted the Regulator to examine and alter pool rules on several occasions. In 1994, the Regulator responded to complaints by RECs concerning increasingly high and volatile pool prices by imposing an arbitrary pool price cap and requiring National Power and PowerGen to divest themselves of a number of power plants, deconcentrating the market.[50]

New Generation. Restructuring has made it possible for independent, private, combined-cycle gas generating plants to obtain licenses and sell into the pool. Approximately 8,500 MW of new capacity of this type has been added. In order to obtain construction financing, essentially all these plants have signed 15-year CFDs with RECs, guaranteeing their sales price as long as they generate into the pool. Seeking even greater assurance of supply and price, many RECs have acquired ownership shares of these "independents."[51] It is estimated that 75 to 90% of all power sold in the pool is governed by long-term contracts.[52]

Problems with Price Cap Regulation. The initial terms of the price caps, set at the time of privatization, necessarily forecasted the ability of NGC and the RECs to reduce costs. Over time it became apparent that the Regulator underestimated the ability of these firms to reduce costs, enabling these firms to earn substantial profits. As consumers and government officials have become increasingly concerned that prices are not dropping as fast as profits have increased, the Regulator has intensified his regulatory scrutiny of prices and costs.[53]

The apparently lucrative niche occupied by the RECs and other factors have made them prominent takeover targets. Successful and unsuccessful bids have come from U.S. utilities, conglomerates, and other companies in the U.K. power sectors.

Retail Price Comparisons. A number of researchers have compared the prices produced by the British system to prices before privatization, prices in the rest of Europe over the same periods, and estimates of what prices would have been in the United Kingdom absent privatization. On the one hand, this research shows that restructuring unquestionably increased the efficiency of the generating sector. Labor productivity in this sector has grown much, much faster than productivity in the British economy as a whole.[54]

On the other hand, the combination of pool-related competitive issues and transco and REC regulation have yielded retail prices that are lower than they would have been without privatization, but higher than comparable prices under other alternatives in the United Kingdom, as well as higher than comparable prices in other EC countries. Summarizing the post-restructuring balance of consumer and investor benefits, Sussex University researcher Gordon McKerron wrote: "To date, the benefits to consumers in price terms has been limited, and the chief beneficiaries have undoubtedly been shareholders and senior executives in the [generation, transmission, and distribution] companies...The excessive profitability of the industry is creating stress in legitimacy and could, if sustained much longer, lead to a significant reshaping of the regulatory system so that it will be seen more clearly to represent the interests of consumers."[55]

Impact on Public Interest Objectives. Prior to privatization, the system's commitment to the protection of low-income customers and universal service was a matter of political control. The same was true of the system's activities in energy efficiency, fuel diversity, environmental protection, and research and development. These aspects of electric utilities are intended to serve the public interest, and frequently cannot be justified to private owners on the basis of their impact on utility profitability.

The terms of the public licenses given to suppliers, the gridco, and the RECs call on them to maintain commitments to public interest objectives. However, most observers believe that there has been a significant reduction in the level of effort devoted to these objectives. One observer wrote, "under the current system, the prospects for energy efficiency are dismal" and the prospects for cleaner power plants "likewise dismal."[56] A 1992 Parliamentary Inquiry into the consequences of privatization found insufficient performance against every public interest objective, with particular criticisms of consumer protection, energy efficiency, and research and development.[57]

Chapter 7

NOTES

1 For example, a was willing to pay p^, but instead pays p*.

2 When estimated demand exceeds power bids submitted, system controllers must balance supply and demand by adding a capacity component to pool prices. (See Ruff (1996) p. 13–14 and the discussion on transmission governance and Chapter 9. This introduces importance governance and long-run issues.)

3 In practice, most power plants cannot change their output so fast that they can bid any quantity they choose for each hour. Instead, they must bid a sequence of quantities that is technically feasible for their plant. This introduces mathematical complexities described in by researchers such as Johnson, Oren, and Svoboda (1996) p. 3.

4 Graves and Read (1996) remind us that, even in tight power pools, buyers and sellers do not all see the same price. Prices are still based on sellers' and buyers' costs.

5 Evans, 1996

6 See Fernando and Kleindorfer (1996) p. 13 and Hartman and Tabors (1995) p. 25 on this point. Aggregators play an especially important role under retail access. See Chapter 10 *infra*.

7 Hogan (1996), p. 27–28 discusses and illustrates CFDs in detail.

8 See Brennan, et al (1996) p. 55.

9 Fernando and Kleindorfer (1996) p. 29; Brennan, et al (1996) p. 54ff. and Hunt and Shuttleworth (1996) ch. 12.

10 Hogan, Hitt, and Schmitt (1996) p. 2.

11 Graves and Ilic (1996).

12 For an extremely detailed description of ISO alternatives, see the Pacific Northwest Utility Conference Committee (1996). Regarding ITAs, see Ashleyad Delgado (1996). For discussions of regional differences, see The Harvard Electricity Policy Project (1996).

13 Professor Joskow makes this point in his comments before the Technical Conference concerning ISOs, FERC Docket No. RM95–8-000, January 26, 1996.

14 See Dunn and Hibbard (1995).

15 See, for example, Harvard Electricity Policy Group (1996).

16 George Lobsery, "State Regulators Question Independence of System Operators." *Energy Daily,* October 23, 1996 p. 1.

17 Hogan, Hitt, and Schmidt (1996) p. 21.

18 The FERC appears to have authority to approve these agreements because they are a form of interstate transmission agreement.

19 Order 888, Mimes, p. 279–286.

20 These rate (or more properly, revenue) proceedings will be somewhat unfamiliar, as few proceedings have been devoted to transmission systems alone. If the procedures follow cost-of-service rate making, the main issues to be confronted include (a) an examination of out-of-pocket costs to ensure they are reasonable; (b) appropriate depreciation policies and rates; (c) the valuation and usefulness of the existing capital stock (the "used and useful" prudence issues); and (d) the appropriate level of allowed return on equity for transmission system assets. The latter has been examined in the Electricity Users Group (1996).

21 See the comments of one participant in the Harvard Electricity Policy Project (1996) p. 5.

22 See, for example, Fernando and Kleindorfer (1996), Gross and Balu (1996), and the Harvard Electricity Policy Project (1996).

23 See Box 7-1 *supra.*

24 Many transmission owners object to the term ancillary services because it might be inferred to mean services that are optional, rather than integral to the provision of transmission. I use the two terms interchangeably purely for rhetorical reasons.

25 When researchers at the U.S. Department of Energy's Oak Ridge National Laboratory compared the list of ancillary services utilities that had developed around the United States, they found a wide variation in nomenclature, definitions, and measurement techniques. See Kirby, Hirst, and Vancoevering (1995) ch. 5. In Order 888, the FERC noted that commentors proposed a wide variety of service ranging in number from a handful to 38 separate services.

26 Order 888, Mimeo, Section IV.D.

27 In Kirby and Hirst (1996), the authors examine a sample of 12 utilities' ancillary service costs. In the sample, the range of ancillary service costs varies over a range of +300%–100% of the mean values shown.

28 NERC has formed the Interconnected Operating Services Working Group (IOSWG), a task force composed of transmission owning utilities, distributor utilities, generators, and other market participants. This group and many other researchers and market participants are expected to report to the FERC by the end of 1996 with "technically, sound practices for the management of unbundled ancillary services." (Electric Power Research Institute, IOSWG Meeting Notice, August 1996).

29 The largest single barrier to transaction allocations involves the technical interdependence and non-linearities in the transmission grid as a whole. A number of researchers note that computers' ability to solve nonlinear system problems has expanded enormously, leaving a strong potential for useful work in this area.

30 The dialog concerning competitive ancillary services is the debate over vertical deintegration of the bulk power market in microcosm. While no one disagrees that unbundling and competitive procurement of non-natural-monopoly services is theoretically optimal, if there are economies of scope, or transactions costs between the ISO and the providers of ancillary services and the ISO outweigh the benefits of competition, then nothing has been gained by contracting out. Similarly, regulatory scrutiny of the competitive purchasing practices of the ISO may be easier and more effective than scrutiny of services produced in-house by the ISO, but it is also possible they may not. Finally, the cost characteristics of the provision of many ancillary services is not known, much less the complementarities between these services and other generation products. Attention must therefore be given to whether competition remains effective and independent of the ISO in each unbundled ancillary service marketplace.

As an example of the latter, many observers have noted that the need for reactive power is extremely location-specific. Reactive power cannot be economically shipped, it must be produced very near where it is needed. Thus far, it is difficult to see how a significant number of suppliers can compete to provide the necessary services.

Cost or Value of Unbundling Service(s)

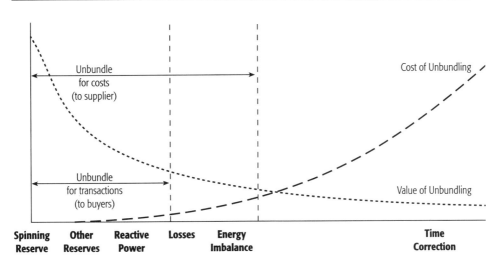

Hirst and Kirby (1995) develop a graphic that illustrates the tradeoffs in unbundling as follows:

This graphic shows that the value of unbundling (the greater efficiency of the marketplace and lower aggregate costs) outweighs the additional transactions and regulation cost for services such as reserves and voltage support (left side of picture). For services at the right side (only time correction is shown), unbundling costs more than it produces in added value. (Time correction is the periodic resynchronization of all area control centers to the same time reference; see Kirby, Hirst, and Vancoevering (1995) p. 24. See Kirschard Singh (1995) for pro-competitive IOS arguments.)

31 Gilbert and Kahn (1995) and Cicchette and Pepys (1996) review foreign resttructurings.

32 California, 1995 Assembly Bill 1890, August 31, 1996, codifying in large part the Final Decision of the California Public Utilities Commission in Docket R.94-04-031 and I.94-04.032.

33 The separation of bilateral contracts and winning poolco bids called for in the California Public Utility Commission's proposal has generated opposition from several researchers who argue that the separation will prevent lowest-cost dispatch of the system. There is much disagreement over the significance of the cost penalty, with many arguing that the value of ensuring that the largest utilities in the state do not foreclose bilateral contracting opportunities justifies suboptimal dispatch. Among much discussion of this topic, see the comments of the California Energy Commission before the California Public Utilities Commission, R.94-04-031 and I.94-04-032, May 28, 1996, the comments of Paul L. Joskow in the Technical Conference Concerning Independent System Operators and Reform of Power Pools Under the Federal Power Act, RM95-8-000, January 26, 1996, Hogan (1995c), Stoft (1996). For useful background, See Porter (1995) and Chandley (1994).

34 Among others, Hogan (1995 a), Cross and Balu (1996), Hartman and Tabors (1996) and Fernando and Kleindorfer (1996) discuss the pros and cons of the two models.

35 During the discussions leading to California restructuring, Southern California Edison's comments before the Public Utility Commission described their view of the equity benefits of pools:

A regional pool, says Edison, "allows the free market to operate in a structure that protects all sellers and customers equally. A regulated bilateral-contract-only regime disproportionately rewards large players that can exploit the system. Thus, a regional pool is the only method that assuredly provides transparency and efficiency, and that solves many of the serious concerns and mitigates many of the serious risks to California's customers, energy infrastructure and economy, raised by the implementation of direct access."

As reported in the *Quad Report*, Consumer Energy Council of America, September-October 1994, p. 5).

36 Brennan, et al, (1996), pg. 58

37 See Fernando and Kleindorfer (1996), p. 13. Hartnan and Tabors (1995, p. 25) elaborate on this point as follows:

> The advantages of the Bilateral Transactions arrangement all derive from the fact that additional players participate in the market, while the operational integrity of the system is insured by the ISO. Aggregators and repackagers will be significant players in the Bilateral Transactions market model. These entities will actively seek out transactions that will bring buyers and sellers (most likely as groups) into contact such that the individual preferences of both groups are uniquely met. The usual welfare improvements stemming from competition will result, as suggested by economic theory and as demonstrated by the restructuring experiences in the telecommunications, natural gas and transportation industries.

> This increase in the number and diversity of market participants induces two subsidiary advantages, one regulatory and one innovational. Because the Bilateral Transactions arrangement engenders greater competition through the entry of a significant number of new participants (generators and aggregators), far less regulatory oversight will be required in the steady state. Competition is always more effective in regulating economic behavior than is agency regulation. Likewise, we contend that the increased level of competition engendered by the Bilateral Transactions Model will lead to significant innovation in both the products and the services supplied in the industry. The experience in the restructured telecommunication industry is illustrious in this regard.

In 1994, British electric power regulators responded to pressure to allow bilateral physical contracting for power outside the pool by investigating the subject in a public inquiry. He concluded that "Perhaps the costs or risks for the immediate parties could be reduced by trading outside the pool. But as yet I have seen no firm evidence of significant gains that could be achieved in this respect". (Office of Electricity Regulation, 1994, p. 19).

38 The authors state their conclusion in more detail as follows:

> We have demonstrated both the volatility of "near optimal" scheduling outcomes for resources not base loaded, and the especially negative consequences of volatility for marginal resources (i.e., resources that frequently determine system marginal costs). Specifically, we have shown that variations in near optimal unit commitments that have negligible effect on total costs could have significant impact on the profitability of individual resources. Consequently an ISO charged with making efficient central unit commitment decisions is in a delicate position of having to allocate profits equitably among resource owners with no economic rationale to back the decision. These effects are inherent when attempting to optimize unit commitment from the perspective of a central operator, because of the near-degeneracy of the unit commitment problem and the presence of many near-optimal solutions.

> The results raise serious questions regarding the feasibility of proper mechanisms to oversee the efficiency and equity of a mandatory centrally committed and dispatched pool. We suggest that centralized scheduling by a mandatory power pool, using models appropriate for solving the integrated and regulated utility's scheduling problem, may be perceived by suppliers as unnecessarily volatile and even inequitable, and hence in the long run yield schedules that do not minimize costs. In particular, our results highlight potential pitfalls in central management of dispatch constraints specified by bidders (Johnson, Oren, and Svoboda, 1996).

> The results of these three researchers have not been independently verified, nor have the problems they observe been a major issue in the U.K. pool. Nonetheless, they pose an interesting counterpoint to the widely held belief that poolco outcomes are transparent and efficient.

39 For additional discussion, see Brennan et al (1996) p. 56ff.

40 Graves and Read (1996) p. 8 note that short-term natural gas pipeline capacity trading did not work well until the market adopted standard terms and conditions. This is an example of pro-competitive standardization of trading rules.

41 See Hogan (1995c) and Joskow (1996).

42 Ilic, *et al* (1996).

43 For an elegant and forceful exposition, see McGuire (1996).

44 University of California Electrical and Computer engineering professors Felex Wu and Pravin Vaaiya construct a theoretical proof that an ISO with a computer program that provides system information in real time can enable multilateral trails that converge on an efficient solution. "It is also shown" the authors write, that "the existing communications, computing and control infrastructure is adequate to support [their] proposed model." Wu and Varaiya (1996). Similar arguments are made by M.I.T. electrical engineering faculty Maria Ilic, Eric Allen, Robert Cordew, and Chien-Ning Ya (1996).

45 Newspage, October 16, 1996.

46 Making reference to the costs of litigation and enforcement—two classics transactions costs—the U.K. electric regulator expressed discomfort with bilateral trading in another country that has restructured its power system, Norway:

> Experience in Norway has demonstrated that despatch driven by the requirements of bilateral contracts between generators and customers is not straightforward and that it can give rise to a large number of disputes between parties. Where contract volumes are disputed, the Norwegian Pool has found it difficult to complete the settlement process in a timely way. Considerable thought would need to be given to formulating a satisfactory mechanism for the resolving of such problems were TOP to be allowed here. (Office of Electricity Regulation, 1994, p. 19)

Source: Adapted from FERC Order 888, Kirby and Hirst (1996), and Hirst and Kirby (1996).

47 MacKerron (1996).

48 Recent press reports from the United Kingdom state that the RECs are claiming that it will be impossible to install the metering, communications, and control systems required to effectuate retail choice for the 22 million small customers who do not yet have it. Simon Holborton, "Regulator Slams Power Companies on Competition," *Financial Times* July 15, 1996.

49 Newberry (1995a,b), Newberry and Pollit (1996), Green and Newberry (1992), Energy Committee (1992), Wofram(1995), Yarrow (1996) and MacKerron (1996).

50 Evans (1996), p. 5, and Wibberly (1996) chronicle the main regulatory changes since privatization. Also see Hunt and Shuttleworth (1996) chapter 12 Appendix B and Brower, Thomas, and Mitchell (1996).

51 Newberry (1995a) p. 13.

52 Brick (1994); Newberry (1995a).

53 McKerron (1995, p. 6; 1996 p. 6)

54 Newberry and Pollit (1996)

55 McKerron (1995, p. 6; 1996 p. 6). Similar, stronger conclusions are voiced by Yarrow (1995). Newberry and Pollitt (1996) explore a number of counter factual scenarios and aspects of restructuring. Recognizing that relatively few gains have been seen by consumers in the form of lower prices, they nonetheless conclude that the net impact of privatization was significantly positive.

56 Brick (1994) p. 4 Also see Owen (1994).

57 Energy Committee (1992) paragraphs 140ff

Chapter 8

Market Performance and Regulation in the Long Run

However well or poorly restructuring works in the short-run, there will come a time when new transmission lines will be needed or generators will seek financing to build more capacity. The transmission system must be operated, expanded, and governed to provide for long-run economic efficiency and reliable service in the presence of competitive generators.

Many industry participants admit that creating and governing an efficient transmission governance and transmission system is, in the words of economist Larry Ruff, "probably the most difficult part of creating an electric market."[1] In spite of this acknowledgment, far less attention has been paid to this issue in most restructuring research than has been paid to the questions of short-run market mechanics. Article after article on restructured power markets examines short-term efficiency in great detail and then addresses the long-term simply by noting that it will be essential for the transmission system to expand efficiently.[2] As economist William R. Hughes and transmission engineer Richard Felak concluded after an extensive review, "Pricing for long-run efficiency is the last frontier of transmission pricing. It is time for the industry and the theorists to give it much more attention than it has received to date."[3]

To focus on the long run, it must be assumed that the system will continue to operate on a daily basis with adequate efficiency and reliability. Whatever the particulars of the marketplace, it is also assumed that spot prices are effectively available and visible by location and time across the grid. Market participants are able to watch the trend in spot prices, as well as the price of futures contracts, from publicly available sources. The terms of CFDs and other market arrangements between competitive sellers and buyers will be confidential, but sellers' offers will be widely self-publicized, and the commercial information services available to buyers and sellers in many other commodity markets will seek to serve electric market participants as well.[4]

NODES, ZONES, AND THE SIGNAL TO BUILD GENERATION

As demand starts to regularly outstrip supply in any one portion of the system, locational prices in the area will begin to rise. The network operator will find it increasingly difficult to ship power from available surplus generators into the high-demand region. If the price difference is due to inadequate transmission into the node and/or inadequate generation near the node, the phenomenon will be local; if it is due to a regional shortage of generation, high prices will also be observed at other nodes.

The accuracy, timeliness, and level of geographic detail in the published locational price information is of obvious importance to prospective generators and greatly affects the mechanics of operating and policing ISOs and transcos. Accuracy and timeliness have been addressed largely via specific, industry-wide mandates for real-time information posting on electronic bulletin boards.[5] The issue of geographic detail, however, has touched off a vigorous debate between analysts who argue that every major node (generator, transmission line junction, or substation) should have a spot price and those who argue that pricing ought to be simplified into zones that are bounded by constrained interfaces.[6]

Nodal pricing advocates point out a number of shortcomings with zonal pricing. First, they argue that zonal prices may not send accurate price signals either within an unconstrained zone or in other zones. For example, even without congestion in a zone, new intrazonal flows can affect available capacity across the borders of the zone and within other zones.[7] Absent a very complex pricing scheme, however, a zonal price will not reflect these extra-zonal costs.

Within the zone, the efficiency of zonal pricing depends on the way in which the generation market is structured and the ability to game or exercise market power over the intrazonal dispatch. If the market is not set up carefully, prices may become more volatile than necessary, generators may be able to make excess profits by gaming the system more easily, and opportunities to reduce costs by allowing customers to reduce their demand will be attenuated.[8] According to nodalists, all of these possibilities mean that regulators must play a larger role in policing a zonal market and its ISO.

Nodalists also point out that investment planning will become more difficult with zonal pricing. The natural signal to create more transmission between two nodes is repeated congestion between the two points. Transmission planning in the presence of zones requires a more complex translation between present patterns of congestion and new addition options. This, too, will increase the need for strong regulation, nodalists argue. Similar arguments apply to the related concept of a transmission congestion contract (discussed at the end of this chapter) and the need to create new zones.[9]

In defense of zones, others have argued that the added complexity of administering hundreds of different nodal prices outweighs the efficiency disadvantages of zones.[10] Zonal advocates note that the pool in the United Kingdom and most other real-world systems have rejected nodal pricing as being too complex—although the issue remains contentious and unsettled six years after the establishment of the U.K. national grid.[11] If, for example, zonal pricing encouraged too many new generators to enter within a low-price zone, transmission constraints would likely develop inside the zone, leading to the creation of a new zone.

Zone advocates argue that the main areas of congestion in the United States are and will continue to be across interfaces that cannot be expanded. Hence, the main purpose of zonal differences would be to encourage generators to locate anywhere within the boundaries of the high-priced zones and to focus the attention of the transmission system on relieving these constraints, where possible.

Finally, zonalists make the pragmatic observation that revenue reconciliation (to ensure that transmission systems earn their required revenues), the recovery of stranded costs, and other regulatory functions will be tied—in an accounting sense—to the present boundaries of transco systems. Zonal pricing schemes lend themselves more easily to accounting for these regulatory functions than do nodes, which frequently lie on the border between two systems.[12] Although it is difficult to gauge the severity of these concerns, it is wise not to underestimate the attractiveness of an approach that regulators can more easily understand and implement.

GENERATION CONTRACTS AND MERCHANT PLANTS

As noted in Chapter 7, power market spot prices are inherently quite volatile. When demand begins to constrain supply, volatility increases more than the average pool price because supply constraints usually do not apply during the low demand parts of the day. Hence, it is only in the few high-demand hours of the day that total generation is short and prices must increase quickly and substantially to reduce demand. The total hours during which price rises above short-term marginal cost to levels that equal or exceed long-term marginal cost will be scattered and unpredictable at first.

This behavior has made it largely impossible thus far to justify the financing of new generating units on the basis of the expected future stream of spot price payments alone. In the United Kingdom, for example, the only sellers offering power not under long-term contract are the two large, multi-plant companies that were formed by the breakup of the old system.[13] Even when plants are guaranteed the right to sell at spot prices, financiers in the United States and the United Kingdom have required that the plants have long-term contracts (CFDs) at specified prices prior to obtaining their construction loans.[14] The market signal to new generators that therefore matters most is the willingness of buyers to sign long-term agreements at prices that enable the profitable construction and operation of a new plant. Ironically, this model is strikingly similar to PURPA-era negotiations in the United States between QFs and utilities that were willing to pay capacity payments.

It is worth considering for a moment why financiers might not be willing to take on the risk that future spot price payments over the lifetime of the plant will repay their investment with a profit. The obvious reason is that future capacity payments will depend on the the total capacity of available plants relative to total demand in each hour. No one plant builder can control the decisions of an unregulated group of builders. If too many builders build, there is excess capacity and the pool, finding no need to ration demand, pays sellers only short-term marginal costs (i.e., the cost of fuel and other variable expenses, with no capacity payments component). The only way out of this Catch-22 is to lock in what are effectively guaranteed capacity payments via a long-term contract with buyers.[15]

Buyers' motivations for entering into contracts are the mirror image of sellers'. Sellers fear their inability to control a supply glut; buyers fear a marketplace in which too little energy is available, and prices rise—even, perhaps, only temporarily. They may also observe that, when many other buyers lock in supplies with long-term contracts, supplies may not be available at all during times of shortage, even at high prices. Fears of this nature prompt many wholesale buyers to continue to enter into long-term contracts in the natural gas and oil markets, although not nearly as consistently as they did prior to the emergence of strong spot and futures markets.[16] Undoubtedly these buyers are seeking to avoid what economist Phillip Verleger, Jr. noted during a period of very high oil spot prices: "Buyers who need stocks right now are having to pay an arm and a leg and probably promise their first born" to guarantee delivery.[17] Even without price spikes or availability concerns, price certainty allows buyers to do better planning and budgeting, and insulates them from inconvenient or financially destabilizing cash demands.

The advantages of long-term contracts to sellers and buyers make it likely that a long-term market for contracts—or capacity or price guarantees—will exist for some time.[18] As is now the case in the United Kingdom, and as was the case in the United States following PURPA, new plant decisions will be driven primarily by the willingness of buyers to sign contracts.[19]

There are clear signs on the horizon that some generation firms may, in time, raise the capital to build plants that are not largely contracted for. Plants of this rather different nature are known as *merchant plants*, and are a rapidly emerging area of interest. It is still too soon to tell whether merchant plants will become a permanent feature on the generation landscape, but, in any event, it is unlikely that they will play a very large role in most markets for quite some time, if ever. The enduring prevalence of contract plants makes it likely that access to transmission at prices that are certain in addition to as low as possible will remain an abiding concern in the generation industry.[20]

GRID EXPANSION: ANALYTICAL AND PLANNING ISSUES

There are two ways to look at the underlying forces for grid expansion. One is simply to view it as something driven largely by the location and size decisions of new or replacement generators and load growth.[21] The second view springs from the observation that, in a spot pricing world, the value of transmission is simply the value of generation price differentials between two nodes or zones.[22] These views are both useful, but neither is sufficient to illuminate the analytical, planning, and economic challenges that transmission systems face under restructuring.

In the abstract, spot price differentials seem to be an economically ideal signal of the need for transmission. However, the ideal must reflect the practical considerations that alert price and affect transmission systems in complex ways. For example, spot prices may be high but little different in two adjacent, congested parts of the system.

In practice, the impetus for upgrades may transition back and forth between these two modes of expansion. In places and at times when generation differentials persist (suggesting that generating capacity is consistently cheaper in one place than another but cannot be sold sell to the cheaper area), transmission expansion is merited. When the differentials are not consistent over time and space, local matching of demand and supply will be more common, and grid expansion will not be driven by generation price differentials so much as by the need to reinforce regional systems to provide for reliability in the wake of growth.[23]

The use of spot prices alone as a signal to construct capacity is unlikely to yield a system with capacity sufficient to meet adequacy standards described in Chapter 3. A scarcity of capacity drives capacity price higher, which is in the interest of power plant owners. However, scarce capacity (relative to unpredictable peak loads) hightens the chance that reliability will suffer. Because the owners of power plants bear a very small portion of the cost of power outages, they do not face the correct incentives to build as much capacity as society needs.[24] Most regulatory authories have recognized this by requiring poolco operators or other network authorities to pay generators a capacity payment over and above spot prices designed to induce "adequacy."[25]

Spot prices will also be a bit difficult to use when examining the long-term tradeoffs between generation and transmission investments. To justify a large new transmission line on the basis of time-varying locational price differences, it will be necessary to expect the differences to persist over many years. Indeed, investors' unwillingness to fund a transmission project on the basis of spot price differences should be much like their unwillingness to finance generators without long-term contracts, but worse. First, there are two varying spot prices, one at each end of the prospective upgrades, instead of one spot price earned by the generator. Second, typical transmission lines last 40 years, whereas it is difficult to forecast spot prices even a few months out and easy to build a new generator in a handful of years.

The concept of transmission expansion driven by generator demands encounters different sorts of conceptual and mechanical difficulties. The transco/ISO combination will have no control over the demand for additional transmission service from new generators. this demand will follow generators finding customers to sign contracts. As discussed above, this is likely to follow something of a boom-bust cycle: When volatility trends up, sellers and buyers will rush to sign contracts; when there are capacity surpluses, no one will want to build.

This is just what competitors are expected to do, but it places an extraordinary planning and capital budgeting burden on the transmission system. Transmission planners must be prepared to face a stream of changes in generator decisions which makes systematic planning and budgeting vexing and possibly chaotic.

Difficulties in planning and building the system will translate into capital budgeting and then transmission ratemaking uncertainties and changes. Because the cost of transmission upgrades must be recovered from all users, they become part of transmission prices one way or another.

Unpredictable changes in long-term transmission prices are bitterly disliked by generators because they make long-term contracts more difficult to write and they greatly elevate business risk. This, in turn, makes new investment capital for generators more costly and raises the price of generation. The independent generators' transmission pricing lament, "Just tell me what it will cost," will perhaps be answered more routinely and fairly in the new era—but possibly with an even less desirable answer than before.

A final consideration applies to both views of transmission expansion. In Chapter 4, it was noted that many upgrades to the transmission system provide reliability benefits that are shared by customers and generators stretching over enormous regions. Almost every significant addition to the transmission system—even something as exclusive as a single line traveling to just one plant—usually provides *some* benefit to the system as a whole. With the exception of this sort of line, however, it remains very difficult to determine how to attribute specific reinforcements of the larger system to specific bilateral transactions. The problem is exacerbated by the fact that most transmission upgrades are economical if done in large increments—perhaps 1,000 MW—that are three or four times the size of the average new power plant. If only two new plants trigger a need for the upgrade, who should pay for the extra, initially dormant part of the upgrade?

Some idea of the magnitude of these effects can be gained by looking at the United Kingdom: The National Grid Company (NGC) has divided its capital outlays into three categories, "load-related," "non-load related," and "transmission services." In simple terms, these three categories correspond to investments attributable to accommodating new generation, replacing and reinforcing the existing system, and investments to improve and optimize system reliability and efficiency, respectively. It is particularly in the third category that difficult-to-attribute new projects are undertaken. NGC's forecasts for these three outlays during the 1996–2000 period are shown in Table 8.1. These figures indicate that less than half of the capital costs of maintaining the system can be allocated to new generation as a whole. Moreover, only a small proportion of the load-related outlays is attributed to individual new generation transactions.

Table 8.1. Estimated Grid Capital Outlays in the United Kingdom

Grid Capital Category	NGC Estimated Outlay, 1996–2000
"Load-Related"	£ 560
"Non-Load-Related"	£ 507
"Transmission Services"	£ 131

Source: Offer (1996) Table 4

As seen in Chapter 4, vertical integration naturally addresses this problem. It does not matter whether the investment is charged to generators or transmitters because they are the same company. However, as discussed in Chapter 6, disputes over transmission costs began to arise when QFs and later IPPs operating under PURPA required transmission expansion by IOUs and publics. Now, the long-term economics of generation and transmission expansion have not changed; the primary difference is that the transmission system is entirely independent of all generators and is obligated to expand the system efficiently.

Evidence of all of these tensions and problems is clearly visible in the United Kingdom, where the transmission system (NGC) has been a wholly independent entity since 1990 and is responsible for planning and expansion. (See Chapter 7 for more details about restructuring in the United Kingdom.) In its early years, NGC was deluged with numerous requests for connection, which it planned and budgeted to accommodate. This speculative generation rush, triggered by excitement about the nation's massive restructuring, ultimately proved to be short lived. As a result, NGC's *actual* construction expenditure during the years 1993 to 94 to 1996 to 97 was £862 million, which was almost 50% less than its original projection of £1,620 million.[26]

Since transmission rates were based on required revenues, including capital outlays, generators and power buyers were understandably not pleased to discover that they collectively had paid hundreds of millions of pounds more than would have been necessary had there been perfect grid construction foresight. Ironically, this episode taught NGC that the market signals provided by competitive generators—ie., requests for connection—were *not* to be trusted as a basis for system planning and budgeting. Instead, and with the full approval of its regulator, the NGC now performs its own independent forecasts of the generation market and likely connection requests and costs.[27]

NGC and its regulator have also struggled to determine how much of the costs of transmission upgrades are attributable to specific requests for service, as opposed to adding the costs to the rates of all transmission system users. In U.K. electric parlance, the dispute includes who to charge for the cost of a new line from the existing system to the new generator (the "spur") or the interconnection at the power plant's border (the "substation"). NGC has generally proposed that almost all of the costs of connection should be shared by all users, with only the costs of the connection itself borne by the generator. The Regulator has asked NGC to change this policy to include charging each user for its radial lines as well.

The magnitude of the charges involved illustrates the low proportion of capital outlays that are charged to individual transactions, even among "load-related charges." Of NGC's total revenues of £1.1 billion, less than £110 million is charged to individual generators. If the cost of spurs were to be included, this would raise direct charges only by about £30 million.[28]

When considered in the abstract, the traditional, integrated, iterative study of generation and transmission system expansion options described in Chapter 4 would clearly seem unsuited to a world of unregulated generation.[29] After all, this is a centralized planning model that requires a single regulated monopoly to make investment decisions. Captive customers will pay for the mistakes of the regulated firm and its regulators, as well as for their successes.

However old-fashioned this may seem, all indications are that this centralized, regulated analysis and planning function will continue for quite some time. The extensiveness of common and joint capital costs in the transmission system and the need to make generation-transmission tradeoffs will leave the system's owners, planners, and regulators together in a regulatory fishbowl that is quite similar to traditional planning processes.[30] Short-run concerns about ISO independence, discussed in Chapter 7, complement these long-term drivers for continued regulation.[31]

GRID EXPANSION: ROLES, INCENTIVES, AND REGULATORY BODIES

It is now clear that the timing and extent of investment to accommodate generators, alleviate congestion, and reinforce the system will depend critically on whomever bears these responsibilities and their regulator.[32] The long-run objectives of transmission governance are clear. In order for competition to work in the long run, the grid must: (a) expand to accommodate all economically rational new generators, (b) provide for robust long-term generation competition, (c) invest adequately in transmission upgrades that maintain reliability, (d) appropriately trade off investments in more generation with more transmission; and (e) efficiently plan, budget, finance, and construct the grid.

Luckily for the British, these objectives could be applied to an independent, nationwide grid company regulated by the national government. NGC holds a license (analogous to a U.S. franchise) that allows it to own and operate the grid on the condition that it expand the grid efficiently and carry out the planning and administrative duties outlined above. The U.K.'s electricity regulatory agency monitors and approves specific rules, procedures, and transmission rates for NGC. As a result, the United Kingdom's electric utility industry is blessed with a relatively simple and straightforward division of responsibilities.

In the United States, federal authority to *require* transmitting utilities to enlarge is extremely limited (see Chapters 4 and 6). To a significant extent, it is subject to *de facto* veto by state public service commissions and/or state or local siting and permitting authorities. FERC Order 888 requires transmitting utilities to give all users of the system treatment that is comparable to the treatment given to the owner of the system. However, the Order provides no guidance as to what to do when the system cannot accommodate any new transactions and the transmitting utility is unwilling or unable to build more transmission.[33]

FERC did indicate that it will give greater deference to open access transmission filings made by bona fide Regional Transmission Groups (RTGs, see Chapter 6). Because the FERC has made transmission expansion planning a condition of becoming an RTG, the FERC cleverly created an incentive (in the form of greater deference) to voluntary associations of transmission owners that pledge to do "planning to make expansions that are economically justified from a regional perspective."[34] The FERC also made involvement of state planning and siting authorities a condition of RTG approval.

One group's recent request for approval for status as an RTG illustrates the FERC's approach to promoting efficient grid expansion without mandates. In 1995, The Western Regional Transmission Association (WRTA) applied to the FERC requesting status as an RTG. The FERC approved this application only on the condition that WRTA "provide for the development of a single coordinated regional transmission plan, and that [WRTA] members must 'commit to adopt and promote this...plan before individual state regulatory and siting authorities."[35] The plan had to include the long-run needs of all transmission users, whether or not they were WRTA members, effectively requiring expansion for a bona-fide long-term demander. (Figure 8-1 is a diagram of WRTA's proposed regional transmission planning process.)

Figure 8-1. WRTA Regional Transmission Planning Process

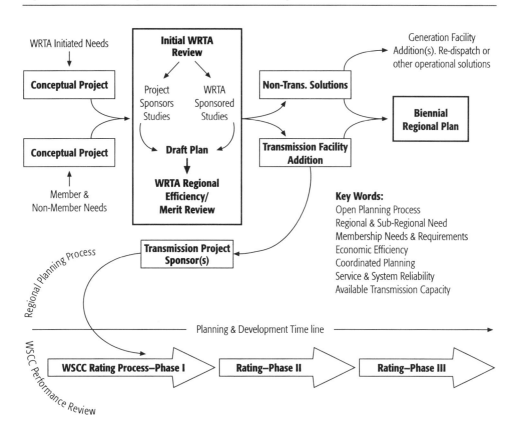

The WRTA responded to FERC's conditional acceptance with a compliance filing that the FERC accepted, thus granting RTG status to WRTA. In its Order granting acceptance, the FERC noted that WRTA "does not require that members *commit* to the RTG's coordinated regional plan." The [WRTA's compliance] filing argued that FERC's requirement that an RTG "adopt and promote" a consensus regional plan "should not be read to require a member to delegate to WRTA the ultimate authority to determine what investments the member must make. Thus, while a member has an obligation to provide service, WRTA cannot require a member to implement a specific proposal."[36]

This example suggests that FERC's RTG requirements and other policies do not entirely substitute for an obligation to site and invest in more transmission.[37] Coupled with the widespread acknowledgement that modern electric markets are not bound by state borders, this situation is prompting a re-examination of the authority for siting and transmission expansion.[38] The discussion of this issue in Chapter 4 showed that siting authority has been a matter of contention for many decades. However, electric restructuring has intensified the interest in, and the need for, regionwide transmission planning and governance.[39]

It is too early to tell whether future federal legislation will shift the balance of authority on grid expansion to the federal government or leave it where it is. It is also possible that all affected parties, including the states, will voluntarily combine state authority into regional entities. At least one state public service commissioner, Chairman Richard Cowart of the Vermont Public Service Board, has proposed that a regional regulatory body be formed with one regulator appointed by the governor of each state in the region.[40] A variety of industry groups, researchers, and others have urged that regional regulation of transmission be seriously examined.

It is probably obvious that the mechanics of establishing a lasting regional regulatory body are daunting. The body must establish a charter, administrative and regulatory procedures, pricing and expansion standards, and internal decisionmaking rules, not to mention the mechanics of funding a significant regulatory edifice. Daunting as these problems are, and the more severe and lasting the tensions restructuring place on state and federal regulators, the more likely it is that federal or voluntary action will redraw the lines of authority over transmission.

OWNERSHIP AND REGULATION: OPTIONS AND INCENTIVES

Thus far, it has been assumed that the grid and/or ISO will carry out its planning responsibilities equally well irrespective of the precise manner in which it is organized or regulated. It is now time to consider whether differences in these features of the restructured transmission system might be the most important aspect of all. In other words, which specific ownership and regulatory approach is likely to best incentivize and police the grid if the transco or ISO is a for-profit, regulated company? Should the regulation be traditional cost-of-service or should it take some other form? Would public ownership or not-for-profit status be better?

Some of the main options and tradeoffs in this difficult subject area can be illustrated with the aid of three simplified scenarios for grid ownership and regulation—as shown in Table 8.2. Across the top of the table from left to right are three options for ownership of transmission: (1) the present structure (i.e., mostly integrated IOUs); (2) a required spinoff of all transmission into independent, private regulated transcos; and (3) spinoff of all transmission into publicly owned, self-governed companies. Because about 30% of transmission is now owned by public agencies, the first two scenarios would probably be more accurately characterized if public ownership was included, but, for discussion purposes, only monolithic scenarios are considered.)

The rows of Table 8.2 present three main options for regulating transmission, including the ISO if there is one. The topmost row is traditional cost-of-service or COS regulation. Under this alternative, the rates and terms of service for transmission would be set much as they are by today's regulators. Whatever the exact form of the transmission pricing tariffs, they are set to earn the owner total revenues that are equal to the owner's total costs, including what regulators believe to be a fair return on investment.

Table 8.2. Categories of Grid Ownership and Regulation Alternatives

↓ Form of Regulation:	Ownership Alternatives		
	Present Integrated Owners of Generation and Transmission	**Independent, Investor-Owned Regulated Grid**	**Public Ownership of Grid**
Traditional			
Price-cap Regulation			
Public Ownership & Control			

☐ **Case to be discussed**   ☐ **N/A**

The second row of the table is price-cap regulation, which is a form of *incentive* or *performance-based regulation* (PBR). PBR is a family of new regulatory approaches that stem from several key observations that were first made by economists during the 1960s. First, economists noted that regulators and regulated firms had "information asymmetries" that allowed the regulated firm to profit from its inevitably superior knowledge of cost and profit opportunities. Second, economists recognized that the mechanics of price and profit determination created incentives that were not entirely intended, and cannot be easily eliminated.[41] The well-known Averch-Johnson-Wellisz effect, by which utilities overcapitalize relative to other inputs, was one such incentive discovered buried within COS regulation.

These discoveries led economists to ask whether different forms of regulation could place just the right profit incentives on regulated, for-profit firms. If so, then rather than having to protect against undesired incentive and information effects, perhaps the incentives created by the regulatory process itself would lead the firm to do what regulators desired. This new approach became known as *incentive regulation*, and more recently as *performance-based regulation* (PBR).

There are many clever and sophisticated versions of PBR for utilities. (Appendix 8A describes several of the major alternatives in this fertile and fascinating area of research.) PBR schemes have become very customized—one rate making formula is used for telecommunications, for example, while a very different one might apply to a gas pipeline.[42] For the purposes of Table 8.2, it is assumed that PBR takes its best-known form, so-called *price cap regulation*.

Under price caps, a utility is given a maximum price that it can charge for specific groups of its services as well as minimum terms, conditions, and quality standards that it must meet. Unlike COS regulation, the utility's profits, investment, and rate base are intentionally *not* regulated in order to give the utility no limit to its potential upside. However, every four years or so, the price caps are reset, and regulators inevitably look at profits earned in order to decide where to set the new cap. The intent of price cap regulation is to incentivize the regulated firm to reduce its costs and boost efficiency by making such behavior profitable.

Box 8-1. Transmission Price-Cap/Incentive Regulation in the United Kingdom

Transmission regulation in the United Kingdom (UK) has evolved considerably since it was first implemented six years ago. At present, it is a combination of price cap and targeted "profit-sharing" regulation. The two approaches work in tandem: The Regulator sets allowable prices, but he also sets a number of performance targets. If the targets are not met, the National Grid Company (NGC) must give up some of the revenues it earned; if they are exceeded, NGC is allowed to bump up its prices enough to recover a bonus payment determined by formula.

The process of setting price caps begins by dividing the NGC's total revenues into two categories, capped and uncapped. The revenues earned by NGC in lines of business other than transmission are not subject to caps, nor are the revenues collected from specific new generators for the purpose of constructing and operating facilities that are dedicated to specific transactions ("shallow" facilities, i.e., substations and spurs). The latter are not capped because NGC cannot predict the number of generators that will require future interconnection, nor the costs of building lines to generators whose locations are not yet known.

The remainder of the service provided by NGC is subject to a maximum allowed total revenue charged per unit of peak demand. The level of peak demand on which the maximum is set is determined in advance in order to remove any incentive for the NGC to stimulate peak demand. The Regulator sets this maximum via the following procedure:

1. Levels of demand are estimated for the next price cap period;

2. Operating costs are estimated for this level of demand;

3. Capital costs (including the costs of debt and equity) are estimated over the period:

4. These two estimates are aggregated into total estimate future-period revenue needed.

5. The regulator examines the present level of prices and profits, and the likely range of future cost efficiencies.

6. The regulator sets the new price (average revenue per kW of demand) equal to the old price, escalated at the retail inflation index and reduced annually by the "X" factor.:

New Price = Old Price + Inflation Adjustment – X% productivity adjustment.

When the UK first restructured its industry in 1990, it began with an estimated average revenue and an inflation adjustment, but no X factor. In 1992, the Regulator concluded that profits and prices were so high that an adjustment was needed, and included an X factor of 3% per year starting in 1993. In a recent (August, 1996) review, persistent high profits prompted the regulator to increase X to 6% a year. Hence, if inflation is less than 6% a year during the next four years, the real maximum allowable price for transmission in the UK will diminish.

Although price-cap regulation creates a global incentive to reduce costs, the Regulator and many others in the UK have come to believe that certain kinds of costs require more targeted savings incentives. Accordingly, the Regulator has established target levels for NGC's costs of providing reactive power, transmission system losses, and the capacity component of the pool price. As a "sharing" mechanism to encourage reductions in each of these costs, the Regulator sets a target level for each and then pays a bonus or exacts a penalty if NGC's actual costs are over or under the target. The sharing formula is:

Allowed NGC Revenue = Actual Revenue + (Actual Cost − Target Cost) x (Sharing Factor)

For example, if the target costs of reactive support are $10, actual costs are $8, and the sharing factor is 50%, NGC receives 50% of (10 − 8) = $2, or $1 in extra revenue.

Source: Office of Electricity Regulation, "Transmission Price Control Review of the National Grid Company, Fourth Consultation." August, 1996.

In the United Kingdom, the Regulator has applied price cap regulation to grid pricing, using a highly customized approach (see Box 8-1). In brief, the Regulator looks at forecasted and historic grid operating and construction costs, NGC's profits, and the breakdown of NGC earnings by service offering. He then sets maximum prices for each transmission and ancillary service for the next four years, as well as the specific terms and conditions for each service. To provide even more tailored incentives, the Regulator also sets target levels for items such as NGC's cost of providing voltage support, with a formula for giving more or less profit depending on whether actual costs are above or below target costs. The process is a little bit simpler than a typical U.S. utility rate proceeding, with much of the difference due to the absence of adversarial administrative procedures.

The third row of Table 8.2 is public ownership and control. Public ownership is an ownership rather than a regulatory approach per se. In some states, publicly owned utilities are subject to state rate regulation or similar procedures. In other instances, control over transmission prices will occur via the exercise of political control over the public owner.[43]

Table 8.3 also displays several of the key differences in the incentives faced by grid owners and the responsibilities of regulators for the scenarios set out in the Table. The scenarios in column (1) include transco owners that also own generators. As seen in Chapter 7, the short-run incentive to favor self-owned generators has led to the development of ISOs and widespread concerns over ISO independence. The same concerns motivating ISO independence—an absence of favoritism toward self-owned generators or cost shifting between generation and transmission—apply just as strongly in the long run. Regulators will want to ensure that the expander has treated all new capacity offers comparably; that the transmission upgrades required by each capacity offeror were evaluated correctly; and that the sum total of generation and transmission expansion options is the most efficient scenario for system expansion.

Oddly enough, this process bears a striking resemblance to the seemingly outdated Integrated Resource Planning process that was described in Chapter 1. Indeed, if integrated utilities subject to IRP were required to take bids for all new capacity, and the utility was required to bid its own affiliated generators rather than build more integrated generation, regulator scrutiny of this IRP process would be quite similar to regulators' responsibilities in the column (1) scenarios of Table 8.2.[44]

Under COS regulation, the projected and actual costs of the expander must be evaluated and audited to ensure that the utility's costs were prudently incurred. One advantage of price-cap regulation is that this contentious process is not necessary.[45] However, in its place, regulators must periodically examine the quality and availability of transmission services to ensure that transmission is available to all who want it. It remains to be seen whether this approach will lead to fewer disputes over service availability than cost oversight led to disputes over costs.[46]

Table 8.3. Main Incentives and Attributes of Grid Ownership/Regulation Options

↓ Form of Regulation:	Ownership Alternatives		
	(1) Present Integrated Owners of Generation and Transmission	(2) Independent, Investor-Owned Regulated Grid	(3) Public Ownership of Grid
Traditional	More intense regulatory oversight required to guard against favoring self-owned generators	Incentives to add to rates base do not distinguish between broad-based reliability and transaction-specific capacity additions	
Price-cap Regulation		Economic incentives to add capacity depend on incremental revenues at capped prices verses incremental cost	
Public Ownership and Control			• Traditional pros and cons of public ownership and management • If public owners also own generation, same incentives and problem as in column one

Whatever the form of grid ownership, there is a direct or indirect danger that a failure to expand the grid will constrain the availability of transmission, making transmission more valuable and the grid owner wealthier. This is precisely why the grid is regulated or publicly owned, but neither is perfect at removing the incentive to make transmission scarce and costly. It is thus important to consider whether the built-in information asymmetries and incentives in each alternative help or harm efficient grid investment.

COS regulation is widely acknowledged to provide relatively weak incentives to control costs and strong incentives to add to rate base, so long as rates allow for the recovery of investment.[47] This much-criticized incentive to "pad" rate base here acts as a counterweight to the incentive to limit investment; what regulators must watch for here are capital investments that cost too much for the capacity addition they provide.[48]

Price cap regulation's incentives to add capacity generally work in the opposite direction. In the process of computing the maximum price, the regulated gridco has an incentive to overstate needed construction revenues and get a correspondingly higher cap. Once the cap is set, it is in the interest of the grid to spend as little money as possible to maintain adequate levels of sales. Under price caps, regulators must therefore guard against overforecasting up front and underspending later. They must also continually guard against service quality deterioration.[49]

Public ownership is difficult to analyze in economic terms; the incentives of the public agencies that might own and operate grids and the possible mechanisms of public grid governance are extremely diverse. Moreover, there has been comparatively little systematic evaluation of the impact of public ownership versus private-ownership-with-regulation.[50] The empirical literature does not support either a simplistic view that public ownership is always inefficient nor does it favor the opposite view.[51]

Although some public choice experts might disagree, one would expect that a public grid would not be permitted to exploit transmission scarcity for its own gain, giving it little incentive not to expand. Public investment in the United States is probably criticized more often for overexpanding and "empire-building" than for building too little. As with COS regulation, in this case, such an incentive works in the right direction.[52]

Interestingly, one theme that emerges from many of these studies is the advantage of unifying the ISO and transco in a single, independent entity. Because this entity holds assets, it can be incentivized by being allowed to earn more or less profit—or if it is public, to build more or less empire. Information gained from dispatching and/or operating the spot market can be used for planning transmission more accurately. In particular, this improves the ability of the transco regulator to make the transco balance the cost of system upgrades with the periodic capacity payments earned by the pool when capacity is scarce. The simple fact that the transco's revenues begin to exceed costs due to the realization of large capacity charges should help signal regulators that either the generation market is not providing adequate supply, the system is not accommodating expansion, or something else worth looking into is going on.[53]

Apart from this somewhat common belief, there is no consensus about what is the best form of grid ownership or governance in the United States. Each of the main alternatives, including often-criticized traditional regulation, has its adherents, for the reasons just noted and for a variety of specific, practical reasons having to do with regulatory information, procedures, and subtle incentives.[54] There is agreement that divestiture of transmission would make governance easier and probably more successful whatever the specific approach, but absent divestiture it is unclear how well any of the hybrid approaches that the industry will be forced to take will work.

TRANSMISSION CONGESTION CONTRACTS

The difficult challenge of incentivizing and regulating transcos and ISOs has led many economically-inclined analysts to ask whether the greater use of competitive markets in transmission might itself be a substitute for transmission regulation. If transmission could somehow be turned into a competitively traded good, market forces might engender adequate investment and efficient prices. "Without well-defined property rights," Professor Hogan writes, "the alternative would be to rely solely on the monopoly grid owner to expand the grid and send everyone the bills."[55]

This premise has led to the development of a concept Hogan calls a *Transmission Congestion Contract* (TCC). TCCs are tradable rights to receive the costs of congestion between two nodes on a transmission system. These rights would be sold, by the transco or ISO, thus providing a hedge against transmission price increases. If they work, TCCs are the long-awaited answer to generators' desire for long-term fixed-price transmission.

Hogan (1995a) p. 17 admitted that the characteristics of transmission networks made questionable the idea of a measurable unit of transmission capacity:

> Unfortunately, the natural presumption of the existence of such a well-defined physical transmission capacity right confronts three seemingly insurmountable problems. First, assignment of such rights implies control of the use of the physical transmission of power in a way that would reverse the separation of ownership from control that is an essential feature of open access to the transmission network. Second, restricting use of the grid to match the allocation of such rights would either place great demands for constant retrading in a short-run secondary market or would compromise the ability to achieve the least-cost dispatch. To obtain the economic dispatch, the system operator needs control over a sufficient range of flexible plants and control over the full transmission grid. Third, and more arcane, except in the case of a substantial under-allocation of what would seem to be natural transmission rights, an under-allocation designed to avoid any possibility of confronting any transmission constraints, there is no well-defined physical capacity that can be allocated and assured. Due to the many interactions across locations, the "capacity" of the network is not amenable to any easy definition and the ability to move power between locations cannot be assured. The capacity of the network at anytime depends on the configuration of the inputs and outputs.

Hogan's answer to these problems is to define TCCs as financial contracts that pay the holder the congestion cost differences between two nodes. Although the research results are still unsettled, some research by Hogan and others suggests that it is theoretically feasible to create this kind of tradable right.[56] A TCC of this nature is a natural counterpart to a CFD, which hedges price for a downstream buyer.

TCCs received a powerful boost when the California Public Utilities Commission ordered regulated California IOUs to develop a pool and an ISO that offered TCCs.[57] The decision has not been implemented to date.

TCCs and the idea of marketizing transmission is fascinating, but first principles suggest that the idea will, at best, be a partial solution. As previously discussed, a large proportion of transmission system capital outlays are associated with replacement and broad-based grid reinforcements. These investments are not directly associated with specific congestion relief, and they are unlikely to be recovered by TCCs, were they ever to come to pass.

Hogan notes that "due to economies of scale and network effects," there are likely to be instances where the sum of the benefits of relieving transmission constraints as reflected in the benefits received by the holders of TCCs won't be sufficient to justify all of the grid reinforcements and upgrades needed to provide reliability. Hogan calls this the "residual role for monopoly investment."[58]

In one of the most extensive and rigorous technical analyses of grid planning under competitive generation, a team of M.I.T. researchers headed by Professor Maria Ilic examined TCCs and other grid expansion incentive schemes. The team concluded that TCCs might be problematic for the reasons Hogan identified, namely that available transmission capacity might prove too difficult to allocate. In addition, they emphasize Hogan's second caveat that a strong role will continue to exist for centralized planning and investment:[59]

> A dynamically efficient gross-complementing between enhancing the transmission grid and the efficient creation of system service generation can only be done at the interconnected system level. This argument is in the essence of our suggestion that without active (minimal) information provided at the systemwide level, the system may either become technically unreliable in the sense that the reliability of service may either drastically reduce or the systemwide dynamic efficiency may deteriorate. Because of this it becomes important to develop means of protecting the cost of system support in a reliable way.

A final issue concerning TCCs is the impact they will have on generation and transmission market power and their attendant political acceptability. TCCs can be seen as the right to earn profits from a monopolist's unwillingness or inability to expand an essential facility. For obvious reasons, the industry has not previously sold off the rights to earn profits from essential electric utility facilities without accompanying regulation of prices and/or profits.

Hogan assumes that buyers of TCCs or sellers of generation that desire a guaranteed locational price differential will buy TCCs as price hedges. But suppose a third party begins buying TCCs and then attempts to block any expansion of the transmission system through legal or economic means. This would be a perfectly rational strategy: the more congested the system becomes, the more valuable the TCCs.

It is probably the case that regulators and all elements of the transmission segment would be involved in comporting TCC rights and ATCs and policing the overall market. This example is raised merely to illustrate the fact that TCC markets may raise practical questions that have not yet been understood, much less addressed.

POLICING COMPETITION IN THE GENERATION MARKET

The benefits of restructuring to consumers and to the nation as a whole stem directly from the cost efficiencies and service expansion a competitive generation industry promises to create. If unregulated generation markets become dominated by a very small number of firms, they may have the ability to exercise what economists call *market power* (the ability to profitably raise price by "a small but significant and nontransitory amount" without other competitors forcing prices back down). Insufficient or ineffective competition among generators in a market will yield profits and prices that are higher than under effective competition, which would harm consumers and undo the benefits of restructuring.

Many of the various means of acquiring market power in electric generation markets have been discussed in previous chapters without explicitly labeling them as such.[60] Transmission is regulated in order to prevent its owner from abusing its monopoly status, and regulators and policymakers are going to extraordinary lengths to ensure that the transmission system does not become a source of unfair competitive advantage to any generators. These can be viewed as efforts to prevent market power related to the vertical relationship between G and T.[61] In this section, these critical issues are put aside and the discussion is limited to issues arising in the generation sector itself—horizontal competitive issues.

The history and cost characteristics of the integrated electricity industry suggest that generation concentration following restructuring is a reasonable concern. When the U.S. generation sector was completely deregulated in the 1920s, generation became a highly concentrated industry. Recent experience in the United Kingdom, where the government privatized the state-owned electricity monopoly into only three competitors, has also demonstrated the importance of effective generator competition. Many researchers and the U.K.'s own electricity Regulator have concluded that this arrangement conferred market power upon the three generators. Accordingly, the Regulator called for the divestiture of 6,000 MW of generation by two of the three dominant firms.[62] Regulators in California have required the largest utilities in their state to sell off some of their generators in order to ensure that—as part of that state's restructuring—the generation market is competitive.[63]

A review of the recent literature on scale economies in generation (Chapter 4) found that the era of individual *plant* economies had passed, but that there was no basis for concluding that the lowest-cost size of generation *companies* had diminished. Indeed, legitimate cost savings may be realized by expanding and diversifying one owner's portfolio of generators over many geographic areas and unit types. When economic growth and electricity sales are strong in one region, for example, they may be weaker in another. Similarly, when the price of one fuel is high, or availability is reduced, other fuels may be cheaper or more plentiful. All these effects help a generation company blend earnings from temporarily more profitable units with those from temporarily less profitable ones, smoothing out overall company earnings.[64]

Additional advantages may stem from mass-purchasing operating supplies or new capital goods for large numbers of plants. From everyday experience it is clear that it is not unusual for suppliers to give quantity discounts, or premium service to large customers.[65] Maintaining a fleet of generating plants undoubtedly entails incurring some costs that do not vary much with the number of plants owned; the more units these costs can be spread across, the lower the average cost of service.

These considerations have led some to predict that restructuring will cause (or is causing, others say) a wave of mergers that will ultimately produce substantially fewer generation firms than exist today. Almost ten years ago, a Wall Street investment banking firm made headlines by predicting that utility deregulation would cause the 220 or so then-existing utilities to consolidate into 50 much larger firms within five years—a slogan made famous as "fifty within five."[66] Although merger activity has not proceeded in nearly such rapid terms, a recent wave of similar predictions is noteworthy not so much for its focus on numbers of firms as on the size of the transactions.[67] Management consultant Michael Weiner (1996) boldly predicts that "a half a dozen or so utilities will end up as mega-generators, controlling perhaps 80% of total generation capacity."[68] In December, 1992, Public Service Company of Indiana and Cincinnati Gas and Electric Company formed an 11,000 MW new utility, Cinergy. Less than four years later, its President asserted that he "is still not large enough to compete effectively with American Electric Power, Duke, Southern, or the Tennessee Valley Authority, the large players in our area. We believe in scale and are currently looking for another company to merge with in order to continue to reduce costs and grow our business."[69] Economists such as Sally Hunt and William G. Shepard also predict that generation consolidation will become an increasingly important issue.[70]

In view of the benefits of some concentration and the dangers of too much restructuring, policymakers face a tremendous challenge. If generation companies become too large, they may cease to compete fiercely, or even collude to raise prices. Sharing planning information and making pricing and transmission information widely and rapidly available—actions required for fair, short-term competition and efficient grid expansion—may further facilitate tacit or explicit collusion.[71]

The United States is fortunate to enter the era of restructuring with an industry that is unquestionably the most diverse in the world (Chapter 5). Depending on the boundaries of the emerging regional generation markets, there are often many firms in these markets today.[72] Where this will not be the case, restructuring must begin by establishing a sufficient number of sufficiently small competitors, relative to the size of the market, to foster competition.

Starting with a competitive generation market, growth in horizontal market power will probably arise primarily from mergers or acquisitions of generator assets. Furthermore, entry into generation markets by new rivals must be (or become) difficult. Consequently, policing horizontal market power becomes primarily an issue of merger and acquisition policy, with constant attention paid to the conditions of entry.

Table 8.4 summarizes the main types of antitrust regulation that apply to electric generators before and after restructuring in the United States including jurisdiction over horizontal mergers by generators or integrated utilities. The first two rows of the table treat the roles of the public agencies responsible for policing competition in all industries. These agencies, principally the U.S. Department of Justice, the Federal Trade Commission, and the attorneys general of individual states, are reponsible for enforcing the principal antitrust statutes in the United States. These laws (including the Sherman Act, the Clayton Act, and the Federal Trade Commission Act, and related federal and state statutes) make it illegal to attempt to monopolize any industry, conspire or collude to raise prices or limit output, or engage in a merger whose effect "may be substantially to lessen competition, or to tend to create a monopoly."[73]

Enforcement of these laws proceeds in any of several ways:

- **Actions Initiated by Federal Agencies.** When a firm that owns generators seeks to merge with another firm, federal procedures require a notification filing before federal authorities.[74] Federal antitrust agencies then examine for the possibility of harm to competition and may seek to halt or modify the merger. Although an informal negative opinion or the threat of prosecution is often enough to persuade the merging parties to alter their effort, a binding alteration can be made only if federal prosecutors take the case to the Federal Trade Commission or to federal court and succeed.[75]

- **Actions Initiated by Private Parties.** The Clayton Act allows individuals or companies who believe they have been harmed by anticompetitive practices forbidden under the Act to seek compensation in federal court. If the allegedly harmed party can demonstrate actual harm, the law allows that party to collect three times the value of the economic harm suffered (so-called "treble damages").

- **Actions Initiated by State Agencies.** Many state governments charge their attorneys general with the responsibility of protecting consumers against unfair competition or anticompetitive practices.[76] These agencies can also initiate claims for treble damages.[77]

In any of these instances, upon a finding of market power, courts may act by preventing a merger, ordering the divestiture of certain assets or lines of business of the merging firms, by preventing certain contracting or pricing practices, or by applying other remedies.[78]

The third and fourth rows of Table 8.4 show antitrust responsibilities that currently reside in U.S. electric regulatory agencies. The Federal Power Act (FPA) and most state public utility statutes contain language that gives the regulator broad responsibility to prevent utilities from exercising market power, harming consumers, or engaging in anticompetitive conduct. This is, after all, the primary purpose of regulating franchise

Table 8.4. Protection Against Market Power in Electric Generation, Pre and Post- Restructuring

	Pre-Restructuring	Post-Restructuring
Federal Antitrust Statutes	• Hart-Scott-Rodino pre-merger notification, with opportunity to challenge merger • Opportunity to initiate action at FTC or in courts anytime market power is believed to exist	• Unlikely to change from pre-restructuring policies
State Antitrust Statutes	• Can initiate actions under federal antitrust laws and sometimes state consumer protection statutes	• Unlikely to change from pre-restructuring policies
Federal Electric Regulators	• Prior approval of all merger of most IOUs, with authority to impose some conditions that prevent market power increases. • Prior approval of sales generating units or lines subject to FERC jurisdiction	• To be determined
State Electric Regulators	• In many states, prior approval of merger of utilities subject to state commission jurisdiction, with authority to impose some conditions • Prior approval of construction, sale, or purchase of generating units of utilities	• To be determined

monopolists in the first place. Beyond this general authority, Section 203 of the FPA gives the FERC additional specific responsibility for reviewing and approving *in advance* mergers of electric IOUs or sales or purchases of electric generation of transmission by the same.

The standards used by the FERC to evaluate whether to permit such actions generally parallel the analyses used by antitrust authorities. Most state PSCs have taken a similar approach. However, it should be noted that the FERC's review of mergers has sometimes gone beyond a pure inquiry into the competitive effects of the merger.[79] In addition, there has been some disagreement about whether the application of the specific methods of competitive analysis used by the FERC, which were designed for non-utility industries, are accurate for utilities and are properly applied.[80]

In a recent Notice of Inquiry, the FERC reevaluated its merger policies and procedures in light of utility restructuring.[81] The FERC's new policy was prompted by twin objectives of ensuring sound competition and reducing regulatory red tape, based on the impacts of the proposed merger on sound competition, the ability to protect the parts of the industry that remain regulated, and electricity prices. The Commission proposed a new five-step evaluation procedure intended to shorten the review process and conform more closely to traditional economic analyses of mergers.

Because restructuring will almost certainly result in most or all generation no longer being subject to state or federal regulation, prior regulatory approvals of generator sales or mergers may no longer be required. Hence, the major shift in antitrust governance triggered by restructuring will not be in statutes that continue to apply to generators, but rather in the methods of monitoring and law enforcement. Protection against excessive concentration, collusion, and unfair practices by unregulated generators will become entirely responsive, as is most antitrust enforcement today.

Some observers argue that an abandonment of prior approval will leave electricity markets insufficiently protected, while others argue that the present system of overlapping and redundant responsibilities is excessive and inefficient.[82] Ultimately, the extent to which restructuring will reduce the lengthy reviews that are presently required depends on how policymakers balance the need to allow unfettered competition in generation with the need to provide for adequate antitrust protection.

Appendix 8A

A Brief Introduction to Performance-Based Regulation for Electric Utilities

An enormous variety of regulatory approaches may be characterized as performance-based or incentive regulation (PBR). Indeed, because every form of regulation considers performance, and contains incentives, strict application of the term is meaningless. However, in the United States, the term has become synonymous with forms of regulation other than cost-of-service (COS) rate making.

Of the many forms of PBR, two major versions are important for electric utilities: price caps and generalized PBR. The former occasionally includes revenue caps, while the latter encompasses a variety of terms used in the industry today, including *sharing schemes, customer service incentives, profit-sharing,* and *performance bonuses*.[83] A third approach, optional tariffs, is not widely used today but has helped to produce interesting literature.[84]

PRICE-CAP REGULATION

Under this approach, a regulator sets a maximum (but not required) price for each service offered without examining or regulating the utility's investments, rate base, or profits—except when the caps (maxima) are set for the first time. Price-cap regulation gives utilities an incentive to reduce costs as much as possible—100% of all cost reductions beyond those assumed in the capped price flow to the utility as added profit. Revenue caps are similar, but they set a maximum allowed total revenue for the utility.[85]

Under the simplest price cap approach, a regulated franchised utility could reduce the quality of services to the lowest possible level that customers would be willing to bear and still pay the maximum price. Hence, price caps have always been implemented along with relatively complex quality standards that regulators police. If a utility fails to meet these standards, a penalty is applied.

Even with quality standards, it is generally believed that it is not necessary to give utilities 100% of the profits they earn by cutting cost in order to provide adequate cost-cutting incentives. As a result, price caps have generally been implemented with formulaic adjustments that have the effect of passing on some of the cost savings to customers, rather than the utility keeping all cost savings as profit (from whence the term *sharing or productivity factor* originates). The typical approach in price cap regulation is to index next year's allowed maximum price to the rate of inflation and an expected level of cost reduction that must be shared with customers. If the rate of inflation is measured with the consumer price index (CPI), a typical simple formula is:

$$\text{Price (next year)} = \text{Price (this year)} + \text{CPI} - X$$

where X is the sharing factor. Hence, if inflation is expected to be 5% next year, cost reductions are expected to be 6%, and half of any cost savings are to be returned to consumers, next year's price should first be increased by 5% for inflation and then reduced by 3% to rebate half the expected cost savings. If the utility fails to make its targeted cost savings, its profits will suffer, which is the whole idea.

Price-cap regulation came into widespread use in the United States in the late 1980s as a means of regulating telephone companies, which were beginning to undergo deregulation and diversification into unregulated lines of business. Price-cap regulation has since been applied to integrated electric utilities, gas pipelines, and local and regional phone companies. These applications have produced a decade of experience that must be regarded as mixed. Of the many problems and criticisms leveled at price caps, several stand out as significant.

The first criticism relates to the initial setting of the capped price and other factors in the formula. Because this is typically done when a regulated entity changes from a prior mode to price caps, there is little basis for determining the proper initial price or sharing factor. By definition, some of this problem goes away as the regulator gains experience with successive price caps.[86] However, the regulator must continually guess the entity's future ability to increase productivity and reduce costs. Some critics argue that this is a much more difficult burden to place on regulators than policing profits—particularly in industries where technological change is rapid and unpredictable.

This leads to the second major criticism, which is that price cap regulation has a heads-I-win-tails-you-lose character to its outcomes. If regulators guess too low on the sharing formula, a price-capped entity may earn very large profits, while captive consumers experience no price reductions. On the other hand, if regulators are too severe, the regulated firm will suffer losses and could go bankrupt, which is not a permissible outcome. Hence, regulators can err on the side of greater profits, but not in a direction that benefits consumers.

This problem is exacerbated by seemingly inevitable surprise events that have large—but unforeseen impacts—on utility costs. Suppose, for example, a hurricane causes extensive damage to a utility system. The cost of repairing this damage was not figured into the price cap, so regulators and the utility must agree on a special exception that raises allowed prices to account for these unforseen costs. Although everyone agrees that this sort of thing is necessary, some argue that this kind of adjustment is one-sided. That is, it is unlikely to occur if the utility stumbles onto a windfall that greatly reduces costs and hikes profits, but will occur if prices must rise.

In practice, this asymmetry has proven to be troublesome. When price-capped utilities begin to report large profits, the pressures mount on regulators to immediately revise the sharing factor.[87] Of course, if the sharing factor is continually revised to keep utility profits at levels that are considered fair and reasonable, then price cap regulation becomes the equivalent of traditional regulation. Hence, political considerations have prompted price cap regulation to resemble COS regulation to a surprising degree.

Another criticism of price caps involves incentives for regulated firms to reduce services or service quality. Because the firm being regulated is the only monopoly provider of services, most customers will continue to buy at the regulated price even if quality suffers. If reducing quality cuts costs, the regulated firm can earn greater profits.

For example, Maine regulator Barbara Alexander writes:[88]

> However, most utility commissions have struggled with how to retain sufficient oversight of customer service and reliability during the term of the performance-based regulation. Early PBR decisions contained no special provisions for maintenance of customer service, as commissions reasoned that they would rely on their existing rules and investigatory authority to address any problems that later arose. Many commissions have since found that this approach is sufficient. Particularly in the western U.S., states are scrambling to address deteriorating telephone service quality that occurred after performance-based regulation was approved.

The practical implications of this incentive to reduce quality is that regulators must carefully set standards for quality, monitor utility performance, and penalize poor quality in a meaningful way. Developing quality indices that are accurate, available to regulators, and effective has proven to be a major challenge for regulators in telecommunications and electric power.

Finally, the establishment of a price cap generally requires a separation of the firm's services into two baskets. One basket of services is unregulated and uncapped, while the other is subject to a price cap that is based on the weighted average price of the regulated basket. These initial steps, while necessary, introduce a variety of problems.

First, to the extent that the utility reports its costs and profits by lines of business, it has every incentive to shift costs from the unregulated to the regulated lines of business. This incentive to shift costs to captive customers exists equally under COS regulation of a diversified firm, but in that case it is expected that regulators inspect all costs in detail and disallow those that are not attributable to efficient regulated service. Price-cap regulation is intended to relieve regulators of the need for policing costs in detail, but to do so for a diversified firm invites trouble.[89]

Second, the regulated basket of products is also subject to manipulation that is designed to increase prices and profits without adding to consumer benefits. For example, suppose that a utility decides to offer a new regulated service in the middle of a period in which the weighted average price of the prior selection of services was fixed. What weight should be put on the new service? By redesigning or renaming its services, the utility may be able to effectively evade the price cap without appearing to do so.[90]

The experiences of U.S. and U.K. regulators who have tried price-cap regulation have varied. The U.K. Regulator has confronted many of these criticisms, particularly concerns over "excess" profits earned by capped firms.[91] Even though the regulator has been forced to increase the sharing formula repeatedly—most recently, only one year after it was established—he continues to conclude that price caps are superior to other forms of PBR. Some U.S. regulators agree, although the center of gravity in the United States appears to be shifting more toward generalized PBR.

GENERALIZED PBR AND PROFIT SHARING

Generalized PBR is essentially the establishment of a formula that sets forth the profit (as opposed to maximum price) that a regulated entity is allowed to earn. Under this approach, regulated price is determined by adding up costs and profits (which sums to revenues) and dividing by sales. Typically, profits are first set to a modest or baseline level; performance targets are then used to increase or reduce profits above or below these levels, depending on performance. For example, suppose one performance target for electric utilities is the energy efficiency of their average power plant. A PBR formula might look like this:

Allowed Profit = Baseline Profit + (Actual Efficiency – Target Efficiency)
x (Sharing Factor)

For example, if actual average efficiency was 50%, and the target efficiency was 45%, the utility would receive additional profits as set by the sharing factor. Other performance targets for electric utility PBRs include electric fuel costs, power plant heat rates (a measure of energy efficiency), average prices for other utilities in a sample, success implementing demand-side management programs, complaint rates, employee safety rates, and levels of reliability.[92]

As another somewhat mechanical, formula-based approach, PBR suffers from some of the same criticisms as price caps. Just as it is necessary to guess future cost levels for the regulated service basket under price caps, it is necessary under PBR to sometimes make much more detailed predictions of optimal service targets and associated profit bonuses. The requirement that the PBR regulator pick a group of performance targets from among literally thousands of possibilities is analogous to the price cap regulator's choice of the weights on the basket of regulated, price-capped services—and as in the latter case, there are sometimes many ways that utilities can manipulate their reported costs or performance to their advantage.[93]

In addition, PBR schemes are subject to some of the same profit and readjustment asymmetries that burden price caps. First, regulators must formulate an appropriate baseline level of performance and profit, representing average acceptable performance, as well as the appropriate sharing factors.[94] If profits are too high, consumers will clamor for price reductions; if profits are too low, the utility will clamor for a new formula. As with price caps, unforeseen events can force a recalibration of allowed profits or prices or discarding the PBR formula altogether. To prevent the latter, utilities in California have proposed a "Z factor" that allows them to raise or lower prices according to the occurrence of unanticipated costs that are outside their control, including such things as changes in environmental laws, the costs of adjusting to restructuring, natural disasters, and other occurrences.[95] This factor is added to the formula shown above, following the sharing factor, and is zero unless unusual events occur.

A final pragmatic issue concerning both price caps and PBR is the administrative and expertise burdens that they place on the regulator. Although both schemes arguably reduce the cost of regulation for the regulated firm, they require regulators to undertake different and more sophisticated analyses than they have in the past. Without increases in resources, regulators may be hard-pressed to do a good job at PBR or price-cap regulation.[96]

There are the many additional theoretical and practical issues associated with the design and administration of incentive regulation plans. Table 8A.1 summarizes several of the main attributes, problems, and literature on price cap and general PBR regulation. For additional reading on PBR, see Joskow and Schmalensee (1986), Baron (1989), Brown, Einhorn, and Vogelssang (1991), the Edison Electric Institute (1993, 1995, 1996), Weisman (1993), Marcus and Grueneich (1994), Weisman and Sappington (1994), Hill and Brown (1995), Pfeifenberger and Tye (1995), and Comnes, et al (1995).

Table 8A.1. Main Forms of Utility PBR or Incentive Regulation-Overview

Form	Incentive and Information Advantage	Selected Applications to Date	Main Responsibilities of Regulator	Problems and Observations	Selected References
Price-cap Regulation	• Inherent incentives to reduce cost	• AT&T's long-distance prices in the U.S. after divestiture (1984) Electric Transmission and franchised distribution prices in the U.K., 1990–present	• Ensure service quality is high • Guard against under-capacity • Reset prices periodically	• Deterioration of quality service is a concern • Concerns over excess profits cause a convergence with cost-of-service regulation • Difficult to decide bundle of service on which to set maximum price • Shifting of pricing towards most captive customers	• Miller (1995) • Isaac (1991) • Xavier (1995) • Offer (1996)
Customized PBR/Profit-Sharing	• Incentives to improve performance or reduce cost are narrowly targeted • Less need to police cost and prudence, as above	• Some use by U.S. electric and gas utilities • U.K. Transmission price contain some specific incentives	• Determine specific PBR scheme • Measure results and recalibrate	• Difficult to set measurable targets that cannot be manipulated by regulated firm • Unintended side effects may be perverse	• Alexander (1996) • Comnes, et al (1996) • Hill and Brown (1995) • Marcus and Greunich (1994) • Edison Electric Institute (1996) • Navarro (1996)

251

Chapter 8

NOTES

1 Ruff (1996) p. 21.

2 For example, Gross and Balu's "Synthesis Paper on Wholesale Markets, Power Pool Proposals" (Gross and Balu, 1996) barely mentions long-term grid investment; likewise the conference summary of the Harvard Electricity Policy Group's "Special Session on ISO Governance." (Harvard Electricity Policy Group, 1996). McGuire's (1996) brilliant defense of bilateral decentralization says, in the last paragraph of the paper, "Important matters not addressed here include contract enforcement, non-performance, contingency power generation, and all long run matters such as investment in generation and grid capacity." (P. 7) (and similarly, see Chao and Peck (1995) p. 15). An equally common approach is typified by this passage from University of Wisconsin Professor Fernando Alvarado (1996), p. 35:

> Because of the continued monopolistic role of the transmission grid, its expansion will probably have to continue to be regulated. Criteria for network expansion should be based on the determination of the common benefits that are attained by expansion. That is, optimal expansion of regulated transmission systems would be done much as it is today, based on the least cost overall expansion of the system taking into consideration not only the cost of expansion itself, but also the expected value of the long term marginal benefits to all participants. The benefits accrue as a result of the expansion of the OLB [grid] to reduce security costs due to congestion, and also as a reduction of energy costs of transmission.

Finally, Backus and Baylis (1996), p. 38, predict that the transmission system in the United States will engage in strategic overbooking and other forms of gaming, presumably evading regulators to do so. The authors allege that "current U.S. thought on deregulation have yet to embrace [transmission system] gaming and have not considered overbooking."

3 Hughes and Felak (1996) p. 56. For an important exception, see Einhorn (1990).

4 In addition, sellers are likely to face certain required conditions on their operation, probably reflected in state or federal licenses. See Chapter 9 infra.

5 The point here is that there is widespread agreement that real-time posting is a proper and adequate *approach* to timeliness and accuracy. A number of difficult technical issues complicate the posting of specific real-time information. One of the largest such issues is the definition of available transmission capacity for a large number of locations in a network. Because flows between two nodes can alter availability capacity between other nodes, ex-ante postings of available capacity become inaccurate as soon as some posted capacity becomes used. In a companion order to FERC Order 888, Order 889 required that utilities post the availability of transmission capacity and other system data needed to effect bilateral trades. (Docket No RM95-9-000, April 24, 1996).

6 For example, utilities in California have proposed that prices be posted not for every node in the state, but rather for four very large zones that cover the state and are bounded by already-congested interfaces in their system. California Energy Commission (1996) p. 22

7 Hogan (1996b) p. 23 gives an example of this effect in a simplified network.

8 These points are discussed in the context of the California proposal for zonal pricing in California Energy Commission (1996) pp. 24–32.

9 Specifically, transmission congestion contracts (TCCs) are defined with respect to two nodes. If they are defined with respect to two zones, problems will occur when regulators determine that congestion patterns have changed enough to redraw the boundaries of zones. In general, redrawing or subdividing zones could greatly disrupt a market whose many contracts and arrangements were made under the original zonal definition.

10 This argument, including the points following, is made by Fernando and Kleindorfer (1996) p. 34–35.

11 At present, the United Kingdom has 14 transmission pricing zones. A recent review of transmission pricing by the U.K. regulator found "little support" for nodal pricing among those who commented on the subject "on the grounds that it would create volatile and uncertain prices." Nevertheless, the Regulator stated that he is "not necessarily convinced by the counter arguments to nodal pricing," but he declined to require a change away from zones. (Offer, 1996, p. 55–56)

12 The following passage from Fernando and Kleindorfer (1996) p. 34 expands on this point:

> Transmission service providers will remain regulated entities for the foreseeable future. The complexities of revenue reconciliation, revenue requirements, comparability reviewers and capability assessments are going to be difficult enough in zonal pricing, reset annually or semi-annually. They would appear to be almost impossible under the added complexity of nodal resets. But if transmission prices are to be fixed for a reasonable length of time, it should be clear that the required scenario averaging across time will not benefit much from the added complexity, of having to do this averaging at each node. As a further problem in regulatory complexity, if issues of shareholder and customer cross-flows are raised, these will be more difficult to sort out in a nodal pricing environment than under zonal pricing, where zones and recoverable embedded cost can be clearly identified with respect to native customers and ownership boundaries.

13 Newberry (1995a) p. 12

14 In the United Kingdom the contract is a CFD; in the United States a physical bilateral contract is more common. It is possible to foresee that strong, diversified generating companies or other sources of capital may decide to build "merchant plants" to sell entirely into the spot market. The first of these plants was announced in the United States in early 1996, and may signal a trend. However, experience in the United Kingdom has demonstrated the opposite behavior, i.e., increased integration between generators and distributors.

15 In a study of investment in oil markets, which also exhibit great short-term price volatility, Verleger (1993) observes that rational investors ought not to be discouraged by volatility. However, Verleger and others cite two possible pathways by which volatility might adversely impact investment. The first, exposited by Pindyck (1991) and Dixit (1992), uses option theory to demonstrate that volatility causes rational investors to delay their investments later than they would if prices were stable at the same average level as volatile prices. The second pathway by which volatility impacts the marketplace is via increased transactions costs—adjusting to changing cash flow and budgeting needs, for example. Verleger notes that forward contracts and futures are designed to reduce the adverse impacts of volatility.

16 See Verleger (1987) constraining oil market contracting. Colleagues Frank Graves and James A. Read, Jr. (1996) similarly note that the use of long-term contracts for natural gas supplies has diminished considerably in the wake of open-access gas transportation and the emergence of gas spot and futures markets.

17 Martha M. Hamilton, "Crude Oil Prices Surge After U.S. Attacks on Iraq." Washington Post, 9/4/96, p. C1.

18 See Hogan (1996a), p. 26 and Jaffe and Felder (1996). Another more technical explanation for the continued likelihood of capacity contracts is their value as financial instruments. Graves and Read (1996) demonstrate that power capacity will probably be priced using approaches from options theory. Briefly, they note that a unit of capacity can be defined as a series of options for the purchase of energy at a fixed future price and time—analogous to common financial call options. With a number of strong assumptions, traditional "Black-Scholes" options pricing approaches can be applied to capacity pricing. This has the intuitive result, for example, that

greater volatility of energy spot prices increases the value of capacity. Hence, an option-theoretical view of capacity also suggests that buyers and sellers will value contracts as a means of smoothing volatile price streams, and more so if the supply or demand side of the market goes through rapid "boom-bust" cycles that exacerbate the natural volatility of electric network prices.

19 As of July, 1994, 80% of all power sold in the United Kingdom was governed by a long term pricing agreement. Offer (1994) p. 5. Interestingly, roughly the same fraction of total electrical energy sales are sold under firm contracts or from integrated capacity in the U.S., which at present has no operating merchant plants and therefore 100% bilaterally contracted plants.

20 This helps explain generators' preference for zonal pricing in the United Kingdom note 11 *supra*, which is less accurate and less volatile, but also less likely to be substantially changed over time.

21 These decisions will depend to a significant degree on the locational nature of transmission prices and the responsiveness of electricity users to location and time-based price signals. Hence, there is the expected long-run circularity observed in markets: generator and user decisions will depend on the price signals sent by transmitters (as well as other external factors), which will depend partly on generator and customer decisions.

22 Professor William Hogan: "Avoiding sustained locational cost differences defines the economic rationale for investing in transmission." (Hogan, 1996a, p. 29).

23 This is quite similar to what was observed in the National Power Surveys and other national planning studies of the 1960s and 1970s. Even without considering siting approval issues, many apparent regional price differentials were not large or persistent enough to justify interregional transmission lines. Many utilities therefore planned and built more or less to achieve company, regional, or pool balance. On the other hand, along the West Coast of the United States price differentials between the Pacific Northwest and Southern California were sufficiently large and persistent to justify the construction of three large lines down the Coast.

24 See Jaffee and Felder (1996).

25 For example, this payment is part of the uplift payment in the United Kingdom' power pool, Appendix 7A.

26 NGC's overbudgeting for transmission capital is not entirely attributable to requests for generator hookup. According to a study by a consultant to the UK regulator, £210 million was directly attributable to generators who applied for service but later canceled their request.

27 Offer (1996) p. 23. NGC uses a form of scenario-based planning not unlike integrated resource planning procedures used in the United States for many years.

28 Offer (1996) p. 5, 52.

29 For starters, the concept of generation adequacy (Chapter 2) would happen to appear to be largely obsolete. In competitive power markets it is assumed that long-run price signals will engender sufficient generator capacity to allow the system to function well in the future—acknowledge that spot price plays a much larger role matching demand to supply then previously. Nevertheless, many policy makers and industry officials would prefer that adequacy be monitored in order to spot unforseen problems before major shortage or surpluses develop. This has prompted policymakers to return adequacy requirements this far, and to reconfigure capacity reporting procedure to try and count unregulated generators, who traditionally have not been required to report capacity or output to the government. For example, in 1996, The Energy Information Administration of The U.S. Department of Energy turned a Task Force to examine later reported requirements for traditional and non-traditional utility participants.

30 Formal mathematical analysis of the planning problem by some of our most advanced engineering- economic researchers does not alter this conclusion. Following a formal, highly mathematical examination of the issue, a team of researchers from the Massachusetts Institute of Technology led by Professor Maria Ilic, et al (1996) p. 42 concluded that:

This is a very complex question because it involves tradeoffs between making long(er) commitments to generation-based services, on the other hand, and commitments needed to enhance the grid, on the other. In light of this a particular need emerges for developing economic incentives capable of comparing the cost of out of merit generation to the cost of enhancing the grid to avoid this inefficiency. This algebra is tricky, the payoffs are seen over (what used to be) planning horizons:

E.g., over ten years it may be more cost-efficient to build a FACTS [transmission] device than to persistently operate out of merit generation needed to avoid transmission grid congestion. When assessing the process of creation of systems control services over the time horizons such as months and years, this issue must be addressed very carefully. Because of market uncertainties the analysis will definitely result in some measures of risk taking. The regulators should clearly define responsibilities of various players over the longer time horizons.

31 In addition to the public benefits of fair and robust competition, the electric grid provides certain public benefits that have traditionally been the responsibility of regulators and regulated companies. Some of these responsibilities, such as the responsibility for reliability and fuel diversity, will likely devolve to the grid or the ISO. See Part 3 *infra*.

32 For an early discussion of the incentives of transmission monopolists to constrain total transmission capacity, see the Transmission Task Force Report (FERC, 1989), ch. 5, 6.

33 For transmitting utilities that hold an obligation to serve native load under state law, the Order permits these utilities to reserve capacity for their own system's future use (though they must allow it to be used for short-term transactions prior to the time it is needed). Other than this important exception, all requests for transmission service must be considered equally.

34 RTG Policy Statement, Docket RM93-3-000, July 30, 1993 mimeo p. 17.

35 Order Accepting Compliance Filing, Western Regional Transmission Association, 71 F.E.R.C. Par. 61,523, June 9, 1995.

36 Order Accepting Compliance Filing, Western Regional Transmission Association, 71 F.E.R.C. Par. 61,524, June 9, 1995.

37 One additional complication: The emergence of RTGs is the only apparent change in the planning role of regional reliability councils. Whereas these voluntary groups were (and in some cases still are) the primary regional planning form they are displaced and/or duplicated by the RTG planning role. Reliability councils appear to be dividing between some who want to become ISOs or poolco operators, others who seek to set adequacy rules for system in their region, and others who favor other new roles in the energy industry.

38 For a selection of recent views, see Curtis (1992), Stuntz (1992), Stalon (1992), the Harvard Electricity Policy Project (1996), and the Proceedings of the Fourth Biannual NARUC-DOE Electricity Forum, Santa Fe, NM, October 21, 1996.

39 The historical difficulty of finding utilities willing to build, and state regulators willing to approve, regional transmission for broad grid purposes was discussed in Chapter 3.

If U.S. firms continue to own both G and T, there will be a long-run counterpart to the need for ISO independence and an unfortunate disincentive to undertake broad-based transmission system improvements. Diversified IOUs may rationally prefer to spend their money on unregulated generator investments rather than regulated transmission system upgrades. This is especially so for expansions that improve regional reliability, but do nothing to improve their particular prospects as generation competitors.

According to Gary Hunt (1990), manager of a municipal utility member of the New England Power Pool, the onset of competition has already caused the traditional willingness of large integrated systems to expand transmission to disappear. Hunt notes that the phenomenon is largely due to competitive rivalry between large systems and between systems and new entrants. In addition, the disappearance of pool-planned units that many systems owned has removed an obvious incentive for broad regional transmission reinforcements that many systems and their regulators would support.

40 Regional regulation is a recurring topic of discussion among state regulators, in print and in private, such as in the "1991 State Regulators' Forum," *Public Utilities Fortnightly,* Nov. 1, 1991, p. 26.

41 See Joskow and Schmalensee (1986), Brown, Einhorn, and Vogelsang (1989), and Olson and Costello (1995).

42 See Comnes (1994) and Patrick (1995).

43 Shepard and Gies (1974) Chapter 1 describes public ownership regulation in useful detail.

44 IRP processes in some states very closely resembled this process for supply side resources (for example, see the IRP process set forth in the final Order in Docket 89-239, Massachusetts Department of Public Utilities, 1990). Another key feature of IRP is equitable treatment of demand and supply-side options. This IRP feature would not necessarily be replicated in column (1) scenarios, depending on how the organization conducting the short-term market accepted bids for demand-side resources.

45 To set the initial price caps, a cost-of-service examination much like a traditional rate determination is sometimes employed. Also, if regulators come to believe a price-cap regulated firm is earning excessive profits, they may examine the firm's cost in the traditional fashion—if they have the authority to do so.

46 When applied in practice, COS and price-cap regulation have exhibited a tendency to converge, i.e., to adopt some of the other's attributes. Price-cap regulators are not supposed to readjust caps in response to changes in profits—the essence of traditional regulation—but this is precisely what has occurred in the U.K. electric market since P.C. regulation was implemented in 1990 (See Offer (1996)). Conversely, traditional regulation has attempted to become more incentive-based, allowing some costs to be passed through or indexed and profits to be increased or decreased based on performance objectives. Nevertheless, the inherent character of the two approaches remains different. For more discussion of this point, see Appendix A.

47 The criticisms have surfaced periodically for years, e.g. in Kahn (1970) V. I. p. 26. U.S. Department of Commerce (1987) and the National Regulatory Research Institute (1991), ch. 2.

48 See Kahn (op cit) and Sherman (1989) p. 211 for two treatments of these issues. Also see Costello and Cho (1991) ch. 2.

49 See Kihlstronard Levhari (1977), Cabral and Riordan (1990), Lewis and Sappington (1992), Fox-Penner (1992), and Miller (1994).

50 Vickers and Yarrow (1989) ch. 1. Also see Vennard (1965), Tolletson (1996), Randolf and Ridley (1987), Kaufman and Dulchinos (1987), Wilcox (1971), and Shepard and Gies (1974).

51 Ibid, section 2.5. Also see the reference in Jones, Tandon and Vogelsang (1990) and Shepard (1965).

52 Some observers feel that nonprofit or public status for the grid will result in inadequate incentives for efficient service; others argue the reverse. See Harvard Electricity Policy Project (1996). Others, such as Newberry (1995a), argue for public ownership.

53 Newberry (1995c) goes so far as to propose public grid ownership; see also the discussion in Backus and Baylis (1996). PNUCC (1996) also notes the advantages of consolidating these two functions; so does California utility activist James Caldwell:

> Transmission assets owned and operated in a regulated "common carrier" mode is perhaps the ideal model. In fact, separation of ownership from operating control presents thorny governance issues to keep short term operation and long term investments incentives properly balanced. However, existing variegated ownership patterns and contractual relationships, and vaguely defined "transmission rights" held by non-owners mitigate against expecting a ("Big Bang") sale of all transmission assets at first market formation. Governance must be designed to accommodate this existing situation or no progress will be made in a reasonable amount of time. In the U.S. however, divestiture of the transmission assets of integrated utilities into separate companies is a difficult financial and legal undertaking.

54 While COS regulation has certainly borne its share of criticism, it would not be fair to say that there is widespread rejection of this approach in favor of PBR, either by academic experts or by practitioners. COS regulation's simplicity, occasionally useful incentive properties, and regulators' extensive experience ameliorating its shortcomings in specific, well-developed ways has given this form of regulation surprising durability and even occasional high marks. For example, Joskow and Schmalensee (1986) and Schmalensee (1989) lean in this direction; more recently, a team of researchers at Lawrence Berkeley National Laboratories studied a sample of actual PBR programs used in the United States during the past few years and concluded that "unless PBR plans are aggressive through their setting of minimum terms,...PBR may be no better than existing cost-of-service/rate of return regulation." (Comnes, et al, 1996, p. 16).

55 Hogan (1996b), p. 35.

56 Hogan (1992), Peck and Chao (1995), and Bushnell and Stoft (1996) suggest that Hogan's financial TCCs are feasible, though often under somewhat restrictive conditions; Oren, et al (1994) cast some doubt on the proposition.

57 Policy Decision, Docket R.94-04-031/I94-04-032, December 20, 1995, p. 40.

58 Ibid.

59 Ilic, et al (1996) p. 49.

60 Market power tends to be explored using the same concepts of *vertical* and *horizontal* we applied to industry structure in Chapter 4. This section primarily discusses horizontal antitrust issues in the generation segment; Chapters 7 and 8 discuss the prevention of vertical market power (by properly structuring the ISO and pool, for example) and regulation of transmission itself for the purpose of preventing "horizontal" market power in the transmission segment. Horizontal and vertical market power issues pertaining largely to retail trade and the distribution sector, will be discussed in the following two chapters.

61 For a detailed summary of transmission-related antitrust considerations, see Barnes (1990); other useful articles include Fairman and Scott (1977), Pace (1984), and Green (1990).

62 See Green and Newberry (1992), Newberry (1995a, b), MacKerron (1996), Newbery and Pollitt (1996), Yarrow (1996), and Offer (1996).

63 "PG&E Puts For Sale Sign on 3,059 MW of Capacity." *Energy Daily,* October 23, 1996, p. 1. Also see Borenstein, *et al* (1996).

64 Management and strategy consultants now counsel utilities to view their generators as a portfolio of assets, much like a mutual fund represents a portfolio of equities. (See, for example, Graves and Read (1996), Weiner, et al (1996) and Hashimoto, Hansen, and van Geyn (1996), as well as Budhraja (1995)).

65 Weiner, et al (1996) notes that "the generation sector will split into two camps: those companies that achieve unprecedented cost reductions *through scale economies in operations and fuel procurement;* and those...that employ radical new generation technologies..." (P. 15, emphasis added).

66 Richard Gilluly, "Shearson Lehman Predicts Utility Marriages But Some Don't Agree," *The Electricity Journal*, July 1988 p. 6.

67 By historical standards, the number of actual and proposed mergers is far from record levels. The 1920s saw hundreds of mergers in a typical year; even as late as the 1960s, there were as many as 40 mergers in a year, according to Breyer and MacAvoy (1971) p. 156 [citing Edison Electric Institute figures]. Depending on whether one counts merger offers that were withdrawn or denied, five to fifteen mergers a year have been occurring in the past few years. (Schreiber, 1996). Also see the Economics Resource Group (1996) (note 70 *infra*) and the discussion in Frankena (1996) p. 3.

68 Weiner, et al (1996) p. 15.

69 "Mergers and Unbundling: Countertrends in Electric Utility Restructuring?" Conference Proceedings, Economics Resource Group, Cambridge, MA, March 1996.

70 Hunt (1996) p. 79; Shepard (1996a, b); Baylis and Backus (1996). See also the discussion of the FERC's merger policies *infra.*

71 These issues are explored in the context of another deregulated industry, motor carriers, in Tye (1987).

72 Of course, this stylized fact does not substitute for a careful examination of competitive conditions in each regional or subregional market. For example, Shepard (1996a, b) argues that some regional markets are already dominated by several firms. In each of the large utility mergers examined by the FERC during the past ten years, economists have analyzed pre- and post-merger horizontal market conditions to determine whether the merger had the potential to increase market power.

73 The Sherman Act, 15 U.S.C. § 1-7, bans monopolies and attempts or conspiracies to monopolize. The Clayton Act, 15 U.S.C. § 12-27, 44, prohibits anticompetitive trading practices and merger or acquisitions that may create market power. The Federal Trade Commission Act, 15 U.S.C.§ 41-57a, creates the commission and prohibits certain methods of competition and deceptive practices.

74 This premerger notification requirement arises from the Hart-Scott-Rodino Antitrust Improvements Act of 1976 15 U.S.C. § 1311-1314; 18 U.S.C. § 1505. It applies only to mergers that meet a complex series of conditions including the buyer and seller, size classes, the percentage of voting control of the company required and other factors. [See 16CFR 801-03.]

75 These actions are most often taken to challenge a proposed merger, but may also involve pricing, trade, or contracting practices believed to injure competition.

76 See appendix A of the Horizontal Merger Guideline of the National Association of Attorneys General, March 10, 1987, Washington, D.C..

77 15 U.S.C. § 15c (a) (2). See Holmes (1988) section 8.09 for further discussion.

78 See Holmes (1988) Section 5.02 for a list of the extensive set of remedies allowed under law.

79 The commission's present policy is to examine six factors originally developed in 1967: the effect of the merger on cost and rates; the contemplated accounting treatment; the reasonableness of the purchase price; the possibility of coercion of the buyer; and the merger's impact on state and federal regulation. [Commonwealth Edison Company and Central Illinois Electric and Gas Company, Docket No. E-7275, 36 F.P.C. 927.] These criteria have been applied and modified in a series of subsequent proceedings, including Utah Power and Light 45 FERC ¶ 61, 095 (1988)], Northeast Utilities 58 FERC ¶61,070, and El Paso Electric 68FERC ¶ 61, 181(1994)] to name a few. See Moot (1991) and Franfena and Owen (1994). Among the other changes, the FERC has focused most strongly on cost and rate impacts, competitive impacts, and the impacts of the merger on regulation.

80 The procedures for merger are set forth in the U.S. Department of Justice and Federal Trade Commission's Merger Guidelines, 4 Trade Reg. Rep. (CCH) ¶ 13, 104 (April 12, 1992); also see Salop and Simons (1994) and Warden (1982). Their applicability to electric utilities (in theory and as practiced by the FERC) is discussed in, among other works, Joskow (1995), Frankena and Owen (1994), Warden (1996), Wall (1995), Shepard (1996 a,b), Peirce (1996), and Frankena (1996). Also see discussion accompanying the following note.

81 Merger Policy Under The Federal Power Act, Notice of Inquiry, RM96-6-000, Order No. 592, Dec. 18, 1996.

82 For the former view, see Penn (1995). Discussion of the subject as a whole is extensive in the Federal Energy Regulatory Commission's recent inquiry into its merger approval policies, *In the Matter of Merger Policy Under the Federal Power Act,* Docket No. RM-96-6-000.

83 These schemes go by a variety of names, including shared savings, profit-sharing, sliding-scale customer service incentive, performance bonus, and others. See Edison Electric Institute (1993).

84 Much of the recent interest was spawned by Vogelsang and Finsinger (1979); the concept is discussed further in Brown, Einhorn, and Vogelsang (1989) and Baron (1989). Of particular interest, one of the few proposals for an incentive-based *long-run* transmission pricing approach involves an optional tariff approach; see Einhorn (1990).

85 In practice, this requires an additional step beyond computing maximum price—namely the calculation of an assumed level of sales. Proponents of energy conservation note that, under price caps, utilities have an incentive to encourage electricity usage because this increases their profit. Revenue caps provide the opposite incentive, as additional sales above the target level can reduce profits. Revenue caps are a form of decoupling that has been implemented and in some cases discarded by U.S. regulators and is still used to set transmission prices in the United Kingdom today (see Box 8-1). For discussions on the pros and cons of revenue caps and decoupling, see Costello (1996).

86 This is precisely the argument made by the United Kingdom regulator in response to criticism of the initial levels of prices after deregulation. After (1996) p. 13

87 See, for example, "Can The Regulators Hold Them?" *The Economist,* June St., 1991, and Studness (1995).

88 Alexander (1996) p. 47.

89 For example, Cabral and Riordan (1989) write: "A clear advantage of price cap regulation is that it reduces the substantial administrative cost of rate-of-return regulation. Under current practice, any proposal by AT&T or a local carrier to change prices is subject to an expensive review process…Under price cap regulation, such a review is triggered only by proposed increases…above the price cap." (P. 93).

90 Loube (1995) and Isaac (1991) examine the behavior of one electric utility operating under price-cap regulation which chose to spin off a generator, thus throwing off the basis of the original price cap. Isaac concludes that it is impossible to determine whether this was a willful manipulation of the regulatory process largely because regulators do not have the information to determine when such manipulations occur. Isaac also found that the regulators in this case were sensitive to public pressure over high profits, which led them to change the price caps prior to the agreed-upon revision date.

91 See Steners (1995) and Xavier (1995) for discussions.

92 See Comnes (1995), ch. 3.

93 Alexander (1996), Woolf and Michals (1995), Marcus and Grueneich (1994), Comnes, et al (1996), and Navarro (1996).

94 Typically PBR formulas specify a range of performance levels and baseline profits rather than a single level. The range in which performance targets may fall without triggering an increase or decrease in profits is sometimes called the "deadband." See Navarro (1996) and Comnes, et al (1995) for discussions of sharing mechanisms and initial profit levels. (Navarro in particular claims that baseline profits are set too high in most cases). Mergers and Vickers (1996) also criticize sharing formula on these and other grounds.

95 See Comnes, et al (1995) ch. 3.

96 Hill and Brown (1995).

Chapter 9

Customer Choice, New Retail Services, and the Distribution System

No aspect of restructuring has captured the imagination of the public and policymakers, nor sparked such violent disagreements, as the idea of consumers selecting their own electricity suppliers. At one hearing on the idea, a manufacturer reportedly stood up and said, 'For everything else I use in my factory, I take competitive bids. Why can't I do this for electricity?'[1] Legislators see it as an extension of the fundamental engine of American economic progress, industries see it as a means of lowering costs, environmentalists see in it new opportunities for clean energy, and independent generators see it as a monumental new market for sales directly to users.[2]

The idea has also ignited vociferous disagreement. Some in the utility industry, its labor organizations, and its regulatory community are concerned that retail choice will diminish reliability and service quality, with few offsetting benefits. Many environmental groups worry that restructuring will increase air pollution, reduce energy efficiency, and stunt the growth of renewable energy. Finally, some consumer advocates fear that restructuring will mean the end of affordable electricity service for low-income families, as well as rising rates and poorer service for residential customers as a whole.[3] As one utility labor leader recently summed it up, "Even the advocates of [retail access] say it is a bad idea whose time has come."[4]

A significant fraction of the controversy over retail competition ("retail choice" or "direct access") actually concerns two consequences of the idea rather than the idea itself. The first consequence is that some utilities may see the value of some of their generation assets greatly diminished by competition—a problem now referred to as *stranded costs*. The second feared side effect is that the variety of public goods now provided by utilities, such as special provisions for low-income customers, will fall victim to retail deregulation. This chapter and the next focus on the issue of retail competition itself; Chapters 12 through 16 examine the public interest aspects of restructuring in greater detail.

CAN POWER BE TRANSPORTED TO END USERS?

Many early discussions about retail choice emphasized that it is utter fiction to think that the power from one generator can be directed to a single user at the other end of the grid.[5] Indeed, this idea was laid to rest in Chapter 2. Yet, despite this physical impossibility, some key features of what is usually regarded as retail competition are achievable in the power industry. Direct access is an unusually complicated form of retail competition—presenting a host of largely unanswered regulatory and economic questions—but it is a real form of competition nonetheless.

The mechanics of competitive retail shopping for electricity begin with the wholesale markets described in Chapters 7 and 8. The prerequisite for direct access is a well-functioning wholesale market and its associated institutions—i.e., an open-access transmission system, a grid operator or ISO, well-defined ancillary services, and so on. This is an extremely critical assumption because, as was seen in the previous four chapters, complete wholesale competition is a brand new policy with many details yet to be resolved. Intuitively, if the wholesale portion of an industry is not competitive, shoppers in retail stores are almost certainly not seeing economically efficient price and supply conditions.

Whether the wholesale marketplace is pool-based or bilateral, the entry points for retail markets are the substations that connect the bulk power network to lower-voltage distribution systems. The electrical features of the distribution system are much like those of the transmission system. Like the high-voltage grid, it must maintain frequency synchronization and voltage support within certain tolerances in order to function at all. It must ensure that power received at its entry points equals power demanded by all retail customers on the system at every moment. Like the grid's area control centers, distribution systems have dispatch and control centers that monitor the network twenty-four hours a day, responding instantly to imbalances and network failures.

The primary engineering difference between the bulk power grid and the distribution system is that the latter is designed with less flexibility and redundancy, particularly near the customers' meters. In a transmission system, there is a system topology and control hierarchy that keep all of the system fully operational in the wake of one or more major component failures. In simple terms, transmission grids are planned and operated to continue to function with the unplanned loss of one or two large lines.

Distribution systems have some redundancy, but as a whole they are one-way, one-path networks that branch into successively smaller limbs as they progress toward the customer. There is much less ability to adjust this system to accommodate a line failure or other problem, short of fixing the failure itself. Although the distribution system is controllable, it is usually at a level that affects several square miles of the system, not at a level that can adjust the grid individually for each customer.[6]

As examined in Chapters 6, 7, and 8, competitive generation requires open- access transmission. This, in turn, requires that prices be set for transmission and ancillary services as best as is possible. The prices paid for these services reimburse the owners of the transmission system for their costs of providing service, including the difficult-to-allocate common goods needed to keep the grid functioning.

In order to allow retail purchase, there must be similar open-access charges for the distribution system. Just as with the transmission system in which the cost of transporting power varies with all other flows on the system, the same is true in the distribution system. This raises a distribution system ("disco") transport pricing problem that is almost as complicated as bulk power transmission pricing. In brief terms, how can a disco give a consumer a price for shipping power to her home, when disco's costs of providing this "transport" depend on unpredictable patterns of demand for all of her neighbors, as well as the prices disco is paying each moment for power and ancillary services received at its connection point to the bulk power network?

In distribution systems, these problems are exacerbated by the large numbers of disparate individual customers, the small size of the transactions, and other attributes of distribution systems.[7] The average distribution utility has about 40,000 customers. Although it is probably *theoretically* feasible for a distribution utility to run a computer that continuously measures the interdependencies of all of its customers and computes efficient hourly distribution rates (spot prices) at each customer's meter, it is far from practically feasible, particularly for residential customers.[8] Instead, a handful of utilities have gotten as far as measuring marginal distribution costs down to good-size districts within their systems for an average period such as a year. For example, the Pacific Gas and Electric Company has refined its costs to reflect differences in 201 "Distribution Planning Areas" that contain an average of more than 20,000 customers each.[9]

Even if it was possible to price each disco transaction separately, the system and its users would still have to contend with distribution system outages, which are a thousand-fold more common than bulk power outages. When part of the distribution system goes dark, retail transport to the dark sections obviously stops, and prices in many other parts of the system may be driven up or down dramatically as flows rapidly readjust themselves. Other features of the distribution system also contribute to greater volatility in marginal distribution costs than are observed in many bulk power systems. One study of distribution marginal costs found that, during just a few hours of each year, distribution costs jump to levels higher than \$2/kWh—more than 20 times normal prices. These studies also show that distribution costs vary greatly by geographic location and year; the latter because distribution system upgrades gradually manifest large economies of scale but must usually be made in large increments.[10] And all this does not begin to touch the long-run issues of disco expansion and generation/transmission tradeoffs, which are examined below.

Economics teaches that the rates for distribution should ideally equal the marginal costs each transaction imposes on the system, with provisions made for ensuring adequate total revenue. For the foreseeable future, however, retail "transportation" tariffs are going to be the same sort of crude approximations of marginal costs that today's "bundled" retail rates represent. In other words, irrespective of large differences in the actual costs of retail transport to different customers, retail transport rates will almost surely be set for a small number of "customer classes" such as residential, commercial, and industrial.[11]

Retail competition per se is not the main determinant of short-term pricing efficiency in the distribution segment of the industry. With or without retail competition, disco users will pay for distribution services, i.e., transport across the distribution system and related disco network support. The level of effort and detail each LDC puts into disaggregating its costs by location and equitably allocating the system's joint and common costs are the critical factors that determine the economic efficiency of the distribution segment of the industry. This will continue to be a matter determined by the LDC and its owner or regulator—perhaps independently of whether that utility offers direct access. Direct access itself does not guarantee that the economic efficiency of power distribution will improve.[12]

THE MECHANICS OF RETAIL ELECTRIC SALES

Assuming, for illustration purposes, that Disco offers open-access distribution rates that are reasonably accurate, as just described. If Consumer can shop for retail power, why shouldn't he see what Acme Company can offer and how the transaction can work?

Acme cannot control the frequency or voltage of power at the delivery point; this is set by the grid for everyone, and Disco's job is to maintain the common standard. In addition, Acme cannot control the reliability of the bulk power system (control is in the hands of the area control center or the ISO) nor can it control how rapidly Disco will come out to repair the wires when a tree limb knocks power out in the neighborhood. If the bulk power marketplace is set up properly, it will have sufficient redundancy and controlability to enable grid operators to keep bulk power reliability high. Accordingly, if Acme's plants suddenly fail, it is likely that Acme's customers will feel no physical effects, as is the case today. When an outage affecting Acme's customers in one distribution area occurs, Acme's plants may be working fine, but they will temporarily have to shut down or (more likely) will instantly switch to "selling" to someone else on the network until the line to its customers are repaired, at which point Acme will then switch back to selling to its customers (more on this below).

The agreement with Acme will thus likely revolve primarily around the price of service, which may vary by time of day and season, and other financial and pricing terms. It may contain price guarantees, offer discounts or promotional rates, and there may be a number of associated products or services, including frequent-flyer style reward programs to encourage brand loyalty to purchasing from Acme.[13]

This is a somewhat different form of retail purchase than a visit to the local hardware store or dry cleaner. First, reliability—the single most evident aspect of what consumers see as the quality of electricity service, is virtually 100% outside the control of any competitive supplier. Reliability requires that the marketplace be set up so that the grid interacts with all suppliers and that the disco works efficiently to maintain local reliability. The link between Acme and system reliability is reflected (along with many other factors) in Acme's price.

Under retail competition, buyers essentially shop for an efficient but entirely invisible upstream operator—someone who aggregates supplies in clever ways, strikes the hardest bargain with generators or fuel suppliers, builds and operates generators more cheaply than others, and is good at playing by the rules of the pool or the ISO. Almost everything they do to differentiate themselves as competitors is invisible to its customers and not felt in the physical aspects of the supply. Instead, it is felt by customers in the prices they are offered. As an example, if the bulk power marketplace has been set up effectively, generators that are not reliable will have to buy backup power and other ancillary services from the grid (see Chapter 7) more frequently than others. Because these services may be quite expensive, an inefficient plant operator's costs and sales prices will soon rise relative to its competitors, and shoppers will presumably reward the cheaper generators.

Incidentally, there is no reason why Acme has to own any power plants. Acme may simply be a company that contracts with independent generators, the grid, distribution systems, and retail customers, buying and selling to make a profit. A variety of terms have evolved to describe these types of companies, many of which already exist today; four common names are *power marketers, aggregators, supply coordinators,* and *marketcos.* Some observers predict that most retail suppliers will not own plants—the very antithesis of vertical integration—so that they can be free to buy from whomever is cheapest at the moment. However, there is also evidence that strong marketers will be closely associated with large, diversified generation companies.[14]

The ownership of power plants by deregulated retail sellers is one of a host of issues related to the administrative or transactions costs of efficient retail supply. As described in Chapter 4, the boundaries of firms (i.e., the set of products firms offer and the extent of vertical integration) are determined by the cost penalties or savings involved in owning versus contracting for services. Therefore, if it is cheaper for Acme to own generators rather than to contract for them, it will do so, and vice-versa. These determinations will revolve not just around the estimated costs of power plants, but also around the costs of striking agreements between gencos and marketcos, the complexity of these agreements, poolco or ISO rules, and so on.

Recall from Chapter 3 that no matter how many units of service are signed up for with Acme, or how usage may vary from hour to hour, the system must always remain in immediate supply-demand balance. Even if Acme is not required by its agreements with customers to supply everything they need at every moment, somewhere there must be a supply of balancing power to make sure balance occurs.

Every retail seller will therefore be working with the grid and the disco to account for every unit of power its customers are using in every instant (more practically, each fraction of an hour). There must be a system spot market or backup power source that the grid or the disco can call on to balance total load and generation, and interconnected operating services must be supplied regionally by the network and locally by the disco. If an Acme customer takes more than the amount he or she contracted for with Acme, that customer will pay an "imbalance" penalty. If Acme fails to meet its commitments to the network, it will likely be liable for penalties.[15] This helps explain why the Connecticut Department of Public Utility Control stated flatly in a 1994 investigation of retail access, that "...the Department believes that retail wheeling tariffs or contracts should contain provisions that would allow the local utility to exercise some degree of control over loads served by others on its system to maintain the quality and reliability of service."[16]

Some idea of the complexity of these arrangements is illustrated in Figure 9-1, which reproduces the rules proposed by the Pacific Gas and Electric Company (PG&E) in 1994 for implementing retail choice. In these rules, PG&E used the term "supply coordinator" to refer to retail sellers. As the figure shows, retail sellers have a number of complex responsibilities to the operators of the transmission and distribution systems in order for the network to work.[17] Retail sellers will have to do many of the things that discos do now, sometimes as part of an integrated utility, and they will be required to follow extensive procedures and rules set by ISOs, transcos, discos, and state and federal regulators.[18] Meanwhile, the distribution system will probably increase its regulated activities.

Rather than make individual customers go through all of the steps in Figure 9-1, supply coordinators (or whatever they're called) will make most or all of the arrangements that are necessary to get power to consumers' homes. These firms, which will contract with gencos, the grid, and the disco, will provide retail customers with "one-stop shopping." They won't be able to control transmission or disco rates, but they can take care of all the scheduling, rules, and billing paperwork and might consolidate all payments into one monthly bill.[19] If they are efficient in their procurement, administration and use of the grid, they can compete on the cost of the total package offered. Nevertheless, power bills after retail choice will have to be more complex than they are today in order to allow customers to evaluate competing offers. Residential power bills will most likely include one or more payments to a local distributor, additional payments to the state or regional network, a separate payment to one or more generation companies, and other taxes and charges, including those arising from public interest programs.

Because this is an important and confusing subject, it is worthwhile to recap:

- No, power cannot be selected and shipped from one generator to a specific customer.

- Individual generators will have almost no control over the quality or reliability of electric service, with or without retail choice.

- Yes, individual customers can sign contracts with generators which mainly involve prices and quantities. Additional non-power services will also be offered.

- Economic efficiency gains in the distribution segment of the industry depend on the quality and accuracy of distribution price-setting, not on retail competition per se.

- Marketers or other "middlemen" will simplify billing, but retail choice requires someone to make complex, separate arrangements with generation, transmission, and distribution companies.

NEW RETAIL ARRANGEMENTS

Right now, it is legally possible for distribution systems or integrated utilities in many states to sell many products other than power. Two common examples are home appliances or energy conservation services. In many places, utilities provide such services under the watchful eyes of state regulators. Some utilities lease or provide discounts on home appliances; others provide energy management services to large customers. A number of utilities sell both electric and gas, and rural electric cooperatives offer their members insurance and other products.[20]

Figure 9-1. Pacific Gas and Electric's Proposed Rules for Retail Suppliers[1]

Role of the Supply Coordinator

When a supply coordinator is selected by a direct access customer, the coordinator becomes a single point of responsibility for meeting an eligible customer's supply needs. Customers may also act as their own supply coordinator.

Responsibilities:

The supply coordinator would:

- Arrange for all supply to meet the customer's load
- Arrange transmission and distribution grid services and comply with established grid procedures
- Meet current standards for reliability, control area service[2], and operating requirements
- Match supply with customer's load, and account for losses
- Schedule supply sources with the grid operator

These are the same responsibilities marketers and utilities must comply with in today's wholesale market.

What Supply Coordinators Are Billed For

The supply coordinator would receive a bill from the host utility grid operator for regulatory approved cost-based charges for:

- Grid services[3]

To the extent that the following services are not provided by the supply coordinators themselves, a bill from the host utility or other suppliers would be received for:

- Generation supply
- Matching of generation supply to customer load
- Generation supply reserves
- Other power services

And, if necessary, a bill for the settlement of accounts from a settlement agent[4] for:

- Unscheduled power deliveries (imbalances)

---

1 Source: Direct Access: The First Three Years "Working Draft" Handout, Sept. 13, 1994.

2 Control area services include backup supply, voltage regulation, operating reserves, etc.

3 For example, supply power factor charges, supply interconnection charges, grid congestion charges, scheduling fees, automatic generation control, if necessary, etc.

4 This would be the host utility or an independent entity with responsibility for settlement of accounts. Parties would have audit rights to the settlement process.

For a variety of reasons, utility involvement in products other than pure electricity service has waxed and waned over the decades. At least until 1925, utilities themselves were responsible for introducing and marketing virtually all of the early electric home appliances—lamps, refrigerators, electric fans, radios, irons and washing machines.[21] Although this tradition continues in some utilities, as a whole, the utility industry gradually withdrew from direct sales of electricity-using goods. One evident reason, at least among IOUs, was the increased regulatory scrutiny that diversification triggers. In addition to satisfying the requirements of the Public Utility Holding Company Act, regulated utilities engaged in unregulated ventures had to regularly assure that they were not shifting costs from unregulated to regulated business lines by engaging in harmful "self dealing."[22] These concerns generally led to utilities focusing on promotional schemes aimed at increasing sales of electricity via increased sales of appliances by others, rather than utilities themselves selling the appliances. In the 1980s, investor-owned utilities went through a short-lived wave of diversifications into non-electricity businesses.[23] At the same time, the trend toward utility-sponsored energy efficiency programs renewed some utilities' involvement in downstream equipment, although sometimes not as a profit center.[24]

If it does nothing else, the prospect of retail competition has largely severed this regulation-induced hesitation toward actively experimenting with new product and service offerings. Marketcos are not price-regulated and generally will not be subject to the Holding Company Act. They will be roughly as free as any other retailer to sell whatever bundles of products and services, in addition to electric power, they find profitable. What might these services be?

REAL-TIME PRICES AND NEW RETAIL SERVICES

Although it is difficult to predict the new bundles of products that will ultimately prove successful, Table 9.1 summarizes the *kinds* of service that are likely to be offered. This extremely broad array may be grouped into the three categories shown on the three rows of the chart. The first category (row 1, Table 9.1) involves systems that alter or control customers' electricity use, or otherwise monitor or change power system usage patterns.

Prior to the 1970s, electricity rates varied with volume or size of load, but rarely by time period. When oil price shocks, inflation, and high construction costs drove rates up in the 1970s, the industry and its regulators began to experiment with time sensitive pricing as a means of reducing demand—particularly peak-period demand, which is the most costly to serve. A number of experiments were conducted in which residential, commercial, and industrial customers were given several price levels for different parts of the day.[25]

Among industrial customers, time-sensitive rates were generally well-received; many factories found that they could reschedule at least some usage to lower-cost periods by rearranging production.[26] Conversely, many residential ratepayers objected to the practice based on a perception that it was unfair to charge dramatically higher prices to those whose schedules did not permit shifting the time of their maximum power use.[27] This led regulators to limit time-based price increases to residential users, a move industrial customers opposed.[28]

Table 9.1. Likely Categories of Retail Services Sold by Electric Retailers

Type of Service	Present Status	Examples of Expanded Offerings Following Retail Choice
Services linked to monitoring or changing electricity demand based on system prices or conditions	• Rapid growth of real-time pricing systems offered by discos • Utilities and unregulated manufacturers testing and selling some control systems	• Home and office real-time pricing systems sold and serviced by third parties • Backup power supplies
Services that power retailers may find profitable to offer	• Some public and IOUs offer products (generally appliances), subject to scrutiny of regulators and antitrust officials	• Other fuels and fuel purchasing services • Telephone communications services • Price guarantees or hedges for power costs • "Smart" appliances or onsite power generation
Services that make use of the distribution company's infrastructure	• Generally set by state regulator policies and approvals	• Installing communications systems that use power lines or conduits • Metering and billing services

Early experiments with "time of use" pricing used one or two pre-determined periods of price increase based on estimates of cost differentials in these periods. As communications and computing systems have gotten stronger and cheaper, the industry has moved to the point where the same hourly prices system operators use for the bulk power markets can be communicated to large customers at little expense.[29] (Of course, the actual prices users pay include transmission and distribution charges, but so far these have rarely been time-varying). Users who are able to use these price signals to shift loads and save money are better off, and society uses lower-cost resources to generate electricity.[30]

Figure 9-2 shows a pilot real-time pricing control setup that is now operating in a large hotel in New York City. This system reads hourly prices provided by the local utility and uses pre-programmed computers to turn off various air handlers, fans, and other equipment based on detailed calculations of cost savings realized if the periods of equipment use can be shifted back or forth by one or two hours. For example, during a high-cost period, the speed of some fans might be reduced and then restarted at an elevated speed during a low-cost period. Continuous measurement of temperature and air quality inside the building can ensure that the internal climate remains healthy and comfortable, overriding the cost-based controls if necessary.

Figure 9-2. Simplified Diagram of a Real-Time Pricing Control System in a Large Hotel

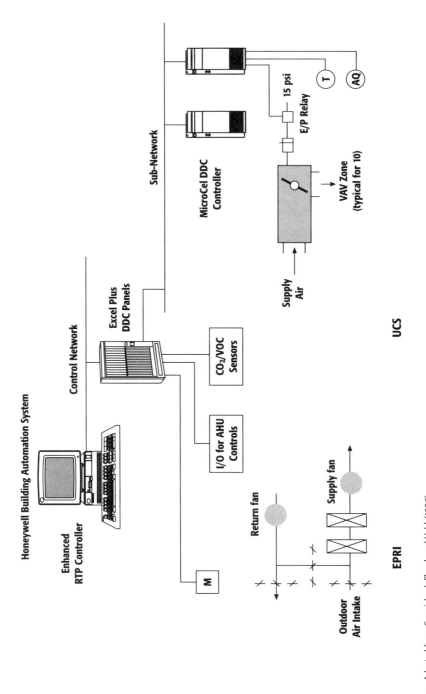

Source: Adapted from Carmichael, Flood, and Vold (1996)

Figure 9-3. Actual Daily Reduction In Energy Use and Cost in Hotel RTP Pilot

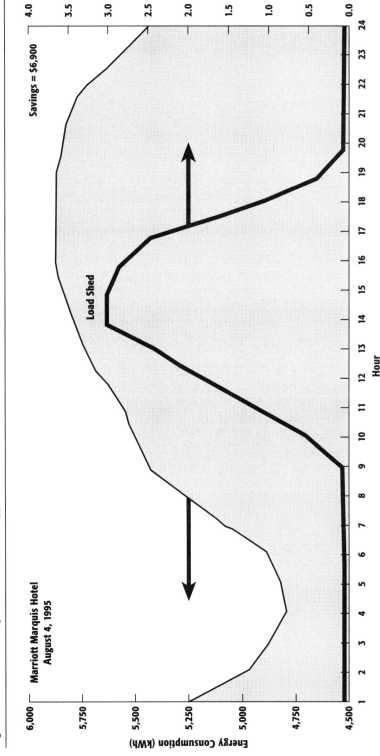

As real-time prices rose from under 10 cents/kWh to a high of 300 cents/kWh, the RTP Control System reduced loads below what they could have been without RTP. The result was less use during high-cost periods and a savings of $6,900 in electricity costs for the hotel this day.

Source: Adapted from Carmichael, Flood, and Void (1996)

Figure 9-3 shows the difference in the pattern of this building's use of power before and after installing the real-time control system for one actual day during 1995. As the hotel's actual prices rose to a peak of $3.00/kWh, the control system successfully reduced usage during high-cost periods by the shaded amount shown, yielding a total savings of $6,900 for this one day. Undoubtedly the altered operations of the building equipment reduced the comfort of some of its inhabitants, but most experts believe that substantial savings can be achieved with little or no change in comfort if real-time systems are carefully engineered and operated.

Similar cost-lowering circuitry can be installed in homes, factories, stores, and other establishments. In addition to enabling lower use in reponse to price, the new equipment may be able to use a communication network to remote-control home appliances, remotely read utility meters, monitor security alarms, and so on. The terms "smart home" and "smart buildings" refer to structures that have installed appliances and internal wiring to allow for such control. These innovations, which appear in limited applications and pilot programs today, are rapidly expanding across the United States.[31]

If power retailers begin to sell real-time communications and control services, it is possible that they will also begin to sell or lease the "smart appliances" that these systems can control. With reduced regulatory oversight, these activities could be taken over by new entities and expanded into many more types of equipment under far more varied sales or leasing arrangements. It is conceivable, for example, that a power retailer could own, install, and operate the air conditioners in large buildings, or own and operate roofing surfaces that double as photoelectric power generators.[32] When these arrangements include power production or storage equipment in addition to demand-altering appliances, they lead to the concept of a "distributed utility," discussed briefly in Chapter 2.

Time-varying electric prices present a tremendous new opportunity for electricity retailers and related businesses. In this context, the role of retail competition is merely to produce economically accurate hourly prices. If they can be produced any other way with no offsetting disadvantages, then many of the benefits of retail competition could be provided to customers without what most folks think is the core of retail choice, i.e., shopping for one's own seller.

Power market futures contracts, options, forward contracts, and other new kinds of power purchase and financial contracts will also play a role in the new retail products—with or without retail choice per se. If electricity prices vary unpredictably over time, experience shows that markets will develop products that ensure against or transfer price risks between parties. These products will probably not be used directly by individual households, but utilities, power markets and other sellers and middlemen will use them extensively to create products for retail sale.[33]

These ideas and new products underly a proposed alternative to retail competition which Professor William Hogan has dubbed *efficient direct access*.[34] Hogan and others note that, if a wholesale market is correctly set up to produce accurate hourly prices, and these prices are made available by traditional franchised monopoly discos in place of the simple retail tariffs they use today, then every retail customer will be able to achieve the equivalent level of efficiency as in a robust competitive retail marketplace without the hassle of shopping and contracting for suppliers. (See Chapter 10 for a comparison of efficient direct access to actual retail choice.)

OTHER RETAIL SERVICES

The second category of expanded retail products in Table 9.1 are products and services that do not use utilities' physical infrastructure or involve electricity, but may be effectively sold by power retailers. Some of these products are closely linked to the in-home or in-office control systems that also control power. For example, a recent *Wall Street Journal* article reported that an electric utility had entered into a joint venture with a telephone company that not only produced systems that controlled appliances, but also provided burglar alarms, medical monitoring and many other possible services over a wireless communications network.[35] Even farther from electric power, utilities may launch credit cards (as did AT&T) or market goods or services that are only tangentially related to their core business.

The final category in Table 9.1 encompasses services that make use of utilities' infrastructure, but do not involve electricity per se. In many parts of the United States, utilities are installing communications (fiber optic) cable in the trenches that they hold exclusive or limited rights to use. Other utilities are exploring communications networks that make use of their utility poles or selling metering and billing services to other companies.[36] Because some of these new products involve communications systems, some utilities see a synergy advantage in owning the communication system that will provide power on non-power related services.[37]

These category-three products and services can, for the most part, be offered only by the disco and transco segments of the industry, which own the electric network infrastructure. In many cases, advance regulatory approval is required before utilities can provide these services and regulators will undoubtedly scrutinize the non-power businesses to ensure that costs are not shifted. In other words, the growth of this category of services depends strongly on how they are viewed by state and federal regulators in the context of massive industry change.

Going back to the earlier example, Acme's decision to own power plants was driven by the transactions cost savings of integration vis-a-vis the benefits of competitive outsourcing. The same own-versus-buy tradeoff applies to Acme's decisions to offer any of the unregulated services in Table 9.1.[38] If adding a service reduces costs or provides a marketing advantage, sellers can be expected to pursue the addition. Many believe that cross-product synergies (or scope economies) will force most successful retailers to sell all of the major end-user fuels (e.g., natural gas, electricity, propane, kerosene, and other fuels), as well as energy management services.[39] Returning to a point raised by environmentalists in the 1970s,[40] many argue that future retailers will not sell electricity or gas, but simply the cheapest means of providing heat or other services. The idea that the retailing of gas, electricity, and other fuels and services will be joined has been christened "convergence" or "the BTU marketplace."[41]

PRICE-DIFFERENTIATED RELIABILITY AND SYSTEM EFFICIENCY

The preceeding example of buying retail power from Acme illustrated that reliability had no connection to Acme's reliability as a generator (other than a small, indirect contribution to price). Factoring in real-time pricing and communications and control hardware, this statement must be modified in a complex but important way.

Proper real-time prices are an indication of the degree of capacity scarcity on the regional system at any given time. Experience has shown that some power customers would actually prefer to disconnect themselves from the power system during periods when prices are high, reducing their costs along with their "reliability" (assuming reliability in this sense includes the number of minutes during which a customer chooses not to have power, as well as involuntary losses of service).[42] Customized retail arrangements that include real-time prices, therefore, allow willing customers to partly control their reliability. They cannot use these custom arrangements to *increase* their own reliability above the levels the system provides, but they can lower their reliability (indirectly increasing it for others) and save money for themselves and the system as a whole.[43]

This is a remarkable improvement over the present power system because it is a win-win situation for everyone. Power users who do this voluntarily are better off because they have lower bills that must be worth the inconvenience of their self-chosen disconnections. Every other power user in the region is also better off, because by disconnecting during high-price periods, peak prices are reduced and power system reliability increases for everyone who does not disconnect. By pricing electricity more accurately in each time period, those who value power highly during high-price periods can continue to use it. Those who do not value power highly are prompted to shift their use to periods when power is cheaper for everyone.

Some idea of the enormity of the potential for savings this rationalization provides appears in a paper by Professors Samuel Oren and Dennis Ray who construct a hypothetical utility with eight types of customers, each with a group load of 300 MW, a group average willingness to disconnect, and a group average cost of enduring a random outage. Under the authors' simplified assumptions, the total generation costs of serving the entire fictional system are reduced by as much as 80% when classes can choose from a carefully-developed menu of self-curtailment options.[44]

There is no guarantee that anything approaching these sorts of savings can be produced in the real world, but neither is there doubt that real-time pricing, reliability self-selection, and other RTP-related services will produce substantial opportunities for efficiency improvements. Remarkably, these efficiency improvements shared by customers of all sizes, even if they do not themselves get or use RTP, provided costs are not shifted into these customers' rates.[45] Equally remarkably, these costs do not *require* retail choice, nor are the estimated cost savings highly dependent on the nuances of wholesale or retail market organization. A better time structure of prices will allow enormous efficiency gains so long as the prices are more accurate than today's annual flat rates.

DISTRIBUTION SYSTEM GOVERNANCE

The industry's distribution systems will remain regulated whether or not retail choice is implemented. However, the issue of who has the authority to order or allow retail access may not yet be fully settled. At present, the FERC has taken the rather complex position that (1) The Energy Policy Act of 1992 prohibits it from ordering or approving retail access or choice, but (2) if anyone else orders or approves it, the *rates* that transcos or discos charge for retail transport are FERC-jurisdictional.[46]

If legislators and the courts ultimately uphold FERC ratemaking authority over disco transport, the governance of discos will be uniquely complex. State regulators will undoubtedly want to continue to approve most regulated disco actions; state and federal regulators will have to coordinate as they set rates for transport (at the federal level) and sale. (To avoid making the rest of this discussion as complex as the real situation may actually be, vague reference will be made to the disco's regulator from now on without further specificity.[47])

The next step is to examine the issues involved in regulating distribution with and without retail choice, assuming in both instances a well-functioning competitive wholesale market. The first scenario—continued exclusive franchises and regulated disco sales—is similar to today's industry, with the exception that certain changes underway today will continue. The bulk power market will more fully reflect open-access transmission, and local distributors will more commonly offer RTP and other service enhancements. States and municipalities will continue to govern their local systems, perhaps transitioning to performance-based regulation (PBR). If the wholesale power marketplace is functioning well, each disco and its regulators will have a spot-price benchmark against which any disco's power costs can be compared, even if the disco chooses to own and operate most of its own capacity.

In the retail choice scenario, regulators' first task is to set regulated prices for the "transport" of power from the system's bulk power gateway to each retail destination zone.

Other than the differences in topology and adjustability, discussed at the outset of this chapter, the difficulties in setting such rates are similar to the difficulties involved in setting unbundled transmission prices in the bulk power system which are discussed in Chapters 7 and 8. To review briefly, there are four types of issues:

1. **Non-Discriminatory Poolco or Bilateral Access to Distribution.** Although it is probably easier to mimic whatever is done at the regional level, there is the option of setting up a distribution-system poolco or bilateral trading regime. Either way, the system must be non-discriminatory, and, in particular, must not favor generators affiliated with the disco. In the bulk power market, this has produced great pressure either for complete G-T divestiture or for the creation of an ISO. Although disco independence has not received nearly as much attention, a few practitioners have called for divestiture to include the distribution system, as well as transmission.[48]

2. **Distribution Network Services.** As in the bulk power grid, not every cost of operating the disco system can be attributed to transportation. Hence, disco-level integrated operating services must be defined and procured efficiently. Just as the ISO will have to make sure the bulk power network maintains sufficient spinning reserve and voltage support to keep the bulk power network up, the disco must buy or self-supply reactive power and other network services. Reactive power needs and real power losses, in particular, can be quite significant in distribution systems. Thus far, there are few options for getting reactive power other than through the facilities discos themselves own.[49] At least for starters, services like this will be provided by the disco, and charged to all users according to procedures approved by state regulators.

3. **Revenue Adequacy.** Disco rates for transport and ancillary services must be set so that the disco receives enough revenue to cover its costs plus a fair return. As with transmission rates, theory and actual data suggest that charging measurable marginal costs does not produce anywhere near enough revenue. Additional funds must be collected either by allocating non-marginal costs to each customer or seller by class or some other way.

4. **Optimal System Expansion.** The disco must be given incentives to expand efficiently and to make appropriate tradeoffs between expanded distribution lines and generation alternatives. For example, suppose that it was true that the installation of a small generator at the end of a long disco line was more efficient than adding more transport capacity to the line. Does the disco have the incentive to pursue this least-cost solution if it is simply a wires company?[50] Such questions have not yet been resolved.

Chapter 3 discussed the common difficulties of siting transmission lines. Discos have smaller-scale versions of these same problems; by some accounts their siting problems can be as difficult as those for large lines. Adding distribution capacity in urban areas may mean digging up streets or putting up transmission poles in alleyways, along roads, or across parks; any of these activities can spark vehement local opposition. New disco lines can also be expensive, e.g., land and land rights in some urban areas are very costly, and the refusal of certain property owners to grant easements can necessitate expensive, circuitous wiring routes.

Today, this is all done with the mindset that the disco has an absolute obligation to serve all customers with no discrimination. Moreover, discos earn a regulated return on their lines, and they also benefit by selling more of their own power. At least in theory, all this gives the disco the incentive to press forward with maintaining system adequacy. To some extent, regulators will continue to impose these obligations if retail choice is adopted.

As examined in Chapter 3 the controversy over how to charge customers for transmission system expansions has raged for many years in the United Kingdom as well as the United States. In simple terms, there is usually a fundamental choice between charging specific growing customers for the cost of the expansion versus "rolling in" expansion costs into the prices paid by all customers. The problem was exacerbated by the fact that optimal expansions are usually much larger than the one or two transactions that necessitated the upgrade.

Many of the same problems apply to disco system expansions. If an office building expands, and the local network needs reinforcement, who should pay for this expansion? If the reinforcement makes the entire network more reliable, how is the value of more reliability for one user versus the value of more capacity by another divided among users?

In principle, there is nothing new about these problems, nor are they tied to retail competition per se. Just as it is today, it will probably remain the case that regulators will determine the fair allocation of distribution system capital costs and reflect these in LDC transport tariffs. Although the incentives exist today, it is possible that retail choice creates an added incentive to appeal to a commission to try and reduce particular customers' transport costs. For example, it may seem less unusual for a regulator to customize retail transport rates to individual users (particularly large ones) on the basis of transmission cost studies than it is to alter full-delivered retail prices.

Given the technical nature of the power grid, many of these issues involve both transmission and distribution. Even if there come to be no issues of cost allocation or legal jurisdiction involving network costs, discos will have to participate in and transact with the bulk power network as well as with state regulators. As with most of today's utilities, discos, acting as transporters will be the servants of two masters: (1) federal regulators and regional entities such as ISOs that set network rules and rates, and (2) state regulators who rule on whether discos acted prudently with respect to the amount and cost of each service taken from the wholesale market.[51] This will be true even if all discos are wholly independent and limited strictly to transportation, both of which are somewhat extreme assumptions.[52]

THE DISCO AS SELLER OR RELIABILITY PROVIDER OF LAST RESORT

Whether or not retail choice is allowed and the disco offers regulated transport rates, it will, in some fashion undoubtedly, be required to provide the same sort of power service it provides today under regulated rates to consumers who choose not to shop. Few question the necessity of requiring this function—indeed, it is a technical necessity, much like requiring the ISO to provide immediate balancing services at wholesale. This function adds greatly to the task of regulating discos.

There are two basic approaches to the provider-of-last-resort function. One approach is to place all customers who do not choose to select a retail provider into a group whose power needs appear to continue to be provided by the existing disco—from the vantage point of these customers. However, rather than supply the power itself, the disco will take bids from independent generators to supply the needs of its group. The costs of transport and other services will be added in and the price to this group will roughly equal the average costs of supply. In other words, the disco will remain a cost-of-service regulated retailer to this group, but its power will be purchased on the wholesale market.

Under the second approach, the disco is not required to purchase all power for this same group, but may instead supply some or all of it from its own integrated plants. Prices will again be cost-of-service regulated. The key difference between the two approaches is that the first requires 100% purchase (deintegration), whereas the second allows integrated supply.

If regulators permit the second approach, they will face the challenge of balancing the disco's responsibilities as seller with its responsibilities to provide open access to all other retailers who are "shippers" on its system. These two roles can present diametrically opposed incentives for the disco and its regulators. On the one hand, the disco-as-seller may wish to aggressively court sales and retain customers. Regulators may prefer this because greater sales may lead to lower costs or less financial risks being passed on to the sales-customers who undoubtedly strongly prefer the same. At the same time, the disco must provide a "level playing field" that allows all other sellers to compete fairly against each other and against the disco-as-seller.[53]

These challenges have not escaped the attention of state regulators. In the state regulatory proceedings that have examined retail competition, there has been extensive discussion about the need for a residual sales function, as well as the need for fair and efficient disco access.[54] The discussion has considered, among other things, the rules under which customers can leave and return to the regulated or "core" sales service, the prices for core services, and the flexibility the discos should have over pricing core services.[55] For example, the California Public Utility Commission's first order leading to the state's recently adopted retail choice plan stated:[56]

> To allow eligible consumers to choose without restriction between the regulated price for bundled utility service and the price offered by the generation services market may severely reduce the utility's ability to plan for, and reliably serve, its remaining customers. Absent modifications to the compact's traditional duty to service, consumers may make choices about electric services which they find economically attractive, but which are undesirable with respect to the broader goal of allocating society's resources efficiently. That is, the consumer may choose to leave the utility franchise because the *regulated price* the utility charges is higher than the price offered in the generation market for comparable service, despite the fact that the utility's *marginal cost* of providing the service is actually less. This Commission has aggressively developed policies to discourage what is referred to as "uneconomic bypass" in each of the industries we oversee. As such, uneconomic bypass by no means represents a novel challenge to California's utilities, or to this Commission. Indeed, our policies designed to deter uneconomic bypass in California's electric services industry have enjoyed considerable success. Yet despite this success, we are nonetheless concerned about two troubling aspects of those policies. With the threat of bypass on the rise again, we believe it is appropriate to revisit those policies....

> With these considerations in mind, we propose that the following principles comprise California's policy governing its investor-owned electric utilities' duty to serve in a new era marked by consumer choice through direct access:

> • The utility may compete to retain direct access consumers based on disaggregated prices and services.

- The utility may negotiate prices with direct access consumers which diverge from tariffs for generation and generation-related services (e.g., coordination and system control services, backup services, etc.). Those prices will receive streamlined regulatory scrutiny and enjoy expedited approval procedures. The terms and conditions governing pricing flexibility will be established as part of the service unbundling proceeding we propose. At a minimum, we propose to allow the utility to freely negotiate prices with direct access customers as long as those prices do not exceed current tariffs or fall below the company's marginal cost.

- Utility shareholders will contribute to the full recovery of revenue shortfall and receive the gains resulting from any price discounts the utility offers consumers. We will establish a revenue or price cap framework to govern utility operations related to direct access service. Like caps applied in other industries (and proposed by California's utilities in their PBR initiatives), the framework will include a formula which explicitly accounts for inflation, productivity, and events beyond utility control (the so-called "Z factor"). In addition, the framework will include a symmetric band, comprised of a floor and ceiling, governing utility earnings. The ceiling will limit the aggregate amount utility shareholders may earn, while the floor will limit shareholders' exposure to revenue losses due to increased competition.

- Direct access consumers will contribute to the recovery of the uneconomic portion, if any, of the utility's generating assets resulting from our new competitive framework. Direct access consumers will make that contribution in the form of a "competition transition charge" assessed as part of the demand charge. We will open a proceeding to establish the mechanism governing the competition transition charge.

It is critical to recognize a key component of our proposal. The utility is *not* at risk for the total revenues customers eligible for direct access currently contribute to the utility's revenue requirement. On the contrary, the actual amount at risk is a fraction of that contribution. Under our proposal, the utility is at risk *only* for those revenues tied to the *economic* portion of the utility's generating assets, and any overhead tied to the delivery of generation services. We do not propose to place at risk the *uneconomic* portion of the utilities generating assets.

The second approach was taken by the Rhode Island legislature, which recently passed a retail competition bill under which the residual supplier role must be carried out by purchasing generation from the wholesale market:[57]

In recognition that electricity is an essential service, each electric distribution company shall, within three (3) months after retail access is available to forty percent (40%) or more of the kilowatt-hour sales in New England, arrange for a last resort power supply for customers who are no longer eligible to receive service under the standard offer and not adequately supplied by the market because they are unable to obtain or retain electric service from nonregulated power producers. The electric distribution company shall periodically solicit bids from nonregulated power producers for such service at market prices plus a fixed contribution from the electric distribution company. Acceptance of bids by the electric distribution company and the terms and conditions for such last resort service shall be subject to approval by the Commission. The bids requiring the lowest fixed contribution from the electric distribution company shall be accepted.... All fixed contributions and any reasonable costs incurred by the electric distribution company in arranging this service shall be included in the distribution rates charged to all other customers.

Other states are exploring various other approaches to the provider-of-last-resort issue. The important point is that no utility restructuring scheme has as yet required *all* customers to stop buying bundled service, leaving an important and sometimes large sales role for the disco.[58]

The tension between systems' needs to cooperate on grid planning and the emerging desire to compete (see Chapter 8) will intensify further under retail competition.[59] Already, disco utilities are hesitant to share planning information about their systems with other discos. With increasing frequency, regulators are being asked to review utility filings under confidentiality agreements, pitting the public's right to observe the governance of the distribution monopoly with the need not to compromise the viability of the disco-as-seller by revealing its business information to competitors.[60] Meanwhile, planning to serve the needs of customers who may come and go from the regulated system, but are nevertheless guaranteed service, may be quite difficult.[61]

Another delicate area of regulatory balance concerns the customer information that traditional distributors hold on each customer account. Many potential retail sellers argue that this information confers a marketing advantage on the disco-as-seller or its unregulated marketing affiliate. In the few states in which retail choice has been implemented, regulators have required discos to share certain customer information with all interested potential sellers. This, in turn, raises concerns with consumer protection advocates, who sometimes fear that sensitive customer information is being too widely disseminated.[62]

DISCO INVOLVEMENT IN UNREGULATED VENTURES

Beyond the tensions between its sales and transport roles, disco involvement in unregulated businesses raises an additional set of concerns. As noted earlier in this chapter, the evolution of technologies and markets will unquestionably encourage retail electric sellers to explore integration into communications networks, control systems and controllable appliances, other fuels, energy and fuels management, and other retail services. Synergies may be involved in allowing the regulated disco to sell some of these services.[63]

Allowing a single firm to own regulated and unregulated lines of business that trade with each other may present incentives for the firm to shift costs from the unregulated to the regulated portion of the business. Shifting of costs from the unregulated businesses to the costs of the disco operation is economically harmful in two ways. First, shifts of this nature cause distribution rates to be higher than true costs, harming all who purchase from the disco. The second harm falls to the marketplace for the unregulated good. Here, the disco's ability to shift costs would allow it to charge a price lower than its true costs, sending a false economic signal to the marketplace. Moreover, if other unregulated sellers cannot compete with the utility's artificially low price, they may be driven out of business, and fair competition will suffer.[64]

As discussed in Chapter 4, one of the purposes of the Public Utility Holding Company Act of 1935 was to protect against this sort of competitive harm. Many state public service commission statutes also contain provisions empowering regulators to examine transactions between discos and their affiliates. Regulators may lack authority to prevent a disco from entering an unregulated business, but they will likely retain the ability to disallow the costs of affiliate transactions in regulated prices and to study the synergy benefits of integrating disco and other retail services. Regulators will have to balance the unique benefits of disco vertical or multi-product integration with the risks of affiliated transaction abuse and the cost and quality of enforcement efforts. Of course, this tradeoff must be tempered by the scope of the regulators' specific authorities as well as other important practical considerations.

DISTRIBUTION GOVERNANCE: SUMMARY

All of this provides a picture of distribution governance and transactions costs that is perhaps even more complicated than many of today's ratesetting, planning, and investment approval processes (Figure 9-4). Discos must set rates for transport and ancillary services, recover adequate revenues, treat all retail sellers comparably, plan and expand the system efficiently, obtain siting approvals, be extremely careful with affiliate transactions and ventures in unregulated areas, aggressively pursue unregulated ventures when there are unique synergies, serve as a seller of last resort, maintain reliability and quality of service, maintain revenue adequacy and accommodate public interest programs. Regulators must watch over all of these activities, balancing the roles of the disco and the jurisdictional authorities of state and federal agencies.

Chapter 10 goes on to explore whether traditional or performance-based regulation is better suited to this rather imposing list of tasks. Whatever the answer, it is apparent that retail choice places a number of unprecedented and little-understood responsibilities on the regulators and owners of disco systems. Moreover, the scope of regulation and litigation at the distribution level may remain large or even *increase* following retail choice—a truly paradoxical outcome.[65]

Figure 9-4. Summary of Governance Challenges in the Distribution Segment Following Retail Choice

BULK POWER MARKET

↑   ↑

TRANSACTIONS   TRANSACTIONS

↓   ↓

DISCO as Seller of Last Resort	DISCO as Transporter
Issues Include:	Issues Include:
• Form of regulation to incentivize least-cost supply	• Setting efficient disco transport rates
• Planning uncertainties, including customers who suddenly leave or return to system	• Revenue adequacy
	• Incentives and ability to expand system efficiently
• Fair dealings with providers of disco transport and competitors	• State regulator approval of bulk power market costs
• Allocation of common costs between sales and transport customers	• Jurisdictions of regulators and their coordination

↑   ↑

Affiliate Transactions

↓   ↓

DISCO as Participant in Unregulated Power and Non-Power Lines of Business
Issues Include:
• May allow for important cost or marketing synergies
• Creates opportunities for cost-shifting and competitive harm

Chapter 9

NOTES

1 This anecdote was relayed to me by the Honorable Gregory Conlon, California Public Utilities Commission, in an unpublished conversation, 1995.

2 Useful surveys of the debate can be found in the National Regulatory Research Institute (1994) (or for a shorter version, Sikkema and Hill (1995)). Also see "The Battle Over Retail Competition: A Dialogue." *The Electricity Journal*, June 1994, and "Wired Up and Looking for a Fight," *Governing The States and Localities*, July 1994.

For well-known works making the case for retail competition, See Kennedy and Baudino (1991), the testimony of Kenneth L. Lay, Chairman and CEO, Enron Corp., before the subcommittee on Energy and Power, Committee on Commerce, U.S. House of Representatives, May 15, 1996, Anderson (1996), and the California Public Utilities Commission Order Instituting Rulemaking and Investigation, R.94-04-031/I.94-04-032, April 20, 1994 (The so-called "Blue Book" proceeding), along with the associated staff report, "California's Electric Services Industry: Perspectives on the Past, Strategies for the Future," Division of Strategic Planning, California Public Utilities Commission, February, 1993.

One widely-acknowledged impetus behind industries' and some regulators' desire for retail choice is the existence of significant differences in the rates of adjacent utilities. Although regional disparities have long existed in the industry (Federal Power Commission, 1964), only recently have adjacent utilities had large cost disparities. (Drazen-Brubaker and Associates, 1992). Although these disparities are higher than in recent history, many argue that they are a temporary phenomenon irrespective of retail competition (Cohen and Kihm (1994), Kihm and York (1994)).

3 See the Edison Electric Institute (1989, 1992) and Jurewitz (1994) for examples from the utility industry, the International Brotherhood of Electrical Workers (1996) and the California Coalition of Utility Employees (1994) for labor organizations' view, Cavanaugh (1994) and Cohen and Kihm (1994) for environmentalists' views (also see Dave Kansas, "Power Industry's Wheeling Get New Opposition," *Wall Street Journal* March 7, 1994). Regulators' opposing views are illustrated by Michael Martz, "Regulator Voices Doubts On Retail Electric Competition," *Richmond Times- Dispatch*, April 10, 1994, p. 1. For representative consumer community views, see Oppenheim (1989) and "NCLC Skeptical about Retail Competition," *The Quad Report,* September/October 1994, p. 4.

4 Dave Poklinowski, President of IBEW local 2304, quoted in "The Deregulation Battle Goes On," IBEW Journal, August 1995.

5 For example, See Edison Electric Institute (1989).

6 Other than remote switching and relaying, the primary controls on the distribution system are changing settings on transformers and remote-controlled reactive power devices. For customers whose demand is large, utilities or the customers themselves often install equipment controlled by the power company that effectively adjusts the local grid at the point of entry into the large customer. This equipment is installed to provide better circuit protection, power quality, reactive power (voltage support), also referred to as power factor correction. See Rustbakke (1983) chs. 10, 11 and Panisi (1992).

7 This point is recognized in Kennedy and Baudino (1991) p. 42 as well as in Edison Electric Institute (1989b). Also see the discussion on grid planning *infra*.

8 Real-time metering and other residential service issues are discussed in more detail below. As a practical matter, the difficulties are not in metering, but rather in measuring the impacts of small "retail wheeling" transactions across multi-megawatt power grids. Edison Electric Institute (1989b) points out that the *smallest* bulk power transaction is on the order of 25 MW, equivalent to roughly 5,000 households. Although improved computation might eventually lower this by an order of magnitude or more, it is nonetheless unrealistic to think individual marginal cost impacts can be computed for individual retail transactions.

9 Energy and Environmental Economics (1996), p 1–5 and Energy Information Administration (1995c).

10 See Energy and Environmental Economics (1994, 1996) for additional information.

11 In fact, in most of the states that have entered into retail wheeling pilot programs, service is offered under a written tariff much like sales tariffs. These tariffs are offered to all customers within a specified class. Other utilities have filed or are computing pro-forma tariffs in many states. For one survey, see the testimony of Mark Drazen before The Alberta Public Utilities Board, Drazen-Brubaker and Associates Project 3770, June 1993, Appendix B.

12 In addition, under retail choice, power marketers will seek to sell to a geographically disparate group of individual customers. To achieve economies of scale in its own purchases to supply these customers, such a marketer will aggregate its load. However, the distribution charges required to transport power to this customer group may vary considerably by customer. The marketer must either tailor charges to each customer or include average distribution charges in bills to all customers. The latter approach is identical to today's approach, except that today's utilities aggregate all loads in a single geographical region.

13 One utility, Cleveland Electric Illuminating, has already announced such a program called the "2002-Energy and Beyond Awards Program." Points are given based on electricity purchases, which may be accumulated and used to "purchase" new electrical appliances. The Ohio Public Utilities Commission has given the program conditional approval. See "Marketing Program Policy Statement," Public Utilities Commission of Ohio, Columbus, OH, Nov. 6, 1996. According to the *Wall Street Journal*, another power marketer is preparing to release a program called "Energy Bucks" that gives new customers frequent flier miles. (Peter Fritsch, "Tired of Phone Wars?" *Wall Street Journal,* April 16, 1996, p. 1).

14 Eschewing ownership of assets would be consistent with the "virtual utility" concept (Awerbuch, Carayannis, and Preston, 1996). Evidence in the other direction comes from the long-distance telephone industry, in which the three largest firms own considerable assets. In addition, the largest U.S. power marketer, Enron, is one of the three largest owners of independent power capacity and is integrating further into generation ownership. Other utilities that claim to be pursuing aggressive retail strategies are also acquiring rather than shedding assets (see evidence on recent merger activity, Chapters 5 and 8).

15 The presence of loop flows and system loss increases that cannot be attributed to individual transactions will pose interesting challenges for the allocation of imbalance penalties. These challenges will increase as the number of small (i.e., retail) transportation transactions rises. See Henney (1996)

16 Decision, Docket No. 93-09-29, September 9, 1994 Mimeo p. 63. Also see The Edison Electric Institute (1992) p. 31ff.

17 In Figure 9-1, PG&E owns and operates both the transco and disco parts of the system. In places or under scenarios in which the two parts are owned separately, the rules will resemble those shown in the figure, but will likely become still more complicated.

18 Apart from the rules required to ensure conduct that keeps the grid functioning efficiently, retail sellers will probably face licensing requirements in many states and some public interest responsibilities. Both topics are discussed below.

19 It is impossible to determine how long it will take the industry to develp its accounting procedures and regulations to the point where a retail choice customer receives only one monthly bill. See King (1996).

20 I have visited appliance stores or "energy centers" (in which new energy-using products are demonstrated) owned by Duke Power, the Pacific Gas and Electric Company, and the Southern California Gas Company. Several utilities have leased appliances to large customers, such as Boston Gas. (Also see Dennis Moran, "Alternative Financing for Promoting Gas Technologies," *Gas Energy Review*, the American Gas Association, May 1996.) As of 1991, Mississippi Power operated a 26-store retail chain of appliance stores under the name Electric City. ("Mississippi Power is Cutting Back Unregulated Retail Appliance Unit," *Electric Utility Week*, April 22, 1991).

21 See Platt (1992) ch. 5 and Hughes (1983). The latter reports (p. 223) that Commonwealth Edison opened its first "Electric Shop" in Chicago in 1909, and that Edison sold appliances door-to-door as well.

22 See Fox-Penner (1990e). A number of high-profile affiliate transactions cases heightened regulators' awareness of this issue; for example, see "The Baby Bells Misbehave," *Business Week,* March 4, 1991.

23 Many factors contributed to the move away from diversification. In some cases, the financial results were disappointing; in others, regulators admonished utilities to "stick to their knitting" rather than focus attention on unregulated businesses. In addition, the management literature of this era emphasized that successful companies engaged in businesses that made use of their "core competencies." In the utility industry, this was interpreted as suggesting that utility diversification should be into activities closely associated with electricity, such as the more traditional diversification into appliance leasing. See Public Utilities Reports (1981) and Shapiro (1986). For two suggestive press accounts, see Frederick Rose, "Utilities, Flush With Cash, Enter New Fields." *Wall Street Journal*, July 1, 1996 p. 6, and Leah Beth Ward, "A Reckoning for the Utility High Rollers," *New York Times*, July 29, 1990, p. D4.

24 As a resident of Massachusetts, I used to purchase energy-efficient equipment from a warehouse operated jointly by Massachusetts investor-owned utilities. The warehouse was part of these utilities' regulator-approved energy efficiency programs.

25 The most expansive study of time-varying prices during this period was the Electric Utility Rate Design study, a ten-volume opus commissioned in 1974 by The National Association of Regulatory Utility Commissioners and involving the Edison Electric Institute, The Electric Power Research Institute, and several consulting firms. Task forces in the study focused on the following ten topics:

1. Analysis of Various Pricing Approaches;

2. Elasticity of Demand for Electricity;

3. Rate Experiments for Smaller Customers;

4. Costing for Peak Load Pricing;

5. Ratemaking;

6. Measuring Advantages of Peak Load Pricing;

7. Metering;

8. Technology for Utilization;

9. Mechanical Controls and Penalty Pricing; and

10. Customer Acceptance.

Numerous Study and Progress Reports were issued over the following several years. For a useful overview of this period, see Shepherd (1983). The industry also began experimenting with rate levels that shifted between seasons of high demand (summer or winter) and low-demand, sometimes referred to as "seasonal differentials." As with all traditional ratemaking, the differential was based on the estimated average differences in the costs of serving customers in different seasons. In many jurisdictions these differentials survive to this day.

26 See, for example, Christensen Associates (1988).

27 The impacts of time-varying prices and other aspects of retail access on small customers are discussed in more detail in Chapters 10 and 11.

28 See (Kennedy, 1978), Cargill and Meyer (1971), and Malko and Simpson (1978), Hirshberg and Aigner (1983), Park and Acton (1984), and Aigner and Hirshberg (1985).

29 Real-time metering is much more expensive than traditional electric meters, which simply accumulate power use over many, many hours (e.g., one month). Real-time meters store the amount used each hour in electronic memory. In addition to measuring use, however, a communications network is needed to send real-time prices to each house so that users can look at price change or program devices to respond to them. The cost of providing this communication, metering, and control system is presently estimated in the range of one thousand dollars per meter, making it too costly for most residential customers at present. (Richard Wakefield, Cassazza Schultz and Associates, Arlington, VA, unpublished presentation, 1996). However, technological advances will continue to reduce the cost, and the communications network may be used for many services other than power prices. If multiple other uses can be found for the communications part of the system, widespread RTP may be far more viable among residential customers. See Cavanaugh (1995).

30 Leaving aside externalities and inaccurate pricing, the economic efficiency impacts of real time pricing are unambiguously positive. It can be shown that no electricity customer who voluntarily shifts his or her electricity use in response to accurate costs can be worse off. Thus, the total adjustment to spot prices must be an improvement (See Oren and Ray, 1996; Taylor and Schwarz (1996)). Similarly, it can be shown that shifts of electricity production from high- to low-cost periods reduces the total cost of supplying electricity, and maintains or increases the total production efficiency of the system. Measurements of the total economic welfare implication of real-time pricing must incorporate, among other things, the possibility of inaccurate prices, the impact of altered electricity use on externalities not reflected in price or otherwise accounted for, and the fairness of the changes in electricity bills to all customer classes.

For more information on RTP, see Siddigui and Woodley (1994), Energy and Environmental Economics (1994), and Christensen Associates (1995 a, b).

31 See KiaShantè Breaux, "Gadgets Ease the Story of Electric Rates." *Wall Street Journal*, July 5, 1996, p. B6. For an example of a large, evolving utility experiment, see Crane and Leonard, (1991). See Wharton, Hanser, and Fox-Penner (1997) for a recent survey.

32 See Newcomb (1994).

33 See Hunt and Shuttleworth (1996) ch.11.

34 See Hogan (1994 a, b, 1995 a).

35 Christian Hill, "CellNet Signs Pact with Joint Venture of Ameritech, Wisconsin Energy Corp.," *Wall Street Journal*, June 3, 1996, p. 35. For another revealing discussion, see Herbert Cavanaugh, "Utilities Are Entering An Alarming Business," *Electrical World*, January 1995 p. 63.

36 These activities are surveyed in Cavanaugh (1994), who begins by noting:

Electric utilities have the most critical private-industry need for "real-time" communications in the nation, placing great reliance on extensive internal communications systems to monitor, control, coordinate, and protect their operations. For this reason, advanced utility-owned and -maintained telecommunications systems rival, and in many cases exceed, the systems operated by commercial communications providers.

Also see the proceedings of Telcom Strategies for Utilities, International Business Communications, Washington, D.C. September 16, 1996. Weiner, et al (1996) goes so far as to suggest that the value of energy saved by energy management systems (see the following sections) is sufficient to allow power utilities to purchase communications utilities.

37 For a thoughtful discussion, see Mitchell and Spinney (1996) and the references therein.

38 See Hunt and Shuttleworth (1996) p. 33. Some of the services in Table 9-1 involve the distribution infrastructure itself and therefore require ownership arrangements or transactions with regulated discos. This raises governance issues discussed at the end of this chapter.

39 See Costello, Burns, and Hegazy (1994), Oren and Ray (1996), and Mitchell and Spinney (1996) for useful discussions of this point.

40 See Lovins (1976) and Sant, (1980).

41 One recent and typical conference presented by Pasha Publications was titled "BTU Marketing and Networking," and introduced itself as follows:

"All I need is heat, light, motion and steam," is how Simpson Paper Energy Director Jackson Mueller sees the evolving energy market. Gone are the days when he buys each fuel from a different supplier. The paper manufacturer is only one of a host of new customers seeking more and more services from their energy suppliers. They are buying BTUs. Fewer industries are tied to a single fuel, and competition within the electricity industry has created a new breed of mega-marketers positioned to play and profit with interfuel dynamics. Buyers want to maximize use of facilities and fuel to get the best energy deal. Sellers want to sell more than one fuel to provide the maximum value to the customer. They refer to themselves as "energy marketers," "the Energy Store," "refiners of natural gas" or simply "energy managers," dropping any reference to a single fuel in favor of names that show their versatility and fuel mix. Increasingly common are deals such as this: Aquila Power delivers gas to a New York power generator and takes its payment in kilowatts it transports and sells to a wholesale power user in New York City. This conference is intended to help those who want to understand the relationship between fuels and how to profit from making fuel selections or combining fuels in a single series of transactions. Gas marketers can rub elbows with power brokers and spot coal dealers and vice versa. The inherent volatility of gas and electricity will create opportunities to arbitrage fuels, locations and timing. A competitive electricity market will increase the competition and interplay between gas and electricity as well as coal–still the principal fuel in electricity generation. Whenever possible, gas and electricity will move interchangeably from one location to another to seek higher values.

Steven Parla and Vince Salvato, two stock market analysts at the CS First Boston Corporation, describe future successful retail energy conglomerates as follows:

The successful energy service company will need to have a national presence, physical assets to complement financial capability, a high degree of sophistication in trading and derivatives, strong back office operations, and an entrepreneurial culture. Key attributes will include:

- Reliability and reputation.

- Strategic assets.

- Full coverage in all physical commodities.

- A complete spectrum of products and services.

- State-of-the art risk management capabilities.

- Nationwide transport management expertise.

- A large and diverse customer base, allowing for maximizing economies of scale and identifying arbitrage opportunities.

- The ability to provide financing to both energy suppliers and customers.

- Highly sophisticated information technology capabilities, including back office functions.

> Although these capabilities entail a substantial commitment of people, culture, and time as well as certain physical assets–full-service BTU *marketing is not a capital-intensive business.* Since virtually all energy businesses that preceded it were, and given the "trading" and service nature of the energy service business, it appears quite different from what most energy companies consider themselves to be (Parla and Salvato, 1995).

Similar arguments can be found in Smith (1995).

> Interestingly, this business-based literature may be contrasted with public interest literature on the future of energy efficiency programs and markets after restructuring. We examine this in more detail in Part III; here we merely note that many of the "energy service companies" (ESCOs) who began as sellers of energy conservation technologies are well-positioned to become multi-faceted businesses that add fuel brokering and trading and financial products to their traditional lines of business. See Part III, Goldman and Dayton (1996), and Newcomb (1994).

42 See Douglas (1996) for a discussion of emerging technologies that will actually be able to provide *higher* reliability for certain customers or parts of the distribution system.

43 A power user can improve reliability above system-provided levels by adding backup power sources, such as batteries, small generators, or "uninterruptible power supplies." These pieces of equipment have always been unregulated end-user technologies.

44 Oren and Ray (1996) Section III. The example requires an estimate of the costs borne by customers following an unplanned outage, as well as the reductions in unplanned outages prompted by self-chosen reliability schedules.

45 The Massachusetts Energy Efficiency Council (MEEC) notes that several practical factors limit the benefits of RTP to residential customers. First, most will not have real-time metering or control for some time due to the expense. Second, many customers dislike price volatility and prefer dealing with stable prices (on this point, see "A Real-Time Day in The Life of Ivan Sanborn," unpublished manuscript by William LeBlanc, Barkat and Chamberlin, Boulder, Co., 1988). Finally, MEEC notes that RTP does not address certain market failures or externalities (on this point see part III). Comments of MEEC before the Massachusetts Department of Public Utilities, DPU 96-100, May 24, 1996.

46 Order 888, Docket Nos. RM95-8-000 and RM94-7-001, Mimeo, p.400 ff and Appendix G. A similar, equally sticky issue is the boundaries of the distribution system for ratemaking purposes. Also see Costello, Burns, and Hegazy (1994) ch. 4, and "Federal/State Jurisdiction" (mimeo). The Alliance for Competitive Electricity, Washington, D.C., April 1996.

47 For additional discussion of state and federal jurisdictional issues and retail choice, see the comments filed in Federal Energy Regulatory Commission Docket RM95-8-000 and RM94-7-001 (1995).

48 See, for example, Cavanagh, (1994) and Brockway (1994).

49 Large commercial and industrial customers can often "supply" reactive power by altering the use patterns of their large equipment. This is a useful but idiosyncratic source of VAR support.

50 Cohen and Kihm (1994) section IIIB. argue that retail competition will reduce the ability to make efficient tradeoffs.

> This last question is paticularly interesting because some observers predict that technological change will make small-scale power sources economical to operate on distribution systems–the so-called distributed utility concept (Weinberg, 1993; EPRI, 1996). If distribution rates are set accurately, the correct economic signals will encourage efficient tradeoffs between distributed generators and larger lines. If distribution rates are set on a highly simplified basis to reduce transactions costs, the net impact relative to internalizing the tradeoff within the utility (which occurs in RTP/PBR) is uncertain.

51 The issue becomes still more complex when one considers that many discos will not be regulated by state commissions, but most will be subject to federal or regional rules. See Schoengold (1995) for an extensive discussion.

52 In the words of the Connecticut Public Service Commission, "Coordination of state and FERC policies on transmission pricing and cost recovery will likely be necessary to allocate costs properly and establish retail wheeling rates." (Decision, Docket No. 93-09-29, DPUC Investigation into Retail Electric Transmission Service, Sept. 9, 1994, Mimeo p. 66).

53 Hunt and Shuttleworth (1996) p. 68

54 Two of many examples of extensive deliberations on disco governance can be found in "Restructuring New York's Electric Industry: Models and Approaches," Phase II Final Report, case no. 94-E-0952, September, 1995, Chapter Four, and Report of the Advisory Committee on Electric Utility Restructuring," Public Service Commission of Wisconsin, Madison, WI, October, 1995, Chapter Five.

55 Frame (1993) also discusses the disco's need for pricing flexibility.

56 California Public Utilities Commission Order Instituting Rulemaking and Investigation, R.94-04-031/I94-04-032, April 20, 1994, p. 42–46.

57 Rhode Island Utility Restructuring Act of 1996, 96H8124, August 4, 1996, p. 27

58 A March 1996 proposal by Sheldon Silver, Assembly Speaker for the State of New York requires the New York Commission to "certify that all customers will retain access to services at reasonable prices" before retail access is allowed, but is unclear on the remedies available to the Commission. Massachusetts proposed retail access rules include a required "basic service" provided under regulated rates from disco purchases (Proposed Rules 220 CMR 11.00, March 15, 1996). Retail Choice legislation enacted in New Hampshire simply states: "Electric service is essential and should be available to all customers. A utility providing distribution services must have an obligation to connect all customers in its service territory to the distribution system. A restructured electric utility industry should provide adequate safeguards to assure universal service. Minimum residential customer service safeguards and protections should be maintained. Programs and mechanisms that enable residential customers with low incomes to manage and afford essential electricity requirements should be included as a part of industry restructuring" (N.H. House Bill 1392, 374-F:2.V, 4/16/96).

59 Planning difficulties are discussed by the Edison Electric Institute (1992), Jurewitz (1994), in the Decision of the Connecticut Department of Public Utility Control, Docket 93-09-29, September 9, 1994, Section II. J, Kihm and York (1994), and in the references cited in note 46 *supra*. Kennedy and Baudino (1991) p. 45 argue that planning differences will be minor.

60 See Jurewitz (1994) p. 65.

61 See Costello, Burns, and Hegazy (1994) p. 75 and Franze (1993) Section II.

62 For representative discussions by regulators, see "Restructuring New York's Electric Industry: Alternative Models and Approaches," Phase II. Final Report, Case No. 94-E-0952, Sept. 1995, p. 50 and The Public Service Commission of Wisconsin (1995) p. 107. Also see Schultz (1996)

63 See Costello, Burns and Hegazy (1994) p. 71 and the discussion of vertical integration economies in Chapters 1 and 4 *supra*.

64 Oren and Ray (1996) ch.IV, Fox-Penner (1990), and Hunt and Shuttleworth (1996) p. 59, 68 discuss potential anticompetitive harms arising from vertical integration or cross-ownership. The Federal Energy Regulatory Commission has acted to police affiliate transactions on many occasions. It has promulgated generic codes of conduct for oil pipelines, natural gas pipelines, and power markets, and it has examined specific transactions. (See, for example, Order on Rehearing, "Inquiry into Alleged Anticompetetive Practices Related to Marketing Affiliates of Interstate Pipelines," RM87-5-001, December 15, 1989 for a generic policy and Order Noting and Granting Interventions and Rejecting Rates, Docket No. ER9-40-000 ("Terra Comfort Decision"), September 7, 1990. Also see the discussion in Chapter Five concerning the Public Utility Holding Company Act of 1935.

65 Utility consultants Steven Kihm and Greg York (1994) argue directly that state regulation will increase rather than decrease under retail choice. Former economics professor and utility executive John Jurewitz (1994) agrees, saying that "nothing could be further from the truth" than the proposition that retail access will diminish regulation. Also see Cornelli (1996) and Dunnard Rossi (1996)

Chapter 10

The Costs and Benefits of Retail Choice

Now, with a sense of the mechanics of retail competition and the roles of the disco and regulators in mind, it is time to do a more complete comparison of retail choice and its alternatives. To simplify the comparison as much as possible, two very general organizational forms for the retail or "downstream" segment of the industry are examined.[1] Both scenarios begin by assuming that there is broad, effective competition in the wholesale power marketplace—competition that is largely unaffected by the presence or absence of retail competition.[2]

The first scenario is a continuation of the current, exclusive retail sales franchises for distribution utilities, i.e., no direct retail choice. However, in this scenario it is assumed that all discos offer real-time pricing (RTP) to all customers who want it, allowing for Professor Hogan's "Efficient Direct Access." It is also assumed that the disco is regulated, using a form of performance-based regulation (PBR) rather than traditional cost-of-service (COS) regulation.[3]

The second scenario is one in which retail choice is offered to all customers. The disco sets regulator-approved, area-specific rates for retail transport, continues to sell to those who choose not to shop, and has the responsibilities and challenges described in Chapter 9.[4] Table 10.1 highlights the two basic scenarios that are examined—"RTP/PBR" and "Retail Choice"—respectively, in slightly more detail in this chapter.

There are clearly many important variations on these two scenarios which will not be examined here. Interested readers are referred to the many pieces of research that examine more numerous and detailed arrangements.[5] In particular, there are many important alternatives for governing the non-competitive roles of the disco, ranging from public ownership to a variety of innovative regulatory schemes.[6] The two scenarios contrasted in this chapter were not chosen because anyone can prove they are the best, but rather because in the coming era, more will be seen of both, and because they are helpful in illustrating some of the less obvious aspects of retail choice.

The difficulty of placing values on many qualitative or difficult-to-measure aspects of industry change requires comparison according to a number of criteria, each of which relates to a generally-acknowledged economic or social policy objective. Here again the list of criteria used in many studies is long and complex. For purposes of the current comparison the list has been reduced to the following five broad categories:

Table 10.1. Characteristics of the Two Retail Scenarios to be Contrasted

Characteristic	"Retail Choice"	"PBR/RTP"
Bulk Power Market	Widespread, effective competition, with open-access transmission and effective markets for interconnected operating ("ancillary") services. Futures contracts and "Contracts for Differences" based on locational price differences are available.	Widespread, effective competition, with open-access transmission and effective markets for interconnected operating ("ancillary") services. Futures contracts and "Contracts for Differences" based on locational price differences are available.
Function of Distribution Utility ("Disco")	(1) Provides transportation for power retailers and associated services under open-access terms and regulated rates; *(2) retains a traditional power sales function for all customers who do not purchase from unregulated marketplace, under regulated rates.	Purchases or supplies from owned resources all electricity demanded by retail customers in its franchise area under regulated rates.** Regulators require discos to make available to all customers who desire real-time prices based on those the disco incurs when purchasing on the bulk power markets.
Power Retailers	Generation companies and power marketers compete to sell to all customers who elect to purchase from market. Retail sellers must abide by federal and state regulations, including all rules associated with transporting power on the transmission or distribution system. However, the prices charged by power retailers are not regulated.	Only the disco may actually sell power in its franchise area. However, "Energy Service Companies" (ESCOs) may enter into unregulated financial arrangements with customers which change customers' costs of electricity service. Power marketers continue to serve the wholesale marketplace, and power marketers, gencos, and ESCOs all transact in electricity-related financial arrangements, such as electricity futures and options. These financial arrangements hedge the risk of price variablity from disco real-time prices.

*It is assumed that PBR is used to govern the disco under retail choice, largely to provide service quality incentives (see discussion in text).

**The scenario assumes that the distributor is investor-owned rather than publicly owned, and further assumes that state regulators govern disco rates using "performance-based regulation" (PBR). Most of the conclusions of the comparison apply to an efficient, performance-based publicly-owned distributor offering RTP to all customers.

1. **New Products and Services.** Does retail choice make it more likely that the market-place will provide services (other than electric power) that provide value to customers and would not be available as easily, quickly, or cheaply than under RTP/PBR? Because customized pricing is one of the most important new services that retailers can offer, and involves financial products, it is examined as a new service, apart from the overall average level of prices.

2. **Electricity Prices and Monthly Bills.** Total monthly bills over the course of a year or more reflect the overall average costs of electricity paid by customers. How much will the average effective prices and costs differ between the two scenarios? Will there be rate reductions or increases, and will they apply equally to all customers?

3. **Reliability, Service Quality, and Customer Administrative Costs.** Under the two scenarios, customers may face different levels of reliability or quality of service. The time and cost required to verify or rectify performance errors may change significantly. In addition, retail choice will require customers to shop for electricity providers; the RTP/PBR scenario encourages shopping for energy service providers. In both instances, evaluating competitive suppliers is uncertain and potentially costly. Is one arrangement more likely than the other to be convenient and/or reliable?

4. **Ability to Incorporate Public Interest Programs.** The essential nature of electric service and the industry's importance in energy and environmental policy have made it a key participant in many programs that serve legislators' expression of the state or national interest. Will the two access scenarios incorporate these programs and needs in a more or less satisfactory manner?

5. **Ability to Govern the Distribution Segment of the Industry.** Chapter 9 examined the challenges faced by discos and their regulators. The choice of two specific scenarios illustrates the governance challenges a bit more concretely, and for completeness this criteria belongs on this list. However, the general issues are unchanged from the previous chapter, and only a little additional discussion is necessary.

What about overall economic efficiency as a criterion? If listed separately, it would encompass all of the criteria listed above—essentially the composite total of cost savings, transactions costs saved and imposed, the value of improvements or declines in service quality and reliability, and so on. It would also be useful to know how sensitive the overall or composite gains are to the ability of state and federal authorities to carry out their roles under different industry scenarios. These fundamental questions of public policy are taken up again at the end of this chapter.

NEW PRODUCTS AND SERVICES

As fully discussed in Chapter 9, retailers as well as regulated discos are likely to offer the three categories of new products and services described in Table 9.1. Some of these products are tied to the technical operation of the power system, while others are related to financial, administrative, or energy services as a whole. In all cases, the decision to offer more products and services is driven entirely by synergies and profits for all sellers other than the disco, whose motivations will be tempered by state or federal regulatory policies.

Figure 10-1. How Retailers Profit Under...

Retail Choice	Real-Time Pricing/PBR-Regulated disco
1. Power Marketer signs up customer, offering any combination of pricing, power-related services, and non-power related services as well. Marketer may also offer to evaluate and install or operate RTP-controlled equipment on customer's premises.	1. Energy Service Companies (ESCOs) sign up customer, offering any form of pricing, power-related services, and non-power related (same). ESCO may offer to install or operate RTP-controlled equipment on customer's premises.
2. Power Marketer goes to bulk power marketplace to arrange customer's generation supply and signs contracts with GENCOs.	2. No need for ESCO to arrange generation.
3. Power Marketer must arrange and pay for bulk power transmission to customer's distribution system. Marketer must also pay system operators or ISO for all ancillary services needed to move power.	3. ESCO need not arrange for physical flows—which remains the responsibility of the regulated disco. Instead, the ESCO purchases a "contracts for difference" (CFD) between the generators it contracts with and power prices at the disco's entry point.
4. Power Marketer must arrange and pay for sub distribution transmission to customer's distribution system. Marketer must also pay system operators or ISO for all ancillary services needed to move power.	4. ESCO does not have to arrange distribution because the regulated disco still sells and transports power to all its customers.
5. Power Marketer collects revenues from customers according to its pricing offer and pays for generation, transmission, ancillary services, distribution, and other charges.	5. The ESCO collects revenue from customer according to its pricing offer and pays for generation, pays the disco's tariff, and pays or receives money according to the CFD it arranged.

In the RTP/PBR scenario, there are no unregulated power retailers per se. Each customer receives a monthly bill based on traditional or real-time rates. The disco operates a system in which all customers who so desire can observe hourly prices and adjust their use accordingly. In the retail choice scenario, the same is true, but power retailers can make arrangements with generators, the transmission and distribution systems, and can offer an endless variety of pricing and service plans, subject, of course, to numerous transco and disco rules.

The rules under which prices are offered to customers are set by the disco and approved by the regulator. It is very unlikely that the regulator will give the disco the same totally unconstrained ability to alter prices that unregulated retailers will have, so it is likely that the customer's ability to customize its price and service plan will be substantially better under retail choice.[7] This applies primarily to large customers, but even small customers may be offered temporary sale prices or special bargains as a group—similar to what is currently happening in the U.S. cellular phone and long-distance marketplace.

Interestingly, in theory, the same variety in pricing can be offered by energy service providers, so long as each customer can receive RTP. Consider, for example, a factory owner in the retail choice scenario who signs a contract with a power marketer for a customized, time-varying price for electricity. The contract includes the cost of the transco and disco transportation tariffs, ancillary services the marketer had to purchase from the bulk power market, and whatever else the marketer herself must pay for in order to provide service.

Under the RTP/PBR scenario, that same factory could sign a contract with an ESCO, a power marketer, or perhaps even the disco itself, wherein the ESCO guarantees the factory a pricing package that could be virtually identical in complexity and customization to those offered by the marketer. How? The ESCO contractually agrees to pay the factory's *actual* power bill to the regulated disco—which could be as simple as today's traditional monthly bill—and charges the factory the prices it wrote into its custom contract. The difference between the payment the ESCO makes on the factory's traditional bill and the revenues it receives from the customized pricing it offers the factory sets up the ESCO's profit incentive. If the ESCO can ensure that the factory will respond to real-time pricing signals, thereby reducing its demand and costs as computed by the power company, the ESCO will pay less to the power company and can pocket some of the difference in profits. To do this, the ESCO may have to set up hardware to read the real-time pricing signals and adjust the factory's demand (as in Box 9-3). It can also earn profits by mimicking what the power marketer does, using time-and- location-sensitive power futures contracts or "Contracts for Differences," assuming they are available for all nodes in the wholesale market. (Remember, both scenarios assume that the bulk power market continues its march toward full, open competition).

To illustrate this difference more clearly, Figure 10-1 traces the steps that the power marketer must go through to make profits under retail choice and the steps that an ESCO goes through to profit under RTP/PBR.

- The power marketer must arrange and pay for actual transmission and distribution, which, of course, is not paying for physical transportation, but is rather paying for the marginal impact of each generation injection at the genco's location and the matching customer's consumption beyond his or her meter.

- In contrast, the ESCO relies on the fact that there is still a regulated disco in charge of buying or otherwise supplying every customer's total needs, purchasing from the competitive bulk power market. Because this is by assumption a competitive market with prices for each location, the ESCO can simply purchase a Contract for Differences ("CFDs," described in Chapter 6) and send the same price signal to the customer.

It may at first seem seem odd that a financial hedge based on locational spot prices can mimic retail access; this really is not odd at all. It is actually more odd to view the power retailer as "purchasing transportation," because this is not really what happens. In any case, it is possible to show that, if transmission and distribution prices are perfectly efficient, the prices power retailers and ESCOs can offer to customers are exactly identical, and produce identical profits in the two different scenarios.[8]

Because no one expects any future industry structure scenario to produce perfect prices or zero market imperfections, net pricing and efficiency differences will depend on the nature and extent of the imperfections, and how they might differ between the two scenarios presented above. For example, retail choice requires reasonably accurate prices for transmission, ancillary services, and distribution transportation. As discussed in Chapter 9, this presents some daunting challenges. On the other hand, efficiency in the RTP/PBR scenario requires the ready availability of accurately-priced CFDs or location-sensitive electricity futures, and that regulated discos purchase power from the bulk power market and operate their systems as efficiently as unregulated sellers.

One attractive feature of the RTP/PBR aproach is that it effectively assures that there will be an active, visible, spot market against which financial instruments such as CFDs can be priced. On the other hand, the RTP/PBR preserves the disco as the sole aggregator of demands in any one area. This may prevent the emergence of a more active market for trading in capacity, load shifting, and other products that could replace the need for today's rigid reserve requirement.

It is difficult to gauge the magnitude of such imperfections with hindsight, much less forecast their importance in the future, but it is important to be aware of them as the market develops. Another way to gauge which of the two scenarios may produce better results is to consider the overall magnitude of the gains to be realized from competition, and then to consider which retail structure is most likely to realize these gains.

This viewpoint requires consideration of what management strategists call the "value chain" and the "core competencies" of suppliers in the two scenarios. Power marketers make their money by skillfully aggregating supplies from many generators, perhaps owning and operating efficient generation, and understanding how to work with the transco and disco systems to ship power cheaply. Their primary focus may be these upstream activities, including owning large portfolios of generators and large power trading operations.

In contrast, the natural initial focus of the ESCO is the customer's usage pattern and costs. The ESCO must begin by convincing a customer that it is worthwhile to let the ESCO take responsibility for the traditional power bill and instead pay the ESCO. This may focus the ESCO on saving money by making the customer more efficient (in addition to trying to save money by financial arrangements in the bulk power marketplace).

There is no reason why power marketers cannot also be good at helping customers save money via demand adjustments and RTP, nor is there any reason why ESCOs cannot be as good as power marketers at bulk power transactions. And, at this point, no one can tell whether there are greater opportunities for realizing economic efficiency gains by more competitive trading in the bulk power market or via new retail products, services and financial arrangements. If it turns out that upstream trading is the more important source of efficiency gains, and these are better realized by power marketers, then retail choice may prove to be more advantageous than RTP/PBR. Conversely, if downstream opportunities are of greater unrealized value and a continued disco supply role leads to fewer overall complications, inefficiencies, and transactions costs, then RTP/PBR may provide greater overall benefits.

Finally, there are services other than electricity and electric pricing plans. Are retailers under the two scenarios more or less likely to be innovative and seek out new power and non-power services?

There are no immediately identifiable impediments to power retailers or ESCOs offering any power or non-power related products in the two scenarios. In both cases, sellers can still knock on doors and offer consumers communications services, smart appliances, smart buildings, fuels management, and whatever else they want to sell. Nonetheless, there are undoubtedly subtle differences that cannot yet be fully understood or predicted. For example, it may be much easier to convince customers to buy electricity, as opposed to marketing the kind of financial arrangements just described, even if the two are identically fictitious in actuality. Power marketers may benefit from owning generating plants; it is difficult to tell whether ESCOs would find any similar synergies in the RTP/PBR scenario. Conversely, ESCOs may be better at marketing onsite services or RTP-based control systems to customers and leaving the bulk power market alone.

Which of the two scenarios—retail choice or RTP/BPR—will work best? It is still too early to tell. It is probably fair to give a slight advantage to retail choice because I have the greatest confidence that a variety of products and pricing plans will emerge under this scenario. The difficulty lies in gauging the size and value of the difference. At the very least, retail choice comports more closely with the conventional modes of buying and selling products in other markets. At this point, both the uncertainties and the range of possible benefits are quite significant.

RETAIL POWER COSTS AND BILLS

Having already discussed differences in the ability of regulated versus unregulated firms to customize their pricing offers, and to add other products and services to the pricing bundle, this section examines likely trends in the overall bills that customers will pay for electricity under the retail choice and RTP/BPR scenarios.

The rates at which retail sales occur are a main preoccupation of traditional regulation or public ownership. The set of prices at which power is sold at retail, or the rate structure, typically has five to ten customer classes. Any customer that takes service in the franchise area and meets the terms of a particular customer class is entitled to service at the rates, terms, and conditions specified in that customer class rate or tariff. Most utilities have two to four classes of residential, commercial, and industrial customers, plus classes for government customers and other identifiable groups.

Whether regulated or unregulated, the average prices a viable firm charges for its products must be expected to ultimately cover its costs, plus a return on investment commensurate with investor risk.[9] Each component of the costs of supplying customers may or may not depend on short-term sales volumes, i.e., be a fixed versus variable cost.

Finally, regulation does not change consumers' price sensitivity: different sorts of customers respond to price increases differently, but everyone responds by ultimately using less as prices rise. Research in regulated and unregulated settings has confirmed that price response depends on the type of customer, time period, magnitude and duration of the price change, and other factors.[10]

Unregulated firms in competitive markets ask themselves, "what set of products can I make which maximize profits, given prevailing prices, my fixed variable costs, and the sensitivity of the market to price increases?" In contrast, regulated firms' pricing decisions are largely in the hands of regulators. The latter must be committed to ensuring that revenues cover costs, but beyond this there are a number of objectives other than profit maximization which help determine rates. These considerations have developed over almost a century of regulated utility ratemaking, and include such things as fairness, an absence of undue discrimination, economic development, universal service, and other social objectives.

Appendix 10A contains a comparison of the principles of regulated utility ratemaking and cost allocation with unregulated pricing. The Appendix explains the basis for the widespread prediction that retail choice will result in increased prices for most residential and small commercial customers relative to large industrial and commercial customers. The Appendix explains that deregulation will make "volume discounts" much more common, and they will have the effect of shifting a greater share of generators' fixed costs onto smaller users.[11] To the extent these cost realignments result in prices closer to marginal costs, economic efficiency will increase.

The likelihood that retail choice may increase small user prices—relatively, and perhaps absolutely—has not been lost on representatives of small consumer and small business interests. Recently, a group of small businesses placed an advertisement in a Washington, D.C. newspaper intending to influence the U.S. Congress against retail choice legislation. Many other consumer groups have expressed concerns in one way or another, although not all such groups oppose retail choice.[12] During the dialog leading to retail choice in California, the director of a prominent California consumer group said:[13]

> We all know that the Public Utility Commission (PUC) on April 20 proposed to revamp the state's electric industry by increasing competition and decreasing regulatory scrutiny. The PUC's proposal offered vague assurances that *all* consumers would share in the benefits of a more competitive industry. We're not so sure.

> We had a glimpse of what's in store for California consumers on September 15, when the PUC approved sweeping changes in rates for all telephone services, effective January 1, 1995. Under the PUC's newly rebalanced rate design, Pacific Bell's residential customers face a 35 percent increase in the cost of basic monthly phone service, and PacBell's small business customers will see a 24 percent rate hike.

This glimpse into the telephone industry's restructuring, including the procedural violations, is revealing because the PUC has also been reciting the mantra of competition with regard to that industry. So, it won't come as any surprise when I predict that the restructuring of California's electric industry will result in higher bills, and quite possibly a lower equality of service, for small users.... It doesn't take an economist with a lot of fancy degrees to figure out that when some customers have market power and competitive choice, and other customers do not, there is going to be a tremendous incentive for the utility to charge its captive customers more in order to offer better prices to those customers who have a choice.

Interindustry comparisons of rate decreases and shifts are are quite common, but cannot be relied on as anything other than an illustration of general principles. Each utility industry has significant differences in its cost structures and its division of authorities and costs between each stage of the delivery chain. The rate impacts of retail electric choice cannot be predicted with any precision from the rate impacts of telephone or gas deregulation. At best, one gets a rough sense of the two key trends at work, one to lower the average of all prices to all users and the second to shift prices relatively upward for small customers. Without considerable analysis, there is no basis for assuming that the numerical magnitudes of these effects in other restructured industries applies to the power industry.

Remembering this strong caveat, the dual impacts of overall price declines and relative price increases can be seen in a recent study of the natural gas industry during the period following open-access in gas pipelines. According to a study by the American Gas Association, between 1987 and 1995, the prices for natural gas at the wellhead and the price of natural gas transportation (interstate as well as local) both declined by roughly the same amount, about 26% in real terms. Average prices for the unregulated commodity (gas) and the services that were continuously regulated throughout the period (gas transport) both declined as a whole. Hence, increased competition in what might be viewed as the analog of the bulk power market prompted savings in the regulated transport segment of the industry as well.

The phenomenon of cost shifts toward smaller users is equally evident. As Appendix 10A predicts, the total price decline to large industrial customers was twice as large as the average small-customer decline (33% and 16%, respectively). Because total average costs declined 25% overall and residential declined only 16%, it is evident that residential prices fell much less than others. However, the net effect of the two phenomena was to reduce inflation-adjusted residential rates.[14]

These gas industry results do not imply that electric retail choice will cause residential and small business prices to go up overall. This depends on the overall composite impacts of all aspects of retail choice which are reflected in price. However, in order for small user average prices to decline, overall prices must decline significantly across the industry, so as to offset the allocational shifts retail access will bring about. Because prices in competitive markets tend to track marginal costs, it is important to assess the magnitude of the decline in marginal costs in the two scenarios. Most importantly, will retail choice force costs and prices down more or less than RTP/PBR when both are coupled to a competitive wholesale marketplace?

ADMINISTRATIVE HASSLES, RELIABILITY, AND THE QUALITY OF SERVICE

At first glance, it may seem odd to lump administration, reliability, and service quality within one broad criteria—even in this highly simplified discussion. However, in the views of most customers, these three concepts are closely related, and it is difficult to talk about any one in isolation.

As used in this discussion, administrative costs are the aggregate costs required by customers to evaluate, shop for, purchase, and pay for electricity, as well as the costs of altering service, rectifying billing errors, and reporting and correcting service deficiencies. For many customers, these costs are measured in hours devoted to these tasks and perhaps aggravation rather than dollars. Administrative costs also include sellers' costs of marketing, contracting, collecting revenues, paying vendors, and servicing customers' changes and complaints.

Service quality is an expansion on reliability, which was defined in Chapter 3. Interestingly, until recently, the utility industry had no numerical indices of service quality other than those used for reliability. Most public service statutes and the charters of most publicly-owned utilities call for high-quality customer service, and most utilities have adopted credos or vision statements that pledge allegiance to the notion. Nevertheless, until recently there was no formal measurement or benchmarking of quality other than via reliability measures.

In recent years, utilities and market researchers have worked to translate service quality into measurable, monitorable dimensions. The common types of criteria that are emerging from these efforts are as follows:

- Measurements of stranded administrative processes, such as the average number of minutes a customer must wait on hold before a customer service representative answers a call;

- Average rates of complaints received or customer dissatisfaction, as reported in surveys or received at complaint bureaus; and

- Indices of reliability, such as the average number of power outages experienced by the average customer during a year.

For purposes of the current discussion, the nexus this criteria represents should now be clear. Administrative costs for all parties go up considerably when the network is unreliable, or when service quality does not meet customers' expectations. When all goes well, customers' administrative costs tend to center on selecting suppliers and paying bills; when things go wrong, the time and cost of restoring quality service becomes a major concern. With this in mind, it is now time to review each of these areas under the two scenarios.

Service Quality. In Chapter 9, it was noted that one of the remarkable aspects of retail choice is that most of what are currently considered quality attributes of electricity service per se will not be controllable by the retailer. Under either of our two scenarios, but especially under retail choice, the provision of quality service will require the cooperation of unregulated retailers and the distribution company.

Obviously, the disco's incentives to provide quality service will depend on how well regulators measure quality service and reward or penalize the disco's performance. Measuring and providing incentives for quality service is one of the most important elements of PBR, and is described in more detail in Appendix 10B. For purposes of the current comparison, it is difficult to conclude that the service quality provided by the disco will differ significantly between the two scenarios. Once again, the main difference in disco service quality is the dual-role conflict of the disco, which may well be something that state regulators can monitor and perhaps even turn to their advantage.

As to the quality provided by unregulated service providers, the competitive marketplace is, of course, the ultimate judge and jury. However, each of the scenarios has an unregulated sector, and it is not clear whether power marketers or ESCOs have a service quality edge. As noted in the example of Acme power retailing, the only way the power retailer can control such things as voltage, waveform quality, or reliability is to install and operate a variety of devices at the customer's site. These are precisely the sorts of things that ESCOs do as well; indeed it is the *main* thing ESCOs do to make profits. In both scenarios, service quality from the unregulated providers will probably be a function of the robustness of completion among providers.

Administrative Costs. From the standpoint of most customers, the administrative costs of obtaining electricity are practically nothing today, and most customers have very little contact with their power companies. Most customers would be hard pressed to name another seller of anything on whom they spend comparable amounts of money and yet have so little contact. When all goes well, the oversight that a public service commission exercises keeps customers content with service and prices, and many consumers' roles are limited to paying a monthly bill. When the power does go out, there is no question about who to call or hold responsible.

In the Acme power example, Acme would not have the primary control over our quality of power service other than self-chosen interruptible pricing options if Acme offers them. It was also noted that Acme might not actually own any power plants or lines, and it may be located anywhere in the United States since it is not tied to the geography of the local or regional grid. Because Acme may not be a local company, the customer's contact may be via advertisements and over the telephone.

The RTP/PBR scenario provides "one- stop accountability" for service quality problems. For most customers and most problems, there is a single place to go to resolve billing disputes, reliability problems, power quality problems, and so on.[15] Of course, the issue of service quality extends well beyond the question of whether the customers' lights are involuntarily off. Like all other sellers of services, regulated or unregulated power retailers will have to answer customers' questions, send bills, process payments, change service options, and attend to a host of other factors. To do so, they must cooperate and transact constantly with the disco, the transco, the ISO, and many other parties that most customers will never know are there.

Factors such as the essential nature of power service, the fact that transmission and distribution will be closely watched by regulators, and the momentary control demands on the system could lead to vexing administrative disputes, with customers stuck in the middle. *Forbes* magazine writer John R. Hayes chronicled ("Blackout." *Forbes,* Nov. 4, 1996, p. 346) one such recent predicament in the telephone industry:

> In April, with no warning, Frontier Corp., the phone company based in Rochester, N.Y., cut off all outgoing long distance and incoming 800 calls to hundreds of Los Angeles businesses. Customers are still seething.

> "I'm sure we lost quite a lot of business," says Michael Stadler, chief financial officer of Los Angeles-based S&M Moving Systems, which does employee relocations for Intel and others. "If they can't get us, they call somebody else."

> What had S&M done wrong? Nothing. It was an incidental victim of a billing dispute that began in 1993 between two telephone companies: West Coast Telecommunications, purchased last year by Frontier, and one of its customers, Santa Monica, California based Fibernet. Claiming that Fibernet was behind on payments, Frontier pulled the plug on it. Fibernet's customers were the ultimate losers.

> Now headed for court, the dispute between Fibernet and Frontier has caught the attention of the industry and of the California Public Utilities Commission. The question: what obligation, if any, did Frontier have to Fibernet's customers?

> If the business service in question were anything but a traditionally regulated utility, the answer to the question would be "none." S&M, after all, had no contractual relationship with Frontier. The moving company was paying Fibernet to deliver service, and if Fibernet failed to deliver, the beef was with Fibernet.

> But public utility commissions do not see themselves as enforcers of contracts. As regulators, they like to think of themselves as guardians of the public welfare. Hence the California PUC may step in and dictate to companies like Frontier how they should treat end users such as S&M.

> Count on plenty more disputes like this one as telephone and electric companies evolve from regulated utilities into players in a free market.

In another instance, I recently spoke to an electricity control room operator who scheduled a power marketer transaction that had to be altered for reliability reasons. The control room called the power marketer, only to receive the marketer's phone answering machine. Left with no one to talk to, the controller unilaterally altered the transaction.

These anecdotes do *not* constitute proof that administrative costs and hassles will be worse under retail choice than RTP/PBR. However, they seem to square with widespread customer perceptions and the fact that transactions will simply be much more complex under retail choice, leaving more room for error.

A final element of transactions costs is the time consumers spend shopping for and evaluating alternative retail offerings. Every one knows that it takes time and patience to do comparison shopping for anything. In a world in which power prices fluctuate 24-hours-a-day, finding a supplier who will be cheaper or better than others is going to be a daunting task that many customers will want to simplify or avoid entirely. This is particularly true for small residential and commercial users, for whom the cost differentials between alternative suppliers may be quite small.[16]

The unwillingness of many small consumers to bear the cost of shopping for retail power may well be economically rational, and has been observed in other deregulated markets such as telephones and natural gas.[17] In effect, the number of small customers who choose not to shop is an indication of a perception that the benefits of competition do not outweigh the costs of shopping under present rules and perceived market conditions. However, even where this is true it cannot be concluded that non-shoppers do not benefit from the advent of retail choice. If retail choice or RTP/PBR is properly implemented, the industry as a whole should become more efficient, reducing the aggregate cost of supply. Assuming proper implementation, non-shoppers will be worse off only if the industry raises prices to non-shoppers and lowers them to shoppers. However, because non-shoppers are buying at regulated rates, such cost shifts should be observed and moderated by regulators.

THE ABILITY TO INCORPORATE PUBLIC INTEREST PROGRAMS

The following chapters of this volume examine a variety of public goods the electric utility industry has played a key role in providing in the past. These goods include universal service and the protection of low-income customers, protecting the environment, and various national energy policy objectives.

With few exceptions, the legal obligations to provide these goods were imposed on utilities that either owned generation or sold at retail. As might be expected, the patchwork of legal authority over members of each segment of the industry (Table 4.1) makes the imposition of any obligation by state or federal law quite difficult. The fifth row of Table 4.1 shows the tremendous variation in the legal obligations to do integrated resource planning—an obligation closely linked to public interest objectives.

Unfortunately, retail choice makes the allocation of responsibilities like this more difficult. From the outset, discussions of retail deregulation have raised the thorny question of which segment of the new industry, if any, should bear future public interest obligations. Because these obligations are by definition related to public goods, the principles that guide efficiency in private-goods markets offer only limited insight. Moreover, because the physical, economic, and political nature of these public goods is each, very different, generalized solutions are essentially impossible.

The following chapters examine each main public interest obligation and the proposals for reducing or allocating these obligations to participants in the new power industry. For the purpose of comparing retail choice and the RTP/PBR scenarios, the overall conclusion that emerges from the discussion can be summed up as follows: from the standpoint of administrative and political achievability, the RTP/PBR scenario has minor advantages in sustaining social programs—whatever the level of obligations policymakers impose on the industry.

The reasoning underlying this preliminary conclusion is quite simple. The industry is much more complicated under retail choice; tracing the identity of power flows and power sellers is decidedly more complex. Any obligation imposed on the generators or sellers of power may have to be traced across hundreds of thousands of small transactions scattered across the nation, as opposed to the convenient centralization within regulated distributors. For better or worse, because virtually every U.S. household receives service from a single, readily identified distributor, the utility industry has been made a convenient means of enforcing social obligations.

DISTRIBUTION SYSTEM GOVERNANCE

The conflicts and challenges facing distribution companies and their regulators are examined at length in Chapter 9. If this discussion is placed in the context of the two scenarios, there is a significant difference between RTP/PBR and retail choice.

Many of the distribution governance problems described in Chapter 9 apply to both scenarios, but one sort of conflict is stronger under retail choice. If the distribution system retains the sales franchise, the conflicts between the disco-as-seller and disco-as-transporter are eliminated. Because the disco sells all retail power in its area, regulators no longer need to strike the delicate balance between giving the disco the pricing flexibility and overall economic strength to remain a reliable seller of last resort and leveling the playing field for all retailers other than the disco.

The main tradeoffs for this considerable simplification of the regulators' task revolve around the other net costs and benefits of retail choice described in this chapter. In addition, the RTP/PBR scenario does not simplify or remove the regulatory challenges associated with policing the unregulated activities of distributors. It may lessen the pressures on distributors to engage in unregulated ventures, and it may make it easier to construct access or information policies that enable other services to be provided. Also, under retail access, state regulators will probably play a significant role in licensing or certifying the capabilities or practices of retail sellers, whereas under RTP/PBR, this is not necessary. (In the latter case, however, regulators may oversee some aspects of the markets for power-related financial instruments.) For all of this, the calculus of costs and benefits is devilishly difficult, and no theory or model is sufficient to predict the size of the net differences between the two scenarios—only their likely character.

THE OVERALL NET BENEFITS OF RETAIL COMPETITION

Table 10.2 summarizes a comparison of the two rough retail scenarios according to the equally rough main criteria that are being employed. At this level of generality, it is impossible to conclude that one scenario produces greater economic welfare than any other, particularly if a value is attached to "fairness" (however defined) and other public-good considerations.

On this last point, a somewhat fierce division of opinion has erupted between economists, regulatory agencies, and other industry participants. With regard to the difference in economy-wide savings between retail scenarios—given strong wholesale competition—two schools of thought have emerged. One school argues that the additional savings from retail choice will be substantial; the other argues that the additional savings will be small or negative, and are outweighed by the various potential disadvantages of retail competition described throughout this and the previous chapters.

The pro-retail-choice school of thought often cites evidence from other deregulated industries, where competition has almost always reduced overall costs. Proponents of the RTP/PBR scenarios respond to these examples by noting first that they are not usually a good measure of the difference between the particularly refined scenarios the power industry must implement, including the fact that bulk power deregulation is expected to produce significant benefits in either case. They also point out that deregulation of other industries and electric power in the United Kingdom and other countries has produced record numbers of small-customer price increases, problems with service quality, reliability, competition adequacy, and other issues that call the ultimate success of some deregulation proposals into question.

The school of thought that is skeptical-to-negative about the net benefits of retail choice emphasizes that it will be very difficult to distinguish long-run savings from the shifting of costs onto other customers or investors. For example, utility analysts Robert McCullough and Ruben Brown concluded that retail competition in California would result in small cost reductions but larger cost shifts, resulting in 13% higher prices for residential and commercial ratepayers in the early years of retail choice.[18]

Several studies have attempted to quantify the net benefits of wholesale competition, but only a handful have attempted to separate the net benefits (if any) of retail competition relative to a scenario such as RTP/PBR. One study of retail choice that has gained much attention (completed by the Citizens for a Sound Economy (CSE) Foundation in May, 1996) computed even larger aggregate benefits of $110 billion a year—over half the cost of America's total power bill.[19] Obviously, savings of this magnitude exceed likely cost shifts toward small customers, making them net gainers even if their rates remain high relative to large users.

Many researchers with long industry familiarity find this conclusion unsupported and unreasonable. Dale Phariss, an analyst with the National Rural Electric Cooperative Association, examined the CSE study and wrote:[20]

The CSE conclusions are based on assumptions that are both inaccurate and misleading. The...report sidesteps proper economic methodology, which, if applied to its own statistics, would show that residential and small-business customers would save little or nothing at all, while large industrial consumers would get most of the benefit.

In a study sposored by a broad coalition of consumer, low-income, rural, small-business, and other groups, economist Mathew Kahal concluded that the CSE results "rest upon a flawed analysis and a superficial understanding of present industry trends...[and]... is based upon misconceptions and conceptual errors." Kahal found that when the errors in the CSE are corrected, " the long-run savings evaporate" and "there is a risk of harm to some consumers." Investor-owned utilities issued a statement expressing similar views.[21]

The second school of thought argues that these large estimated savings violate economic common sense. When the Federal Energy Regulatory Commission estimated the economy-wide net benefits of wholesale completion under Order 888, it arrived at a figure of about $5 billion per year—about 2.5% of electric power bills.[22] This figure is very much in line with 30 years' of studies of improved system efficiencies (discussed in Chapter 4), although there is lingering dispute over whether such figures include all transactions and environmental costs.[23]

Unleashing the competitive energies of the national wholesale marketplace's 3,000 buyers and perhaps 1,000 sellers will, in FERC's opinion, save $5 billion a year. Does it follow that by keeping the same group of sellers, but enlarging the group of buyers from 3,000 discos to hundreds of thousands of power marketers and individual customers, the savings will suddenly leap to $100 billion?[24]

Most economists find it unrealistic to think that present or future wholesale buyers will fail to compete strongly or smartly enough to overlook savings 20 times as large as they are achieving. As the discussion in this and the previous chapter makes clear, it is not realistic to assume that retail choice will bring forth legions of new products or services that the RTP/PBR industry cannot effectively produce. And finally, regulators will continue to govern the marketplace in many ways under both scenarios, so this can hardly be the source of such large savings. The view of retail competition skeptics is well-expressed by British researcher Gordon MacKerron (1996) p. 5, speaking of the introduction of retail choice in the U.K. which is presently scheduled for 1998:

> It is very difficult to imagine that it will be worth the while of potential suppliers to compete for more than a very small proportion of the 22 million potential customers, and clear that a genuinely competitive system will lead to very significant new transaction costs in metering or profiling and in the setting up of complex new trading and settlement systems. The process is currently guided by what seems close to a blind faith that a competitive solution will automatically minimize costs: it seems at least possible that a less radically competitive system could provide similar benefits with less complexity and up-front costs.

Table 10.2. Summary Comparison of RTP/PBR and Retail Choice Scenarios (All evaluations subject to wide error bands)

Characteristic	RTP/PBR Scenario	Retail Choice
Marketplace Variety of New Products and Services Offered	Good and possibly better for new services on customer premises	Good and possibly better for new upstream innovation; possibly better overall as well
Changes in Overall Customer Power Bills and Economy-Wide Savings	Less shifting of costs to residential and commercial customers and possibly less overall savings, though probably not by much. Savings likely for all users	Greater shifts of costs to residential and commercial customers. Possibly higher overall level of savings. Net impacts on residential and commercial customers uncertain, especially if quality of service and administrative costs factored in.
Power System Reliability	Possible slight improvements over retail choice	Possible slightly diminished relative to RTP/PBR
Service Quality Other Than Reliability	No difference between scenarios?	No difference between scenarios?
Consumers' Administrative Costs	Lower than under retail choice, particularly when problems arise	Higher than under RTP/PBR
Governing the Local Distributor	Somewhat less difficult than under retail choice, though still difficult	Somewhat more difficult than under RTP/PBR

Nor would such solutions necessarily involve more regulation: arguably the complexity of the issues involved in setting up completion—including obligations to supply, special terms for the elderly and infirm, and the general issue of removal of cross-subsidies leading to protest form rural and/or small consumers—will actually provoke more active and complex regulatory intervention than would a modified form of the traditional monopoly franchise.

One of the most balanced views of the net benefits of retail choice is provided by a team of researchers from Resources for the Future, a Washington, D.C. research foundation. Over the course of a series of studies, one of which the author was involved in, this group has expressed the view that the net benefits of retail choice are probably small and center mostly on the availability of new products and innovative arrangements, not price reductions.[25] This conclusion was succinctly summarized by Resources For the Future, economists Douglas Bohi and Karen Palmer (1996), who wrote:[26]

If markets are sufficiently competitive, the retail model of restructuring is likely to produce a greater array of products and services and lower [overall] electricity prices, but the wholesale [RTP/PBR] model may yield lower transactions costs and better encourage transmission investment. Which model is better? We don't know yet.

Retail choice may be better for the development of new products and services, particularly if the unrealized values are more upstream than down, and it is undoubtedly better for undoing the economic inefficiencies inherent in retail ratemaking. On the other hand, RTP/PBR may be better at realizing downstream opportunities, it is almost certainly easier and cheaper to administer, and it may also be an inevitable and logical endpoint for electric industry restructuring. To repeat Bohi and Palmer's final thought: "We just don't know."

Appendix 10A

Retail Prices and Cost Allocation With and Without Regulation

Unregulated firms set prices with the single, overarching objective of maximizing profits. Regulators, with the authority to set prices, have a more complex set of objectives. Table 10A.1 summarizes the objectives of public ratesetters in the eyes of three leading authorities on rate policies, Professors Bonbright, Danielsen, and Kamerschen (hereinafter "BDK").[27] Associated with each of these objectives is an attribute of an ideal regulated structure of electricity rates.

According to BDK, there are three main objectives for utility ratemaking: economic efficiency, fairness, and administrative simplicity. Whereas the first objective is shared with unregulated industries, the latter two are not, and they have the potential to cause regulated prices to differ significantly from those set by a competitive market with similar products and costs.

The third major objective shown in Table 10A.1 involves return on investment or level of profit. Regulators must allow regulated firms to earn enough profit to compensate investors, but not more than an "appropriate" level. Discussions of the methods of deciding the correct return on equity or absolute profit for a regulated firm occupy hundreds of books and thousands of hours of debate in utility rate proceedings each year.[28]

However, our purpose here is to understand differences in cost allocation and prices in deregulated retail markets, and for this purpose profit level differences are not of primary importance.[29]

The final consideration shown in the table is administrative simplicity and transparency.

Table 10A.1. Bonbright, Danielsen, and Kamerschen's View of Ratemaking Objectives and Utility Rate Structures[1]

Ratemaking Objective	Regulated Price Structure Attribute
Economic Efficiency[2]	Classes of customers and pricing structure for the class encourages efficient use of resources
Economic Efficiency	All costs are reflected in rates, including all externalities
Fairness	All joint or common costs are apportioned "fairly" across customers who benefit there from
Fairness	No undue discrimination towards any customer or customer class
Economic Efficiency	Promotes innovation and other "dynamic efficiencies"
Capital Attraction	Rates and allowed profits produce efficient amounts of capital at costs equivalent to compensate investors for expected risk-return profile
Balance Between Administrative Simplicity and Adherance to All Other Principles	Rates are reasonably stable and predictable for both regulated firms and customers
Balance Between Administrative Simplicity and Adherance to All Other Principles	Regulation produces rates that are understood, believed fair and reasonable, and can be effectively administered

Source: Adapted from Bonbright, Danielsen, and Kamershen (1988) p. 382ff

1 This table is adapted from the discussion on p. 382 ff of BDK (1988). The order of the entries has been changed for expositional convenience, and no prioritization is implied by the order of listing in this table other than to reflect BDK's conclusion (p. 389) that the alignment of costs and prices (insofar as this is possible) is the single most important "measure of reasonable rates."

2 BDK (1988, p. 385) call this objective "customer rationing," i.e., "rates that are designed to discourage the wasteful use of public utility services while promoting all use that is economically justified in view of the relationships between the private and social costs incurred and the benefits received." This definition corresponds closely to the more widely recognized term shown in the table.

This straightforward consideration applies to unregulated as well as regulated markets, though much more so under regulation. Consumers, buyers, and even the employees of sellers must be able to understand and keep track of the terms and conditions of prices and services offered by sellers.

PRICING PHILOSOPHIES AND COST ALLOCATION

"Without doubt," BDK write, "the most widely accepted measure of reasonable public utility rates and rate relationships is cost of service." (p. 389). BDK are undoubtedly correct—there is little disagreement that rates should reflect costs. However, electric utility networks have many cost elements that are very long-lived, and give service to many types of users at once. These facts give rise to many definitional variations on cost: short-run marginal, long-run marginal, average, and so on. Which cost do BDK refer to?

Economic theory teaches that efficient prices are always set equal to the marginal cost of serving each customer.[30] However, the ratemaking objectives shown in Table 10A.1 suggest that utility ratemaking is not based solely on economic efficiency (i.e., marginal cost), but *is* based primarily on "cost".[31] Hence, rather than using the economists' standard of marginal cost, regulators devise their own definition of cost and base rates on costs as they allocate them under their definition so as to equate total costs and revenues and thereby avoid a revenue shortfall that might occur under marginal cost pricing. This methodology has become known as Fully Distributed Costing (FDC).[32] More importantly, the use of a cost-based pricing standard with regulator-determined allocations of all recorded costs makes cost allocation the central exercise in regulated ratesetting.

Universally, this method of ratemaking allocation begins by defining customer groupings or *classes,* with similar non-transferable attributes. In most utilities, the main customer classes are residential, commercial, industrial, and governmental, joined by agricultural users or other groups in many cases. The next step is to divide all costs into two categories: those that can be allocated specifically to each unit of consumption and each transaction and those that cannot. Very crudely, this is the difference between variable costs, which vary with the output of the firm, and fixed costs.[33]

In the electric utility industry, there is generally a somewhat obvious distinction between the inputs whose consumption rates rise and fall with the rapidly-varying nature of power demand, such as fuel, and most other costs, which vary only gradually with the total size of the firm. Table 10A.2 shows the percentage breakdown of costs in various categories for all electric IOUs in 1994. Column (2) shows the percentages of total cost in that category for the total IOU segment of the industry; column (3) indicates the extent to which this cost element varies directly with the level of sales during a typical year, and therefore can be assigned directly to each transaction and customer with no need for an indirect allocation scheme. As column (3) shows, a substantial amount of a utility's costs— typically 50% to 80%—do not vary directly with sales over the course of a year.

With few exceptions, regulators ascribe to each customer class those variable expenses that can be clearly associated with that class. For example, suppose a utility decides that it is going to change the meters that are used for its commercial class customers, but no others. The cost of these changes would almost always be allocated by regulators to

Table 10A2. Aggregate Cost Structure, 1994 Investor-Owned Electric Utilities

Cost category	Expense, 1994 (millions of dollars)	Proportion of total cost	Fixed or variable
Steam Power, Total Operation Expenses	29,780	0.283	variable
Steam Power, Total Maintenance Expenses	4,422	0.042	fixed/variable
Nuclear Power, Total Operation Expenses	8,513	0.081	fixed/variable
Nuclear Power, Total Maintenance Expenses	2,827	0.027	fixed/variable
Hydraulic Power, Total Operation Expenses	271	0.003	fixed/variable
Hydraulic Power, Total Maintenance Expenses	173	0.002	fixed/variable
Other Power, Total Operation Expenses	931	0.009	variable
Other Power, Total Maintenance Expenses	187	0.002	fixed/variable
Other Power Supply Expenses Purchased Power	29,213	0.278	fixed/variable
Other Power, System Control and Load Dispatching	251	0.002	fixed
Other Power, Other Expenses	309	0.003	fixed
Total Transmission Expenses	2,069	0.143	fixed
Maintenance Expenses, Total Distribution Expenses	5,933	0.056	fixed
Total Customer Accounts Expenses	3,546	0.034	fixed
Total Customer Service and Informational Expenses	1,956	0.019	fixed
Total Sales Expenses	232	0.002	variable
Total Administrative and General Expenses	14,515	0.138	fixed
Total Expenses, 1994	105,128		

Source: Energy Information Administration, Financial Statistics of Major U.S. Investor-Owned Electric Utilities 1994. Washington, D.C.: U.S. Government Printing Office

commercial rates and no others. Similarly, fuel costs are allocated to each customer class in accordance with its energy use. It is the allocation of the joint and common costs associated with fixed investments that is difficult and somewhat arbitrary.

ALLOCATION OF COMMON COSTS

Deciding how to allocate the remaining (and usually large) fixed costs across the customer classes is the task BDK call "perhaps the most complex and difficult part of modern rate theory." (P. 427). This is the crucial point at which regulators' judgement and methods for apportioning costs come into play—usually within bitterly contested rate proceedings. Economists Stephen Brown and David Sibley (1986, p. 44) observe that:

This is not a straightforward task and is the source of many of the most muddled, lengthy and unsatisfactory proceedings in regulatory history. The root of the problem is that prices must be set for each service offered by the firm but substantial amounts of cost represent facilities used in common by several or all services and cannot be allocated in a clear cost-related way to any single service.

Under the fully distributed cost method, the regulator is essentially free to pick any method of cost allocation that results in a complete allocation and reasonably balances the regulator's objectives and responsibilities.[34]

Modern ratemaking practice uses what is known as three-part pricing—a customer charge, energy or fuel charge, and a capacity payment. In greatly simplified terms, the customer charge is equal to all of the fixed costs that are associated with adding one more customer to the utility's system, even if that customer uses very little power. For each class, this is the cost of setting up and keeping track of a customer account: sending utility trucks to the house, installing and reading the meter, sending out monthly bills, and so on. None of these costs vary much with the amount of power used each month or year, but they do vary by customer type, so there is a relatively non-controversial way to allocate them.

Energy charges are similar to the variable costs discussed above, and they can be allocated more or less accurately to each customer class. The remaining costs of the utility, generally associated with the utility's generation, transmission, and distribution fixed plant, are allocated via any of several methods. The most common allocation method is known as the peak responsibility method, where each class' demand at the time of the utility's overall peak is used to allocate fixed costs. The idea underlying this method is that the utility's total fixed assets are sized to safely and reliably accommodate the system's highest total use. If peak use dictates the size of fixed costs, then one's contribution to peak ought to be one's fixed cost share.[35]

The net result of this division of costs into categories, creation of customer classes, and allocation of costs to class is a set of average prices that ideally achieve objectives like shown in Table 10A.1. For the United States in 1993, the distribution of average rates is shown in Figure 10A-1. These total rates reflect the aggregate of the *directly attributable* fixed and variable costs as well as the *allocated* share of fixed costs that regulators have assigned, on average, to each class.

UNREGULATED AND RAMSEY PRICING

Cost allocation does not play the same central role in unregulated power markets as it does under price regulation. An unregulated power marketer (assumed to be serving the same customer classes as above) is interested in maximizing its total profits.

Assume that this firm serves the same customer classes as the regulated firm, and that customers cannot shift between classes. Also assume that unregulated firms have roughly the same collection of fixed costs and variable costs.

The unregulated firm begins to set price by allocating the variable costs to each user class. As noted above, any other assignment would clearly violate common sense and the principles of economic efficiency. However, when it comes to the fixed costs, the firm asks itself a simple question: How can I set prices for each class above the baseline levels of allocated variable cost so as to earn the highest possible revenue, covering my fixed costs and making the largest possible profit?

The prices that meet these conditions were first mathematically determined by mathematician Frank Ramsey in 1926, and they are widely known as "Ramsey Prices."[36] Ramsey proved that the profit-maximizing prices to several user classes in the presence of fixed as well as variable costs depended on the price sensitivity, or *price elasticity,* of each user group. The less price-sensitive the group, the higher its profit-maximizing price.

The price sensitivity of electricity customer classes has been carefully studied over many years, and there is little controversy about the approximate magnitude of typical class price elasticities. Residential electricity consumers are the least price-sensitive, generally because electricity is not a large fraction of their household budgets, electricity is essential or near-essential for many uses, and household schedules often do not allow much flexibility with respect to electricity use. Commercial users are moderately price-sensitive; they can adjust some demand, and electricity costs can be an important expense element for many stores and offices. Finally, and for a variety of reasons, industrial users are the most sensitive group of all.[37]

If unregulated power retailers have significant fixed costs, there is every reason to believe that they will use Ramsey pricing principles to set prices for retail electric power. Residential customers will pay the highest prices because they are least price-sensitive, commercial rates will be lower, and industrial rates will be the lowest of all.

In contrast, regulators have often considered and usually rejected Ramsey pricing. Among other things, regulators have found them difficult to implement, "unfair," or otherwise inconsistent with the ratemaking principles in Table 10A.1.[38]

Figure 10A-1. Average Revenue per Kilowatthour by Sector, 1994

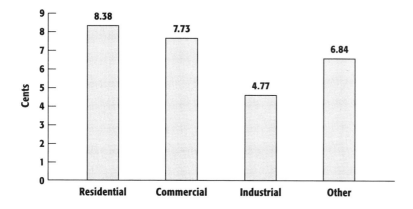

U.S. Average = 6.91 Cents

Notes: Data are final. Other includes sales for public street and highway lighting, other sales to public authorities, sales to railroads and railways, and interdepartmental sales. The average revenue per kilowatt-hour of electricity sold is calculated by dividing revenue by sales.

Source: Energy Information Administration, Form EIA-861, "Annual Electric Utility Report."

Appendix 10B

Regulating the Quality of Electricity Service

In well-functioning markets, customers make tradeoffs between price and quality, and the marketplace provides a variety of multidimensional qualities and prices.[39] When regulated entities are involved in the production chain, product variety and the overall level of quality in the marketplace are strongly influenced, directly or indirectly, by regulation's form and function.

As noted earlier, the quality of service experienced by electric customers will depend on the quality provided by regulated transmitters and distributors of power, as well as on the ability of the regulated and unregulated stages to work together effectively. Former New York State Public Service Commissioner and Professor Eli Noam likens quality provision to a chain that is only as strong as its weakest link.[40] This quality interdependence is exacerbated by the dual roles of the disco and perhaps the transco (assuming incomplete vertical deintegration), which may diminish the disco's desire to cooperate.

Beyond these considerations, regulation itself has become a source of concern over quality. Price-cap regulation, which is probably the most common form of PBR, has an intuitive, built-in incentive to reduce quality in the regulated network. For a regulated monopolist whose costs are linked to quality, the reasoning is simple: price caps entitle the utility to pocket all of the difference between allowed price and cost. If customers will continue to buy sufficient amounts of the regulated good even when the utility cuts quality and cuts costs, reducing quality can become profitable, at least up to a point.[41]

Although economic theory predicted this all along, interest in quality enforcement has grown as regulators heard increasing complaints about quality declines following price cap regulation. Former Maine Public Service Commission staff director Barbara Alexander wrote:[42]

> "Early PBR decisions contained no special provisions for maintenance of customer service, as commissions reasoned that they would rely on their existing rules and investigatory authority to address any problems that later arose. Many commissions have since found that this approach is insufficient. Particularly in the western U.S., states are scrambling to address deteriorating telephone service quality that occurred after performance-based regulation was approved."

MONITORING AND ENFORCING QUALITY

The science of measuring and incentivizing quality is new and evolving rapidly. Thus far, three main regulatory approaches are apparent. The first is to increase the monitoring of service quality dimensions within cost-of-service regulation via periodic filings or proceedings. The results of these proceedings lead to reporting requirements, required service standards, and perhaps reductions or increases in allowed profits when regulators examine rates and costs. This approach was adopted in New York, where the Public Service Commission issued a 1991 order requiring each regulated utility to adhere to certain service standards and to report its performance against these standards in an annual report.[43]

The second approach is similar to the first, but embeds quality standards in a PBR framework. PBR lends itself to this approach because the essence of PBR is the specification of a set of performance standards to which regulated profits are tied. Nowadays, it is unusual to see a PBR scheme for a regulated gas, electric, or telephone local carrier that does not contain some quality feature. As an example, Central Maine Power (CMP) was ordered by its regulators to adopt a quality incentive scheme as part of its PBR. Table 10B-1 shows the five quality criteria and benchmark levels of the CMP scheme. For each criteria, an annual point score based on service performance is awarded and aggregated across all five criteria. Based on the aggregate score, an earnings penalty ranging from zero (all benchmarks met or exceeded) to $3 million is applied to CMP's earnings.[44]

The third approach to quality enforcement is to provide direct incentive or penalty payments for easily monitored and well-specified performance variance episodes. In the United Kingdom, electricity distributors' franchise agreements require that they pay penalty fees directly to customers when they fail to meet certain targets, such as remaining on-time for service calls.

Although some precedents have been set by these approaches, the techniques for measuring and rewarding quality are still in their infancy.

Chapter 10

NOTES

1 Many analyses of retail choice examine a larger number of more detailed scenarios, including Hunt and Schoengold (1995), the Public Service Commission of Wisconsin (1995), Hunt and Shuttleworth (1996), and Brennan, et al (1996).

2 Bohi and Palmer (1996) argue that this is not an appropriate assumption because the bulk power markets will be significantly affected by the presence or absence of retail competition. These authors argue that retail competition will help the bulk power marketplace in some ways, but on balance will make it work less well in the long run by making it more difficult to justify investment in the transmission system. We return to this particular point briefly under the heading of the reliability impacts of retail choice below. In general, however, and without disputing Bohi and Palmer's point, we abstract from any differential impacts of our two scenarios on the bulk power market.

3 The reader is entitled to know why PBR is a part of this scenario, and which of the almost infinite varieties of PBR do we have in mind. As to the former, the advantages of moving to PBR from traditional regulation are discussed in Chapter 8). In this particular case, we have assumed that the wholesale power market is broadly competitive. Even if this means that many discos remain integrated with their gencos, which appears increasingly unlikely, the use of bulk power purchased competitively is sure to become a much more common aspect of every disco's supply strategy. If we view the role of discos largely as purchasing agents on behalf of their franchise customers, PBR may be particularly suited to giving the proper regulatory incentives to purchase efficiently. This argument is made most forcefully by Cavanagh (1994, 1996).

As to the latter question, we simply assume that the disco faces performance requirements upon which a portion of its allowed profits are set. We assume this is done intelligently and effectively, so discos do a reasonably good job of minimizing the total bills of their customers subject to regulatory and other constraints. We do not specify this idealized form of PBR in more detail than this; in practice, the ideal form may vary somewhat according to each disco's circumstances and local regulatory policies. For more information on the performance standard aspects of PBR for discos, see Appendix 10B.

4 In the retail choice scenario, we also assume that disco earnings are determined via a PBR approach rather than via traditional ratemaking. We return to this point in our discussion of service quality incentives below.

5 To cite just a few of many important efforts, Public Service Commission of Wisconsin (1995), Hunt and Shuttleworth (1996), Bohi and Palmer (1996) and Brennan, et al (1996).

6 In particular, public ownership of transmission and distribution is receiving renewed attention as part of the overall restructuring debate. See, for example, Schweitzer (1996).

7 Note that a disco or an affiliate of a disco may offer a wide variety of financial products that may hedge or otherwise alter the effects of disco prices. See the discussion of disco pricing flexibility in Chapter 9 *supra* as well as Hunt and Shuttleworth (1996) p. 70.

8 Also see Professor Hogan's discussion of "Efficient Direct Access" (Hogan 1994a). The identical point is made in Hunt and Shuttleworth (1996, p. 87) quite concisely, where the authors call what we have called RTP/PBR the "optimization" scenario and retail choice the trading or market scenario.

An informal proof of this proposition is as follows: Suppose, in both scenarios, there is a perfectly competitive price for generation G in every time period and perfectly formulated transmission and distribution transport tariffs (T and D respectively) for each period as well. These transport tariffs include all ancillary services.

Suppose that disco regulators also set prices perfectly, with zero transactions costs. Then retail prices R will be the sum

(1) $R = G + T + D$

for every period in a retail price location for which tariffs T and D apply.

In the retail choice scenario, the power marketer's profits per unit of power sold in each period are:

(2) Marketer = Sales price–$(G + T + D)$

The marketer sells at whatever price she can and then must pay the sum of G, T, and D to providers to deliver the power. The ESCO's profits under an agreement in which the ESCO pays the customer's power bill and profits from the difference between its own sales price to the customer and this customer's bill is, for each unit sold in each time period:

(3) ESCO = Sales price–Retail Price R

However,

(1) $R = G + T \times D$

So:

(4) ESCO = Sales Price–$(G + T + D)$
which is the same equation as profits for the marketer.

The preceding proof can be modified by letting G be the average market hub price or price of an electricity future at the market hub. Power marketers or ESCOs may then profit by finding or making power "below market" and purchasing this cheaper power.

In the case of marketers, this causes a direct cost savings as the marketer's cost of supply is now G + T + D,

where G is the cost of below market power. In the case of an ESCO who could make or buy power for a unit price G, the equivalent profit could be made by reselling this at wholesale. In both cases, profits are G–G.

9 This refers to an industry with approximately equilibrium capital stock. If technological change, altered demand patterns, investor or regulator mistakes, or other factors create excess or insufficient capacity, investors may earn more or less than the risk adjusted cost of capital until the industry reaches capital equilibrium.

10 See Appendix 4A for additional discussion and references.

11 Residential customers also have demands that are steady over time. These demands are cheaper to serve; customers with the better "load factors" will also receive discounts.

12 See Colton (1996) and Energetics (1995) for two useful surveys.

13 "How Restructuring Affects Customers" Speech by Audrie Krause, Executive Director, Toward Utility Rate Normalization, Northern California Power Agency Annual Meeting, Sept. 22, 1994

14 These results also illustrate the tension between the sales and transport roles of the distributor. Commenting on the fact that regulators allowed transport price increases for residentials much larger than those allowed for industrial users, the study's authors (Tucker and Davis, 1996) noted:

> At the same time, the competition in large-volume applications with low cost oil and coal placed special pressure on T&D charges to industrial users and electric utilities. By squeezing out cross-subsidization of small-volume users by large-volume ones that had long been a hallmark of regulated ratemaking, the natural gas industry has been able to respond to the competitive pressure and even increase its market share substantially in large-volume markets.

In other words, regulators were sufficiently concerned about keeping LDCs viable to allow them the price flexibility to keep bypass at reasonable levels.

15 When homes or businesses install real-time control systems, and something goes wrong, it will be necessary to determine whether the problem is in this network or the customers' system.

16 This is not to deny that the marketplace will find many ways of simplifying shopping or making it entertaining. For example, there is reportedly an Internet site that searches and compares telephone calling plans at no cost to the user. Mike Mills, "On The Web, Finding a Carrier to Fit Your Calling Pattern." *Washington Post*, 10/29/96, p. C1. Also see Weiner, et al (1996) p. 25.

17 In U.S. natural gas markets that allow retail choice, most residential and small commercial customers have thus far chosen to remain with their utility. See, for example, Residential Choice Off to a Slow Start in New York," *Gas Daily*, 6/12/96 p. 1 and Agis Sapulkis, "Natural Gas, Unnatural Selection," *New York Times*, 10/23/96 p. D1. In California, commercial customers shifted from zero to 10% retail choice in just two years (California Energy Commission, 1994).

18 "The Price of Energy Industrial Customers", *Electricity Daily*, V.2 no. 87, May 5, 1994. Also see Cohen and Kihm (1994), Bohi and Palmer (1996), Corneli (1996).

19 Maloney, McCormick, and Sauer (1996).

20 Pharris (1996) p. 1.

21 Kahal (1996) pp. 57–58, and "Statement of the Edison Electric Institiute on the Citizens for a Sound Economy Foundation Electric Competition Study," Edison Electric Institute, Washington, D.C., May 30, 1996

22 The precise range of FERC estimated savings is $3.8 to $5.4 billion (Order 888, Docket No. RM95-8-000 and RM94-7-001, April 24, 1996, Mimeo, p. 3).

23 See Chapter 12 infra.

24 The Regulatory Assistance Project, a federally-funded project that assists state regulators, made a similar observation:

> If wholesale competition is functioning well, the utility will already be acquiring all cost-effective supplies, and there is very little chance that retail customers will find resources that offer additional system cost savings under a retail wheeling framework. If the utility is taking advantage of wholesale competition for new resources, then it is unlikely that new supplies identified by customers will beat the utility's marginal supply costs very often or by very much, and as a result retail wheeling will yield little, if any, economic benefit. (Regulatory Assistance Project, 1994).

25 See Bohi and Palmer (1996), Brennan, et al (1996), and Bohi (1996). Similarly, Hunt and Shuttleworth (1996, p. 70) note that the main difference in economic benefits between wholesale and retail choice "is on the customer side." By this they mean that real-time pricing, which they assume is available only under retail choice, enables large customer savings. It follows from this argument that the efficiency differences between RTP/PBR and retail choice are minor.

26 Bohi and Palmer (1996).

27 This table refers to the objectives of regulated pricing policies, not to the objectives of regulation itself. In addition to pricing, regulation must remain attentive to efficiency incentives (intentional and otherwise), information asymmetries and other potential real-world defects, transactions costs, and other sometimes subtle economic, legal, political, and cultural forces that may play a role in regulated outcomes.

28 See the discussion and references in Appendix 4A and Chapter 15, as well as Bonbright, Kamerschen, and Danielsen (1988), and Kolbe, Read, and Hall (1984).

29 It is reasonable to abstract from profit level differences because the theory of utility regulation suggests that differences in profits should be relatively small between regulated and deregulated scenarios at *equilibrium levels*. The overall profit levels of regulated firms are supposed to be set at levels that mimic competitive markets, and which are sufficient to attract just enough capital to keep the industry at efficient long-run levels. Competitive markets are supposed to function with the same long-run objective.

This argument refers only to the total level of profits, and does not imply that the profit gained from each group of customers or transactions will not differ significantly between regulatory and deregulatory scenarios. We focus attention on the costs and prices faced by customer groups in this section partly because the gap between these two dictates the distribution of profits across transactions.

31 Professor Kahn (1970, V. I., p. 65) writes, "the central policy prescription of microeconomics is the equation of price and marginal cost. If economic theory is to have any relevance to public utility pricing, this is the point at which the inquiry begins." Kahn advocates the use of short-run marginal cost (defined in a particularly precise manner, p. 71), including externalities but excluding long-run fixed costs, unless competition is present. Fixed costs should be recovered using Ramsey-like prices (see discussion following), if necessary (p. 77ff).

32 BDK, Kahn (1970), and Malko, Smith and Uhler (1981) discuss many limitations on the use of marginal costs alone as the basis of ratemaking. The former conclude their review by noting that:

One may summarize the foregoing criticisms of the proposal to base utility rates on short-run marginal costs by saying that, in giving sole consideration to one very limited though important function of prices, that of securing the optimum utilization of whatever the plant capacity exists at a particular time, it would surrender other functions of even greater importance including, particularly, that of the long-run control of the demand for and supply of utility services. Mainly for this reason, we take it, the proposal has won less support than otherwise, even from those economists who are most impressed with the shortcomings of full-cost standards of ratemaking.

The main limitations BDK refer to include (predictably) fairness, inability to measure costs, and revenue shortfalls. Shepherd (1983) contains an extensive discussion of the use of marginal costs and time-sensitive pricing in utility rate structures.

33 See BDK (1988) ch. 19, Brown and Sibley (1986) Section 3.4.

34 "A fixed cost is the cost of the smallest (least expensive) batch of inputs that the firm can buy if it is to be able to produce any output at all.…In addition, the total cost of such inputs does not change when the firm changes its outputs by an amount that does not exceed the…production capacity. Any other cost of the firm's operation is called variable because the total amount of that cost will increase when the firm's output rises." (Baumol and Blinder, 1991, p. 429).

(For interpretation and application of these concepts in a regulated context, see Kahn (1970, V. I, p. 77), Joskow (1975), Cicchetti, Gillen and Smolensky (1976), BDK (1988) ch. 19 and Anderson and Bohman (1985).

Practically speaking, the longer the time period, the fewer costs are fixed. Over a period of years, firms can sell off their fixed assets and adjust the magnitude of just about any cost to better match firm output. Accounting rules specify the conceptual distinction between fixed and "current" assets, and generally give rise to a distinction between costs that vary over the course of a year with sales levels and investment outlays that do not. (Smith and Skousen, 1977, p.65).

35 Because FDC does not adhere to economic efficiency principles, the economic literature is widely critical. Among many examples, see Kahn (1970, V. I., p 151), Brown and Sibley (1986) p. 59, and Baumol, Koehn, and Willig (1987).

36 See BDK (1988) ch. 19 and Munasinghe and Warford (1985).

37 See Ramsey (1926), Baumol and Bradford (1970), and Bromard Sibley (1986).

38 See Appendix 4A for additional discussion and references.

39 See, for example, Tye and Leonard (1983), and the testimony of Paul F. Levy and David M. Weinstein, Maine Public Utilities Commission, Docket 92-315, February 1993.

40 See Tirole (1992) Chapter 2.

41 Noam (1991).

42 See Noam (1991), Fox-Penner (1992), Alexander (1996).

43 See Barbara Alexander, "How to Construct a Service Quality Index in Performance-Based Ratemaking", *The Electricity Journal,* April 1996. Interestingly, Professor Noam (in 1991) writes that "It was greatly feared that a more competitive and decentralized environment would lead to serious service degradation because the local exchange companies would be starved for investment funds. But though many people still firmly believe that these fears have become a reality, there is little evidence to support this view." (Eli M. Noam, "The Quality of Regulation in Regulating Quality: A Proposal for an Integrated Incentive Approach to Telephone Service Performance," in Michael A. Einhorn, Ed., *Price Caps and Incentive Regulation in Telecommunications,* Boston, MA: Kluwer Academic Publishers, 1991.) It is beyond our scope to consider whether Noam is right and Alexander is wrong or that Noam did not have adequate data in 1991 to conclude that quality concerns had arisen.

44 State of New York Public Service Commission, Order Adopting Standards on Reliability and Quality of Service, Case 90-E-1119, July 2, 1991.

45 See International Business Conference (1996).

Part III

PUBLIC INTEREST AND THE ELECTRICITY INDUSTRY

Chapter 11

Universal Service and Low-Income Ratepayers

The acclaimed documentary *Hoop Dreams* follows the lives of two teenagers from low-income families in the Chicago area. In a portion of the film, one teen's apartment has its power disconnected for nonpayment of the household power bill. Using battery-powered cameras and light from the apartment's hallway, the film briefly records a frightful period in which the family attempts to operate in near-total darkness.

These unforgettable scenes help illustrate a deep-seated public aspect of electricity: universal access at affordable prices. Similar views are held, to varying degrees, with regard to such other necessities as food, medical care, clothing, shelter, education, and other utility services. Hence, by virtue of its role in consumption, electricity joins an otherwise disparate set of private goods over which a public interest is asserted, for reasons entirely apart from the cost characteristics of the industry.

Many of the necessities on this list have been the subject of government attention for hundreds of years. In contrast, electric service is barely a century old and hardly began life as a necessity. Early utilities grew up in large and moderate cities, but generally could not profitably extend distribution lines to sparsely-populated areas. Envious of their urban counterparts and excited by a number of rural electric pilot projects, rural communities clamored for electricity at affordable rates.

A political turning-point for universal service occurred in 1932, when presidential candidate Franklin D. Roosevelt told rural, regional audiences that "[electricity] is no longer a luxury. It is a definite necessity...It can relieve the drudgery of the housewife and lift the great burden off the shoulders of the hardworking farmer...."[1] Roosevelt himself became aware of this issue when he noted that power rates for his small cottage in Warm Springs, Georgia were four times those of his apartment in Manhattan.[2] Roosevelt made good on his views by establishing the Rural Electrification Administration, the Tennessee Valley Authority, and the Bonneville Power Administration in addition to forging the investor-owner industry as it exists today.

Due to the efforts of—and often, competition between—the public and private industry, a remarkably high proportion of U.S. homes and businesses have electricity service.[3] In the United States, lack of access is much more an affordability issue than a physical inability to connect to power lines. As the fraction of American structures that remain wired for power is not expected to drop due to restructuring, future access and affordability issues are closely linked.

Today, universal service in electricity far exceeds the provision of other necessities, even by other regulated utilities. Telephone saturation has been level at about at 94% for many years, (and it is closer to 70–80% in low-income households), while only 91% of U.S. homes have indoor plumbing.[4] And unfortunately, millions of Americans today lack adequate medical care, shelter, or both.

UTILITY AND GOVERNMENT PROGRAMS TODAY

Starting in the 1960s, universal service transitioned from an issue driven primarily by rural access to an issue centered on urban low-income affordability. The large, sudden energy price increases of the 1970s caused added financial hardships for hundreds of thousands of low-income families, prompting a wave of concern among utilities and political leaders.

Although energy prices have dropped in real terms, energy costs remain a major health, safety, and even survival concern today. The typical low-income household in the United States spends $1,100 a year on energy, which is almost as much as the average U.S. household spends.[5] Energy takes one of the largest shares of low-income household budgets, and with low-quality housing stock and little capital there is often little choice but to use energy inefficiently. Energy price increases that many Americans may not notice can force low- and fixed-income households to choose between food and warmth.[6]

At the federal level, the two largest consumer energy programs in history were enacted, and remain in force today. The U.S. Department of Energy's Weatherization Assistance Program, now supplemented by state and local utility programs, has installed energy-savings measures in about 4.4 million of the nation's 29.1 million low-income households since its inception in 1976.[7] The Low Income Home Energy Assistance Program (LIHEAP), established in 1979, provides about $1.0 billion to state agencies that disburse the funds to more than 5.5 million low-income households each year to help pay for the cost of home energy. The latter program, which can provide emergency cash assistance directly to households facing power disconnections, greatly assists public agencies and utilities when disconnection looms.[8]

During the same period, most states adopted low-income/universal service ("LI-US") programs that applied to regulated LDCs. Reflecting the tenor of the times, the 1975 California legislature required utilities to offer a "lifeline rate"—a rate at or below the cost of supply for a customer's first 300 kWh of power and 75 therms of gas. The legislation stated that "light and heat are basic human rights and must be made available to all people at low cost for basic minimum quantities." Utilities understandably complained that selling power below cost was inefficient, harmful, and far less preferable than selling at marginal cost and providing cash subsidies to low-income households.[9]

The more durable LI-US state regulatory legacies fall into two categories—limits on disconnections and special low-income billing practices.[10] Typically, residential tariffs set by regulators prescribe detailed conditions under which utilities may terminate service for nonpayment, including such things as prior notice, notification of third parties, absolute prohibitions in winter months, and prohibitions on disconnection if members of the household are ill or disabled. Low-income billing practices include budget billing (levelized year-round bills), payment plans capped at a percentage of household income, and low-income discounts. In many areas, the implementation of these policies has produced a complex but often successful network that includes specialists within utilities, social service agencies, and state and local energy agencies, all directed toward reducing power bill nonpayment and disconnections.[11] State regulators often prohibit utilities from denying credit or service to low-income or financially insecure customers.

In recent years, however, state utility commissions have begun to use discount utility rates as a more cost-effective alternative than other low-income programs. Utility regulators reason that it is less expensive to provide affordable utility rates to low-income households in the first place, than to provide service at higher rates, incur the costs of providing working capital to cover unpaid arrears, spend money on credit and collection, and to still write-off a substantial part of billed revenue to certain customers as uncollectible. By offering such affordable rates, utilities have found that they actually can increase the revenues they collect and decrease the expenses they incur in the process of collection, to the benefit of all involved, including nonparticipating customers.[12]

The total costs of these programs are difficult to measure because they draw on a number of utility resources. According to U.S. government compilations, gas and electric utility low-income programs had direct costs totaling $425 million in 1994.[13] In addition, utilities spent about $90 million on energy efficiency programs targeted directly at low-income groups, supplementing federal and local programs.[14] Acknowledging that these figures do not include many hard-to-measure intangible and indirect costs, this amounts to roughly one quarter of one percent of total electric power industry annual revenue.

RESTRUCTURING AND LOW-INCOME PROGRAMS

Most modern LI-US programs are designed around the local franchised distributor. Wholesale competition or the real-time pricing-based scenario explored in Chapters 9 and 10 do not create an immediate requirement for change in this area. However, retail choice will allow customers and sellers to select each other, and if these sellers are unregulated they obviously may elect never to serve low-income customers, or may choose to stop service in whatever manner the (unregulated) retail power sales agreement provides.[15] Provisions for universal service are likely to be an integral part of a retail choice scheme, at least until it can be determined that competitive retailers will voluntarily achieve acceptable levels of universal service and other LI-US polices.

The few states that have thus far permitted or are exploring retail choice have demonstrated an acute awareness of these issues. California's restructuring legislation states in its preamble that "It is the further intent of the Legislature to continue to fund low-income ratepayer assistance programs,…in an unbundled manner…." and the legislation mandates no reduction in LI-US expenditures from 1996 levels.[16] New Hampshire's parallel legislation contains an even more explicit policy declaration:[17]

> A restructured electric utility industry should provide adequate safeguards to assure universal service. Minimum residential customer service safeguards and protections should be maintained. Programs and mechanisms that enable residential customers with low incomes to *manage and afford* essential electricity requirements should be included as a part of industry restructuring.

According to experts at the National Consumer Law Center, "only a few jurisdictions… have set a course that does not include funding direct protections for low-income customers as part of the restructured electric industry."[18] Moreover, the Telecommunications Act of 1996 contains a clear-cut federal requirement for universal service in telephony, demonstrating that recent federal legislative sentiment on this issue remains strong (see Box 11-1).

OPTIONS FOR UNIVERSAL SERVICE WITH RETAIL CHOICE

State and federal legislators have been unwavering in their support for universal service, but they have provided little guidance as to how to achieve it under retail choice. In Chapters 2 and 9, it was noted that the distribution utility automatically and immediately becomes the physical supplier of last resort by virtue of the instantaneous operation of the power grid. Once a home is connected to the grid, any power retailer other than the local disco may shut off service from its generators, for bill nonpayment or any other reason. However, the lights will stay on in that "disconnected" home until the home is physically disconnected.

This situation raises two significant and immediate questions. First, should the power that must be provided in these situations come from the disco itself, or should discos have the authority to requisition power from other retailers? If the latter, then the disco must find at least one unaffiliated generator and pay it to provide emergency service wherever needed. Either way, it may be quite difficult to plan for this supply—the disco may have little or no notice of the disconnection, and, therefore, no time to notify its suppliers of this special power need.

In addition to direct provision of power by the LDC (if it owns generators), two models have been proposed. In the first, all retail sellers of power must contribute a small amount of power to a universal service pool of kilowatt-hours, or possibly bear retroactive assessments for such power. In another, the LDC or a regulator auctions the right to supply LI-US power needs to the disco under regulated terms. A third proposal would assign the obligation to serve (or to pay for service to) a portion of LI-US needs to retail sellers equal to each seller's proportion of total market share (excluding LI-US customers). These proposals are largely untested, except that the third approach is used to apportion responsibilities for providing insurance to "uninsurable" or "high risk" consumers in some markets.

Establishing the principle that a competitive industry is obligated to provide for universal service, and then establishing the administrative structures to enforce such an obligation, is not an insurmountable task. As noted above, in the various insurance industries universal service is provided through various sorts of "assigned risk pools." The property insurance industry has developed FAIR Plans (Fair Access to Insurance Requirements) to promote universal service. The obligation of hospitals to serve the indigent is made an explicit quid pro quo for the provision of certain federal construction grants. The obligation of non-profit health care providers to provide indigent health care is made a quid pro quo for the grant of certain public perquisites, including federal, state and local tax benefits. While one may question the success in reaching "universal" service in any of these industries, developing mechanisms in furtherance of meeting the need for widespread service in non-price-regulated can hardly be called new.

Box 11-1. Universal Service in the Telecommunications Act of 1996

The Telecommunications Act of 1996 articulates seven "principles" of universal service:

1. *QUALITY AND RATES*—Quality services should be available at just, reasonable, and affordable rates.

2. *ACCESS TO ADVANCED SERVICES*—Access to advanced telecommunications and information services should be provided in all regions of the nation.

3. *ACCESS IN RURAL AND HIGH COST AREAS*—Consumers in all regions for the Nation, including low-income customers and those in rural, insular, and high cost areas, should have access to telecommunications and information services, including interexchange services, that are reasonably comparable to those services provided in urban areas and that are available at rates that are reasonably comparable to rates charged for similar services in urban areas.

4. *EQUITABLE AND NONDISCRIMINATORY CONTRIBUTIONS*—All providers of telecommunications services should make an equitable and nondiscriminatory contribution to the preservation and advancement of universal service.

5. *SPECIFIC AND PREDICTABLE SUPPORT MECHANISMS*—There should be specific, predictable and sufficient federal and state mechanisms to preserve and advance universal service.

6. *ACCESS TO ADVANCED TELECOMMUNICATIONS SERVICES FOR SCHOOLS, HEALTH CARE, AND LIBRARIES*—Elementary and secondary schools and classrooms, health care providers, and libraries should have access to advanced telecommunications services as described in subsection (h).

7. *ADDITIONAL PRINCIPLES*—Such other principles as the Joint Board and the Commission determined are necessary and appropriate for the protection of the public interest, convenience, and necessity and are consistent with this Act.

Section 254 of the Act spells out a general mechanism for achieving these goals by calling for a proceeding on universal service by a board composed of regulators from the Federal Communications Commission (FCC) and several state utility commissions, as well as a state consumer representative. The board is required to report to the FCC within nine months, and the FCC then has an additional seven months to initiate rulemaking to provide for universal service.

continued on next page

Box 11-1. Universal Service in the Telecommunications Act of 1996 *(continued)*

The legislation gives limited guidance to the FCC for its rulemaking. First, it makes clear that every interstate telecommunications carrier shall contribute to whatever "support mechanism" is chosen, unless its contribution would be de minimis. States may not enact rules that are inconsistent with the federal mechanism, and must require all intrastate carriers to contribute to state mechanisms much as interstate carriers contribute to federal mechanisms. In addition, carriers are barred from charging different rates in urban and rural areas.

The legislation also instructs the FCC to periodically consider the extent to which various telecommunications services are "essential to education, public health, or public safety" or are "consistent with the public interest, convenience, and necessity," as well as the extent to which such services are being provided by the competitive marketplace. The Commission may add or remove services from its list of "essentials" based on these considerations.

To implement these provisions of the Act, the F.C.C. convened a special "joint board" of federal and state telecommunications regulators. The joint board's first report[29] contained a number of implementation recommendations, but also reflected a number of serious ongoing points of disagreement. For example, board members could not agree on whether to fund universal service via a single, nationwide fund or two funds, one devoted to schools, libraries, and health care porviders. Board members also disagreed about the assumptions and methods of computing the cost of poviding universal service in each region.

---

1 Recommendations of the Federal-State Joint Board of Universal Service, Common Carrier Federal Communications Commission Docket No. 96-45, Nov. 7, 1996.

The second threshold question is how the costs of providing universal service and other low-income protections should be paid for. This brings to the surface longstanding tensions between economists' view of efficient redistribution and the pragmatic actions of policymakers. For decades, economists have pointed out that this particular public interest aspect of restructuring was purely a redistribution of private goods from high- to low-income consumers. If society prefers to do this, economists argue, government should use the methods of redistribution most common in our government, namely taxes and government appropriations.[19]

Regulators and many practitioners counter these arguments with the observation that there are practical reasons to favor redistribution via utility programs. First, the costs of these programs are too unpredictable for government appropriation cycles. Second, franchise utilities are the ideal agent for LI-US funding as well as administration, as these utilities have efficient monthly billing and payment operations. Finally, it is sometimes observed that requiring regulated private firms to provide a governmentally-mandated service may be a useful manner of executing government's wishes, however despised the obligation may be by the regulated firm.[20]

Others argue that the obligation to provide for universal service is not one imposed upon the industry, but rather an obligation that the utility industry has voluntarily accepted as part of its franchise agreement. This obligation is one that serves as the industry's "payment" for the grant of substantial public benefits provided to it. So long as the utilities enjoy the fruits of that exchange, they must abide by the obligations that were bargained for as part of that exchange.[21]

Nevertheless, restructuring is prompting a useful review of these issues and a broadening of possible funding mechanisms. It is now common to hear discussions about the possibility of states imposing a charge on all forms of energy to fund LI-US or other public programs (so called "BTU charges").[22] It is even more common to hear concerns that these "wires" or "system benefits" charges be imposed so they cannot be bypassed by large industrial customers, which may physically bypass the electric LDC by connecting (at very high voltages) to transmission lines or by self-generating power, or by any other customer or customer class.[23]

Following an extensive inquiry, a Working Group established by the California Public Utilities Commission recommended that California adopt a single statewide per-kilowatt-hour charge to fund LI-US programs at all electric and gas utilities.[24] Based on 1996 outlays, this charge would amount to about 0.4% of utility revenues.[25] The Working Group noted that distribution utilities that collect this fee would have to balance outlays and collections, as some service areas had far more low-income consumers than others. Many other consumer, environmental, and public interest groups across the nation also support such a "System Benefits Charge" levied by the distribution utility to fund public-interest programs.

Retail choice customers may trade with their power seller entirely by phone or computer, and the retailer may have no nearby physical offices or staff. However, many of the interactions needed to identify customers in distress, provide legally-mandated notices of disconnection, and discuss options for obtaining public assistance often require physical visits to the customer premises. Moreover, research has shown that the more intensive and customized the interaction with the potential problem account, the lower the revenue ultimately lost by the utility.[26] A process that is more distant or mechanized may thus be less successful than some of the programs utilities and state and local agencies manage today.

At the same time, today's franchise utilities fear that they will retain the responsibility of being last-resort seller without adequate compensation. For example, even if a local LDC is repaid the entire cost of the electricity it supplies for LI-US uses, it is extremely difficult to count the less direct costs of administering these programs—e.g., the time required to interact with social service agencies. These responsibilities drain management and staff time and attention, and new funding and administration procedures may take years to develop.[27]

This fear and the onset of restructuring has prompted utilities to increase their efforts to reduce overdue bill payments and sometimes to more aggressively disconnect customers. The Southern California Edison Company made headlines in 1995 when it announced intentions to disconnect over one-half million delinquent customers because of the pressures it felt from impending retail competition.[28]

Overall, the present state of this issue in utility restructuring is unsettled. Whereas legislators appear to solidly favor the maintenance of universal service and low-income programs, there is no widely-accepted way to do this under retail choice. Distribution utilities will resist being burdened with obligations no other competing retailers must bear, and independent retailers will undoubtedly seek to minimize any responsibilities to pay for or implement LI-US programs. Regulators face unpleasant choices, and advocates for the low-income community fear that fights over responsibilities and funding will have the effect of forcing the industry's level of effort downward, perhaps shifting responsibilities to the weakest possible provider. In the end, it is hoped that policymakers can rapidly develop approaches that are effective, economic, and balance humanitarian, equity and economic efficiency in an acceptable manner.

Chapter 11

NOTES

1 Brown (1980) p. 34.

2 Childs (1980) p. 15.

3 Historical statistics of the U.S., Colonial Times to 1970, Part 2, Table S-108-119. Bureau of the Census, 1976. However, having service as defined by the census doesn't include temporary disconnections, which are numerous. See discussion below.

4 Statistical Abstract of the United States, 1974 (Table 826) and 1992 (Table 884); U.S. Census of Housing, 1981, Table A-1; and Mueller and Schment (1995).

5 Unpublished figures from Joel Eisenbergy, Oak Ridge National Laboratory, June 1996.

6 See Colton (1990), Sheehan and Osterberg (1994), and Burns, et al (1995).

7 "Weatherization Program," U.S. Department of Energy, Office of State and Community Programs (EE-44), November, 1996.

8 Low Income Home Energy Assistance Program, Report to Congress for FY 1994, Department of Health and Human Services, Administration for Children and Families, March 1996.

9 See Dahl (1978), Symons (1978), Koger (1979), and Roll and Lande (1980) for criticisms of lifeline rates on a number of grounds. Burns, et al (1995) shows few remaining lifeline rates. Also see Colton (1996c).

10 See Burns, et al (1995). These two categories encompass a variety of specific products and services ranging from budget counseling to energy audits. Sections 115(g) and 304(a) of the Public Utility Regulatory Policies Act of 1978 contain federal electric and gas disconnection standards, respectively.

11 For examples and discussion, see Burns, et al (1995) and Guyant (1995). Low-income energy program expert Joel Eisenberg notes that there is an "extraordinary variation" in the level of effort across the states.

12 Colton (1995d).

13 As these figures are collected by surveys, there are large definitional and collection discrepancies. Unpublished figures from Joel Eisenbergy, Oak Ridge National Laboratory, June 1996 show approximately $140 million in discounts, $44 million in fuel payment assistance, $80 million in arrears forgiveness, and $50 million in deposit waivers.

14 Brockway and Sherman (1996) Tables 2 and 3.

15 Some researchers find that the marketing of advanced telecommunications systems has exhibited such discrimination. One study by the United Church of Christ Telecommunications Project found that the offer of video dial tone service was systematically denied to low-income neighborhoods with Black and Hispanic residents. See Colton (1996) and Colton (1995c)for legislative proposals for preventing such "redlining" by a competitive electric industry.

16 California Assembly Bill A.B. 1890, Enrolled 8/31/96, Section 1 (d).

17 New Hampshire State Code Section 374-F:3 VI, as cited in Brockaway (1996).

18 Alexander and the National Consumer Law Center (1996), p. 36.

19 Testifying against lifeline rates below cost in California, Economics Professor Paul Joskow said,

> …the answer is not to set a special price for [the poor], but through the legislature…to supplement their incomes…and if society's members decline to do the latter, regulatory commissions should not do it for them. For when we deviate from cost-based rates in pursuit of social objectives, we begin to distort the efficiency with which resources are allocated by giving price signals to consumers which do not properly reflect cost.

20 Barbara Alexander and the National Consumer Law Center comment on this issue as follows:

> There is no reason why the needs of low-income customers and others vulnerable to the loss of electricity, even temporarily, cannot theoretically be handled via the tax system or through a fee or tax assessed on all energy suppliers. The real question is whether an alternative system will be in place prior to the onset of retail competition for gas and electricity. There is an extensive series of financial assistance programs available for basic needs (Food Stamps, Medicare and Medicaid, AFDC, Low Income Housing and Homeless Shelters, Elderly Meals on Wheels and LIHEAP for home heating expenses) that have developed in the last 50 years as part of the social safety net at both the federal, state and local level. During this same time period utility regulation has provided a cushion, modest to be sure in some states, for vulnerable customers. The replacement of the informal and formal protections with a fully funded equivalent delivered outside the utility structure will be painful for state and federal legislators facing significant budget deficits and cuts in all social service programs.

> Furthermore, it is not clear that it would be more efficient to fund or deliver these programs outside the utility structure. All households consume electricity and because low-income households on average consume less than their middle class neighbors, the so-called regressive nature of funding these programs via a kilowatt-hour charge is not necessarily true. Furthermore, a per kilowatt-hour charge is spread among all customers, not just residential customers, which means all customer classes contribute in some manner to the social obligation to maintain service for vulnerable customers. The relatively small kilowatt-hour charge associated with low-income programs is not the source of the significant rate increases of the past several years in any state.

21 One national legal organization has concluded: "Local governments are realizing the unique value of public rights-of-way for which they act as trustee. Public rights-of-way are acquired and paid for through government action, usually the exercise of a jurisdiction's eminent domain powers. Thus, the rights-of-way are the most valuable property rights in the hands of the government...Local governments must receive fair compensation for granting use of the rights-of-way. Otherwise, government is merely subsidizing the businesses of private rights-of-way users...Traditional users of the public rights-of-way were deemed to provide public compensation in the form of universal service and regulated rates....With traditional users of public rights-of-way, compensation for the use of the public rights-of-way was passed on to the end consumer through rate regulation and other public benefits like universal service, rather than being paid directly by the governments, the actual owner of the public rights-of-way." Nicholas Miller and Kristen Neven, "What is the Emerging Role of Local Governments in This New World of Telecommunications," in Cable Television Law 1996: Competition in Video and Telephony at 12–13 (1996: Practising Law Institute).

22 Vermont now imposes such a charge; Colton (1996a) discussed the possibility at length. Colton (1996b) discusses a national universal service fund.

23 Issues in the design of wires charges are discussed in Colton (1996a), the California Low Income Working Group (op cit), the Regulatory Assistance Project (1995). Cavanagh (1996) discussed "system benefits charges", which are distinct from "wires charges" in that they are not linked to the use of distribution wires per se. Instead, these changes are linked to distribution services, which all customers use irrespective of their use of specific physical facilities.

24 Report of the Low Income Working Group to the California Public Utilities Commission, Draft, Sept. 5, 1996.

25 Fessler, (1996).

26 Burns, et al (1995).

27 Recognizing these considerations, researchers at the National Regulatory Research Institute recently proposed a novel approach to utility responsibilities for LI-US programs (Burns, et al, 1995). The NRRI team proposed that the principles of performance-based regulation be applied to the utility's administration of these programs, in effect rewarding utilities for good LI-US performance using a system of incentive payments matched to certain measurable criteria. Also see Colton (1995b).

28 See Colton (1995). Colton, Guyant (1995), and Burns, et al (1995) cite research showing that more aggressive disconnection policies are not the most effective means of recouping overdue utility revenues.

Chapter 12

Environmental Quality

Coal-fired boilers were used to provide most of the heat in large urban business districts in the late 1800s; most artificial light was produced in gas-burning fixtures. As buildings became taller and districts more congested, air pollution problems worsened and the need for additional artificial light became more acute. Historian Harold Platt notes that the first skyscrapers of the 1880s "turned downtown streets into gloomy canyons filled with smoke from morning to night."[1] A spiral of declining air quality and visibility eventually led cities like Chicago to adopt municipal smoke abatement ordinances, but with little effect.

Perhaps it is appropriate that the electric power industry, which has occupied a prominent position in countless environmental issues, was born as the solution to an environmental problem. Thomas Edison correctly surmised that a smokeless, flameless form of light would prove to be an environmental as well as an economic improvement. Edison's observation remains true today: considering the complete life cycle of production and use, electricity often remains the lowest-impact means to a specific end.[2] For example, utility consultant Mark Mills reports that electric lawn mowers produce one fifth the overall pollution of gasoline-powered mowers over the entire production and use cycle (and solar-electric mowers produce even less).[3]

With nearly equal foresight, the Federal Power Commission surveyed the connection between the environment and electric power in 1964—a time when large generating units were reaching their zenith and commercial nuclear power was about to emerge. The Commission noted:[4]

> Environmental considerations will become increasingly significant factors in the location, design, and operation of future thermal-electric power plants... The problem of air pollution and thermal pollution of water will be of particular concern at most future installations. The industry's vigilance in planning these large installations and incorporating the proper measures to minimize pollution will be the subject of increasing public concern. The electric power industry has long recognized these problems and has spent large sums for research and control of air pollution.

These twin themes—environmental improvement through electricification and environmental degradation from the generation of power—have been essential elements of the power industry since its inception. It is well beyond the scope of this volume to address the full breadth and depth of this nexus; many important works attack this task.[5]

This chapter examines the ramifications of restructuring on the environmental costs of electricity supply. This brief treatment barely does justice to the complex and sometimes profound issues involved.[6]

THE ENVIRONMENTAL IMPACTS OF POWER SUPPLY

Table 12.1 summarizes some of the impacts that electricity supply has on the environment. As shown in Figure 12-1, electric power plants emit about two-thirds of the nation's sulfur dioxide (SO_2), the pollutant associated with acid rain and other environmental problems.[7] Power plants also emit one third of all nitrogen oxides (NO_X), a key precursor to urban smog,[8] and about one third of the nation's annual output of carbon dioxide (CO_2), the gas most associated with global climate change. Utility air emissions are also important sources of ozone, other hydrocarbon pollutants, and airborne heavy metals such as cadmium and mercury.[9]

The second category of utility emissions in Table 12.1 involves water and water pollution. Each kilowatt-hour generated in a power plant consumes 0.3 to 1.3 gallons of water, while the amount of water passing through plants for cooling can be as large as 75 gallons/kWh—more than 100 billion gallons per hour for a typical plant.[10] Use of this water results in damage to the ecosystem from changing water temperature around the plant (thermal effects), the release of chemicals into the water, and damage to fish and other species from waterflow in both hydro and conventional plants.[11]

A third category of environmental impacts is unique to nuclear energy. The entire nuclear fuel cycle, from production of nuclear fuel to its ultimate processing or disposal, creates large amounts of radioactive and hazard waste in various forms. Nuclear power plants have elaborate procedures and controls, but nonetheless they do release small amounts of radiation and other forms of pollution. In addition, nuclear plants themselves must be dismantled and stored as radioactive waste, or decommissioned, at the end of their useful lives. Finally, rare but potentially devastating incidents such as Three Mile Island and Chernobyl pose environmental issues of international proportions.[12]

Electric power transmission and distribution pose unique environmental challenges of their own. Power lines often stretch across remote wilderness areas or unique and delicate ecosystems, which may be disrupted by construction and periodic power line maintenance.[13] Power lines also stretch into every home, building, and community, and, in recent years, concerns have been raised about the electric fields produced by power lines.[14]

Because all of these impacts occur through the use of particular types of power plants or lines, the impact that restructuring has on these environmental issues should be largely a function of changes in the types of power plants used, their efficiencies and pollution control capabilities, and the intensity of their use. This is precisely the approach that most analysts have used to gauge these "supply-side" impacts. Table 12.1 suggests that environmental impacts that are most sensitive to fuel and technology choices are those associated with air pollution and nuclear energy. (See Chapter 16.) Transmission systems also will undoubtedly face pressures to expand in some areas, heightening issues that have historically been associated with new lines.[15] This chapter focuses largely on air quality impacts.

Table 12.1. Environmental Aspects of Electric Power Supply(a)

	Aspect	Impacts Relative to Other Industries	Change in Environmental Impacts Due to Restructuring

Generation	• Air Pollution from Power Plants
	  – Sulfur dioxide
	  – Nitrogen oxides and ozone
	  – Air toxins and other pollutants
	  – Carbon dioxide
		
		Large
		Large but not dominant
		
		Generally small (b)
		
		Moderate to large
			
			Potentially large
			Potentially large
			
			Possibly large
			
			Possibly large

	• Water Pollution from Power Plants
	  – Thermal discharges
	  – Waterborne wastes
	  – Water ecosystem disruption
		
		Moderate(c)
		Small (c)
		Potentially large, especially for hydroelectric plants (c)
			
			Uncertain, probably little change

	• Nuclear Impacts
	  – Air and water plant discharges
	  – Nuclear waste transport processing storage and fuel processing Nuclear accidents
		
		Large (f)
			
			Uncertain

Transmission and Distribution	• Disruption of ecosystems during line construction and maintenance (d) • Aesthetics and land-use planning (d) • Electromagnetic fields (e)	N/A	Uncertain, probably little change

Notes to Table 12.1

(a) This table focuses on the supply of electricity. The *use* of electricity often displaces a supply technology with larger or different environmental impacts. See text.

(b) See Ottinger, et al (1991), and the Air and Waste Management Association (1992).

(c) Ottinger, et al (1991).

(d) Plant Sitting Task Force (1970).

(e) Ottinger, et al (1991) Section V. F.9. and Plant Siting Task Force (1970).

(f) Ibid, Section VI. P., Union of Concerned Scientists (1975), and the U.S. Department of Energy-Office of Environmental Management (1995).

Figure 12-1. Utility and Other Emissions of Air Pollution 1995

		Misc.	Other Fuel Consumption	Industrial Processes & Related	Industrial	Transportation	Electric Utilities	Total
[1]	NO_X	374	727	888	3,206	10,625	7,795	23,615
[2]	SO_2	14	599	2,029	3,029	578	14,869	21,118
[3]	PM-10	42,743	529	932	237	722	266	45,429
[4]	CO_2	437,898			420,299	408,596	448,875	1,715,668
[5]	CO_2	483			463	450	495	1,891

Notes: Misc. for [1]–[3] represents wildfires and prescribed burning
[4], [5] Misc. includes total residential and commercial emissions
[5] is in units of millions of metric tons

Sources: [1]–[3] National Air Quality and Emissions Trends Report, 1994 (Chapter 3, pages 3-12, 3-13, 3-15)
[4], [5] EIA Emissions of Greenhouse Gases in the U.S., 1995, Table 6
National Air Quality and Emissions Trends Report, 1995 (Tables A-4- A-8). EIA Emissions of Greenhouse Gases in the U.S., 1995, Table 6.

Electric utilities are the primary emitters of sulfur dioxide (SO_2) and substantial emitters of nitrogen oxides (NO_X), and carbon dioxide (CO_2). They are very minor emitters of certain other regulated pollutants, including volatile organic compounds (VOCs) and 10 micron-or-smaller particulates (PM-10).

Box 12-1. The Clean Air Act and its 1990 Amendments[1]

The Clean Air Act [42 U.S.C. § 7401] governs the emission of several types of pollutants from a range of prescribed sources.

Criteria pollutants are widespread pollutants whose maximum levels are set by the EPA according to health-based standards. The criteria pollutants set forth in the Act are sulfur dioxide, carbon monoxide, nitrogen oxides, lead, particulate matter 10 microns in diameter and smaller (PM-10), ozone, and volatile organic compounds (VOCs).

Hazardous air pollutants, also known as *air toxins,* are pollutants that are closely associated with toxicity for humans, and are regulated in a separate title.

For each criteria pollutant, EPA sets national air quality standards that every region in every state must meet in order to be in attainment for that pollutant. If the standards are not attainded, the region is classified as non-attainment, with gradations that depend on the frequency and severity of violations. As of 1994, approximately 174 air regions, including almost every major U.S. metropolitan area, were in non-attainment for at least one pollutant. States with non-attainment areas are required to submit *State Implementation Plans (SIPs)* that detail the state's regulatory and non- regulatory programs that are intended to bring the state's air quality areas into attainment status. If any state fails to file an adequate SIP, the EPA may develop and impose a federal plan. The 1990 amendments to the Act require that all major emitters of air pollutants obtain "Title V" permits from state air pollution agencies. These permits are to reflect all Clean Air Act requirements for a single plant. New power plants must also undergo a "new source review" under Section 165 of the Act. Title IV of the Act regulates emissions of sulfur dioxide from large electric utility power plants that together account formore than 80% of the nation's SO_2 emissions. The title sets a national total maximum for emissions from this set of plants and requires all plants to develop and file plans for reducing SO_2 emissions in two phases beginning in 1995 and 2000, respectively. The law allows utilities and others to sell or barter federal "allowances" for the emission of SO_2. This innovative "cap and trade" scheme is probably the largest and most successful example of market-based environmental protection in the world to date.

EPA is required to promulgate a number of rules under the CAA amendments of 1990. In early 1996, the EPA proposed rules for NO_X reductions at utility boilers. In November, 1996, the EPA proposed new air qulaity standards for ozone and particulates. If adopted in their proposed form, these rules may have a significant impact on compliance costs for power plants and urban areas in noncompliance.

---

1   Based on Bureau of National Affairs (1991), Lock and Harkawik (1991), and the U.S. Environmental Protection Agency (1990, 1992, 1993, 1995)

AIR POLLUTION REGULATION AND ELECTRIC UTILITIES

In most instances, the regulation of utility air emissions is dominated by the influence of the federal Clean Air Act, which was amended in 1970 and 1990.[16] This enormous law (Box 12-1) has a number of provisions that are particularly relevant to electric utilities and non-utility generators.[17]

First, the law establishes a complex process for determining whether a particular region meets federal air quality standards, or is in "attainment," for a single pollutant.[18] Most urban areas, including almost all of the northeastern United States and the developed West Coast, are in non-attainment for at least one pollutant, the most common being ozone.[19] Because ozone and NO_X are emitted by many sources, and are transported over wide areas, they are proving to be two of the more difficult pollutants to control. Hence, much of the focus of air emission changes prompted by restructuring centers on NO_X and its attendant impacts on ozone air quality standards.[20]

The 1990 CAA Amendments established a new requirement that each major air emitter must obtain a permit for all pollutants governed by the Act. For the purposes of requiring a permit, EPA's determination of what constitutes a major source varies by the attainment status of each region. Furthermore, if a permit is required, the minimum required pollution controls are progressively more stringent and costly for more severely polluted areas.

Finally, requirements applicable to new sources since 1978 do not apply to pre-existing, older plants that the legislation assumed would retire. Many of these plants continue to operate with fewer pollution controls than modern-day units. According to some estimates, older units have emissions rates *that are* as much as ten times as high as today's units, which tend to be more energy-efficient and often *employ* the cleanest generation and control technologies.

These provisions establish important differences between old and new plants. New plants may be required to purchase "offsets"—i.e., to find existing emitters of one or more pollutants and pay the emitter enough to convince the emitter to stop discharging the needed amount of pollution.[21] Several states are working to implement provisions of this sort through state or regional emissions trading systems.

Urban smog is often the result of NO_X pollution (transported long distances) mixed with "local" hydrocarbons.[22] Recognizing this problem, the law specifically established the Northeast Ozone Transport Region, which stretches from Maine to Washington, D.C. along the east coast,[23] and granted authority for other groups of states to form transport commissions. However, in response to eastern states' claim that transported NO_X from the midwest prevented them from attaining the air quality standards for ozone, EPA created a a much larger group, the Ozone Transport Assessment Group (OTAG), which is comprised of most of the states in the eastern United States,[24] to discuss potential regional pollution control programs.

With or without restructuring, these groups and their non-attaining member states face daunting challenges as they struggle to achieve attainment for all pollutants, as the Clean Air Act requires. States whose pollution is "imported" from others resent the fact that economic activity benefiting others requires them to place controls on local industries, possibly hurting in-state economic growth. Conversely, attainment states see little benefit in reducing pollution in their states, only to see the reductions provide primary benefit elsewhere.[25] As it happens, the locational pattern of existing power generators in the eastern and western United States exacerbates these concerns. Roughly speaking, the midwestern, west-central, and southern sections of the country have a higher population of older, coal-burning power plants than the areas with lower attainment status on the east and west coasts. Prevailing winds tend to shift ozone precursors emitted from these plants from the center of the country to the east.[26] OTAG was constituted as a forum for bringing the precursor source and receiver states together to develop equitable, achievable approaches to resolving interregional tensions. This group has become one of the main centers of restructuring/air quality policy development.

STATE UTILITY PLANNING AND ENVIRONMENTAL REGULATION

Long before restructuring became an issue, state air quality and utility regulators realized that their responsibilities toward electric utilities were closely connected. The connection became even more apparent when many states moved to integrated resource or least-cost planning. Under these approaches, utility regulators were required to take into account all of the costs of electric power supply, including any measurable costs of environmental harm—even if these costs were not a part of electric power rates.[27]

Apart from the enormous difficulty of quantifying environmental costs, utility regulators faced procedural challenges as they tried to factor the environment into utility plans. If environmental controls or offsets cost utilities money, regulators could audit the utility's compliance planning and penalize or reward the utility accordingly. However, if a (measurable) environmental harm was caused by a utility emission, how should that harm be reduced or compensated in a utility planning process?

For a time, regulators experimented with several approaches to this problem.[28] In some states, regulators tried to determine the costs of environmental damage and added this cost into utilities' estimates of least-cost supply. For example, suppose a utility found that new gas and coal plants of equal size and quality cost 4 and 3 cents per kilowatt-hour, respectively. Suppose further that regulators could measure the difference in environmental costs, such as the value of the "residual" damage to streams from acid rain, between the two plants. If these costs were greater than 1 cent per kWh, the gas plant would be considered cheaper "for society," and the utility would have been ordered to build the plant.

This approach was roundly criticized by economists and others for a number of reasons. First, scientific and economic analyses of the monetary value of environmental improvements are uncertain and controversial even between experts; reaching a reasoned judgement requires considerable expertise and effort.[29] Second, economists found many misapplications of economic theory, leading them to conclude that the incorporation of measured environmental costs might not increase human welfare.[30] For example, they noted that if only new electric power plants were subjected to externality "adders," other dirtier sources of energy would become relatively less expensive and increasingly put to use. However, state planning proceedings rarely examined any plant other than the next planned unit, leaving the rest of the system uncontrolled. Somewhat presciently, Professor Joskow wrote:[31]

> Adders also change the complex relationships that exist between the operation of existing resources and the selection of new resources: such changes could actually increase the utilization of existing more-polluting resources beyond what it would have been absent the adders. This is especially likely in inter-connected electric power systems that span several states with extensive trade in energy and capacity.

One of Joskow's concerns was addressed in a little-discussed proposal to use these adders when system controllers were performing economic dispatch. The hourly operation of power plants is based on the system controller's determination of which set of plants have the lowest operating cost per unit of power. When making this determination, the operator could add to the variable cost of each plant a per-kWh environmental fee. Thus, for example, a gas plant whose normal variable cost was 2 cents per kWh would be dispatched before a coal plant with 1 cent variable costs if the environmental dispatch adder was 1 cent or more.[32]

Use of dollar-based economic adders in either the planning or the dispatch process has gradually diminished as integrated resource planning has given way to restructuring. However, these ideas lay the groundwork for several recent proposals that are specifically directed at air emissions in a restructured industry.[33]

RESTRUCTURING'S IMPACT ON AIR POLLUTION

Prior to the issuance of FERC Order 888, U.S. electric competition was largely limited to PURPA Qualifying Facilities and other independent generators. Because most of these facilities were new, they faced CAA environmental control requirements that were similar to those for new utility plants. In this limited sense, EPACT-driven wholesale competition did not appear to introduce significant change in environmental impacts. The high efficiency of new plants and their predominant use of clean-burning natural gas were often viewed as a relative plus for the environment.[34]

When FERC proposed Order 888, observers noted that the increased transmission access that the rule was created to engender could prompt substantially more bulk power trading over wide areas. Because the midwestern region of the United States happens to have many underutilized coal-burning power plants with relatively low operating costs, it seemed likely that these plants would be able to sell to a broader market and thus produce more power and pollution. Figure 12-2 depicts the pattern of NO_X emissions and winds projected to move pollutants from midwestern plant sites into eastern Air Quality regions.

In accordance with the National Environmental Policy Act, the FERC issued a draft and final Environmental Impact Statement (EIS) for its proposed rule. The FERC ultimately came to the conclusion that the rule would not have a material impact on air pollution emissions, and therefore decided not to modify its rule or to propose environmental mitigation measures.[35] The FERC found that the increases specifically attributable to its new rule would equal only about 1% of expected NO$_X$ emissions—a difference that was less than what would be caused by other sources of variation such as fuel prices. In addition, the FERC concluded that it had neither the authority nor the expertise to become an environmental regulator, and that the Clean Air Act and other laws provided EPA with the responsibility and the means of addressing the nation's air quality needs.[36]

The FERC's draft EIS proved to be the spark that set off a storm of examination of, and opposition to, the environmental impacts of restructuring.[37] Studies conducted by critics of FERC's analysis found that the Commission had underestimated the impact of its rule on the expansion of transmission capacity, the ability of coal plants to expand output, increased demand, reduced energy efficiency and renewable energy programs, and other factors. In addition, they criticized the FERC for looking too narrowly at the

Figure 12-2. Major NO$_X$ Sources and Pollution Transport Vectors on High Ozone Days in the Northeast

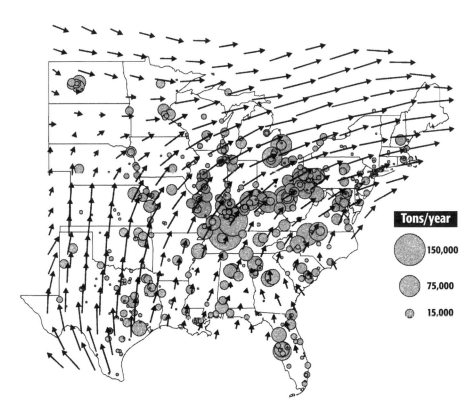

Source: Electricity Restructuring and Environmental Policy: A Review of CEQ, FERC and EPA Actions to Consider the Environmental Impacts of FERC's Open Transmission Access Rule. July 1996.

impacts of its rule alone, rather than at the broader impacts of restructuring.[38] Critics also challenged the FERC's narrow view of its authority under the Federal Power Act. Defenders of the FERC rule alleged that FERC's analysis was adequate and that, in any case, environmental protection was a job for EPA, not the FERC.[39] FERC critics responded by noting that, even if FERC's analysis was accepted at face value, it indicated that the rule would exacerbate non-attainment in many parts of the United States and harm U.S. environmental treaty commitments.[40]

Table 12.2 displays the range of estimates of environmental impacts put forth by the FERC and others during the course of the EIS discussion. With respect to NO_X increases, the table shows that the FERC's estimates are well below those of most other researchers, although it must be noted that several of the studies on this incorporate the effects of retail as well as wholesale competition and the Commission's does not. In any case, these studies show that the range of *restructuring-related* increases in NO_X is of the same size as the *total* increase in this pollutant from utility sources through 2010 without restructuring. In other words, restructuring's rough impact is on the order of doubling the increase in NO_X through 2010.

Governors of non-attainment states, struggling to improve air quality without hampering business growth, found this grim news. Shortly after FERC issued its draft EIS, governors of nine northeastern states adopted a joint resolution calling on the President to ensure that restructuring would not harm air quality in their states. Massachusetts Environmental Commissioner David Struhs further stated:

> We continue to pay the public health bill and suffer the environmental damage caused by the pollution blowing into our region from dirtier plants in the Midwest and Southeast. We want to ensure a level playing field, both for our businesses and our citizens.

These comments reference the fact that controlling smog precursors at the plant source is believed to be much cheaper than controlling precursors in non-attainment areas, which already have stringent controls. One study filed in response to FERC's EIS estimated that NO_X mitigation at plant sites would cost $1.3 billion less than equal mitigation in downwind non-attainment regions.[41]

The FERC EIS debate helped highlight the difficulties of measuring the additional environmental impacts of retail choice beyond those of wholesale competition. Just as it is devilishly hard to quantify the net economic gains from retail choice, it is as difficult to separate retail from wholesale environmental effects. Work in this area suggests that the added environmental costs of retail competition will be subtle and mixed. On the other hand, price competition may be more severe, driving prices below social costs in some cases, thereby encouraging inefficient consumption. All power producers will have a single dominant incentive: to minimize all environmental compliance costs that add no marketing benefits. This case is buttressed by evidence showing that the onset of competition has substantially reduced U.S. utility energy efficiency, as well as the number of renewable energy programs. Other nations that have implemented retail choice have seen these programs all but disappear.[42]

Others counter that retail choice will ultimately improve the environment. They argue that competition will encourage more rapid technical change, and that consumers' inherent desire for cleaner energy will drive the marketplace toward this outcome.[43]

GLOBAL CLIMATE CHANGE AND UTILITIES

The world's scientific community is increasingly convinced that anthropogenic generation of certain "greenhouse gases" (GHGs) is raising the earth's average temperature and destabilizing global weather patterns.[44] Many scientists continue to disagree about the scope and timing of the likely changes, but agree that such changes have potentially massive adverse global impacts, including severe storms, rapid disease diffusion, increases in sea level, droughts, and major ecosystem disruptions.[45]

The largest single GHG is carbon dioxide, which is produced by the combustion of any fossil fuel. As Figure 12-1 shows, electric power producers contribute about one-third of U.S. emissions of CO_2 when they burn coal, oil, gas, municipal solid waste, or biomass. (The combustion of coal produces about twice as much CO_2 as the combustion of natural gas per unit of electricity produced.)[46] Importantly, pollution control technologies that can reduce sulfur and NOx emissions from coal boiler exhausts cannot remove CO_2; no near-term technology appears capable of reclaiming or reducing this exhaust gas economically.[47]

Carbon dioxide is not a pollutant that is regulated in the Clean Air Act. Currently, there are no other prohibitions on CO_2 emissions in other state or federal laws. However, the nations of the world have ratified a treaty that governs national emissions of greenhouse gases—the Framework Convention on Climate Change, or so-called Rio Treaty—which calls upon nations to limit or reduce GHG emissions. Procedures for implementing the treaty are under development in a long series of international meetings and negotiations.[48]

As the international agreement has been evolving, the United States and many other nations have attempted to meet the terms of the treaty by developing "action plans" that are designed to stabilize GHG emissions at 1990 levels by the year 2000. The first United States effort, the 1993 Climate Change Action Plan, contained 46 separate federal and state government programs that were designed to meet the stabilization target.[49] Significantly, these programs were almost all voluntary—no new rules or regulations (emissions or otherwise) were proposed. Of equal significance, the vast majority of the programs were tied, in one form or another, to regulated utilities. For example, several programs sought to increase utility energy efficiency programs, while others relied on utility purchases of non-polluting generating technologies.

It is impossible to predict whether advancing scientific research and world concern about the effects of global climate change will combine to produce national regulations governing CO_2 release, but many analysts foresee eventual national actions that could greatly affect power generators. Observers of air pollution regulation in the United States note that electric utilities are often the first (and sometimes the only) source of a pollutant that is to be controlled.[50] The possibility that electric utilities may sooner or later face limitations on their CO_2 emissions has prompted at least one environmental group and the Interfaith Center for Corporate Responsibility to now consider these emissions to be a potential environmental liability.[51]

Those who are attempting to set international climate change policies view with dismay the impacts that restructuring is likely to have on U.S. GHG emissions. As the fuel choice decisions of generators become motivated entirely by the lowest immediate cost of production, U.S. coal-fired generation may increase for a number of years following restructuring. Projections of the impact that restructuring will have on CO_2 emissions (Table 12.2 suggests that these increases are on the order of 100 million metric tons of CO_2, roughly the amount of CO_2 the U.S. is committed to *decrease* under the Rio Treaty.)[52] Increases of this size, caused by restructuring, will require U.S. policymakers to find substantial additional means of reducing U.S. GHGs if theUnited States is to meet its treaty commitment.

FUTURE PROSPECTS AND UNCERTAINTIES

As the debate over the environmental impacts of restructuring rages, no state or federal legislation has, as yet, been introduced to address possible adverse effects.[53] Foreshadowed by Professor Joskow's criticism of adders and the idea of environmental dispatch, some have suggested that environmental policies and emissions rules should be equalized for all units in a trading region. Others suggest adding provisions to the Clean Air Act to set maximum industry-wide limits on certain pollutants and to allow the trading of emissions permits within these "caps," (as is currently the case for sulfur dioxide).

A third group sees restructuring as an opportunity to integrate public utility and environmental laws at the state level, expand the use of market-based mechanisms, and remove some of the features of environmental rules that are most harmful to industry growth. Calling electric restructuring "the last, best chance to attain the goals of the Clean Air Act," a utility CEO called for a halt to retail competition until environmental issues are resolved.[54]

If this is a chance, it will not be an easy one. In early 1996, a number of state utility regulators met with Vice President Al Gore and Secretary of Energy Hazel O'Leary to discuss restructuring and the environment. The state regulators, many of whom were examining wholesale or retail deregulation in their states, heard exhortations by the Vice President and others to make sure that restructuring does not reduce the nation's substantial environmental progress.

Toward the end of the meeting, one commissioner from a state with many coal-burning power plants addressed Secretary O'Leary. "Madame Secretary," the commissioner said, "with all due respect, we are on a collision course. On the one hand, you are telling us to maintain environmental protection. On the other hand, we are facing enormous pressure to lower utility rates by reducing regulation."

It is possible—perhaps even likely—that restructuring will lead the nation down a path that ends in clean energy and a cleaner environment. It is also likely that the nation will first enter a substantial transition period in which some emissions will increase, perhaps greatly. The ultimate profile of restructuring's environmental impact will be a function of many factors that are as yet unknown. Under any scenario, the result will depend on the pace of technological change, coal and gas prices, transmission expansion and use, and the willingness of state and federal policymakers to address simultaneously two of the most difficult industrial policy issues of our time.

Table 12.2. Comparison of Estimates of Air Quality Effects of Restructuring

	Estimated Annual Charges in Air Emissions as Estimated By...					Total Emissions as Noted, for Comparison Purposes	
	Federal Energy Regulatory Commis-sion	Center for Clean Air Policy	National Association of Regula-tory Utility Commis-sioners	Lee and Darani, Kennedy School of Govern-ment	Resources for the Future	Total Electric Utility Emissions, 1995	Projected Increase by 2010 Without Restruct-uring
Area Covered by Study	U.S.	One large Midwest Utility	U.S.	U.S.	U.S.		
NO$_X$ (thousand tons)	71–127.6	36 –127	896	492	349.9	6,233	572
CO$_2$ (million tons)	2.8–27.9	9–30	327	42.9	113.5	493.8	90

Note: The FERC EIS considers only the effects of Open Transmission Access resulting from Order 888. The remaining studies incorporate other changes prompted by restructuring to varying degrees.

Source: Adapted from Burtraw and Palmer (1996) Table 13-1, Environmental Protection Agency (1996b), Table A-4, EIA (1996) Table 12, EIA (1995) Table A18.

Chapter 12

NOTES

1 Platt (1991) p. 28.

2 Mills, McCarthy and Associates (1994); Analysis Group for Regional Electricity Alternatives (1996).

3 See Edison Electric Institute (19745a, 1994b) and Mills-McCarthy (1994) p. 31.

4 Federal Power Commission (1964) p. 137.

5 For example, see Hollander (1992).

6 Important issues not discussed include the scientific and economic bases for establishing environmental standards, the evaluation of environmental costs and benefits, tradeoffs between different types of impacts and risks (environmental risk assessment), and alternative environmental control strategies.

7 See Harte (1992), Offinger, et al (1991), OECD (1985, 1989).

8 See Holander and Brown (1992), OECD (1985), Ottinger, et al (1991)

9 Office of Air and Radiation (1996).

10 Ottinger, et al (1991), Section V.F.

11 Ibid.

12 Ottinger, et al (1991) Section VI.D.; Hohenennser, Goble, and Slovic (1992).

13 Plant Siting Task Force (1970) and Ottinger, et al (1991) Section V.F.9.

14 Committee of Life Sciences, National Research Council (1996).

15 Ottinger (1991), Edison Electric Institute (1970). Also see Chapter Four *supra*.

16 42 U.S.C § 7401-7626.

17 With several exceptions—notably the Acid Rain provisions in Title IV, discussed below—the CAA classifies power generators by size and other factors, but does not distinguish between utility and non-utility plants. Non-utility plants are sometimes smaller than utility plants, and are preponderantly fueled by natural gas, so they are somewhat less affected by CAA rules. Nonetheless, as utilities and non-utilities increasingly build almost identical new plants, geographic differences in air pollution policies become more important than ownership differences. Compare, for example, Shanker (1990) and Lock and Harkawik (1991).

18 See "Designation of Areas For Air Quality Planning Purposes," Office of Air and Radiation, U.S. E.P.A., 40 CFR Part 81.

19 U.S. EPA (1995b).

20 This was clearly acknowledged by the FERC in its Order 888 (Docket RM95-8-000, April 24, 1996, Mimeo p.660).

21 Notice of Proposed Rulemaking, 61 FR 1442, 1/19/96 *supra*.

22 See 42 U.S.C. § 7506a, Palmer and Burtraw (1996) p. 17ff, and NAPAP (1996).

23 42 U.S.C. § 7511c.

24 See the comments of the Ozone Transport Assessment Group on the Draft E.I.S., FERC Docket No. RM95-8-000, Jan. 10, 1996.

25 This perception may be erroneous in many instances. Midwestern states suffer from the highest acid deposition rates in the country, and include major urban non-attainment areas where emission reductions should produce significant health benefits.

26 See Palmer and Burtraw (1996), NAPAP (1996), and the Joint Comment on Draft Environmental Impact Statement, FERC Docket No. RM95-8-000, Feb. 1, 1996.

27 Eco Northwest (1993).

28 See Eco Northwest (1993), Proceedings of the NARUC Environmental Externalities Conference, Jackson Hole, W4, Oct. 1–3, 1990, Consumer Energy Council of America (1993), and Mauldin (1989).

29 See the Office of Technology Assessment (1994).

30 See Hobbs (1992), Joskow (1992), Burtraw, et al (1993), and the Edison Electric Institute (1994).

31 Joskow (1993).

32 See Gjengendal, Marney, and Johansen (1993), Bernow, Biewald, and Marron (1991), and Eco Northwest (1993).

33 Some environmental experts stress the value of adders as factoring in future regulatory costs. See Cavanagh (1993) and note 52 below.

34 See, for example, Shanker (1992) p. 61.

35 Promoting wholesale competition through open access non-discriminatory transmission by Public Utilities, Federal Regulatory Energy Commission Docket No. RM 95-8-000, Order 888, April 24, 1996 Mimeo p. 666–668 [Hereinafter, "Order 888"].

36 Ibid, pp. 670–676.

37 See Humphrey (1996) and Gary Lee, "Utility Deregulation Could Worsen Air Quality in East, Coalition says", *Washington Post* 2/21/96 p. A4.

38 See the Comments of the Center for Clean Air Policy (2/2/96), the Natural Resources Defense Council (2/2/96), the Northeast States for Coordinated Air Use Management (2/1/96), Joint Comments of the Alliance for Affordable Energy, et al (2/1/96), and the comments of the Project for a Sustainable FERC Energy Policy (2/2/96), all in F.E.R.C. Docket No. RM 95-8-000. Also see Palmer and Burtraw (1996) and Lee and Davani (1996).

39 Comments of the National Rural Electrical Cooperative Association and the American Public Power Association, (2/2/96) and the Midwest Ozone Group (2/2/96). F.E.R.C. Docket No. RM95-8-000. Also see Palmer and Burtraw (1996).

40 Comments of the Natural Resources Defense Council, FERC Docket No. RM95-8-000, Feb 1, 1996. Also see Palmer and Burtraw (1996).

41 Joint Comment on Draft Environmental Impact Statement, Docket No. RM95-8-000, Feb 1, 1996, Figure 9.

42 See the Meeting Summary Environmental Impacts of Increased Competition in the U.S. Electricity Industry, Harvard Electricity Policy Group, April 28, 1994; Cavanagh (1995); Brief of the Conservation Law Foundation in Connecticut Department of Public Utility Control Docket No. 93-09-29, May 3, 1994; Fang and Galen (1996), and the Public Service Commission of Wisconsin Environmental Impact Statement, Docket No. 05-EI-114, October 1995.

43 See Amy Jenness, "Electric Restructuring Might Just Clean Up the Air," *Vermont Business*, June 1996.

44 Houghton, et al (1995).

45 Bruce, Lee, and Haites (1995), Watson, Zinyowera, and Moss (1995); United Nations Environmental Program (1993).

46 Renewable power sources other than biomass do not produce CO_2, and biomass absorbs CO_2 as it is grown. Hence, if new biomass crops are grown at the same rate as they are harvested and burned, there is no net increase of CO_2 in the atmosphere (so-called "closed-loop biomass"). EIA (1996).

47 Other so-called "greenhouse gases" include methane (natural gas), which is increasingly the fuel of choice in electric power generation. Hence, power generators who burn natural gas cause GHG emissions from their fuel supply as well as when they combust the fuel in their plants.

48 International Energy Agencies (1994).

49 President William Clinton, United States Climate Change Action Plan, October, 1993.

50 See the comments of David Owen, Senior Vice President, Edison Electric Institute, in Reuters Financial News Service, Dec. 9, 1996. On this point, the environmental community and utilities may agree. Cavanagh (1995) writes, "From an environmental perspective, the utility sector is the economy's most important leverage point."

51 "Risky Business—The Hidden Environmental Liabilities of Power Plant Ownership." Natural Resources Defense Council, San Francisco, CA, September, 1996. For the opposing view, see "American Electric Power Points Out Shortcomings of New NRDC Report." American Electric Power Company, Columbus, OH, September 17, 1996.

52 The size of the increase depends on the price of gas, coal, and other sources of power in the future, progress on nuclear and renewable energy technological change, and, perhaps on the development of carbon absorption or injection technologies.

53 Representative Patrick Kennedy has introduced legislation to study the air impacts of restructuring. The Center for Clean Air Policy (1995) and others have produced somewhat detailed analysis of specific policy options, and OTAG for the EPA is expected to develop a specific proposal as well.

54 Remarks of Lawrence R. Codey, President, Public Service Electric and Gas, before the Fourth U.S. Department of Energy/NARUC Industry Forum, October 22, 1996. Also see Brennan, et al (1996).

Chapter 13

Energy Efficiency ("DSM") and Restructuring

The arguments for making energy efficiency a matter of public concern begin with the critical proposition that energy is not entirely a private good. There is a public interest, these arguments assert, in using energy efficiently.

Two essential contentions underlie these arguments.[1] The first is that the price of energy paid by most consumers does not reflect the total marginal cost of energy to society as a whole. As previously discussed, because environmental externalities are not necessarily factored into the cost of electric power this is a potential source of discrepancy.[2] Additional sources of discrepancy include the cost of energy-induced economic disruptions or fuel shortages, both of which are diminished if everyone uses less energy. (Discussions of these items appear in Chapter 14). Finally, as discussed in in Chapter 10, electricity frequently is not priced at its own marginal cost because of the presence of fixed costs, regulation of some parts of the electricity industry, and other factors. All of these effects can cause electric power to be underpriced and overused.

For economists, the best solution is to correct deficiencies in pricing, rather than intervening in markets to try and fix a problem with the imprecise instruments of regulation and government programs. In this vein, electricity restructuring itself may help electricity prices come closer to marginal costs, as will market-based environmental programs and other improved public policies. This is particularly true with respect to daily variations in prices and costs—the single largest source of inaccurate price signals. Nevertheless, many believe that energy will continue to be priced below its marginal social cost.[3]

A second basis for utility involvement in energy efficiency programs is quite different, and does not depend much on the mispricing of energy or environmental externalities. Suppose that hundreds of homeowners could save money by insulating their homes, provided that they believed that insulation worked, they could afford the cost of installation, and they could easily find capable installers. Individually, a few of these homeowners may take the time to investigate insulation, shop for installers, and come up with the money to make the installation.

If an organization could assist all of the homeowners with the investigation and selection, charging them a modest fee, perhaps many more would choose to insulate. In effect, the organization becomes a means of pooling and reducing the transactions costs of taking on a large, complex home improvement task. Because there are potentially large economies of scale, one organization could help many homeowners more cheaply per home than individuals could do themselves. It is also more efficient for one organization to provide assistance. (Although there are many competitive firms that would be happy to assist with all these transactions costs, firm selection must be based on reputation and price, which brings it back to the same transactions costs we were trying to avoid.)

This kind of situation describes the "information-market-failure" justification for utility energy efficiency programs, also known as Demand-Side Management programs or *DSM*. Because utilities serve every customer in their territory and are in the business of providing energy, they can be expected to know something about the quality of equipment that uses power and the reputations of local installers and contractors. It is also easier to rely on the DSM advice of power companies, because customers know that the utilities themselves are not profiting from the advice and their activities are policed by regulators. Moreover, utilities ought to be able to spread the costs of acquiring expertise and administering programs over thousands of similarly situated customers, taking great advantage of scale economies.

The transactions-cost argument for DSM arose out of the widespread observation that many electricity consumers seemed extremely disinclined to make investments that saved energy and money. Under the traditional theory of investments, a firm or individual will make an investment outlay today that will earn a present value that is larger than the outlay when discounted at the proper discount rate. Applying this model to energy efficiency purchases, analysts repeatedly found that consumers would not invest unless they earned back all of their investment, plus a profit, in one or two years at most. Meanwhile, many energy efficient purchases that cost more to buy at first paid for themselves many times over in energy savings, but only over a period of 3 to 15 years. Many suspected that transactions costs were the hidden barrier to greater efficiency.

Recent research suggests that consumers' unwillingness to invest in conservation is economically optimal if they are unsure that future energy prices will remain as high as they are today.[4] If this a was a large effect—which many researchers doubt—it would suggest that transactions costs and information imperfections were not important.[5] This controversy suggests that one must be extremely careful about concluding that consumers are acting as if they are irrational or the energy markets have "failed." It also shows that it is difficult to determine the size of the transactions costs and information imperfections that serve as one justification for publicly funded DSM.

UTILITY DSM PROGRAMS—SCOPE AND MECHANICS

Both arguments for utility energy efficiency efforts—overcoming energy mispricing and information barriers—have prompted state regulators in almost every state to order electric utilities to provide a variety of DSM programs. (A few states have placed similar requirements on natural gas utilities, although gas DSM has not been nearly as widespread.[6]) Historically, the programs have taken several approaches:

Residential and Commercial Rebate Programs. To provide their customers with an incentive to install energy-efficient equipment, some utilities have offered cash rebates or reduced prices for particularly efficient brands or models. Programs of this nature provide a double benefit to consumers—they reduce the cost of investing and they reduce the need to evaluate technologies extensively. However, these programs can involve significant utility cash outlays. Utilities typically pay much of the difference between the cost of an inefficient and efficient appliance—and they require considerable utility monitoring to ensure that the appliance, once paid for, is used and maintained properly.[7]

Low-Interest Loans. As an alternative to rebates, many utilities have offered their customers loans for large energy-efficient purchases. Theoretically, a low-interest loan has the same economic effect on a purchase decision as an equivalent rebate—both lower the cost of purchase relative to the stream of benefits received. However, in most cases, the original utility loan programs experienced low customer participation and were difficult for utilities to administer. Recently, some utilities have redesigned and relaunched a new generation of loan programs that are targeted at modern DSM markets.

Information or Design Assistance. To reduce the costs of acquiring accurate information, many utilities provide assistance to customers who must make electricity-using capital investments. This assistance can occur via individual visits, telephone consultation, or at special assistance centers or via broad-based advertising programs. In some cases, utilities target specific customer groups or professionals, such as architects, involved in energy use decsions.

Trade Ally Programs. In some instances, utilities arrange with appliance manufacturers or retailers, providing them with inducements to manufacture or sell energy-efficient models. The boldest example of this approach was a national consortium of utilities that paid $30 million to the winner of a competition to design and produce a "super-efficient" refrigerator (first manufactured in 1994).[8] This one appliance program will reportedly reduce residential energy costs by $1 billion over the next 15 years.

Demand-Side Bidding and Performance Contracting. Because saved energy is saved money, it makes sense to offer to pay customers or Energy Service Companies that can find verifiable ways of saving electricity. Starting in 1987, this idea prompted about 35 utilities in 10 states to solicit DSM proposals from customers. Because of the difficulty of specifying, in advance, the sort of energy saving project that could be measured and monitored, as well as other administrative problems these programs were of limited success. Greater success was achieved when utilities developed "standard offers" that combined elements of DSM bidding with rebate programs.[9] Although DSM bidding has largely fallen by the wayside, it is virtually certain to become a "built-in" feature of the restructured power industry. In particular, the standard offer format lends itself to the competitive retail marketplace for electricity and/or DSM services.

From a very small base, DSM programs expanded during the late 1970s and 1980s to the point where the nation's electric and gas utilities ran more than 3,600 programs as of 1992 with almost 20 million participants and a 1994 total expenditure of nearly $2.7 billion a year.[10] To put this in perspective, this sum is more than twice the United States federal government's total budget for energy efficiency programs and roughly ten times what it is spending to address the threat of global climate change.[11] The results of these programs are quite impressive. According to the Energy Information Administration, these programs reduce U.S. peak demand by 25,000 megawatts—a savings equal to the capacity of 50 modern power plants or about 3% of total U.S. capacity.

One highly positive result of these utility efforts has been the stimulation of a new type of firm, so-called energy service companies (ESCOs). The ESCO industry is closely associated with firms that manufacture energy control systems, some of whom operate branches or affiliates that perform on-site building energy services [See Driesen and Cudahy (1996) and Goldman and Dayton (1996)]. Originally focused on large institutional enrgy users such as hospitals, ESCOs proved to be useful providers of assistance and services as utility DSM expanded Over the past 20 years, ESCOs and their trade allies have grown into a billion-dollar industry that is closely allied with some equipment manufacturers and utilities.[12] As described in Chapter 9, the ESCO industry is in a unique position to profit from services related to real-time pricing as part of utility restructuring.

CONTROVERSY OVER DSM'S EFFECTIVENESS AND IMPACTS

One of the challenging and, to many, objectionable aspects of DSM programs is that they place the utility in the role of purchasing agent for a large, potentially disparate group of buyers—an agent that can subsidize the purchases of some customers with revenues extracted from other captive billpayers. To increase economic welfare, DSM program operators must carefully select energy efficiency investments that customers would not make without the utility's assistance—so-called "free riders." In addition, selected programs must be accurate and efficient with regard to administering a complex set of individualized transactions.

The economic standards that are used to judge when a utility should offer a DSM incentive have been the subject of many bitter disputes between utilities, regulators, and public interest groups. Briefly, utilities generally feel that their DSM programs should not harm customers who choose not to participate. Although utilities recognize a need to foster energy efficiency, doing so in a manner that raises prices and reduces overall sales seems excessive. Conversely, efficiency advocates believe that all measures that lower the total aggregate costs of serving society's energy needs should be encouraged, and that modest transfers between groups of ratepayers or reductions in earnings should not prevent this from occurring.

Because regulators want to ensure that DSM program money is spent wisely, regulatory proceedings governing these programs have traditionally been lengthy and extremely technical.[13] The terms of each program are examined to ensure that "free riders" are screened out as much as possible,[14] the correct level of financial incentive is provided, and that the program is effectively administered. Estimates of the total cost-effectiveness of each program—i.e., the total cost of saving one unit of electricity, including administrative and investment costs—are debated, and the financial impacts on the utility and ratepayers who provide funds but do not otherwise use the program ("non-participants") are hotly debated.[15]

A number of studies have attempted to analyze the overall cost-effectiveness of DSM programs and they, too, have sparked heated controversy. A series of papers started by MIT Professors Paul Joskow and Donald Marron in 1991 claimed that most DSM programs cost more than the value of the energy they saved.[16] These claims were countered vigorously by University of California researchers Mark Levine and Joe Eto, who specialize in DSM program evaluation for utilities and government agencies, including the U.S.

Department of Energy.[17] This debate was never fully resolved in its original form. Now, with the rapid reduction in traditional DSM due to restructuring, the debate has been transformed into one about the future of DSM as a whole. However, as it seems likely that many if not most states will continue publicly-funded DSM in some form, the considerable expertise accumulated to date will be essential for designing and evaluating cost-effective programs in a more competitive environment.

ENERGY EFFICIENCY AFTER RESTRUCTURING

Around 1994, utility energy efficiency programs entered a period of precipitous decline. Environmentalist Ralph Cavanagh wrote, "Utility restructuring per se is in no sense hostile to improved energy efficiency. There can be no question, however, that the industry's overall commitment to efficiency has flagged in the face of uncertainty about the schedule and outcome of...restructuring."[18] Even greater falloffs have been observed in the United Kingdom and other nations that are ahead of the United States in restructuring.[19] Cavanagh labeled the collapse of DSM following Norway's restructuring "appalling".[20]

There is little disagreement about the reasons for this retrenchment. Utilities perceive that they will soon be competing with other unregulated power suppliers and ESCOs. To the extent that DSM programs caused costs and prices to increase to some customers in order to fund DSM programs for others, utilities felt that they could be competitively disadvantaged. Furthermore, as noted in Chapter 9, the advent of real-time pricing and other changes could create an improved free market for energy services, obviating the need for utility programs.

Thus began an intense debate between those who believed that restructuring would sooner or later sufficiently remove both the mispricing and transactions-cost justifications for utility DSM and others who were divided on whether restructuring might eventually do this, but were generally united in believing that jettisoning utility-style DSM at this time is premature.

Much of this debate has centered on two notions of "market transformation" versus "public purpose" DSM. Market transformation DSM generally refers to utility programs mandated by regulators and designed to introduce new information and technologies into a specific portion of the energy services marketplace. The intention of market-transformation DSM is to be self-eliminating: as the target market segment learns about new savings opportunities, it will eventually eliminate the need for public assistance. Meanwhile, new information and new technologies may uncover new transformation opportunities. In contrast, "public purpose" DSM programs address more structural imperfections in the energy services marketplace (i.e., low-income cutomers' reduced access to credit) which are ongoing in nature.[21] DSM programs targeted at these areas are not expected to be self-eliminating, but must continue to search for incremental design improvements and undergo sound periodic evaluations.

In a number of states that are implementing or considering retail competition, the emerging range of DSM policies centers on what some call the "two-track approach:" a continued mandate for the regulated distribution utilities to offer programs, a required shift in emphasis toward market transformation programs, and an agreement to maintain public-purpose programs the market is unlikely to provide.[22] For example, Rhode Island's retail choice regulation requires utilities to spend 2.5% of their revenues on "two-track" DSM, which is slightly more than they spend today.[23] For the moment, it appears that these policies have stabilized the overall size of utility DSM investment, but with a shift in emphasis and a highly uncertain future.[24]

POST-RESTRUCTURING DSM IMPLEMENTATION ISSUES

Restructuring is prompting large shifts in program funding and implementation for the portion of DSM that remains regulated. Three main issues in this area are the funding mechanism, the nature of the entity responsible for planning and administrating DSM, and program methods.[25]

The funding options for DSM are, in general, identical to those discussed in Chapter 11 for low- income programs: require suppliers to fund a pool, collect a "system benefits" or distribution services charge from all users of the distribution system, set a tax on energy sales or use, and other unique approaches. For DSM, the system-benefits charge idea has gained by far the largest currency, and much of the discussion has centered on establishment and management of issues in these kinds of charges.

With regard to the question of who is responsible for planning and implementation, there is a division between those who want a public agency of one kind or another to play a strong role and those who prefer to leave the responsibility with each distribution company, which is where it resides today. There are important arguments on both sides of this issue. On the one hand, statewide or regional administration of programs might provide large-scale economies and ensure that everyone in a large area has equal access to public programs. On the other hand, creating and ensuring efficient management of what is essentially a new public bureaucracy is viewed in some quarters as an anathema. Moreover, many discos have more than a decade of DSM experience; often the local disco is a popular, locally-controlled provider of DSM today.

Implementation-mode issues tend to go hand-in-hand with administration and governance. Most states do not presently have a statewide agency that collects funds and implements DSM programs. If one is created, rather than do its own implementation, it is likely to contract out for services from ESCOs and perhaps existing discos. Some argue that disco's should be prohibited from participating in these markets; others look at the mixed record for DSM bidding and advocate less reliance on competitive procurement and more alliance with LDCs and others.

At this early stage, DSM's future hinges on a host of unpredictable factors. First, the public's continued belief in the importance of energy efficiency as a public good is the *sine qua non* of these programs, and must not be taken for granted. Second the breadth and depth of new markets for energy efficiency is not yet known. Finally, planning and implementing programs will be difficult no matter who bears the responsibilities; effective implementation is essential to continued public and industry support.

Chapter 13

NOTES

1 For additional discussion, see Golove and Eto (1996), Levine, et al (1994), Studness (1993), Sutherland (1991), and Johnson (1992). Hirst and Eto (1995) catalog the original arguments for energy efficiency as (1) defer new power-plant construction and associated rate impacts; (2) reduce environmental emissions; (3) compensate for mispricing; (4) reduce foreign oil dependence; and (5) compensate for the absence of efficiency standards; (6) overcome market barriers; and (7) recognize that utilities have unique capabilities to assist customers with efficiency improvements. With the exception of the first and fourth items on this list, these items are subsumed within the two broader areas of energy mispricing and transactions cost advantages discussed in the text.

2 See Vine, Crawley, and Centolella (1991).

3 Changes in utility industry economics during the past 10 years have acted to reduce the gap between actual electricity prices and prices that reflect marginal social cost. In some parts of the country, electric rates are higher than marginal cost, so today's price already sends a price signal that incorporates some environmental externalities. Electric power plants also continue to get "cleaner" per unit of power, as explained in Chapter 12. See Hirst and Eto (1995) for additional discussion.

4 See Hassett and Metcalf (1993), Metcalf and Rosenthal (1993), Johnson (1993) and Awerbuch and Deehan (1993).

5 Sanstad and Howarth (1994); Hassett and Metcalf (1993); and Sanstad, Blumstein, and Stoft (1995).

6 As with electric utilities, most gas DSM has occurred within integrated resource planning processes, but has been much smaller to date. See Goldman, et al (1993) and Lerner (1994). For dissenting views on gas DSM, see Kretchmer and Mraz (1994).

7 DSM experts Joe Eto and Chuck Goldman note that utility shareholder incentives for DSM performance are another reason to impose high monitoring requirements.

8 See Feist, et al (1994).

9 Goldman and Kito (1994) conduct an extensive survey of DSM bidding auctions. They find that these programs were difficult to administer and more costly than conventional utility DSM programs. Auctions that combine bids for new power plants with bids to save energy were even more complex and costly and less successful. Also see Goldman and Busch (1992) and Charles River Associates (1991), and Goldman, Kito, and Moezzi (1995).

10 Utility DSM spending and profiles may be found in Energy Information Administration (1995) Table 43, Hirst (1994), Gellings and McMenamin (1993) and President's Commission on Environmental Quality (1992). Reportedly, electric utilities account for the vast majority of this spending, according to Nadel, et al (1993).

11 This does not include scientific resrach on the causes and impacts of global climate change, but rather only programs created to reduce emissions related to climate change.

12 Goldman and Dayton (1996) estimate the industry has $450 million in annual revenues and has grown 25% a year in the past two years. In "Performance is Key in Energy Contracting," *Energy User News,* 2/26/96, Ed Bas reports that major equipment manufacturers in this field have more than $500 million additional revenue.

13 For perspective on process issues, see Schweitzer and Young (1994), Subbakrishna (1994), and the record of the proceedings in a major states' integrated resource planning proceeding, such as the Integrated Resources Management Docket for the Commonwealth Electric Company, Massachusetts Department of Public Utilities 91-234, 1992–93. Two reference works on the difficult analytic issues involved are Hirst and Reed (1991) and Charles River Associates (1992).

14 A "free rider" is a customer who would have saved equal amounts of energy with or without a utility DSM program, and participates in the program solely to obtain a cash rebate from the utility.

15 See Chamberlin, Herman, and Wikler (1993).

16 Joskow and Marron (1991, 1992, 1993) and Nichols (1994) offer analyses concluding that DSM is not cost- effective. The rebuttal case is presented in Levine and Sonnenblick (1994) Eto, et al (1993) and Eto, et al (1995). See also note 8 *supra*.

17 Also see Cavanagh (1996) and Hirst, Cavanagh, and Miller (1996).

18 Cavanagh (1996), p. 73. Also see Hirst (1994), Messenger (1996), and Centolella (1996).

19 See Gill and King (1994) and King, et al (1996).

20 Cavanagh (1994) p. 17.

21 For example, consumer expert Nancy Brockway writes that "it is well-known that low-income customers face unique market barriers that make it unlikely they will be able to take advantage of most DSM programs..." Brockway (1995) p. 18. Also see, Fisher, Sheehan and Lolton (1995). Brown, et al (1994) note that about 80% of all utility programs aimed at low-income customers were mandated by regulators.

22 See Eto, Goldman, and Kito (1996), Baxter (1996), King, et al (1996), and Goldstone (1996). Also see the Comments of the Massachusetts Energy Efficiency Council in Massachusetts Department of Public Utilities Docket No. DPU 96-100, May 24, 1996.

23 Unpublished communication, Conservation Law Foundation, Boston, MA, 1996.

24 See the Finance, Regulation, and Power Supply Group (1996), Messenger (1996), and Hadley and Hirst (1995).

25 For the most extensive discussion to date, see the two-volume "Funding and Administering Public Interest Energy Efficiency Programs," Report of the Energy Efficiency Working Group to the California Public Utilities Commission, Decision 95-12-063, Aug. 16, 1996. Also see the references in note 18 *supra*.

Chapter 14

Renewable Energy and Fuel Diversity

Fuel diversity refers to the ability to substitute fuels such that energy systems are not highly vulnerable to shortages or price increases of a particular fuel.[1] Coal supplies can be halted by severe weather or railroad strikes; nuclear power has unique safety and availability concerns; and secure oil and natural gas supplies are always an issue. Fuel diversity is essentially an insurance policy designed to keep the economy functioning when sudden problems are encountered with one type of fuel.

Renewable energy, which is inherently decentralized, plays a unique role in providing fuel diversity. Renewable energy sources tend to be less vulnerable to large-scale supply disruptions—particularly supply disruptions that involve foreign powers or disruptions in other segments of the economy.

Renewable energy plays a second public role as well. Most forms of renewable energy have less of an impact on the environment than conventional energy. Hence, renewables contribute to reduced environmental damage as well as fuel diversity. Studies of the environmental footprint of renewable resources confirm this fact, and in the United States and many other nations there is a widely held perception that renewable energy is relatively "clean."[2]

FUEL DIVERSITY AND ENERGY SECURITY

Science has long recognized the unique properties of energy, as distinct from other economic resources. According to the law of conservation of energy, energy is neither created nor destroyed; it merely changes form. Early economists focused on capital and labor as the most important resources. As advanced economies have become increasingly dependent on huge, uninterrupted energy flows, policymakers have recognized that the constant availability of adequate, reasonably-priced energy is essential for national security and economic growth.[3]

Although there are many economic goods that are undoubtedly just as crucial to the economy as energy, the obvious nature of energy dependence and the fact that energy supplies have been threatened in recent memory give rise to a widespread national sentiment that adequate energy supplies are a matter of public interest. The term *energy security* has been coined to refer to this expressed national need.[4]

Among economists, there is genuine debate over the extent to which energy security remains a concern that private markets cannot adequately address. Formally, the issue turns on whether there is a significant negative externality associated with energy shortages, in the form of diminished economic growth for the whole economy or in other shared costs. Those who associate the energy shortages of the 1970s with reduced economic growth and public hardship intuitively believe that such a connection exists, but many economists are not so sure.[5]

Some analysts who believe that competitive power sellers will not want to be too dependent on any one source argue that competition itself will provide adequate fuel diversity. Moreover, they note that the evidence favoring large negative externalities when fuel shortages occur does not justify large, expensive programs. In the context of deregulated power generation, no one is sure just how much the United States should invest in preventing gencos from choosing the best fuel supply from their own standpoint—even if this results in all sellers relying on a single type of fuel that may someday be vulnerable to price spikes or disruptions.[6]

A full treatment of this debate is beyond the scope of this book, but it must be recognized that this underlying debate casts a shadow over electric utility programs dealing with fuel diversity and renewable energy. However, even if these arguments are correct, they do not eliminate the need to consider renewable energy programs, as the latter contribute to environmental objectives as well.

FUEL DIVERSITY TODAY

In regulated industry planning proceedings, regulators usually include an examination of fuel diversity in their review of utility investments. This may occur in a qualitative, judgmental fashion, or alternative fuel price and supply scenarios may be modeled to predict whether a utility system is particularly vulnerable to some aspect of its fuel supply.[7] The result is a tradition of fuel diversification in the utility industry which is seldom rigorously justified, but has generally been viewed as a beneficial feature of the industry.

Figure 14-1 shows the percentages of electric power generated by the major fuel sources in several of the world's major economies. The figure suggests that although the total U.S. power system is not as well-diversified as that of Japan, it is as good or better in this respect as the other major economies shown. Within the United States, regional electric fuel dependence varies greatly. On the whole, the United States continues to be most dependent on coal, which is a comparatively stable fuel, and is reasonably well-diversified with respect to the other major forms of fuel.

Since wholesale competition began in the electric power industry, there has been a significant shift toward plants and fuels with minimum costs, particularly natural gas and coal.

Table 14.1 shows actual and projected capacity additions by U.S. generating companies between 1988 and 2004. Between 1988 and the present, plant additions using sources other than gas have all trended downward, and almost all projected additions are gas and coal.

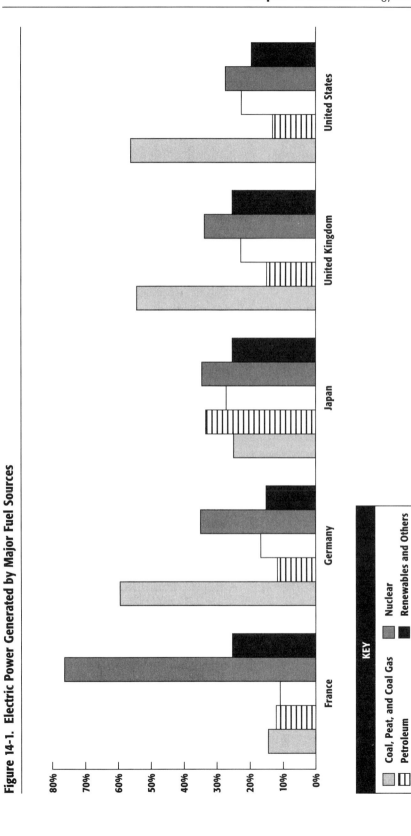

Figure 14-1. Electric Power Generated by Major Fuel Sources

KEY

- Coal, Peat, and Coal Gas
- Petroleum
- Natural Gas
- Nuclear
- Renewables and Others

Source: The IEA Energy Statistics for OECD Countries

Table 14.1. Actual and Planned Additional Capacity (in megawatts)

		Coal	Petroleum & Gas	Nuclear	Hydro-electric	and Other	Total
Actual	1988	1,784	710	2,341	218	168	5,221
	1989	1,967	1,499	3,391	106	3	6,966
	1990	3,063	778	2,300	52	62	6,254
	1991	792	1,759	–	1,073	–	3,624
	1992	498	2,218	–	624	–	3,341
	1993	–	2,217	1,150	92	1	3,460
	1994	540	3,117	–	–	184	3,841
	1995	471	2,095	–	565	55	3,186
Planned	1996	1,173	1,961	–	45	–	3,178
	1997	122	1,376	–	44	–	1,543
	1998	–	2,819	–	140	–	2,959
	1999	660	3,892	–	74	100	4,727
	2000	750	4,779	–	189	14	5,733
	2001	–	4,341	–	–	–	4,341
	2002	500	4,533	–	38	–	5,071
	2003	–	3,081	–	–	100	3,181
	2004	1,325	3,743	–	–	100	5,168

Source: EIA Electric Power Annual, 1988–1995.

It is extremely difficult to measure how much more vulnerable the U.S. economy will become if electric restructuring continues this trend toward less fuel diversity—if at all. Apart from renewable energy proposals, it is equally difficult to determine how a competitive power industry might be required to increase its level of diversity. Should an arbitrary limit be placed on the use of one fuel, in the aggregate? Does it make sense to hold a lottery among power producers and require the losers to employ different fuels? Or should some generators be required to have backup sources or plans for use of a second fuel type?

Most restructuring discussions and enactments have said or done little about fuel diversity (outside the realm of renewable energy). Perhaps this is because studies of the environmental impacts of restructuring focus on renewables and the extent to which restructuring will favor the expansion of coal versus natural gas power plants, and little else. In any event, the rest of this chapter follows this course and turns to the prospects for renewable electricity in the emerging industry.

UTILITY RENEWABLE ENERGY PROGRAMS TODAY

There are six primary methods of generating electricity from renewable resources.

1. Biomass crops can be grown and used as utility boiler fuel;

2. Geothermal energy wells can be used as a source of heat for utility turbines;

3. Wind can be used to turn turbine-generators;

4. Water can be used to turn turbine-generators;

5. Solar energy can be concentrated to boil fluids and operate a turbine: and

6. Photovoltaic ("PV") cells can convert sunlight directly into electricity with no moving parts.

Each of these forms of energy depends on the availability of naturally-occurring, inter-mittent energy. On average, most forms of renewable energy do not generate power as cheaply as conventional power plants. (The most common exception is hydroelectricity, which can be inexpensive at appropriate sites.) However, technological innovation has steadily reduced the costs of non-hydro renewable electricity (Figure 14-2), and if this trend continues as expected, renewable energy will increasingly be cost-competitive in many applications.[8]

Recognizing the public benefits of renewable power, the Public Utility Regulatory Policies Act of 1978 (PURPA) accorded special status to unregulated generators that use renew-able resources. These "small power producers" are guaranteed the right to sell to utili-ties at the utility's full avoided cost.[9] Many of the long-term renewable contracts signed by utilities call for payments at prices that are substantially above today's wholesale power prices. This enables renewable energy producers that could not operate profitably at today's prices to continue to operate under their contracts.[10]

This combination of factors has led to significant utility and IPP involvement in renewable electric production. Renewable capacity (other than utility hydroelectric plants, which are well- established) has grown from near-zero levels in 1980 to an estimated 10,500 MW in 1994, according to data from the Edison Electric Institute and the National Renew-able Energy Laboratory (NREL). According to NREL, an average of almost 2% of U.S. electric generation comes from renewable generators connected to the power grid.[11]

One example of an innovative, utility-based renewable energy program is a "buyer's club" for solar-electric (photovoltaic, or PV) power cells. This program is a joint project of the federal government and a group of 70 large and small public and private utilities. Using its purchasing power, this group hopes to purchase and install, 49 megawatts of PV capacity across the country during the next five years.[12]

In addition to purchasing from renewable energy suppliers, utilities have recognized that, in certain specific contexts, renewable electric generation sources provide benefits that equal or exceed present costs. In some cases, dispersed renewable sources allow utilities

Figure 14-2. Cost Reductions for Wind Energy and Photovoltaic Technologies, 1980–94

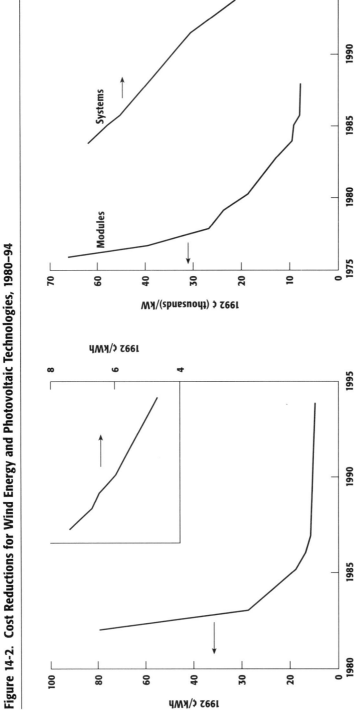

Notes: The cost of wind and photovoltaic (PV) systems and generated electricity have declined sharply. The figure on the left shows data for wind turbines installed in California (which accounts for most turbines in the United States). The figure on the right shows overall U.S. PV module costs and complete PV systems installed at the PVUSA site in Davis, California. To convert PV system costs to an approximate cost of generated electricity, divide the system capital cost by 20,000 to get ¢/kWh. Expanded scales show that costs continue to decline sharply.

Sources: Wind data are from Paul Gipe, Paul Gipe and Associates, Tehachapi, CA, "Wind Energy Comes of Age in California," Dale Osborn, personal communication, April 1994. PV data for modules only are for U.S.-based production and were provided by George Cody, Exxon Corporate Research and Development Laboratory, personal communication, February 1993; Paul Maycock, PV Energy Systems, Inc, January 1993. For complete PV systems, data are for installations by U.S. PV manufacturers under th PVUSA project at Davis, California, and were provided by Dan Shugar, Advanced Photovoltaic Systems, Inc, personal communication, June 1994. OTA (1995)

to provide service to remotely located customers for less than the cost of conventional supply.[13] In other instances, the low-emissions properties of renewables justify their costs, sometimes leading to "externality adders" as described in Chapter 12.[14] In still other cases, utilities believe that their customers want them to incorporate renewable energy into their overall supply plan for diversity or environmental reasons, and to expand their familiarity with renewable supplies in order to prepare for the likely growth of renewable energy.[15]

The decentralized nature of renewable energy is expected to fit particularly well into the "distributed utility" of the far-distant future.[16] (See Box 14-1) As noted in Chapter 1, as the cost of electricity storage declines, the power grid will eventually become more akin to an oil pipeline. During and after this transition, intermittent energy sources such as renewables will be more valuable than they are today, and will naturally fit the electrical architecture of the network. Many see a natural evolutionary course between the use of renewables in locations where they are cost-effective today to more common applications near more conventional urban customers in the distributed-utility era.[17]

RENEWABLE ENERGY AFTER RESTRUCTURING[18]

Price-driven competition in the electric power industry has become fierce, and all indications are that retail competition will intensify this trend among many suppliers. Because renewable power is price-competitive only in small niches, and LDCs cannot be relied upon to continue their past renewable energy activities, renewable proponents fear that renewable energy deployment will stall or reverse itself under restructuring. Indeed, there are already signs that this is occurring—e.g., the 1996 bankruptcy of Kenetech Windpower, the largest U.S. wind turbine manufacturer.[19] According to the National Renewable Energy Laboratory, utility generation from renewable energy reversed a steady upward trend and dropped approximately 30% in 1995, while total industry-wide generation increased roughly 2%.[20]

To forestall backward movement, one segment of the renewable community has proposed that all utilities that sell power, whether regulated or unregulated, be required to get a fraction of their supply from renewable sources. This proposed requirement became known as the "renewable portfolio standard (RPS)." The name was chosen to highlight the notion that the sellers of power should hold diversified generator portfolios, much like investors hold diversified investment portfolios, to shield them from price or supply risks.[21]

The American Wind Energy Association (AWEA) has proposed a renewable portfolio requirement equal to 2% of total sales in the year 2000, rising to 4% of sales by the year 2010.[22] No state has yet adopted AWEA's proposal, but one proposed electric restructuring bill in the 94th U.S. Congress contained a similar requirement for all power sellers in the nation.[23] Figure 14-3 shows that the regional distribution of renewable generation is very uneven today. A 2% purchase standard would require several regions to dramatically increase renewables purchases, while others already exceed the requirement.

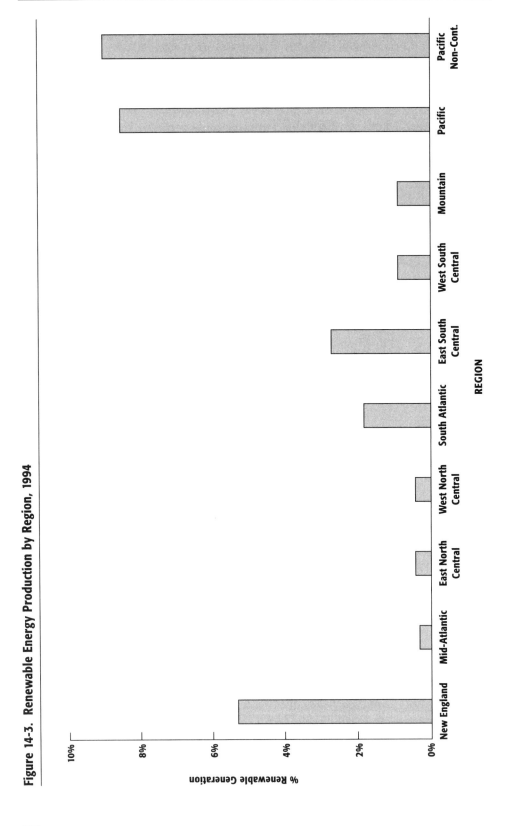

Figure 14-3. Renewable Energy Production by Region, 1994

One of the administrative challenges inherent in a portfolio standard is the labeling and tracking of renewable kWh. Each kWh generated in the future power marketplace may be blended and traded dozens of times before it is ultimately purchased at retail. There is presently no way either to certify that a unit of power comes from renewable resources or to track the course of such units. This problem could be remedied at little cost by national issuance of tradeable renewable generation certificates.[24]

As it has been proposed, the portfolio standard aggregates all renewable energy sources for the purpose of meeting the standard. Because retailers can be presumed to look for the cheapest way to meet this mandate, they will probably buy whichever form of renewable energy is most cheaply available. Some renewable advocates observe that this will lead to no diversification or development within the six forms of renewable energy. Accordingly, they argue that the RPS is less preferable than the conventional diversity-based renewable programs.

Another approach favored by many renewables supporters is to implement programs supporting cost-effective renewable energy installations funded by system benefits charges.[25] This approach is similar to the most likely post-restructuing approach to energy efficiency programs described in the previous chapter. For example, the state of California's restructuring legislation calls for system benefits charge collections of about $40 million per year. These funds are to be distributed according to procedures developed by the California Energy Commission and administered by a public agency. As with energy efficiency programs, it will be necessary for regulators to continually monitor and evaluate these programs.

GREEN PRICING

Survey research shows that a clear majority of Americans believe that the environment is an important public resource.[26] Surveys also show that many consumers claim they are willing to pay more for goods produced with less harm to the environment. This includes energy from less-polluting sources, or so-called "green energy."

The fact that a segment of the population claims to want clean electricity has given rise to claims that a competitive electric market will allow specialty retailers to sell renewable energy at a higher price than other power because it is environmentally superior. This idea, known as "green pricing," is being pursued today in a number of regulated utility pilot programs around the United States.

Preliminary experience with these pricing programs is showing that some customers will participate, but usually far fewer will pay significant extra amounts.[27] For example, in Public Service Company of Colorado's program, as of 1994 almost 85% of those surveyed supported the program, but only 10% of eligible customers signed up and paid an average of $1.90/month more on their bills. Still, this program netted more than $2 million for renewable energy.[28] In total, the Natural Resources Defense Council computes that the five green pricing programs presently operating account for only 2 megawatts of renewable generation—a tiny fraction of the renewable generation stimulated by PURPA.[29]

Some argue that green pricing programs are a superior means of addressing the public need to foster renewable resources because these programs allow those consumers who value the public benefits of renewables to be the only ones who pay for these benefits. They also fit naturally with the idea of unbundling separate retail utility products in the new industry. Based on the economic theory of public goods, there is no reason to believe that these voluntary pricing schemes will produce the optimal level of investment in an accurately-determined public need.[30] On the other hand, policymakers may determine that the level of support provided by these programs is adequate, which would render green pricing perhaps a primary means of supporting renewable energy in the restructured industry.

CONCLUSION

With or without restructuring, the prospects for renewable energy depend on the value that state and federal authorities place on fuel diversity, environmental quality, and the value of reduced energy import dependence. Restructuring's impact on renewable energy will occur inside this broader, shifting context.

On the one hand, restructuring will disrupt almost all proposals designed to foster renewable energy, On the other hand, some retructuring poposals have even stronger government mandates for renewable energy than exist in most areas today. In addition, restructuring is unlocking creativity and variety in the offering of "green" power products to today's retail customer.

The net impact of all these changes on renewable energy and its contribution to rational needs is uncertain. Restructuring may propel renewable energy into a new era of customer-driven, distributed resources or backward into its traditional niche applications.

Table 14.2 reviews the three main policy courses that are evident today for renewable energy programs. To summarize, RPS favors different forms of renewable energy differently, but has gained significant support. System benefits charges dovetail with the most likely treatment of universal service and energy efficiency programs, but place traditional burdens on regulators or regulated administrators. Green pricing will be a natural outgrowth of competitive power markets, and requires little or no government intervention, but its impacts may be too small or skewed to provide sufficient public benefit.

Table 14.2. Summary of the Most-Discussed Renewable Energy Options Under Restructuring

Option	Description	Advantages and Disadvantages
Renewable Portfolio Standard	Requires all sellers (retail or whole-sale, but not both) to purchase or generate a minimum percentage of their supply from renewable sources	Tends to favor the lower-cost renewables over others; may be difficult to imple-ment, particularly if states wish to keep renewables in-state; must be mandated and policed by state or federal regulators
System Benefits Charge	All users of electricity in a state pay a charge to their local LDC or to a statewide group; funds are disbursed according to a plan approved by a public body	Consistent with likely approaches used for other public goods and in other regulated industries; requires traditional large administrative procedures to collect funds and make and implement decisions
Green Pricing	Competitive sellers of power volun-tarily offer renewable electricity to buyers who want it, probably at higher prices	A natural part of the future power mar-ketplace; size and success of programs uncertain relative to public need

Chapter 14

NOTES

1 Substitution need not occur in the physical sense to provide diversity benefits. For example, merely holding a diversified resource portfolio can provide insulation against price increases in a single fuel.

2 For research quantifying the environmental impacts of renewable energy, see Aiken (1996), Barnes (1994) Section III. Regarding public perceptions of the environmental impacts of renewable energy, see Farhar (1993) and "America speaks out on Energy: A survey of 1996 Post-Election Views." Sustainable Energy Coalition, Takoma Park, MD, Nov. 11, 1996.

3 See, for example, United States Department of Energy (1987), the National Academy of Sciences (1986), and Mills- McCarthy and Associates (1995).

4 Ibid

5 See Feldman (1996) and Bohi and Toman (1996) for two recent discussions.

6 For one of the few analytic treatments that bears on this issue, see Dowlatabadi and Toman (1989). These authors find that advanced natural-gas power plants, possibly with coal gasifiers if needed, are economically optimal under a wide range of fuel price scenarios. It should also be noted that the amount and location of fuel storage is an equally important element of insurance against fuel disruption costs and is also frequently man-dated in some form by regulatory agencies.

7 Many integrated resource planning or public service statutes make reference to fuel diversity. For example, in its 1991 filing to obtain permission to build a new gas-fired power plant, the Florida Power Corporation noted that natural-gas fired power would complement a system that otherwise used coal, oil, and nuclear energy. The Florida Public Service Commission agreed, citing fuel diversity benefits. Order determining the need for a proposed Power Plant, 92 FPSC 2:658.

8 Analyses of the future competitive prospects for renewable energy have been the subject of some debate even absent electric restructuring; for representative views, see Resource Data International (1995), Kassler (1994), the office of Technology Assessment (1995), Johansson, et al (1993), and Jane Ellus, "Why Promote Renewable Energy?" OECD Observer, August, 1996.

9 See 10 CFR Part 292 and Fox-Penner (1996).

10 Resource Data International (1995).

11 Porter (1996). Many applications of renewable energy occur in remote locations not served by the grid. Unlike conventional power sources described in Chapters 3 and 4, renewable energy does not always require a grid.

12 Energy Partnerships for a Strong Economy, F95 Program Plans. U.S. Department of Energy, Office of Energy Efficiency and Renewable Energy, Fall 1994.

13 For a description of utilities using PV systems to supplement conventional power at remote substations and other parts of their systems, see Shugar and Orens (1992) and Bzura (1992).

14 See *supra,* note 1.

15 This includes a once-active role in collaborative research and development in renewable energy technologies. Much of this remains, but much is also falling victim to utility cost-cutting pressures. See Blumstein (1993, 1996) and the California Energy Commission (1995).

16 For additional discussion on the distributed utility concept, see the Electric Power Research Institute (1992, 1996), Schuler (1996) and Pfeffenberger, Amman, and Taylor (1996). For a useful overview of the relationship between renewable energy and utility distribution systems today, see Barnes, et al (1994).

17 See, for example, the Electric Power Research Institute (1992), Awerbuch (1993), and the proceedings of the 1996 EUP Annual Planning Meeting, Energy Laboratory, Massachusetts Institute of Technology, Jan. 30, 1996.

18 For useful surveys, see Cohen (1994) and Hoffman (1996). Some renewable energy forms, such as wind and geothermal energy, must be located near naturally-occurring resource concentrations.

19 In the *Wind Energy Weekly,* v. 15 no. 700, June 3, 1996, reports that "Kenetech Windpower management, in a statement announcing the filing, attributed the company's problems with cash flow to: 'deregulation of the electric power markets in the U.S. and corresponding declines in the amounts utilities pay for electricity..'." as well as other factors.

20 Porter (1996) p. 1. The decline in the growth of renewable energy has been occurring for several years.

21 In California, the proposal is known as the "minimum renewables purchase requirement." See California Public Utilities Commission decision D.95-12-063, Dec. 20, 1995.

22 "A Renewables Portfolio Standard," American Wind Energy Association, 122 C. St. NW, Washington, DC, April, 1995. Also see Rader and Norgaard (1996).

23 The California Public Utilities Commission adopted a portfolio standard in its 1995 decision with certain requirements to diversify within the portfolio. This approach was overturned by the California legislature, which sent the matter to a working group for additional policy development. See Wiser, Pickle and Goldman (1996).

24 The RPS concept is also challenging to implement in regional power markets if state authorities want to require that renewable producers locate within their states in order to take advantage of the localized environmental benefits of renewable production. National certification does not help this problem much, and the legal issues in implementing a localized system may be significant. See Hempling and Rader (1996).

25 Systems benefits charges are described in Chapters 11 and 13.

26 See *supra,* note 1.

27 Green Pricing Newsletter, The Regulatory Assistance Project, Gardiner, ME, No. 2, May 1995. On this issue, also see the Office of Technology Assessment (1994). Recognizing these limits, one environmental organization has estimated that green pricing well likely lead to only 2MW of renewable capacity—much smaller than aggregate traditional utility program goals. "NRDC Analysis Finds Green Pricing Programs total less than 2MW." Utility Environment Report, April 12, 1996, p. 14.

28 Nakarado (1994). Also see Byrnes, Rahim Zadeh, de Alba, and Baugh (1996).

29 Fox-Penner (1996) and "NRDC Analysis Finds Green Pricing Programs Total Less Than 2MW". Utility Environment Report, April 12, 1996, p. 14. Also see J. Udall, "Power to the People" *Sierra,* January 1997, p. 26.

30 Asking consumers to voluntarily pay for public goods creates free rider problems—consumers who do pay become jealous of consumers who don't, and yet reap almost the same level of benefits as those who do. In addition, the majority who do not pay obviously contribute nothing to the provision of the good. These two effects together reduce the level of public provision below the theoretical optimum in ordinary cases.

Chapter 15

Network Reliability

The technical nature of power grids makes their reliability a textbook public good. One user's enjoyment of reliable service does not diminish another's, and reliability is supplied jointly to all power users. Changes in retail markets will make it possible to choose individualized *lower* levels of reliability than the grid provides, and expensive back-up features will protect many against outages. For the vast majority of users, however, the overall level of reliability is something everyone will experience in common for the foreseeable future.

As noted in Chapter 2, power systems are uniquely vulnerable to widespread outages that are caused by small, local failures. A single outage can easily reach the level of a regional or national emergency. The Federal Power Act provides the Secretary of Energy with broad emergency authority during such an emergency.[1] The northeast power blackout of 1965 affected 30 million Americans and caused an estimated $400 million in damage; the July, 1996 western U.S. outage left two million people in 11 states without electricity for as long as six hours.[2] Both of these events prompted the President to order significant immediate investigations and remedial measures.[3]

Fortunately, large-scale outages are rare. Figure 15-1 shows the trend in "major disturbances" reported to the National Electric Reliability Council by utilities around the nation. The reporting definition of major disturbance has changed periodically, but not enough to affect the conclusion that there has been neither a positive nor a negative trend in large outages to date (there is also no trend in the number of customers affected per outage).[4] DOE rules call for a report: (1) if a 100 MW or more firm load is shed "to maintain the continuity of the bulk power system;" (2) for systems of 3,000 MW peak or more, any incident in which 10% of system load or more is shed for 15 minutes or more; and (3) for systems less than 3,000 MW, any shedding of 50% or more of system load for 15 minutes or more; or (4) any interruption of 50,000 customers or more for three hours or more. Reporting must occur "as soon as practicable," but not more than three days following the incident. (10 CFR 205.) DOE rules also require the reporting of system vulnerabilities, fuel supply shortages, and other conditions that might lead to severe outages.[5]

The recent large outage in the western United States—at a time when restructuring is well underway in California—raised inevitable questions about the link between restructuring and reliability. One magazine's headline asked, "Who's Watching the Power Grid?" and the U.S. Department of Energy was deluged by reporters asking whether restructuring would cause more blackouts.[6] Long-time engineers and electrical workers' union members, fearful that cost-cutting is already taking a toll on reliability, have emphasized this point in numerous public statements.[7] After an extensive review of the subject, the

Figure 15-1. Major Systems Disturbances

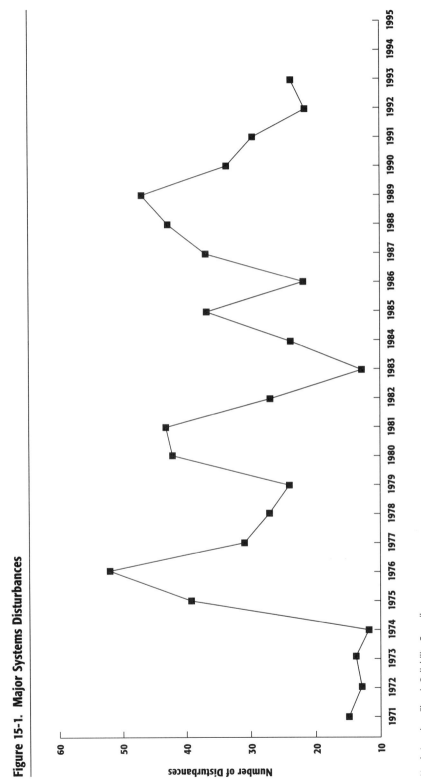

North American Electric Reliability Council
Review of Selected Electric System Disturbances in North America

ordinarily-staid journal, *Electrical World,* bluntly concluded that "no utility engineer at any level claims that reliability will not fall under deregulation." Meanwhile, state and federal policymakers have made it clear that restructuring must not adversely affect reliability.[8]

MEASURING AND MAINTAINING RELIABILITY

Most consumers have an intuitive idea of how to measure reliability—the frequency and duration of outages experienced in their power service. As Table 15.1 shows, these intuitive measures closely match the typical measures used to gauge reliability.

It is important to understand the relationship between *actual historic* reliability statistics and forward-looking reliability *planning* standards. For example, as pointed out in Chapter 3, state regulators often require utilities to plan for a "loss of load probability" of one day in ten years. This standard refers to the ability to meet load under normal weather and operating conditions—not the abnormal conditions that generate most outages. It is really a requirement to have enough capacity on hand to deal with random but not abnormal plant and line failures.[9]

Reliability is the composite result of operation in the short-term and long-term time frames (Chapters 2 and 3, respectively). Reliability first requires adequacy, i.e., planning and building enough aggregate capacity to balance total demand and supply at prevailing prices at every moment. Second, reliability requires centralized, system-wide control so that the entire system can be re-balanced in response to a sudden outage.

The traditional industry ensured G and T capacity adequacy by requiring franchise utilities to maintain enough capacity to serve all demand. Short-term reliability was achieved by voluntary agreements to use regional reliability council standards and protocols and other security-enhancing procedures—all backed by state and federal regulatory authority.

Table 15. 1. Retrospective Outage Reporting

Measure and Definition	Notes
– Customers Losing Service	• Most commonly reported statistic in newspapers
– MW of Load Lost (Shed, Disconnected, Interrupted)	• MW of load shed not a measure of adequacy or load unserved
– Duration of Outage	• Usually refers to period from beginning of outage to complete restoration of service to all customers
– "Customer Minutes": number of customers losing service times average duration of outage	• Measure of severity used implicitly by U.S. Department of Energy for reporting purposes
– "Load-Minutes": Number of MW of load disconnected times average length of time each MW was off system	• Measure of severity used implicitly by U.S. Department of Energy for reporting purposes

Partly due to the self-governance of the reliability council, there has been very little regulation of reliability per se at the state or federal level. The Federal Power Act's Emergency authority is extensive, but it takes hold only after an outage or shortage has occurred.[10] States have enforced criteria for adequacy, but until recently, most never carefully measured the sort of data shown in Figure 15-1 for distribution systems. Sparked by quality-of-service concerns associated with cost-cutting and price cap regulation, some states have started measuring and enforcing reliability performance.

RESTRUCTURING'S IMPACT ON RELIABILITY

Opinions on the public interest impacts of restructuring are routinely divided, but nowhere is the division as pronounced as on the subject of power system reliability. Many economists and system planners believe that restructuring will produce higher levels of reliability at lower total costs. Law professor Bernard Black nicely summarizes this point of view:[11]

> Competition seems more likely to improve reliability. First, in a competitive environment, more customers will choose interruptible or capped service in return for lower rates, which will improve reliability for customers who want non-interruptible service. Second, power plant outages should be less common in a competitive market than today: a regulated monopolist continues to earn a return on out-of-service assets; an unregulated producer doesn't. A regulated monopolist keeps its customers despite bad service; an unregulated producer does not.
>
> Opponents also worry that if the utility loses its monopoly it won't be able to maintain reserve capacity to meet demand spikes. But peaking service—service available when other power sources are already being used—is a market service like any other. Suppliers will provide peaking service if there is demand for it, just as some firms today specialize in emergency backup computer service. For example, today many businesses, hospitals, etc. maintain almost-never-used emergency back-up generators. One can readily imagine an entrepreneur using this largely wasted resource to sell peaking or backup service to firms that now can't afford it or to a distribution utility.

Those who disagree with Professor Black point to a number of phenomena that are as yet anecdotal and longer term in nature. The first is that regional planning and coordination, which requires cooperation, is suffering at the hands of competition. This concern, which affects system economics in normal times as much as it affects the ability to cope with disturbances, was discussed in Chapter 8.

One aspect of regional planning that has raised concern is the traditional notion of adequate total generation. In the context of retail choice, adequacy becomes an economic rather than a physical concept. Unregulated generators sign contracts that guarantee power to their customers; one would expect them to want to have capacity equal to their sales. However, because the bulk power network must have a residual spot market for balancing purposes anyway, generators need not necessarily build plants equal to their demands. Instead, they may rely on purchases from "the system" to make up deficiencies.

This sort of behavior is entirely rational. If many generators do it, the price of spot power will quickly get bid up, and sellers will find that at some point it is cheaper to build another plant for their own sales. Hence, economic theory suggests that bulk power prices should play precisely the role they play in other markets: sending signals to sellers to expand or contract supply to match demand. In theory, this method of finding the optimal level of supply is much more effective than forecasting future demand and building capacity to match forecasted quantity and price.

Reliability worries creep into this picture only if there is a belief that the bulk power marketplace will not work the way textbook markets work. In unusual hot spells, for example, power needs for air conditioning are very large. If sellers are capacity-short, the spot price will skyrocket. Those who pay real-time prices will conserve all they can, while those who need air conditioning for health reasons will want power no matter how much it costs.

If all goes well, those who need power will get it at costs they find reasonable. But if it is profitable for sellers to remain chronically short of power, and the political process does not find frequent price spikes acceptable, then guidelines for adequacy, something like state regulators and reliability councils have today, may be necessary.[12] Economists Frank Felder and Adam Jaffee argue, based on a recent analysis of power system economic fundamentals, that adequacy rules will be necessary to achieve optimal reliability.

The second reliability concern is that the present owners of generation and transmission are failing to invest adequately in their systems because they question whether they can earn a reasonable return on their investment. Because there is some possibility that restructuring will require that any given utility sell off its transmission or generation assets, and such a forced sale may not compensate the seller for recent investments, some utilities are reportedly concluding that some investments are simply too risky to undertake.

It is important to recognize that, even if this phenomenon is occurring, this may be largely a transitional problem. This issue becomes increasingly important if restructuring occurs very slowly, and prospective generation and transmission owners continue to defer investment year after year. It will also occur if transmission fails due to economic or political constraints. In addition, these reported impacts of reduced investment are occurring at the same time that the FERC has ordered nationwide open access and wholesale power trades are skyrocketing. According to many accounts, this is putting unprecedented strain on a transmission network that generally was not configured for large, long-distance power transfers.[13]

The final reliability concern is the short-run counterpart to the transitional unwillingness to invest. Generators may find that it is less costly to buy power from the spot market occasionally than it is to invest in maintaining generating plants in good condition. Similarly, if adherence to reliability council procedures increases costs, and these procedures are thus far voluntary, anecdotal evidence suggests that some utilities are starting to "cut corners" in ways that may be evident only in the wake of a large outage. As with the deregulated adequacy decision, if markets are working, this may improve the allocation of resources. For example, it could turn out to be cheaper overall to have generators that break more often, with a lot of backup, rather than well-maintained generators and less backup.

RELIABILITY ENHANCEMENT AMIDST RESTRUCTURING

A variety of policies are being developed to protect reliability as the industry restructures. The National Electric Reliability Council (NERC) and the regional reliability councils of which it is comprised are in the midst of a "re-invention" that will likely expand its membership from traditional utilities only to all major segments of the new power marketplace. There is much discussion about the need to give the expanded reliability councils (or some other body) the authority to enforce reliability guidelines.[14] These and several other ideas were recommended by the U.S. Department of Energy following its July, 1996 review of several regional blackouts. DOE's national recommendations are shown in Appendix 15B.

State restructuring proceedings are also paying close attention to reliability. Several states have instituted procedures that require regulated utilities to report actual detailed reliability performance each year, allowing for much greater scrutiny.[15] In state restructuring proceedings, responsibility for maintaining adequacy and reliability is tending to remain with state regulators and the operator of the transmission system (in addition to reliance on regional reliability council procedures). A proposal to restructure New York State's utility system maintained substantial public service commission (PSC) authority over reliability:[16]

> At a minimum, current reliability criteria will remain in place. *For purposes of maintaining reliability, the PSC will have jurisdiction over all generation and transmission facilities* in the State, including those owned by PASNY and IPPs (over which the PSC currently has no jurisdiction). Requirements include:
>
> • PSC approval of the independent system operator; Required training for all employees involved with safety and reliability;
>
> • Prior to the implementation of Customer Choice, the PSC must certify that reliability of the electric system will not be impaired;
>
> • The PSC will have authority to restrict electricity imports to maintain reliability; and
>
> • The State may impose additional conditions on any contract, if necessary, to preserve reliability.

These efforts are laudable but unfinished. If they are successful in providing sound incentives for optimal investment and maintenance, the marketplace will improve reliability. If they miss the mark, the consuming public will experience reduced reliability and additional measures may be necessary.

Appendix 15A

Distinguishing Between Reliability Planning and Experienced Reliability Measures

Building enough generating capacity to serve peak load, or adequacy, is a precondition to the ability to keep the system functioning in the event of sudden failures, or reliability. Adequacy is a precondition for reliability, but not the reverse.

Industry participants often hear it said that the utility industry's main reliability criteria is that outages should not exceed one day in ten years. It turns out that this is an *adequacy* criterion used for planning that is often mistaken for a *reliability* criterion. As discussed in Chapter 3, planning consists of forecasting future peak demands and then iteratively planning generation, transmission, and demand-side management changes that will serve peak load. It is here that planners estimate the probability that load will increase above their forecast, and attempt to build capacity that meets all load, with accidental deficiencies projected to occur in no more than 24 hours over the next ten years (87,648 hours).

This planning exercise assumes that all plants and lines in the plan are working normally. Conversely, reliability problems occur when one or more system components fail, triggering outages in the bulk power or (more commonly) the distribution system.

Reliability is factored into planning in a very different way, namely by examining possible modes of failure and protecting against them. Systems are designed to remain operable under an agreed-upon variety of contingencies that could occur anytime, and in any combination.

Because each system's topology is different and many contingencies are triggered by very unpredictable events (including disturbances in other regions), planners have not focused on translating system reliability measurements into frequency-type measures. Planners can tell which combinations of outages they have tested to protect against— and the list is usually extensive—but they cannot predict how often conditions will violate the many cases they have examined and cause an outage.[17] For similar reasons, they cannot predict how widespread the outage will be if it occurs.

All this makes numerical indices of adequacy and reliability easy to misinterpret. A utility's performance on adequacy can be measured by computing the number of days peak load exceeded available capacity. (Adjustments would be have to be made for unpredictable outages during these peak periods because they do not count against adequacy.)

Adjusting the historic record to eliminate any possible outages due to inadequacy produces a reliability record. This record could be compared to the one-day-in-ten years standard— and this is often done—but this is essentially comparing apples and oranges. The system was not designed to provide any specific frequency of outages due to reliability. It was designed to survive a set of contingencies whose probabilities are not numerically known.

NUMERICAL INDICES OF ADEQUACY AND RELIABILITY

Table 15A.1 shows some of the many indices that are used by planners to design systems for adequacy. The simplest forward-looking measure of adequacy is the very familiar reserve margin, or the percentage by which available capacity exceeds expected peak load.[18] The next most common measure is the e expected number of hours with inadequate power, called Loss of Load Expectation (LOLE).

A number of additional measures exist for measuring adequacy prospectively. The severity of the outage could be measured by the number of customers affected, the number of megawatts interrupted, and/or by the length of the outage. Multiplying customers or load interrupted by the duration of the outage produces a composite picture of severity. (The units of measuring outages are megawatt-minutes or megawatt-customers of interruption.) Table 15A.1 contains the definitions of a number of adequacy indicia.

The nature of reliability is such that it is difficult to create indices that measure reliability prospectively even if reliability performance can be looked back on. Table 15A.1 contains a number of common measures of the frequency, size, and severity of outages. Many utilities and regulatory agencies collect data that can be translated into these measures, and they are broadly useful for taking corrective action if there is a perception among the utility or its regulators or customers that reliability is too low. However, there is little formal connection between these measures and the actions of system planners and operators—not because the record does not matter, but because there are no deterministic models that allow data to feed back into design or operational changes.

An analogous situation occurs in airline safety. Airline crashes are extremely infrequent, catastrophic, and entirely unpredictable events. Yet airline designers continue to design planes to meet extremely stringent safety standards. The key to safety is ensuring that any set of possible contingencies (mechanical failures or operator error) can be controlled by redundant systems or procedures.

When a crash occurs, a long investigation may uncover an error in an assumption about how a part of the system works. This may trigger a change in a part or a procedure. Generally, however, it does not change the fundamental way planes are designed or operated.

Because crashes are idiosyncratic and unpredictable, looking back on the historic record may help contribute to technical modifications in design or operating assumptions. However, airline safety experts have for years argued that historic crash statistics are so sparse that they are not good predictors of future crashes. This is precisely the same problem faced by those who would use reliability data for bulk power evaluations. When events are infrequent and linked to many exogenous forces, including other producers, it is difficult to attribute responsibility to one system on the basis of extremely infrequent, linked, and unpredictable events.

Table 15A.1. Prospective Generation Adequacy Measures

Measure	Definition	Notes
Reserve Margin	Surplus available generating capacity on system at time of peak as a percent of peak	• Most common simple measure, but extremely crude
Loss of Load Expectation (LOLE)	Expected percentage of time over planning horizon that peak load will exceed generating capacity	• Most common measure used in advanced planning studies • Similar to LOLP • Does not reflect severity or duration of outages
Loss of Load Probability (LOLP)	Probability that load will exceed installed capacity	• Similar to (and often confused with) LOLE • Does not reflect severity or duration of outages
Expected Frequency of Outages	Number of outages per unit of time	• Easily understood by most consumers, though confused with historic measurements of reliability • More data-intensive than LOLE
Expected Duration of Outages	Cumulative minutes of outages of a period of time	• More data-intensive than LOLE
Expected Load Lost or Unserved by Those Losing Service During Period of Outage	Amount of energy that would have been used	• Sometimes difficult to conceptualize and explain
Expected Unserved Energy	Load between difference in maximum load during outage and load that would have occurred if unconstrained by inadequate capacity	• Difficult to conceptualize and understand significance • Difficult to compute with accuracy

Appendix 15B

U.S Department of Energy Recommendations to Protect and Improve Reliability As The Utility Industry Undergoes Restructuring[19]

1. The Department of Energy will participate in the review of the causes, responses, and preventability of all future major power disturbances and report upon the results of such review, including any recommendations, in accordance with its authorities and the Secretary's discretion, and in cooperation with the (NERC) and the reliability councils.

2. The Department will review existing federal authority concerning reliability and, if appropriate, propose to amend that authority in the context of comprehensive electricity restructuring legislation.

3. NERC and the Regions should review the recommendations of the NERC's 1991 *Survey of the Voltage Collapse Phenomenon* report and other reports to determine what additional follow-up is needed.

   The Department requests NERC to establish a tracking system in cooperation with the regional reliability councils to track the state of implementation of its recommendations regarding voltage collapse and any reasons for delay in their implementation.

   It also requests NERC to periodically publish reports on the status of implementation of all of its recommendations, as part of a broader security issues reporting system.

4. NERC will continue to define and enforce reliability standards to ensure that a continuous supply/demand balance is maintained and that transmission systems are operated within security limits. It will address and periodically report to the government and the public on:

   - Accountability for reliability. The Department requests NERC to describe how it will ensure that accountability does not become muddled or ambiguous through the changes occurring in the electric industry.

   - Compliance with reliability standards. The Department requests NERC to report to it on how completely its planning and operating guidance is carried out as well as adopted by regional reliability councils. The Department suggests that NERC review its tracking system with regional reliability councils to ensure that it captures the actual state of member-entity implementation of this operating and planning guidance, as well as its adoption by regional reliability councils.

- NERC's ability to keep pace with the changes in the industry. The member entities of the regional councils should provide the councils and NERC with sufficient resources, in-kind support, and deference to enable them to carry out their reliability mission now and during the transition to a more competitive future of the industry. These resources should be sufficient to meet current reliability requirements, anticipate future requirements, and ensure their accomplishment.

NORTH AMERICAN ELECTRIC RELIABILITY COUNCIL

5. NERC should proceed upon its four *Strategic Initiatives for NERC* to ensure that reliability will be maintained. Among the most important of these are the implementation of formal, coordinated regional and interregional security processes (the need for which has been clearly shown by the WSSC outage of July 2 and 3, 1996): further development of ancillary operations services which promote both efficiency and reliability and timely implementation of the NERC Security Process Task Force.

6. The Department suggests that NERC modify its periodic reporting on the reliability of the North American electricity supply and delivery system to include more information on security. The annual and seasonal assessments are fundamentally reports on adequacy. Adequacy is an important component of reliability; however, by the NERC's and industry's own statement of reliability, security issues are increasingly important in assessing the actual state of reliability of the North American electricity supply and delivery system. The incidents of voltage collapse reported by FERC in 1991 and the WSCC outage of July 2 and 3, 1996 underscore the increasing importance of security issues.

7. The Department will work with others to develop and focus the pursuit of a suitable research agenda of policy analysis issues and technological challenges affecting the reliability of the U.S. electricity supply and delivery system. It will create a task force for this purpose in cooperation with other federal agencies, NERC, EPRI, the industry, and others. The agenda will address relevant technology development and and analysis tools, control schemes, operating practices, and data requirements for ensuring reliability under changing industry structure and regulation. It will sponsor appropriate forums to identify technical and policy analysis issues, define research needs, and disseminate results.

8. NERC will continue to define and enforce reliability standards to ensure that a continuous supply/demand balance is maintained and that transmission systems are operated within security limits. It will address and periodically report to the government and the public on:

   - Willingness of market participants to share the information required for reliability assessments.

9. NERC and Regional Councils should initiate an immediate review of the information management issues brought to the forefront by the July 2, 1996 WSCC outage. Some of the areas that will be addressed include:

- Guidelines for reporting disturbances to NERC and DOE;

- Disseminating information throughout the industry;

- Guidelines for invoking NERC Special Events Analysis Teams; and

- Contact with the media. NERC and the regional councils need to develop systems for reporting to the media and the public immediately after a significant event, and thereafter, as understanding develops of its cause, the status of response efforts, and the significance of the event.

10. The Department urges all regional reliability councils to cooperate with NERC in providing perspectives and experiences to help shape appropriate policies regarding under-voltage automatic load shedding, improve understanding of the causes of voltage collapse, contribute to the development of improvements in operator sensors better to predict when such collapse may occur, achieve improvements in modeling, and identify and share the best available operator training package.

Chapter 15

NOTES

1 Section 202c of the Federal Power Act 16 U.S.C. § 791 a.

2 Prevention of Power Failures. Report to the President by the Federal Power Commission, July 1967, Vol. I p. 8. Damage figure is in 1995 dollars, inflated from 1965 dollars. The Electric Power outages in the Western United States, July 2-3, 1996. Report for the President, U.S. Department of Energy, August, 1996. p. I.

3 Ibid.

4 DOE rules call for a report if (a) 100 MW or more firm load is shed "to maintain the continuity of the bulk power system;" (b) for systems of 3,000 MW peak or more, any incident in which 10% of system load or more is shed for 15 minutes or more; and (c) for systems less than 3,000 MW, any shedding of 50% or more of system load for 15 minutes or more; or (d) any interruption of 50,000 customers or more for three hours or more. Reporting must be done "as soon as practicable," but not more than three days following the incident. 10 CFR 205. DOE rules also require the reporting of system vulnerabilities, fuel supply shortages, and other conditions that might lead to severe outages.

5 "Power System Emergency Reporting Procedures," Department of Energy: Office of Emergency Planning and Operations, August 1992, Form OE-417R, OMB No. 1901-0288.

6 *Business Week,* June 17, 1996, "Who's Watching the Power Grid," by Peter Coy.

7 "Is Deregulation Putting America In The Dark?" *Electrical World,* October, 1996, p. 21. Also see "Will Deregulation Short-Circuit America's Electric Power Supply?" International Brotherhood of Electrical Workers, Washington, D.C., August 1996.

8 "The Department [of Energy] is committed to ensure [sic] that the nation's electric power supply remains reliable now and in the future." U.S. Dept. Of Energy (1996) p. iv. "We will not allow any reforms to compromise either safety or reliability." California Public Utilities Commission Order Instituting Rulemaking and Investigation, R.94-04-031, April 20, 1994.

9 Appendix 15A describes the relationship between planning and historic reliability measurement in greater detail.

10 Section 202(a) of the Federal Power Act of 1935 [16 U.S.C. § 824(a)] authorizes "certain actions as are necessary to assure an abundant supply of energy throughout the country," and Section 311 of the Act authorizes the federal government to collect information necessary for this purpose.

11 Black (1994) p. 66

12 This discussion refers to generation adequacy. Transmission adequacy is a key element of the long-run health of competition in bulk power markets, and is discussed in Chapter 8.

13 David Foster, "West's Power Is Stretched Too Thin, Groups Warn". *USA Today,* Aug. 15, 1996. P. 14a. Many studies of the economics of interregional power transfers conducted in the 1960s and 1970s reviewed in Chapter 5 rejected the idea of building many transmission lines between regions.

14 See the U.S. Dept. Of Energy (1996) and *Electrical World* (op cit).

15 See, for example, "Order Adopting Standards on Reliability And Quality of Electric Service, "Case 90-E-1119, New York Public Service Commission, June 26, 1991, and Section 410.460 (b) of the Electric Service Reliability Policy, 83 Ill. Admin. Code Feb 8, 1995.

16 "Competition Plus". Sheldon Silver, Michael J. Bragman, Paul D. Tonko, New York State Legislature, March 1996. p. 8

17 Computer programs of this nature are under development in the utility industry, but are not yet in widespread use.

18 If capacity always exceeded peak, then no adequacy-related outage would ever occur. In practice, the peak load used to compute reserve margin isn't the highest possible peak under any circumstances, it is the highest peak in all but the highest 0.1 days per year. In this way, reserve margin is crudely linked to the slightly more sophisticated one day in ten years LOLE standard. Typical target reserve margins for large U.S. systems are on the order of 15%, though this may drop as the industry restructures.

19 U.S. Department of Energy (1996) pp. 25–33. Recommendations applying to specific reliability councils have been omitted.

Chapter 16

Stranded Costs

Every major shift in government policies imposes costs as well as benefits, including the costs of adjusting to or accommodating new policies. "Stranded costs" refers to two kinds of costs that utilities may not be able to recover in the prices they charge as industry competition expands. Stranded costs are the largest and most readily identified transition cost in electric restructuring.

One type of stranded cost could be an ongoing annual expense that regulated utilities, (e.g., distributors) will continue to be required to incur, but unregulated sellers will not. These expenses include payments for many of the "stranded benefits" discussed in the past five chapters, such as programs for energy efficiency and universal service. As noted earlier, if these costs are allocated only to incumbents' utilities, their costs will rise above those of newer, unregulated sellers and sales will shift to the new suppliers who have lower prices because they do not bear these expenses.

The second type of stranded costs refers to long-lived existing utility investments that may not be recoverable at their book value as the industry restructures.[1] As noted in Chapters 7 and 10, competition in the power industry centers very much on price. Everywhere deregulation is entering the industry, the price of power is facing downward pressure. Large, rapid price reductions in power rates will make it difficult for some of today's utilities to earn enough to cover the costs of investments that were made under cost-of-service regulation. The types of investments that may become stranded fall into three major categories:

1. **Generating plants that are more expensive than today's power plants.** In Chapter 8, it was noted that competition has a tendency to drive prices down to the costs of the lowest-price units available today, or even lower. This particularly impacts existing utilities that have, for one reason or another, built power plants that produce power at costs that are higher than today's wholesale power prices.[2] Environmental cleanup liabilities associated with nuclear and fossil plants may also be unrecoverable, absent policy changes.[3]

2. **Long-term fuel or purchased-power contracts that are more expensive than today's prices.** Historically, public and private utilities alike have responded to their historic adequacy obligation by making long-term, firm investments in, or contracts for, power plants and fuel. Utilities have an obligation to pay for these contracts even though many of them now specify prices higher than those available on comparable terms to buyers today. Contracts mandated by government laws and policies form an important subset of these stranded costs.

3. **"Regulatory Assets."** Regulatory assets are an assortment of regulator-approved "extended payment plans" for certain large expenses. A regulatory asset allows a utility to count on its balance sheet a "promissory note" effectively promising that the utility's future revenues will be adequate for amortization of a specific liability. This technique has been used to "phase-in" gradual rate increases in place of rapid, large ones, much like the refinancing of a loan to reduce payments but greatly extend the payment period.

If deregulation occurs during the middle of the amortization period, utilities may not be able to charge prices high enough, or earning revenues large enough, to continue amortization. Financial analyst William Abrams argues that the accounting rules that allow utilities to count regulatory assets on their books become "problematic" under sharp revenue decreases, possibly triggering large write-offs.[4]

Each of these types of stranded investment has become the subject of significant and contentious debate. First, there is some disagreement as to the size of the investments that competition puts at risk. Second, there is much disagreement as to whether these costs should be provided for as part of restructuring or whether large losses should be imposed on investors and taxpayers by simply ignoring them. Finally, for those who argue that they should not be ignored, several approaches for treating stranded costs are under discussion.

STRANDED COST ESTIMATES

Table 16.1 shows a number of estimates of stranded costs which have been prepared during the last several years. These estimates vary partly because different estimates do not include all items in the three categories just mentioned. Of equal importance, these figures are comparisons of projected future market prices and the future cost of operating a utility, both of which are very hard to predict. The estimated portion of any existing asset that is unrecoverable is sensitive to assumptions, including the future price of fuels, interest rates, the reliability and efficiency of units as they grow older, air pollution rules, state and federal policies allowing competition and setting the prices for transmission, ancillary services, and so on; and many other variables. A variety of methods and approaches can be used to calculate stranded costs, each with its own particularly important assumptions and simplifications.[5]

The key conclusion that emerges from Table 16.1 is that, under a wide variety of scenarios and methods, stranded costs are a first-order public policy issue. The Oak Ridge National Laboratories' lower estimate of $72 billion in stranded costs, for example, is more than three times the estimated stranded costs incurred during the gas industry's recent restructuring.[6] Were all these costs to be paid for out of utility stockholders' earnings, the industry would lose a total of about 38% of its equity and a number of utility companies would likely face bankruptcy.

Among potentially stranded nuclear generating plants, one group—the nation's 109 existing nuclear plants—presents an especially important challenge. These plants represent an investment of about $120 billion and supplied 22% of the nation's electricity in 1994.[7] Many of these plants have average costs higher than today's wholesale price. Indeed, without counting *any* recovery of capital costs, 45% of the nuclear plants in the U.S. have operating costs above 3.0 cents/kWh.[8]

TABLE 16.1. Estimates of Electric Restructuring Stranded Costs (Billions of Dollars)

Estimate Made By	**Eric Hirst and Les Baxter, Oak Ridge National Laboratory**	**Moody's Investor Service**	**Resource Data International**	**Texas Public Utility Commission**
Geographic Area	United States	United States	United States	Texas
Types of Stranded Costs Included in Estimate	All elements, implicitly	All elements	All elements	All elements
Estimated Loss ($Billion)	$72–$102	$135	$73–$163	$11–$15

Source: Hirst and Baxter (1995), Moody's Investor Service (1995); *Electricity Journal,* March 1995; Matlock (1995); Feiler and Seiple (1995).

More importantly, neither the owners of these plants nor the American people can simply shut these plants down and walk away. The cost of closing and safely disposing of these plants, or *decommissioning,* is estimated to be between $300 million and a billion dollars per plant.[9]

For nuclear plants that are expensive to operate, this is forcing utilities to choose between attempting to reduce operating costs, potentially impacting the safe operation of these plants, and incurring massive cleanup costs. Although nuclear plant licenses require strict adherence to safety, and the safety record of most U.S. nuclear plants is good, safety questions are starting to arise. The *Wall Street Journal* recently reported that "Some nuclear critics have suggested that increased competition...may be leading to corner-cutting on safety. The U.S. Nuclear Regulatory Commission's top administrator for the Northeast [region of the U.S.] has said...[one utility] has appeared to emphasize profits at the expense of safety since the late 1980s."[10]

Meanwhile, U.S. Energy Information Administration analyst James Hewlett estimates that 15 percent of all nuclear utilities have unfunded decommissioning liabilities equal to more than 40 percent of their total equity; another 35 percent have liabilities greater than one fifth of their equity.[11] Many observers fear that the U.S. government ultimately may have to step in and help fund shutdown and clean up efforts for utilities unable to meet their obligations. (See Box 16-1).

Another unique category of stranded costs are those associated with specific past governmental policies. Utilities' obligation to purchase all power offered from cogenerators and small power producers (under contracts explicitly approved by regulators) represent one such obligation. Other laws passed in the 1970's placed limits on utility plant and fuel choice, unwittingly adding to today's stranded costs.[12]

BOX 16-1. Nuclear Power Plant Decommissioning As A Potential Stranded Cost

The 109 operating commercial nuclear plants in the United States hold licenses that allow them to operate for 40 years, though there is increasing doubt that most plants will run for the full length of their licenses.[1] At the end of their useful lives, nuclear plants must be *decommissioned* according to procedures set by the U.S. Nuclear Regulatory Commission (NRC).[2] NRC rules require the disassembly and disposal of the highly-radioactive inner workings of a plant, usually by shipping the plant piece by piece to a remote, guarded storage area.[3] Related NRC rules require plant licensees to contribute money each year to a fund to be used for decommissioning at the end of the plant's life.[4]

The amount of money to be contributed to each plant's decommissioning fund over the life of the plant–presently $117 million in 1989 dollars–was calculated based on an early estimate of the cost of decommissioning. However, when the NRC set this amount, no commercial nuclear plant had yet undergone decommissioning. During the past few years, the actual decommissioning of several U.S. commercial plants has begun, and the costs are turning out to be far higher than original projections. A. Clegg Crawford, an executive from the Public Service Company of Colorado (PSC) recently summarized PSC's and other utilities' experience with decommissioning in a speech reported in *Nuclear News* according to the article:[5]

> Crawford went on to discuss some of the pitfalls of decomissioning and decontamination of a nuclear power plant such as PSC's Fort St. Vrain: Actual costs are likely to exceed initial projections by a factor of three or more; costs for disposal of low-level radioactive waste (LLW) escalated at 13.3 percent per year during the period 1980–95, and that trend will probably continue; and postponing dismantlement may mean the loss of the skilled, knowledgeable personnel familiar with the plant needed to do the job.

> The total cost of decommissioning Fort St. Vrain was $350 million, according to Crawford, but only $189 million of that was for decommissioning; the rest was for storing and handling the spent fuel, which is not included in the typical decommissioning cost estimate. PSC built an independent spent fuel storage installation.

> The Shoreham nuclear plant cost $700 million to decommission; the estimate was $200 million. The cost estimate to decommission the Rancho Seco nuclear plant was $250 million; the actual cost for dealing with the spent fuel alone was $500 million.

Decommissioning sums of this magnitude vastly exceed the amounts regulators and utilities have been using as the basis of collecting funds for decommissioning. For example, regulatory staffer John Rohrbach reports that the Yankee Rowe plant's costs were more than three times the size of the trust it accumulated over its 30-year operating life.[6]

BOX 16-1. Nuclear Power Plant Decommissioning As A Potential Stranded Cost
(continued)

The potential for underfunded decommissioning trusts is exacerbated by three related factors. First, decommissioning fund collections are set by most state regulators based on expected forty-year plant lives.[7] While it is possible that some plants will operate this long or longer, no plant has yet reached this age; of the seven U.S. nuclear plants closed to date, the oldest was Yankee Rowe at 30.[8] As EIA analyst James Hewlett notes, premature shutdowns have a particularly harmful effect on accumulated decommissioning funds because this robs the funds of their final few years, during which earnings on the accumulated balance are largest.

The second factor helping to understate nuclear decommissioning liabilities is the increasing cost of interim and final waste storage. Technology consultants SAIC estimate that half of the decommissioning trusts established by utilities do not include interim waste storage costs.[9] According to Hewlett, these costs escalated from $25 per cubic foot in 1970 to $350 per cubic foot today; Crawford expects them to rise to $600–$1000 per cubic foot in the future.[10] Rohrbach (1995) warns that, even if the U.S. government completes the Yucca Mountain waste storage facility by the 1998 target date, utilities with spent fuel or other wastes may have to wait many years for the site to accommodate their shipments.

The final exacerbating factor is a feature of electric utility tax law that, at present, would bar the tax deductability of decommissioning funds. If maintained, this provision may cut deeply into utility earnings and the incentive to set funds aside.[11]

These costs raise the specter of extremely large unfunded liabilities associated with nuclear units long after these plants have ceased to operate. The possibility that utilities facing competition may lack the funds or financial stability to adequately operate or decommission plants has prompted the NRC to propose tighter restrictions on decommissioning funds for utilities. In its first notice of the proposal, the NRC put its concerns in rather blunt language:

> The inability of the licensee to provide funding for decommissioning may adversely affect the protection of public health and safety. Also, a lack of decommissioning funds is a financial risk to taxpayers (i.e., if the licensee cannot pay for decommissioning, taxpayers would ultimately pay the bill.)[12]

These concerns also extend to utilities' ability to pay for interim and final waste storage and other long-lived elements in the commercial nuclear fuel cycle.[13]

---

1 For plants that are operating well, there is also great interest among some utilities in obtaining license extensions of five to twenty years. See Energy Information Administration (1995d) p. 51ff for a recent discussion.

2 The rules were promulated in 1988 in 53 FR 24018, June 27, 1988, amending 10 CFR Part 30 *et seq.,* and amended in 1996 in 61 FR 39278.

continued on next page

BOX 16-1. Nuclear Power Plant Decommissioning As A Potential Stranded Cost
(continued)

---

3   According to EIA analyst James Hewlett (1995, p. 11), in other nations, shuttered nuclear plants are first allowed to sit unused and guarded for a 50-year or longer period while they become less radioactive, allowing the owner to defer the cost of cleanup, accumulate a larger cleanup fund, and work on a plant that is less radioactive and therefore less costly and dangerous.

4   See 10 CFR Part 30 *et seq.* (1988).

5   Gregg Taylor and Allen Zeyher, "The Sustainability of Nuclear Power," *Nuclear News*, August, 1996, p. 75.

6   Rohrbach (1995) p. 56.

7   Hewlett (1995), Rohrbach (1995).

8   Energy Information Administration (1995d) p. 57. Several U.S. utilities are exploring extending the licenses and economic lives of their plants, which has the opposite effect. See Bretz (1994) and Hewlett (1994).

9   Hewlett (1995) p. 8.

10 Ibid and Taylor and Zeyher (1996), *op cit.*

11 Elaine Hiruo, "Tax Issues May be Potholes on Road to Electric Competition." *Nucleonics Week*, October 24, 1996.

12 Advance Notice of Proposed Rulemaking concerning the Financial Assurance Requirements for Decommissioning Nuclear Power Reactors, 61 FR 15,427 (1996). As of this writing the NRC has not issued a final order in this proceeding. Also see Studness (1996), who argues that the costs of decommissioning "should be billed to the distribution customers for whom the plants were built."

13 The Financial Accounting Standards Board (FASB) is studying accounting principles used for decommissioning. According to Ferguson (1995), these standards presently allow utilities to defer these costs, placing them at greater risk following restructuring: FASB is considering changes to these rules that will expand the reporting of potential decommissioning costs in utility financial statements (See Ferguson 1996).

FOR AND AGAINST STRANDED COST RECOVERY

A variety of arguments are being made for and against policies that ensure stranded costs are recovered as the industry restructures. These arguments often begin with a discussion about what has become known as the regulatory compact: utilities are granted exclusive franchises in exchange for an obligation to render adequate service at reasonable prices to all within the franchise. The implication of this compact is that government should not change policies that have the effect of denying compensation to investors who have built facilities specifically to carry out their responsibilities under the compact.

A number of industry figures, some with long experience in the regulated industry, have questioned whether such a compact ever existed. Economist Robert Michaels, who went so far as to do a computer search for the words "regulatory compact," found that the words occurred starting only around 1983.[13] This view draws on the school of thought that has long seen utility regulation as an unneeded form of protectionism for the industry.[14]

Most observers and practitioners find the denial *of an implied bargain* between government and regulated firms in conflict with the history of the industry, as well as common sense (see Box 16-2). Instead, careful study of economic and legal interpretations of the responsibilities of regulators and the regulated indicates that an implicit bargain has indeed existed, but it was never meant to be a full guarantee of recovery for all invested sums. The bargain cannot be presumed to have been so strong or inflexible as to prevent any government policy changes, nor should it prevent regulators from providing positive and negative incentives to utilities via allowed profit levels. In other words, the regulatory compact must not insulate utility investors from all the effects of their own managements' actions, good or bad.

The notion that the regulatory bargain must strike a balance between providing adequate investor returns and discipline for utility managers is important, but sometimes overused. The concept of stranded costs is often associated with the idea that utility regulation has made huge, costly mistakes, and stranded costs are simply paying off past regulators' and utilities misdeeds. Some past investments may well have been mistakes, but as a general proposition this view of stranded costs is incorrect.

The modern history of the power industry has been one of successive generations of more efficient and less costly plants. Had we decided to implement competition in the past, we could have called every older generation of power plants since the 1920s stranded because the costs of each vintage were lower than those of the preceding. Instead, regulators viewed each new vintage as a cost-reducing benefit to be shared by all ratepayers, and the cheaper costs of the new plants were blended with the old. The fact that today's power plants are cheaper than yesterday's is, in itself, not a sign of a mistake.

This notwithstanding, the particular circumstances of the 1970s and 1980s make stranded costs look like mistakes and have engendered significant resistance toward their recovery. Economist Robert Michaels notes that most stranded costs come from the 1970s, when some utilities went ahead with construction plans while others reduced or canceled plants—all with regulatory approval. Should there be recovery, Michaels asks, "only for those utilities that refused to admit their mistakes to the market?"[15] Consumer advocates and others have generally objected to recovery on this basis, as when a Massachusetts consumer activist criticized a recovery proposal by saying, "This proposal... foists the costs of the many uneconomic nukes and other plants the consumer groups have consistently opposed on to consumers."[16] Industrial customer groups have similarly recommended explicitly linking stranded cost recovery to "the degree of efficient planning and management" in the past.[17]

Utilities and others have responded that every utility investment is explicitly or implicitly reviewed by regulators, and that regulators have already had their chance to penalize utilities for past mistakes.[18] (Many regulatory commissions have assessed large penalties on utility stockholders due to plant cost overruns). They further note that policies on competition in the industry should be based on the usefulness of competition on its own, not on a one-time reduction in power rates that could have been achieved at many points in the industry's past, whether or not mistakes have been made.

Another anti-recovery view holds that recovering stranded costs is harmful to competition because it will "reward" or "protect" inefficient producers. This argument assumes that utilities with stranded costs will somehow be allowed to charge high prices to captive

Box 16-2. Whither the "Regulatory Compact?"

In dozens of early public utility proceedings, in which public service commissions awarded exclusive franchises for railroad, natural gas, water, telephone, and electricity service, the obligation to provide adequate service was voiced as an express condition of the franchise grant. In 1916, the Maine Public Service Commission wrote,[1]

> Yoked to, and in perfect step with, this monopolistic right of each company is the duty which it owes to the public. If a company owed no duty to those living within its territory, and could act its own pleasure unrestricted and unenlarged by law or other rule of conduct, the result would be that each such company would, within its territory, have all the authority of a feudal lord, demanding and receiving unmerited and arbitrary tribute, yielding in return those things, and only those things, which his capricious pleasure suggested. Such, however, is not in accord with present knowledge of law, equity, or modern enlightened practice. The enjoyment of the monopoly compels the performance of resultant duties. If a utility would occupy, exclusively, a given territory, it must serve adequately, fairly, fully, this same territory. For the very reason that it is the only one in the field, it is under imperative obligation to serve, within reasonable bounds, all whom it finds within its field.

This conclusion and many others[2] indicate that there is a condition that attaches to the grant of a franchise. However, the franchise's entitlement to recover costs made under its obligations—the essence of the stranded cost recovery issue—has a long history of legal interpretation that reached key conclusions in a series of Supreme Court decisions.[3]

In *Federal Power Commission v. Hope Natural Gas*, the court reached what is still considered the standard for determining the reasonableness of utility investor returns. The court said:

> From the investor or company point of view it is important that there be enough revenue not only for operating expenses but also for the capital costs of the business. These include service on the debt and dividends on the stock…By that standard the return to the equity owner should be commensurate with return on investments in other enterprises having corresponding risks. That return, moreover, should be sufficient to assure confidence in the financial integrity of the enterprise, so as to maintain its credit and attract capital.

This case also established that the reasonableness of investor returns was to be judged by the final result, not by the use of any particular method of rate setting or return calculation.

While *Hope* set a clear standard for reasonable investor returns, the courts have also held that the regulatory compact is not a guarantee of a fair rate of return on *all* investments. In a 1912 opinion, Justice Oliver Wendell Holmes described regulators' need to balance fair returns with protections against excessive or improper charges:[4]

On the one side, if the franchise is taken to mean that the most profitable return that could be got, free from competition, is protected by the Fourteenth Amendment, then the power to regulate is null. On the other hand if the power to regulate withdraws the protection of the Amendment altogether, then the property is nought. This is not a matter of economic theory, but of fair interpretation of a *bargain.* Neither extreme can have been meant. A midway between them must be hit. [emphasis added].

Kenneth Rose of The National Regulatory Research Institute recently completed an extensive survey of the legal and economic history of the regulatory compact. Rose concluded that:[5]

A description of the regulatory compact as historically interpreted, may be as follows: the careful balance between compensatory rates and confiscation of utility property that allows a utility an opportunity to earn a reasonable return on their investment in exchange for providing safe and reliable power at reasonable cost to all customers who request service. This opportunity is held in check by the used-and-useful and prudent-investment tests, as well as from competition from government ownership, fuel substitutes, and self-generation. Another important feature of the compact is the continuous rebalancing that takes place to accommodate changing conditions in the industry. Clearly, some kind of rebalancing is needed again. Retail access broadens the scope of competition, but not the potential magnitude of its effect as compared with other forms of competition that the electric utility industry has historically faced. However, a means of recasting the compact that is consistent with past treatment of assets but does not unreasonably impair the development of competition needs to be found.

Economists J. Gregory Sidak and Daniel F. Spulber added to this controversy recently by arguing that deregulatory actions by governments without stranded cost compensation was a government appropriation of private property (commonly known as a "taking") prohibited under the Constitution's Fifth Amendment.[6] These economists argue that utilities denied stranded cost recovery would have grounds for lawsuits against state or federal governments.

---

1   *Churchill v. Winthroup & W. Light and Power Co,* P.U.R. 1916 F. 752, quoted in Nichols (1928), p. 151.

2   See Nichols (1928) for an extensive review.

3   The important cases include: *Bluefield Water Works & Improvement Co. v. Public Service Commission of West Virginia* (26 U.S. 679, 1923), *Hope* (320 U.S. 391, 1944), and *Duquesne Light Co. et al v. Barasch, et al* 488 U.S. 299, 303 (1989).

4   *Cedar Rapids Gas Light Co. v. Cedar Rapids,* 223 U.S. 655, 669 (1912).

5   Rose (1996) p. 69.

6   Sidak and Spulber (1996).

customers as restructuring continues, harming competition and consumers.[19] "If honored, such claims will substantially raise the cost of bypass," Professor Michaels writes. "...Inefficient utilities with high demand charges will be hardest to bypass."[20]

A number of economists and policymakers have pointed out that this characterization of stranded cost recovery is unfair and inaccurate.[21] Appropriate policies to recover stranded costs should be imposed on *all* customers on whose behalf the cost was incurred. If handled in this manner—such as by recovering the cost with a uniform surcharge on transmission prices—stranded cost recovery will neither reward nor penalize *today's* power producers, and the most efficient producers will win in the future.[22] Indeed, the true argument is the reverse: Incumbents with stranded cost liabilities cannot possibly compete fairly with new entrants who have no such liabilities. In the words of former policymakers William Steinmeier and Linda Stuntz,[23]

> "Costs without a customer" are not an inevitable result of the transition to greater competition. They will not occur if regulators structure the transition to assure efficient competition, and carefully balance the relevant public interests. Efficient competition potentially benefiting all customers will not be achieved if some customers are able to avoid paying their fair share of costs approved, ordered or deferred by regulators under traditional regulation. If such cost avoidance is allowed, it would promote unfair competition among power suppliers, uneconomic bypass, distorted pricing and inappropriate cost shifting.

A number of economists and policymakers have carefully examined policies for recovering stranded costs in a "competitively neutral" manner, including William B. Tye and Frank Graves and Professors Joskow, Kahn, and Baumol.[24] In general, these economists find that properly set non-bypassable fees imposed on all customers who use the transmission or distribution system allow forward-looking competition to proceed based on efficiency.[25]

Yet another argument against recovery is sometimes made based on considerations that might be called regulatory gamesmanship rather than economic theory. Agreeing that customers or new suppliers should not escape legitimate obligations, economist Ken Rose of the National Regulatory Research Institute (NRRI) claims that an announced policy that recovers all stranded costs using competitively neutral charges give utilities no incentive to reduce them. Instead, utilities are encouraged to shift costs into the "stranded" category, knowing that these costs will be shared by all suppliers.[26] To guard against this, Rose suggests announcing a policy that makes utilities responsible for paying at least a small percentage of stranded costs out of their own pockets.

Finally, some argue that utilities should not recover stranded costs because they have been compensated for the risk that utility deregulation could cause in the returns paid to utility shareholders already. For example, analysts at the National Association of Regulatory Utility Commissioners reportedly calculated that 72% of the major electric and telecommunications utilities earned returns higher than the average return of unregulated Dow Jones industrial firms between 1972 and 1992.[27] As utility stocks have been less risky than the market, these results imply that utility stockholders have been overcompensated to date. If indeed overcompensating has occurred, it could be viewed as compensation for the possibility of later deregulation.[28]

Several utility industry and financial experts, including A. Lawrence Kolbe and William Tye, have demonstrated that the "already-been-compensated" argument is incorrect (See Box 16-3). Investor returns much higher than those allowed to date would have to have been paid to investors to compensate them for the risks of losing large portions of their invested principal, should deregulation cause this to occur. Moreover, modern financial theory finds it difficult to accept that utilities could earn consistent and large above-market returns.[29]

Without necessarily relying on the regulatory compact agreement, another group of analysts points out that changes in government policies frequently reduce the value of the past investments of individuals or businesses.[30] Critics of this view as it applies to stranded costs have two rejoinders. First, even if actions like this are sometimes unavoidable, they shouldn't be encouraged. Second, there is a long-term benefit for governments that honor past commitments to investors. By giving investors greater confidence in the wide range of government policies that influence investment and economic growth, investors are encouraged to provide capital at lower costs. This point was articulated in the context of the stranded cost debate in the 1996 Economic Report of the President, which stated:[31]

> But recovery should be allowed for legitimate stranded costs. The equity reason for doing so is clear, but there is also a strong efficiency reason for honoring regulators' promises. Credible government is key to a successful market economy, because it is so important for encouraging long-term investments. Although policy reforms inevitably impose losses on some holders of existing assets, good policy tries to mitigate such losses for investments made based on earlier rules, for instance, by grandfathering certain investments when laws and regulations change.

With little doubt, investors will respond to the stranding of costs by raising their assessment of the riskiness of participating in future electric market investments.[32] This will cause the cost of capital to increase throughout the industry, raising the cost of power to every user. In the 1970s, U.S. monetary policy allowed inflation to reduce the value of investments that pay fixed annual sums, such as bonds. Investors responded to this policy shift by demanding higher interest rates for a variety of investments, an "inflation risk premium" that lasted a decade or more and added billions to the U.S. national debt. At the extreme, it is even possible that investors may have grounds to sue the U.S. government for losses caused by deregulation (see Box 16-2).

The final argument presented for stranded cost compensation is tactical and political, and it has been made with great eloquence by writers at both ends of the ideological spectrum. This argument holds that forcing large losses on utility stockholders in connection with restructuring gives utility managements fiduciary responsibility to fight

Box 16-3. Have Utility Stockholders Already Been Compensated for Stranded Costs?*

One of the arguments used against policies designed to provide stranded cost recovery is that utility stockholders have already been compensated for the possibility that their investments would be stranded, i.e., decline significantly in value. Experts in utility economics and finance reject this argument on the grounds that it violates several foundational precepts of capital markets and utility economics.

Under regulation, utility investors ideally earn a return on their investment equal to the returns investors earn on investments that have comparable risk-reward profiles. A bit of simple arithmetic shows that utility investors would have to have been paid much larger returns than they have earned in the past to compensate them for large stranded cost writeoffs.[1]

Suppose an investor has a choice of earning a certain 12% on a non-utility investment or investing in a utility with a stranded cost liability equal to 33% of its shareholder equity. There are two outcomes possible for the investor in the utility, each with a 50% chance of occurring. Under one scenario, the investor loses 33% of her investment and earns nothing on it; under the second, the investor receives back the principal invested plus a return on the investment.

Under these conditions, how high a return would we have to pay to make it equally worthwhile for an investor to invest in a utility or earn a certain 12% in a non-utility investment? Simple arithmetic shows that the required return on investment each period that the utility does not incur the stranded cost writeoff is 58%—far above what utility investors usually earn. Indeed, in a report warning its investors to stay away from utility stocks, one stock brokerage firm summarized investors' predicament as simply: "Heads you win five cents, tails you lose a dollar."[2]

Economists A. Lawrence Kolbe and William B. Tye have used formal economic theory to demonstrate that the cost of capital typically used by regulators cannot compensate utility investors for large writeoffs imposed by the actions of regulators.[3] The proof requires little more than the definition of cost of capital, which is return on a dollar invested *averaged across the probability of all future outcomes*.[4] This is not to deny that investors have accumulated a growing awareness that policies were likely to change, making future returns more volatile.

Forward-looking allowed returns for regulated firms may well require adjustments to reflect higher risk today. However, these adjustments will not necessarily compensate investors for large restructuring-induced writeoffs.

---

1 More complete elaborations on this example may be found in Kolbe, Tye and Meyers (1993), Chapter 2, and Kahn (1994).

2 "A Better Yield: As Electric Utility Outlook Deteriorates, REITS Improve." Paine Webber Investment Policy, New York, Nov. 10, 1996,

3 Kolbe and Tye (1996), Kolbe and Borucki (1996), and Kolbe and Tye (1995).

4 See Brealy and Meyers (1988), Kolbe, Read, and Hall (1984) and Kolbe, Tye, and Meyers (1993)

*This box was prepared by A. Lawrence Kolbe.

against restructuring. After all, it is the job of corporate managers to maximize stockholder returns; if a proposed governmental change will greatly harm stockholders, managements are in some sense obligated to oppose it.

Because stranded costs are a one-time transition cost, it is merely a question of who pays for them. Pragmatists who want to speed the transition to competition thus often opt to accept stranded cost compensation in order to remove an important source of utility opposition. "Society's best interests are not well-served when utilities believe that they can only recover their sunk investment by resisting competition at every turn," former regulator Peter Bradford writes. "Far better that they should believe the reverse."[33] Conservative economic scholar and pundit Irwin Stelzer agrees: "Solve the stranded cost problems by sharing the burden between departing customers (responsible for investments during the time of the social compact); shareholders (forewarned and responsible for post-compact commitments); and all customers (responsible for excessive social costs)."[34]

All these arguments illustrate that stranded cost policies require extremely careful attention to both the short and long-run incentive effects of recovery policies on customers, investors, and genco managements. Mindful of the resources available to them, regulators find they must balance fairness and government credibility preservation against the possibility that recovery will harm the development of effective competition. The former requires recovery, while the latter requires carefully policed, competitively neutral mechanisms.[35]

POLICY DIRECTIONS TO DATE

Each of the major constituencies has made known its recommended policies on stranded cost recovery. Given the contentiousness of the issue and the money at stake, there is a surprising degree of agreement on many of the broad policy principles.

The industrial user community has recommended the following:[36]

1. Mitigate all stranded costs to the maximum possible extent by selling assets, writing up the value of plants that competition has made *more* valuable, and renegotiating contracts.

2. Measure remaining stranded costs via a "formal proceeding" in advance of recovery.

3. Share stranded costs between shareholders and utility customers, with the sharing proportion based on:

   a. The proportion of the uneconomic assets that were due to management discretion and those that were due to government mandates.

   b. The utility's performance with respect to the mitigation of the mitigable costs.

   c. The quality of the utility's management at the time the uneconomic assets were incurred.

   d. Used and usefulness of asset.

4. Allocate stranded costs to customers based on "cost causation principles."

The electric utility industry's trade association has advocated these principles:[37]

1. ...Utilities should take all reasonable actions available to mitigate stranded costs. Further, state and federal regulatory commissions can assist by providing utilities adequate financial incentives to mitigate stranded costs, including appropriate "incentive" or "performance-based" regulation....

2. ...System costs should be recovered from those customers on whose behalf the costs were incurred; that is, the "stranding" of costs should be avoided by: (a) ensuring to the maximum extent possible that all customers pay their fair share of system costs; and (b) by not allowing customers permitted to switch suppliers to avoid system costs incurred to serve them.

3. ...The traditional obligation to serve must be reexamined and modified in light of new customer choices. The essential symmetry between obligation to serve and the exclusivity of the retail franchise must be maintained; as one premise is modified by regulatory policy, the other must be appropriately and simultaneously modified.

4. ...Regulatory commissions should explore all available methods for mitigation and recovery of stranded costs, and should employ such methods in combinations that best meet each unique situation.

Finally, the research arm of the National Association of Regulatory Utility Commissioners suggests:[38]

1. Consider the use of price caps to give utilities an incentive to relieve and reduce stranded costs, with an explicit link between the size of the cap and the allowed return "built-in" to the cap.

2. Do not commit to the amount of stranded cost recovery in advance. Recovered amounts should be recalibrated as market circumstances become known.

3. Limit recovery to a fixed period and do not include avoidable operating costs and return on investment (as opposed to the recovery of invested principal).

4. If possible, do not recover 100% of costs, as this gives the utility an incentive to over-estimate them.

5. Place burden of proof on utilities seeking recovery to reduce oversight workloads at regulatory commissions.

Despite many differences of substance and interpretation, these disparate interests agree that substantial recovery is appropriate, costs should be mitigated as much as possible, and recovery should occur in a competitively neutral manner.

The differences, though extremely important, are more tactical in nature. This relative agreement on principles but divergence on tactics is evident in the actions and statements of state and federal policymakers thus far. To date, regulators and legislators have almost universally backed the "full" recovery of stranded costs, subject to a number of conditions and principles much like the ones found in major proposals.

The issue first arose in the United States in connection with wholesale power competition implemented by the FERC. In bulk power markets, generating utilities sell at wholesale to distributor utilities. When the distribution utilities become "requirements" customers of a genco, the genco takes on an obligation to serve all that disco's future demand, much like the obligations state laws place on discos for their retail customers. If competition suddenly prompts the disco to stop taking service and switch to a generation company with a lower price, the requirements supplier may be stuck with stranded investments made to serve the departing disco.

Apparently accepting the main arguments in favor of recovery, the Commission held in an early ruling that wholesale customers who "departed" a system could be asked to pay for lost utility revenues. In spite of a later appellate court ruling that the recovery payments were illegal on antitrust grounds,[39] the Commission's landmark open-access rule (Order 888) similarly stated that utilities were entitled to recover the "full, legitimate, and verifiable" costs imposed on them by departing wholesale customers, provided utilities took reasonable steps to mitigate these costs. These costs could be recovered by charging departing customers an "exit fee" equal to estimated total differences in utility revenues incurred due to the customer's departure.[40] In adopting the standard, the Commission noted:[41]

> We learned from our experience with natural gas that, as both a legal and policy matter, we cannot ignore these costs. During the 1980s and early 1990s, the Commission undertook a series of actions that contributed to the impetus from restructuring of the gas pipeline industry. The introduction of competitive forces in the natural gas supply market as a result of the Natural Gas Policy Act of 1978 and the subsequent restructuring of the natural gas industry left many pipelines holding uneconomic take-or-pay contracts with gas producers. When the Commission initially declined to take direct action to alleviate that burden, the U.S. Court of Appeals for the District of Columbia Circuit faulted the Commission for failing to do so. The court noted that pipelines were "caught in an unusual transition" a result of regulatory changes beyond their control.

As we stated in the Supplemental NOPR, the court's reasoning in the gas context applies to the current move to a competitive bulk power industry. Indeed, because the Commission failed to deal with the take-or-pay situation in the gas context, the court invalidated the Commission's first open access rule for gas pipelines. Once again, we are faced with an industry transition in which there is the possibility that certain utilities will be left with large unrecoverable costs or that those costs will be unfairly shifted to other (remaining) customers. That is why we must directly and timely address the costs of the transition by allowing utilities to seek recovery of legitimate, prudent and verifiable stranded costs. At the same time, however, this rule will not insulate a utility from the normal risks of competition, such as self-generation, cogeneration, or industrial plant closure, that do not arise from the new availability of non-discriminatory open access transmission. Any such costs would not constitute stranded costs for purposes of this Rule.

Wholesale marketplace stranded costs can also be created if a group of retail customers or a state or local region forms a publicly-owned utility, transferring the existing retail franchise to a new entity. The new entity is not likely to own sufficient generation to serve its load at the outset and is therefore immediately a buyer on the wholesale market. The utility formerly holding the franchise, and which had invested to serve the franchise load, has investments stranded by this "municipalization." In Order 888, (see Box 6-1) the FERC allowed utilities to charge exit fees for such municipalizations, although not without some internal dissent.[42]

The FERC's method of recovering stranded costs, promulgated in Order 888, had several key elements. First, the Commission held that costs are recoverable only from departing customers via a one-time "exit fee" or via a "surcharge on transmission."[43] The Commission decided that charging those who stranded the cost was preferable to spreading the costs among all system users. The Commission based its reasoning on adherence to traditional cost causation principles in utility ratemaking—i.e., all attributable costs should be charged via rates—and the administrative difficulties of allocating costs to broad customer groups.[44] The transmission surcharge was allowed to be demand- or usage-based.

The Commission examined and rejected requiring utilities to share a portion of their stranded costs, stating a belief that such sharing was improper for prudently incurred costs.[45] To prevent future stranded costs, new wholesale contracts (i.e., those entered into after July, 1994) were required to address these responsibilities directly.[46]

As to the calculation of stranded costs, the Commission adopted what it termed "the revenues lost approach."[47] Under this method, stranded costs are the difference between the revenues the seller would have earned from the stranded contract and the projected revenues a seller would expect to earn on the open market. The Commission favored this method because it did not require an "asset-by-asset review" of costs and was relatively easy to calculate and assign.[48] The length of time the seller could reasonably have expected to keep the customer (the "reasonable expectation period")[49] was to be determined by the Commission, and mitigation efforts were to be recognized in the estimation of achievable revenues following customer departure.[50]

In every state where retail choice has been examined or implemented to date as well as in the public utility context,[51] policymakers have tended to agree with federal policies favoring full recovery of prudently incurred costs. In California's enormously complex restructuring proceeding, the California Public Utilities Commission proposed to allow the recovery of all stranded costs (including regulatory assets and power purchase agreements) over four years, using a "competitive transition charge" (CTC).[52] The Commission proposed guaranteed recovery using a non-bypassable fee on all customers that "we will fairly allocate…to avoid cost shifting…using cost allocation principles."[53] Transition costs were to be mitigated and offset by assets with higher-than-book value. In a departure from FERC's position and in agreement with NRRI and consumer group views, the Commission did not spare shareholders some of the costs of transition. To effectuate sharing, the Commission reduced the utility's allowed rate of return on stranded costs, while assuring utilities of the recovery of the remaining (lower) return.[54]

The Commission's decision was soon modified by legislation that adopted most elements of the original proposal. The legislature required that the non-bypassable charge be assigned to all customers "in substantially the same proportion as similar costs are recovered as of [now]," and provided that there is no shifting of costs onto residential and small commercial customers. Each utility has to file a "cost recovery plan" with the California Commission detailing separate charges for each element of its transition costs.[55]

In a particularly unique fashion, the California legislation allowed for a rate reduction and "full" recovery by "securitizing" each utilities' stranded assets. In brief, this process uses the proceeds from tax-exempt state infrastructure bonds to refinance utilities "uneconomic assets." This refinancing allows each utility holding these assets to fully amortize the assets while still reducing its required revenues and rates.[56] In effect, the entire process roughly refinances each utility's stranded assets at the low interest costs of tax-exempt bonds rather than for the higher costs of utility capital.

In a roughly similar fashion, Rhode Island's recent restructuring legislation authorized full recovery of the major types of transition costs via a non-bypassable fee.[57] Many other states have adopted or are discussing variations on the general policy of using non-bypassable charges to recover mitigated transition costs.[58]

Taken together, these federal and state policy actions seem to paint a relatively uniform picture of largely complete stranded cost recovery. Although it is true that regulators have thus far been largely consistent, this issue is likely to remain a source of controversy as long as large sums are at stake and there is the possibility of shifting or denying recovery of these transitional costs.

Chapter 16

NOTES

1 As defined by Tye and Graves, " 'stranded costs' are defined to be investments or cost commitments made by incumbents in the prior regime of cost-of-service regulation (e.g., "Sunk costs") that cannot expect to earn their cost of capital and/or be recovered from customers under the proposed new rules of competitive access to utility systems." These are costs that new entrants in a purely competitive generation market would not face, and they represent a handicap or burden that the incumbent must bear. "Costs that face stranding include, among others:

1) Assets used for electricity generation;

2) Costs for purchased power and fuel required under long-term contracts;

3) 'Regulatory assets' (expenses deferred to keep rates low temporarily);

4) Outlays for social goals, such as subsidies for low-income users; and

5) Incentives for renewable energy."

"The Economics of Negative Barriers to Entry: How to Recover Stranded Costs and Achieve Competition on Equal Terms in the Electric Utility Industry," William B. Tye and Frank C. Graves, *The Brattle Group,* May 14, 1996, p. 1. William J. Baumol and J. Gregory Sidak, "Pay Up or Mark Down? A Point-Counterpoint on Stranded Investment," *Public Utilities Fortnightly,* May 15, 1995, p. 1.

2 It is possible that some generating units are worth more in market value than their present book value. Chernick, et al (1996), for example, estimates that Massachusetts power plants are worth approximately $700 million more if sold today than their book value. As with stranded cost estimates, these "stranded value" figures are highly sensitive to many assumptions and parameters used in their computation.

3 See Box 16-1 and Ferguson (1995).

4 Abrams (1996) p. 25 The accounting standard Abrams refers to is known as FASB 71.

5 Stranded cost computation methodology issues are discussed in Del Roccili (1996); Cearley and McKinzie (1995); Rose (1996); Hirst, Hadley and Baxter (1996 a, b); and Lesser and Ainspan (1996).

6 See the Interstate Natural Gas Association of America (1996).

7 Richard Kindleberger, "Two More Utilities Approve Plan for Deregulation," *Boston Globe,* December 24, 1996.

8 See Navarro (1996) and Michaels (1994).

9 See Dorn (1996) and the references therein, Hewlett (1995), and Gregg Taylor and Allen Zeyher, "The Sustainability of Nuclear Power," *Nuclear News,* August, 1996, p. 75.

10 Ross Kerber "NRC May Close Two Plants Owned by Northeast Utility, Citing Safety." *Wall Street Journal,* March 11, 1996, p. A16.

11 Hewlett (1995) p. 8.

12 One listing of such laws is found in Davis (1996).

13 Michaels (1995). Economists Ray Hartman and Rich Tabors (1996) write

> The so-called "Regulatory Contract" seems to be a creation of Alfred Kahn and others in response to the most recent experience with "stranded" electric utility assets. It was articulated to combat the perceived unfairness of public utility commission disallowances of nuclear power plant costs.

All three authors are apparently unaware of Justice Oliver Wendell Holmes' clear characterization of the compact using the words "regulatory bargain" in *Cedar Rapids Gas Light Co. v. Cedar Rapids,* 223 U.S. 655, 669 (1912). (See Box 16-2). The existence of the compact was also endorsed by one of the seminal critics of utility regulation, the late George Stigler, in testimony before the Illinois Commerce Commission, Docket, Commonwealth Edison Exhibit 25.0. p. 14.

14 See Chapter Two supra for an elaboration of this view.

15 Michaels (1996) p. 49.

16 Richard Kindleberger, "Two More Utilities Approve Plan for Deregulation," *Boston Globe,* December 24, 1996.

17 Hughes (1995).

18 Peter Bradford, past chair of two states' regulatory commissions and a former member of the U.S. Nuclear Regulatory Commission, has pointed out that arguments that assert that all past utility investments were adequately scrutinized by regulators are vast oversimplifications. Bradford notes that "the notion that $100,000 per year regulators sit like Maxwell's Demon busily sorting $100 million per year ($3.17 per second) into prudent and imprudent piles is silly enough, even if one assumes that all regulators are perfect statesmen." Bradford points to the underfunding of regulatory commissions and the political influence of utilities over their regulators as longstanding and widely acknowledged influences that weaken regulatory oversight. Bradford (1995) p. 12.

19 See Navarro (1996) and Michaels (1994).

20 Michaels (1994) p. 19.

21 See, for exampe, Joskow (1996), Tye and Graves (1996), and Kahn (1994).

22 See Baumol, Joskow, and Kahn (1995), Kahn (1994), Joskow (1996). Some analysts have used the concepts of "static" versus "dynamic" economic efficiency to organize this panolply of arguments. In this context, static efficiency refers to the recovery of today's stranded costs in a competitively neutral manner, harming neither incumbents nor new entrants. Dynamic efficiency is achieved when competitors, looking forward, have the incentives and ability to compete effectively and without market distortions.

23 Steinmeier and Stuntz (1994). This paper is a rejoinder to Gray and Hempling (1994), which reflects other views of these issues.

24 Baumol and Sidak (1995, 1996), Tye and Graves (1996), Joskow (1994), Kahn (1994).

25 Chapter 9 discusses a number of difficulties other than stranded cost recovery that make it very difficult to establish a "level playing field" between integrated incumbents and deregulated retailers.

26 Rose (1996) p. 89. Rose calls the evasion of legitimate stranded costs "uneconomic bypass." To guard against this tendency, Rose calls for industrial user groups allocating a small portion of stranded costs to the utility itself, giving it an incentive to reduce them as low as possible. Studness (1996), Stelzer (1994), and others point out that this particularly applies to above-market contracts, which utilities can sometimes renegotiate (with significant effort and cost).

27 Foley and Thompson (1993).

28 Michaels (1996) makes this point, and offers results similar to those found in the NARUC study of annual returns.

29 Competition in capital markets implies that no investor can obtain and maintain a riskless arbitrage position, i.e. a riskless self-sustaining money machine which a perpetual state of above-market returns would imply. See Varian, "The Arbitrage Principles in Financial Economics," *Economic Perspectives,* vol. 1, no. 2, Fall, 1987, pp. 55-72. The very function of modern capital markets is to equalize the risk-return opportunities among all investment opportunities. It is difficult to see how any industry with the free flow of capital in or out could consistently outperform other industries. See the entries on Capitalism and Capital, Credit and Money Markets in the New Palgrave–A Directory of Economics (Macmillan, 1987).

30 Cearley and McKinzie (1995), Michaels (1994, 1995), Rose (1996) and Brennan, et al (1996) address the similarities and differences between "stranded costs" in unregulated and regulated industries.

31 Economic Report of the President (1996) p.188. The point is repeated in James Q. Wilson, "Don't Short-Circuit Utilities' Claims," the *Wall Street Journal,* August 23, 1995, p. 12.

32 Baumol and Sidak (1995) p. 22.

33 Bradford (1995) p. 14.

34 Stelzer (1994) p. 41.

35 See Brennan, et al (1996) p. 104.

36 Hughes (1995) and Anderson (1996).

37 Steinmeier and Stuntz (1994).

38 Rose (1996) p. 96ff.

39 The FERC first confronted this issue when a utility filed a wholesale transmission tariff in 1991 that required departing customers to reimburse the utility...

> For the costs of any investment by such a company in production, transmission or distribution facilities that are unrecovered by such company as a result of the provision of service under this Tariff.

(Entergy Transmission services tariff, Pembroke (1994) p. 43). FERC allowed for this cost recovery, but was reversed in 1994 by the U.S. Court of Appeals for the District of Columbia (*Cajun Electric Power Cooperative v. FERC,* 28F. 3d 173, 1994). The D.C. Court held that the stranded cost recovery was an anti-competitive "tying" arrangement prohibited under U.S. antitrust laws. Washington lawyer James Pembroke described the court's reasoning as follows:

> Taking a practical approach, the Court likened the stranded investment provision to the sword of Damocles dangling over the negotiating table. If a seller in competition with Entergy for a power sale does not know what its delivered cost will be until a stranded investment hearing is concluded, that seller faces, in the words of the Court, "deal-killing transactional costs and uncertainties." The significant problem for power marketers at the negotiating stage is not necessarily whether there are costs related to stranded investment, or what the level of those costs will be. The significant danger identified by the Court is whether the mere possibility of incurring stranded investment costs will adversely affect competition by precluding buyers from purchasing from suppliers other than Entergy.

> In spite of this reversal, the FERC soon promulgated a proposed rule on stranded cost recovery that called for the [full] recovery of "legitimate, prudent, and verifiable stranded costs" caused by open access (wholesale) transmission (FERC Stats & Regs., 32, 514 at 33, 095).

Following extensive comments, the FERC adopted essentially the same standard in its final rule, Order 888 (Dockets RM 95-81-000 and RM94-7-001), mimeo, p 451 (cited hereinafter in these notes as "Order 888"). The FERC argued that its recovery of stranded costs via exit fees on transmission customers in the context of an open-access rulemak-ing did not constitute the same set of circumstances that led the Court of Appeals to object to its actions in *Cajun*.

40 More accurately, the exit fee is set by the formula:

$$SCO = (RSE-CMVE) \times L$$

where

RSE = Revenue Stream Estimate–average annual revenues from the departing generation customer over the three years prior to the customer's departure (with the variable cost component of the revenues clearly identified), less the average transmission-related revenues that the host utility would have recovered from the departing generation customer over the same three years under its new wholesale transmission tariff (863/In the case of a retail-turned-wholesale customer, subtraction of distribution system-related costs may also be appropriate.)

CMVE = Competitive Market Value Estimate–determined in one of two ways, *at the customer's option:* Option (1) the utility's estimate of the average annual revenues (over the reasonable expectation period "L" discussed below) that it can receive by selling the released capacity and associated energy, based on a market analysis performed by the utility; or Option (2) the average annual cost to the customer of replacement capacity and associated energy, based on the customer's contractual commtiment with its new supplier(s).

L = Length of Obligation (reasonable expectation period)–refers to the period of time the utility could have rea-sonably expected to continue to serve the departing generation customer. We reaffirm that we do not believe that a one-size-fits-all approach is appropriate for determining the length of a customer's obligation. If the par-ties cannot reach agreement as to the length of the customer's obligation, this period is to be determined through litigation as a part of the threshold issue of whether the utility had a reasonable expectation of contin-uing to serve the customer.

41 Order 888, mimeo, pp. 453–454.

42 Commissioners Hoecker and Massey dissented from the FERC's Order 888 decision to allow the recovery of costs due to municipalization, stating that such policy would "second-guess" state retail stranded cost proceed-ings, was inconsistent with the FERC's absence of retail wheeling authority, and involves "unnecessary legal risk." Dissent of Commissioner James Hoecker, Docket No. RM95-8-000 and RM94-7-001, April 24, 1994.

42 Order 888, mimeo, p. 477. 43 44 Ibid, p. 477-490. 44 45 Ibid, p. 490. 45 46 Ibid, p. 492ff. 46 47 Ibid, p. 573. 47

48 Ibid.

49 Ibid, p. 596.

50 Ibid, p. 599. The Commission placed additional conditions on the lost revenue calculation: stranded costs could not exceed the average annual contribution to fixed power supply costs; the most recent revenue figures should be used; market value could be set by the customer's alternative suppliers, and the customer may market or broker the power it will strand to "self-mitigate" its costs impacts. (Mimeo p.598-599).

51 Public power agencies and rural electric cooperatives are not immune from the stranded cost phenomenon. According to reports in the *Wall Street Journal,* the U.S. Rural Electrification Administration has $11 billion in loans that are in default or "classified as problematic," some because wholesale power prices have dropped below the costs of stranded investments. The *Journal* reports that REA will restructure the debts of cooperatives that have stranded costs. Bruce Ingersoll, "U.S. Is Giving Electric Co-ops Relief on Loans," *Wall Street Journal,* October 3, 1996.

52 California Public Utilities Commission Decision No. 95-12-063 (December 20, 1995), as modified by Decision No. 96-01-009, 166 P.U.R. 4th 1, January 10, 1996, Section I. [hereinafter, California Order].

53 California Order, Section V.

54 Ibid.

55 California Assembly Bill AB1890, as enrolled, Article 6, Section 367 (e) (1).

56 Ibid, Section 11. For a useful description, see Richard and Lavinson (1996).

57 Rhode Island Utility Restructuring Act of 1996, 96-H8124, Section 39-1-27.3

58 Alabama House Bill 350, signed in May, 1996 provides for full recovery of mitigated costs. A proposed settlement agreement pending in Massachusetts also provides full recovery.

Chapter 17

Conclusion

Today, signs of turmoil in the electric utility industry are everywhere. Scarcely a day goes by without a newspaper story about a major utility merger or a new unregulated business venture. Constantly fearful of being bought, sold, or simply "rightsized," employees of power companies at every level seem distracted, and sprinkle their conversations with gallows humor and phrases like, "for my next career...."

Without a doubt, changes in industry circumstances, consumer needs, and new economic and regulatory practices call for a modernization of electricity policies in the United States. The issue is not whether the industry should change, but how—and how to address a plethora of uncertainties with the fewest possible major missteps.

With this objective in mind, the main findings of this volume may be briefly summarized as follows:

1. The technology of electric power delivery causes great interdependence between all buyers and all sellers of electricity in a region. This interdependence makes it extremely difficult to measure the costs or benefits of individual transactions, in turn making it quite difficult to set fair and effective rules for competition. New technologies will probably solve this problem, but not for several decades.

2. Because power transmission and distribution will remain natural monopolies, two-thirds of the power industry will remain regulated by state and federal authorities—even if retail competition becomes widespread. Restructuring is not deregulation; it is a shift from utilities that own their own power plants and serve one area exclusively to utilities that are geographically dispersed and specialize in unregulated generation products or still-regulated delivery.

3. Competition among generators to sell to local distribution companies is now standing federal policy. Implementation of these policies is still in the early stages, and much of the competitive rulebook is yet to be fully determined. In particular, it must be determined how to allow all generators to use the transmission systems at fair and efficient prices. Even before the rules are fully set, this "bulk power" competition has helped drive prices down in many parts of the United States. In the long run, preventing re-monopolization (market power) and allocating increasingly scarce transmission capacity appear to be the main unresolved challenges.

4. Thus far, competition has been at the wholesale level only—individual buyers have not been able to choose suppliers. If local distributors turn into power "transporters," deregulated power producers will be able to provide competitive power sales to individual consumers. Several states have already decided to adopt this policy of "retail choice."

5. Retail choice deregulates the price and availability of power for individual customers, but substantial administrative and regulatory supervision will be required to establish and oversee it. Retail choice will create new product and service bundles for consumers, such as the possibility of buying burglar alarm monitoring service, electricity, and gas all from a single company. However, electricity service may be somewhat confusing to buy, and prices to small businesses and households may not decline much if any.

6. Retail choice removes much state and local control over power generators and threatens to adversely impact many of the public interest features of the present industry. Programs that ensure universal service and protection of low-income consumers will be continued under regulatory supervision, with many important details still to be determined. Restructuring will also cause significant increases in air emissions in the near term, leading to calls for new federal controls or other corrective policies. Energy efficiency and renewable energy programs are also continuing in most cases, although with different emphases.

7. The net total impact of these changes on the nation's welfare is subtle, complex, and of indeterminate quantity. There is little evidence that restructuring will substantially lower average U.S. power prices, which have long been among the world's lowest and have fallen steadily during the last decade. Restructuring will prompt a *relative* shift toward lower power prices for large business users and relatively higher prices for small businesses and households, but this effect also may not be large. It is most likely that the noticeable changes, negative as well as positive, will be in the availability of new products and services and the quality and reliability of service. For a substantial time to come, the benefits of restructuring may well reside in the eye of the beholder.

The electric power industry is leaving its "iron era" and entering the information age. Gone are the days when the power industry's physical assets were its only assets, information was costly, and electricity service was a simple, purely local concern. Today's financial and power markets and the information technology they thrive on could scarcely have been imagined 50 years ago, when utility regulation was born as a compromise between government ownership and private monopoly. Coupled with the transformation of the modern corporation into a "virtual organization," the local, vertically-integrated utility is clearly a thing of the past.

Yet the thought lingers that this industry a critical part of our economic and military infrastructure, and has long been married to a host of political and social interests. The traditions of local control and intersystem rivalry that confounded federal efforts to "rationalize" the industry for decades cannot be expected to go away overnight, if ever. The Department of Energy wrote in 1977 that:

From the historical experience of power systems it is clear that the realization of the potential benefits of the large-scale technology of power systems has been limited by social and institutional factors. Among these, no factor stands out so prominently as the desire of independent systems to remain independent and to maintain autonomous control over their operations.

These influences are certain to reassert themselves as part of the substantial and essential portions of the industry that remain regulated. This should not merely temper expectations about the extent of the unfettered marketplace that is being created—it should also alert all to search for economic solutions that are broadly acceptable. In the restructured power industry, regulation, coordination, and competition will find it difficult but essential to coexist.

Already there are signs of trouble on the horizon. In the United Kingdom, in spite of universally acknowledged improvements over pre-deregulation conditions, a steady series of regulatory changes and high industry profits have made electric policy a significant political threat to the majority party. In the United States, consumers show a desire for greater competition, but there are also growing service quality concerns and dissatisfaction with a loss of community control.

The potential for consumer backlash is joined by two additional imponderables. First, the success of competition will ultimately hinge on the equitable accessibility of transmission lines and new generators. If the siting and expansion of power lines and environmental permitting become the linchpins of competition, state and federal regulators will hold increasing control over the resources essential for competition. Meanwhile, energy companies will continue to grow to a size and scope reminiscent of the 1920s. The combination of these elements in an industry that is notorious for lengthy regulatory and legal battles will create a potentially explosive mixture with the power to propel or repel change unpredictably.

However dramatic the idea of electric deregulation, it will be far more revolutionary to marry regulation and competition in a peaceful and productive fashion. This is a task worthy of Adam Smith's insights, the wisdom of ancient Solomon, and the political skills of a Franklin D. Roosevelt. With a healthy dose of each, we will fashion a power industry that best serves the nation for the rest of the era of power system technological interdependence.

BIBLIOGRAPHY

Bibliography

Abel, J.E. 1984. "Determining an Optimum Capital Structure." *Public Utilities Fortnightly* (24 May): 24–28.

Ackerman, E.T. 1995. A Primer on Performance-Based Regulation. Presented at the Joint NACUC/EEI Seminar, April 19.

Acton, J.P. and B.M. Mitchell. 1979. "Evaluating Time-of-Day Electricity Rates for Residential Customers," Rand Corporation Report, R-2509-DWP. (November).

Adler, A.R. 1994. "Introduction: The Role of Technology in Serving the Customer's Needs". In *The Electric Industry in Transition,* 215–218. Arlington, VA: Public Utilities Reports, Inc. And New York State Energy Research and Development Authority.

Ahern, W.R. 1985. Position of the California Public Utilities Commission's Public Staff Division on the Regulation of Utility Diversification in California. Presented at CPUC public hearings on utility diversification.

Aiger, D. and J. Hirschberg. 1985. "Commercial/Industrial Customer Response to Time-of-Use Electric Price: Some Experimental Results." *Rand Journal of Economics.* 16 (3): 341–355.

Air and Waste Management Association. 1992. *New Hazardous Air Pollutant & Laws Regulations: Their Impact on Industry, Government and the Public.* Pittsburgh, PA.

Aitken, D.W. 1995. "Wind Power and the Environment: An Essay" Proceedings: Wind Power 1995. Washington, DC: American Wind Energy Association.

Alchian, A. and H. Demsetz. 1972. "Production, Information Costs, and Economic Organization." *American Economic Review* 62: 777–795.

Alexander, B.R. and the National Consumer Law Center. 1996. "Consumer Protection Proposals for Retail Electric Competition: Model Legislation and Regulations." The Regulatory Assistance Project (October).

Alexander, T. 1981. "The Surge to Deregulate Electricity." *Fortune* (13 July): 98–103.

Alvarado, F.L. 1996. Rules of the Road and Electric Traffic Controllers: Making a Virtual Utility Feasible. Presented at the Symposium on the Virtual Utility, 1–2 April, Saratoga Springs, NY.

American Gas Association, 1987. *Avoided Versus Actual Costs of Purchased Cogenerated Electricity.* Arlington, VA.

American Public Power Association, 1996. *Public Power: Annual Statistical Section.*

American Public Power Association and National Rural Electric Cooperative Association. 1996. "Joint Petition of the American Public Power Association and the National Rural Electric Cooperative Assocation for a Rulemaking Proceeding to Revise the Commission's Standards Applicable to the Merger of Public Utilities Under Section 203 of the Federal Power Act." Docket No. RM 96-8-000. (17 January).

Analysis Group for Regional Electricity Alternatives. 1996. *Technology Integration and the Environment Under Competition*. Cambridge, MA: The Energy Laboratory, Massachusetts Institute of Technology. (4 June).

Anderson, R. and M. Bohman. 1985. "Short- and Long-run Marginal Cost Pricing: On Their Alleged Equivalence." *Energy Economics*. (October).

Anderson, S.C. 1987. "An International Comparison of the Impact of Safety Regulation on LWR Performance." Massachusetts Institute of Technology Energy Laboratory, MIT-EL-87-006.

Arrow, K. 1975. "Vertical Integration and Communication." *Bell Journal of Economics* 6:173–183.

Arthur D. Little, Inc. 1979. Financial/Economic Implications of S. 1991. *The National Power Grid Study: Technical Study Reports*. Volume II. Washington, D.C.: U.S. Department of Energy, Economic Regulatory Administration, Office of Utility Systems, 353–371.

Asbury, J.G. and S.B. Webb. 1980. "Decentralized Electric Power Generation: Some Probable Effects." *Public Utilities Fortnightly* (25 September): 21–24.

Atkinson, S.E. and R. Halvorsen. 1984. "Parametric Efficiency Tests: Economies of Scale and Input Demand in U.S. Electric Power Generation." *International Economic Review* 25: 684.

Averch, H., and L.L. Johnson. 1962. "Behavior of the Firm under Regulatory Constraint." *American Economic Review* 52: 1053–1069.

Awerbuch, S., E.G. Carayannis and A. Preston. 1996. "The Virtual Utility: Some Introductory Thoughts on Accounting, Technological Learning & the Valuation of Radical Innovation."

Awerbuch, S. 1984. "Industrial Maturation, Market-to-Book Ratios, And the Future of Electric Utilities." *Public Utilities Fortnightly* (2 Feb): 3–11.

Awerbuch, S. 1993. *Valuing Radical Innovation: The Case of the Distributed Utility*. Wilmington, DE: Utility Photovoltaic Group. (October).

Ayres, I. and J. Braithwaite. 1992. *Responsible Regulation: Transcending the Deregulation Debate*. New York: Oxford Univ. Press.

Bailey, E.E., and A.F. Friedlaender. 1982. "Market Structure and Multiproduct Industries." *Journal of Economic Literature* 20: 1024–1048.

Baldwin, C. J. 1966. "Reverse Sharing Is Primary Benefit from Power Pooling." *Power Pooling* 70 (April): 59–61.

Barkenbus, J.N. 1987. "Roles In Electricity: A Brief History of Electricity and the Geographic Distribution of Manufacturing." Oak Ridge TN: Oak Ridge Associated Universities, Institute for Energy Analysis:(July) Electric Power Research Institute, EM-5298-SR.

Barnes, E. G., 1990. "Antitrust Considerations of Transmission Ownership." In *The Transmission Symposium*, 255–285. New York: Public Utilities Reports, Inc., and The Management Exchange.

Barnes, R. and R. Gillingham, and R. Hageman. 1981. "The Short-Run Residential Demand for Electricity," *Review of Economics and Statistics* 63 (November): 541–551.

Barnes, et al. 1994. *The Integration of Renewable Energy Sources Into Electric Power Distribution Systems*. Vol. 1. National Assessment. Oak Ridge National Laboratory. ONRL-6775 V1.

Bauer, D.C. 1996. Social Equity Issues and Electricity Industry Restructuring: A One-on-One Debate. Presented at the 3rd. DOE-NARUC National Electricity Forum. Oak Ridge National Laboratory.

Baumol, W.J., and Alan S. Blinder. 1991. *Economics: Principles and Policy.* San Diego: Harcourt Brace Jovanovich.

Baumol, W.J., M.F. Koehn and R.D. Willig. 1987. "How Arbitrary Is 'Arbitrary'?—or, Toward the Deserved Demise of Full Cost Allocation." *Public Utilities Fortnightly* (3 Sept.).

Baumol, W.J. 1983. "Minimum and Maximum Pricing Principles for Residual Regulation." In *Current Issues in Public-Utility Economics,* 177–196. A.L. Danielsen and D.R. Kamerschen, eds. Lexington, MA: Lexington Books.

Baumol, W.J., C. Panzar, and D. Willig. 1982. *Contestable Markets and the Theory of Industry Structure.* San Diego: Harcourt Brace Jovanovich.

Baumol, W.J. 1977. "On the Proper Tests for Natural Monopoly in a Multiproduct Industry." *American Economic Review* 67 (December): 809–822.

Baumol, W.J. and D.F. Bradford. 1970. "Optimal Departures From Marginal Cost Pricing." *American Economic Review.* 60: 265–283.

Baumol, W.J., P.L. Joskow, and A.E. Kahn. 1994. " The Challenge for Federal and State Regulators: Transistion from Regulcation to Efficient Competition in Electric Power. Appendix to Comments of Edison Electric Institute on FERC's Mega-NOPR (9 December).

Baxter, L.W. 1996. "Proposals for the Future of Energy Efficiency." U.S. Department of Energy under contract number DE-AC05-96OR22464: (May 1996.)

Baxter, L.W. 1996. "Alternative Transition Approaches." *The Electricity Journal* (August/September).

Becker, G.S. 1983. "A Theory of Competition Among Pressure Groups for Political Influence." *The Quarterly Journal of Economics* 98 (3): 371–399.

Behling, B.N. 1938. *Competition and Monopoly in Public Utility Industries.* Urbana: University of Illinois Press.

Berg, S.V., and J. Tschirhart. 1988. *Natural Monopoly Regulation: Principles and Practice.* Cambridge, England: Cambridge University Press.

Berlin, E., C.J. Cicchetti, and William J. Gillen. 1974. *Perspective on Power.* Cambridge, MA: Ballinger.

Bernow, S., B. Biewald and D. Marron. 1991. "Full-Cost Dispatch: Incorporating Environmental Externalities In Electric System Operation." *The Electricity Journal* (March).

Bernstein, M.H. 1955. *Regulating Business by Independent Commission.* Princeton, NJ: Princeton University Press.

Berry, W.W. 1982. "The Case for Competition in the Electric Utility Industry." *Public Utilities Fortnightly* 110 (6): 13–27.

Besanko, D., S. Donnenfeld, and L.J. White. 1987. "Monopoly and Quality Distortion: Effects and Remedies." *The Quarterly Journal of Economics* (November).

Billington, R., and R.N. Allan. 1988. *Reliability Assessment of Large Electric Power Systems.* Boston: Kluwer Academic Publishers.

Blackmon, B.G., Jr. 1986. "Electricity Futures: The Next Step for Bulk Power." *Public Utilities Fortnightly* 117 (9): 14–20.

Blumstein, C. 1996. *Public Goods Research, Development and Demonstration for California's Electricity and Natural Gas Consumers.* University of California Energy Institute.

Bobbish, D. 1992. "From Gas to Electric at FERC: Will It Be Deja Vu All Over Again?" *The Electricity Journal* 5 (June): 52–63.

Bohi, D. and K.L. Palmer. 1996. "Relative Efficiency Benefits of Wholesale and Retail Competition in Electricity: An Analysis and a Research Agenda." Golden, CO: National Renewable Energy Laboratory (March).

Bohi, D.R., and K.L. Palmer. 1996. "Which Model Is More Efficient—Wholesale or Retail?" *The Electricity Journal* (October).

Bohi, D.R. and M.A. Toman. 1996. *The Economics of Energy Security.* Boston: Kluwer Academic Publishers.

Bohn, R., Schweppe, F. and R. Tabors. 1989. "Using Spot Pricing to Coordinate Deregulated Utilities, Customers, and Generators". In *Electric Power Strategic Issues,* Plummer, J. ed. Arlington, VA: Public Utilities Reports, Inc.

Bonbright, J.C. 1941. "Major Controversies as to the Criteria of Reasonable Public Utility Rates." *American Economic Review:* 31 (May): 379–389.

Bonbright, J., and G. C. Means. 1932. *The Holding Company: Its Public Significance and Regulation.* New York: McGraw-Hill.

Bonbright, J. 1961. *Principles of Public Utility Rates.* New York: Columbia University Press.

Borenstein, S., J. Bushnell, E. Kahn, and S. Stoft. 1996. "Market Power in California Electricity Markets." Program On Workable Energy Regulation (March).

Bork, R.H. 1969. "Vertical Integration and Competitive Processes." In *Policy Toward Mergers,* 139–149, J.F. Weston and S. Peltzman, eds. Pacific Palisades, CA: Goodyear Publishing.

Bradford, P. 1995. "A Regulatory Compact Worthy of the Name." *The Electricity Journal* (November): 12–15.

Bradley, S.P., and J.A. Hausman. 1989. *Future Competition in Telecommunications.* Boston: Harvard Business School Press.

Braeutigam, R. 1989. "Optimal Policies for Natural Monopolies." In *Handbook of Industrial Organization,* R. Schmalensee and R. Willig, eds. Amsterdam: North Holland.

Brattle/IRI. 1996a. *Strategic Considerations for Electric Service Unbundling and: Alternatives for Transmission and Ancillary Service Pricing.* Cambridge, MA. (23 January).

Brattle/IRI. 1996b. *Power Pool Bidding and Antitrust Considerations.* Prepared for New England Power Pool. Cambridge, MA. (5 February).

Brealey, R.A. and S. Meyers. 1984. *Principles of Corporate Finance.* 2nd. ed. New York: McGraw Hill.

Brennan, T., K. Palmer, R. Kopp, A. Krupnick, V. Stagliano, and D. Burtraw. 1996. *A Shock to the System.* Washington, DC: Resources for the Future.

Bretz, E.A. 1994. "Nuclear Power." *Electrical World* (July).

Breyer, S. 1982. *Regulation and Its Reform.* Cambridge, MA: Harvard Univ. Press.

Breyer, S.G., and P.W. MacAvoy. 1974. *Energy Regulation by the Federal Power Commission.* Washington, DC: Brookings Institution.

Bringham, E.F., L.C. Gapenski and D.A. Aberwald. 1987. "Capital Structure, Cost of Capital, and Revenue Requirements." *Public Utilities Fortnightly* 119 (8 January): 15–24.

Brockway, N. and M. Sherman. 1996. "Stranded Benefits in Electric Utilities Restructuring—The Electric Industry Restructuring Series." The National Council on Competition and the Electric Industry (October).

Brockway, N. 1994. *A Low-Income Advocate's Introduction to Electric Industry Restructuring and Retail Wheeling.* National Consumer Law Center.

Brockway, N. 1996. *Treatment of Low-Income Consumer Issues.* ORNL Advanced Training. (December). Mimeo. Oak Ridge, TN: Oak Ridge National Laboratory.

Brower, M.C., Thomas, S.D. and C. Mitchell. 1996. "The British Electric Utility Restructuring Experience: History and Lesson for the United States." *The Electric Industry Restructuring Series* (October).

Brown, D.J., T.R. Lewis, and M.D. Ryngaert. 1994. "The Real Debate Over Purchased Power." *Electricity Journal* (September): 61.

Brown, D.C. 1980. *Electricity for Rural America—The Fight for the REA.* Newport, CT: Greenwood Press.

Brown, H.J. 1983. *Decentralizing Electricity Production.* New Haven, CT: Yale Univ. Press.

Brown, L., M. Einhorn, and I. Vogelsang. 1991. "Toward Improved and Practical Incentive Regulation." *Journal of Regulatory Economics* (3): 323–338.

Brown, L., M. Einhorn and I. Vogelsang. 1989. *Incentive Regulation: A Research Report.* Washington, DC: Federal Energy Regulatory Commission. Office of Economic Policy.

Brown, M.A., M.A. Beyer, J. Eisenburg, E.J. Lapsa and M. Power. 1994. "Utility Investment In Low-Income-Energy Effienciency Programs." Oak Ridge National Laboratory (September).

Brown, S.J. and D.S. Sibley. 1986. *The Theory of Public Utility Pricing.* Cambridge: Cambridge Univ. Press.

Bruce, J.P., Hoesung L. and E.F. Haites, eds. 1995. *Climate Change 1995: Economic and Social Dimensions of Climate Change.* Published for the Intergovernmental Panel on Climate Change. Cambridge: Cambridge Univ. Press.

Brynes, B., M. Rahimzdeh, R. de Alba, and K. Baugh. 1996. "Green Pricing: The Bigger Picture, It's Not Just for Residential Consumers." *Public Utilities Fortnightly* (August).

Brynes, B. et. al. 1996. "Green Pricing: The Bigger Picture." *Public Utilities Fortnightly* (August): 18.

Budhraja, V.S. 1995. "Generation as a Business: Facts, Fumbles, Fictions, and the Future," *The Electricity Journal* 8 (July): 36.

Bunn, D. 1995. "Market-Based Pricing and Demand Side Participation in the Electric Pool of England and Wales". In Proceedings: 1996 EPRI Conference on Innovative Approaches to Electricity Pricing. Palo Alto, CA: Electric Power Research Institute. (March) EPRI TR-106232.

Burns, R.E., W. Pollard, T. Pryor and L.M.Pike. *The Appropriateness and Feasibility of Various Methods of Calculating Avoided Costs.* Columbus, OH: National Regulatory Research Institute (June).

Burns, R.E., J. Beecher, Y. Hegazy and M. Eifert. 1995. *Alternatives to Utility Service Disconnection.* Columbus, OH: The National Regulatory Research Institute (May).

Burtraw, D., W. Harrington, A. M. Freeman III and A. J. Krupnick. 1993. *The Analytic of Social Costing In A Regulated Industry.* Washington, DC: Resources for the Future.

Bushnell, J. B. and S. Stoft. 1994. *Electric Grid Investment Under a Contract Network Regime.* Berkeley, CA: University of California Energy Institute.

Cabral, L.M. and M.H. Riordan. 1989. "Incentives for Cost Reduction Under Price Cap Regulation." *Journal of Regulatory Economics* (June).

Cabral, L.M. and M.H. Riordan. 1991. "Incentives for Cost Reduction Under Price-Cap Regulation". In *Price Caps and Incentive Regulation in Telecommunications,* M.A. Einhorn, ed. Norwell, MA: Kluwer Academic Publishers.

California Public Utilities Commission. 1996. CPUC Sets Out Electric Restructuring Implementation Roadmap. *CPUC News.* 415-703-2423 (R.94-04-031/I.94-04-032). March 13, 1996 CPUC-26.

California Public Utilities Commission. 1994. Fourth Round Opening Comments on: Direct Access and Customer Choice Role, Structure and Efficacy. R.94-04-031 (Filed April 20, 1994): I.94-04-032. Sacramento, CA. August 24, 1994.

California Energy Commission. 1995. *Draft Energy Development:* Volume 1, Part II: Research, Development and Demonstration and Electricity Industry Restructuring. Research and Development Committee. (June).

Calzonetti, F.J. 1987. "The Effect of Electricity Prices on Industrial Location." Morgantown, WV: Regional Research Institute, West Virginia Unversity. Resarch Paper 8704 (April).

Calzonetti, F.J., T. Allison, M.A. Choughry, G.G. Sayre and T.S. Witt. 1989. *Power From the Appalachians: A Solution to the Northwest's Electricity Problems?* Newport, CT: Greenwood Press Inc. (June).

Capron, W.M., ed. *Technological Change in Regulated Industries.* Washington, D.C.: Brookings Institution.

Caramanis, M.C., R.E. Bohn and F.C. Schweppe. 1987. "System Security Control and Optimal Pricing of Electricity." *Electrical Power and Energy Systems.* (October): 217–224.

Caramanis, M.C., R.E. Bohn, and F.C. Schweppe. 1982. Optimal Spot Pricing: Practice and Theory. *IEEE Transactions on Power Apparatus and Systems* PAS-101: 3234–3245.

Cargill, T.F. and R.A. Meyer. 1971. "Estimating the Demand for Electricity by Time-of-Day." *Applied Economics* 3: 233–46.

Carlton, D. 1979. "Vertical Integration in Competitive Markets Under Uncertainty." *Journal of Industrial Economics* 27: 189–209.

Carton, D. 1980. "The Location and Employment Choices of New Firms: An Econometric Model with Discrete and Continuous Endogenous Variables." *The Review of Economics and Statistics* LX (3): 440–449.

Carlton, D. and J. Perloff. 1990. *Modern Industrial Organization.* Glenview, IL: Scott Foresman.

Casazza, J.A., et al 1992. "An International View on Competition and Coordination." Paris: CIGRE, 1992 Session.

Casazza, J.A. 1985. "Understanding the Transmission Access and Wheeling Problem." *Public Utilities Fortnightly* 116 (9): 35–42.

Cash, J.L., R. Eccles, N. Nohira and R. Nolan. 1994. *Building the Information-Age Organization, Structure, Control, and Information Technologies.* Boston: Harvard Business Press.

Catalano, P.T. 1981. "Allocating Demand Costs: The Average and Excess Demand Method." *Public Utilities Fortnightly:* 108 (8 October): 58–60.

Cavanagh, R. 1996. *Restructuring for Sustainability: Toward a New Electric Services Industry.* Natural Resource Defense Council (7 February).

Cavanagh, R. 1996. "Electric Services Industries." *The Electricity Journal* 9 (6).

Cavanaugh. 1994. *The Great "Retail Wheeling" Illusion—And More Productive Energy Futures.* One of an ongoing series of strategic issues papers published periodically by E. Source, Inc.: Boulder, CO.

Caywood, R.E. 1956, 1972. *Electric Utility Rate Economics.* New York: McGraw Hill.

Center for Clean Air Policy. 1995. "Dealing with Regional NOx Emission in a Restructured Electricity Industry." Issue Paper #1: (9 August).

Center for Clean Air Policy. 1996. "Emissions Impacts of Competition: Further Analysis in Consideration of Putnam, Hayes & Bartlett, Inc.'s Critique." (17 June).

Center for Clean Air Policy. "New York Case Study: Electricity Restructuring and the Environment".

Centolella, P.A. 1996. *Energy Efficiency & Net Income Neutrality in a Restructured Electric Power Industry.* Presented at the Office of Energy Efficiency & Renewable Energy Seminar on Restructuring of the Electric Utility Industry.

Chadwick, E. 1994. "Results of Different Principles of Legislation and Administration in Europe; of Competition for the Field, as Compared with Competition within the Field, of Service." *Journal of the Royal Statistical Society* 22 (Series A, September): 381–420.

Chamberlin, J.H., J.B. Brown, and M.W. Reid. 1992. "Gaining Momentum or Running Out of Steam? Utility Shareholder Incentive Mechanisms—Past, Present, and Future." Proceedings of the ACEEE Summer Study, v. 8. American Council for An Energy-Efficient Economy, Washington, D.C.

Chamberlin, J.H., P. Herman and G. Wikler. 1993. "Mitigating Rate Impacts of DSM Programs." *The Electricity Journal* (November): 46–56.

Chandler, A.D., Jr. 1977. *The Visible Hand: The Managerial Revolution in American Business.* Cambridge, MA: Harvard University Press.

Chandler, A.D., Jr. 1962. *Strategy and Structure.* Cambridge, MA: MIT Press.

Chandley, J. 1994. Competition and Restructuring in California's Electric Services Industry. Presented at a conference Sponsored by Morrison and Forester (22 July).

Chandley, J.D. 1994. The California Restructuring Debate: Alternative Models For A Restructured Electricity Industry. Presented at the M.I.T. Electric Utility Program Workshop (18 October).

Chao, H. and S. Peck. 1996. Spot Markets in Electric Power Networks: Theory. (30 December).

Charles River Associates. 1992. *DSM Process Evaluation: A Guidebook to Current Practice.* Palo Alto, CA: Electric Power Research Institute.

Charles River Associates. 1985. *Issues in Pipeline Carriage of Natural Gas and Wheeling of Electricity: A Comparative Analysis.* Washington, DC: Edison Electric Institute.

Charles River Associates, et. al. 1986. *Electric Rate Shocks: Origins, Prospects, and Remedies.* A report to the U.S. Department of Energy. With Pfeffer, Lindsay and Associates and QED Research, Inc. CRA Report No. 772.03.

Charles River Associates. 1991. *Acquisition of Third-Party Demand-side Management Resources.* Palo Alto, CA: Electric Power Research Institute.

Charles River Associates. 1988. "The Immediate Consequences of an Oil Supply Emergency for the Financial Markets and Major User Groups." Office of Energy Emergencies, U.S. Department of Energy.

Chase, M.A. 1970. "Power Pools: What to Do After You're In." *Public Power* (February).

Chernick, P., S. Geller, R. Brailove, J. Wallach and A. Auster. 1996. "Estimation of Market Value, Stranded Investment, and Restructuring Gains for Major Massachusetts Utilities." *Resource Insight, Inc.* (17 April).

Chernick, P.L. 1983. "Revenue Stability Target Rate Making." *Public Utilities Fortnightly* 111 (17 February): 35–39.

Childs, M.W. 1980. *Yesterday, Today and Tomorrow—The Farmer Takes a Hand.* Washington, DC: National Rural Electric Cooperative Association.

Christensen Associates. 1995a. "Real-Time Pricing Quick Start Guide: Field Real-Time Pricing." Electric Power Research Institute (June).

Christensen Associates. 1995b. "Fielding a Real-Time Pricing Program—Rennsylvania Power & Light Case Study." Electric Power Research Institute (August).

Christensen Associates. 1988. "Customer Response to Interruptible and Curtailable Rates." Volume 1: Methology and Results. Electric Power Research Institute.

Christensen, L.R. 1986. *Bypass in Regulated Industries: Specific Examples with Emphasis on the Natural Gas Industry.* Prepared for the California Energy Commission.

Christensen, L.R. and W. Greene. 1976. "Economies of Scale in U.S. Electric Power Generation." *Journal of Political Economy* 84 (4): 655–76.

Cicchetti, C.J., W.J. Gillen and P. Smolensky. 1976. *Marginal Cost Pricing of Electricity: An Applied Approach.* The Planning and Conservation Foundation (June).

Cicchetti, C.J. 1975. "The Design of Electricity Tariffs." *Public Utilities Fortnightly* 96 (28 August): 25–33.

Clark, W. 1975. "The Coming Decentralization of America's Electrical System." In *Energy for Survival,* 207–245. New York: Anchor Books.

Clean Air Act. 1992. *Report of the Office of Air and Radiation to Administrator William K. Reilly.* "Implementing the 1990 Clean Air Act: The First Two Years." United States Environmental Protection Agency. Washington, DC. EPA400-R-92-013.

Clemens, E.W. 1950. *Economics and Public Utilities.* New York: Appleton Century Crofts.

Clemons, E.K. 1991. "Information Technology and the Boundary of the Firm: Who Wins, Who Loses, Who Has to Change." Proceedings of the Harvard Colloquium on Global Competition and Telecommunications. Cambridge, MA: Harvard Business School.

Coase, R.H. 1988. *The Firm, the Market, and the Law.* Chicago: Univ. of Chicago Press.

Coase, R.H. 1960. "The Problem of Social Cost." *Journal of Law and Economics* (3): 1–44.

Coase, R.H. 1937. The Nature of the Firm. *Economica* 4: 386–405.

Cohen, A. 1994. "The Political Economy of Retail Wheeling, or How to Not Re-Fight the Last War." In *The Electric Industry in Transition,* 71–91. Arlington, VA: Public Utilities Reports, Inc. & the New York State Energy Research and Development Authority.

Cohen, A. 1994. *Retail Wheeling and Industry Restructuring: Implications for Renewables and the Environment.* Conservation Law Foundation (April).

Cohen, A. and S. Kilhm. 1994. "The Political Economy of Retail Wheeling, or How to Not Re-fight the Last War." *The Electricity Journal* (April).

Cohen, M. 1979. "Efficiency and Competition in the Electric-Power Industry." *Yale Law Journal* 88 (June): 1511–1549.

Colton, R. 1996a. *Funding Stranded Benefits for Low-Income Consumers In a Restructured Electric Industry: A State Data Book* (November).

Colton, R.D. 1996b. *Assessing Impact on Small-Business, Residential and Low-Income Customers: The Electricity Industry Restructuring Series.* The National Council on Competition and the Electricity Industry (October).

Colton, R. 1995b. *Performance-Based Evaluation of Customer Collections in a Competitive Electric Utility Industry.* FSC Papers on Small Users and a Competitive Electric Utility Industry. Current as of November 29, 1996.

Colton, R. 1995c. *Understanding "Redlining" in a Competitive Electric Utility Industry.* FSC Papers on Small Users and a Competitive Electric Utility Industry. Current as of November 29, 1996.

Colton, R. 1995d. *Models of Low-Income Utility Rates.* FSC Papers on Small Users and a Competitive Electric Utility Industry. Current as of November 29, 1996.

Colton, R. 1990. "Energy and the Poor: The Association of Consumption with Income." Boston: National Consumer Law Center.

Committee of Life Sciences. 1996. "Possible Health Effects of Expose to Electromagnetic Fields (Draft)." National Research Council.

Commonwealth of Virginia State Corporation Commission. 1996. *Staff Report on the Restructuring of the Electric Industry.* Volume 1. Case No PUE 950089 (July).

Comnes, G.A., S. Stoft, N. Green and L.J. Hill. 1995. "Performance-Based Ratemaking for Electric Utilities." Berkeley, CA: Lawrence Berkeley National Laboratory. LBL-37577.

Comnes, G.A. and S. Stoft, N. Green. 1995. "Performance-Based Ratemaking for Electric Utilties: Review of Plan and Analysis of Economic and Resource-Planning Issues." Draft Report (September).

Comnes, G.A., S. Stoft, N. Greene and L. Hill. 1996. "Six Useful Observations for Designers of PBR Plans" *Electricity Journal* (April): 16–23.

Comnes, G. A. 1994. Review of Ratemaking Incentive Plans for U.S. Gas Distribution Companies. Presented at the Rutgers University Advanced Workshop in Regulation and Public Utility Economics, Seventh Annual Western Conference, July 6–8, San Diego.

Congressional Research Service. 1976. "National Power Grid System Study—An Overview of Economics, Regulatory, and Engineering Aspects." Washington, D.C.

Consumer Energy Council of America Research Foundation. 1993. *Incorporating Environmental Externalities Into Utility Planning: Seeking A Cost-Effective Means of Assuring Environmental Quality.* A Report of the Consumer Energy Council of America Research Foundation, Washington, DC.

Consumer Energy Council of America Research Foundation. 1990. *Transmission Planning, Siting, and Certification in the 1990s.* Washington, D.C.

Cooper, M.N. 1994. "Protecting the Public Interest in the Transition to Competition in New York Industries." In *The Electric Industry in Transition,* 151–166. Arlington, VA: Public Utilities Reports, Inc. & the New York State Energy Research and Development Authority.

Coopers & Lybrand. 1993. "Electric Municipalization and the Energy Policy Act of 1992."

Copeland, B.L. 1978. "Alternative Cost-of-Capital Concepts in Regulation." *Land Economics* 54 (3): 348–361.

Corneli, S.B. 1996. "Will Customer Choice Always Lower Cost?" *The Electricity Journal* 8 (October).

Cornell, N.W. and D.W. Weebbink. 1985. "Public Utility Rate-of-Return Regulation: Can It Ever Protect Customers?" In *Unnatural Monopolies: The Case for Deregulating Public Utilities*. Pool, ed. Lexington, MA: Lexington Books.

"Corruption of the 'Cogeneration Ethic'." 1987. Letter to the editor in *Cogeneration Journal* 2 (3): 77.

Costello, K.W. 1996. "Revenue Caps or Price Caps? Robust Competition Later Means Healthy Choices Now." *Public Utilities Fortnightly*. 134 (1 May): 28–33.

Costello, K., R.E. Burns and Y. Hegazy. 1994. "Overview of Issues Relating to The Retail Wheeling of Electricity." The National Regulatory Research Institute (May).

Costello, K.W., and S-B. Cho. 1991 "A Review Of FERC's Technical Reports On Incentive Regulation."Columbus, OH: The National Regulatory Research Institute. NRRRI 91–9.

Council On Economic Regulation. 1988. *Competition and Regulation in Electricity Markets.*

Cowan, T. 1990. *The Theory of Market Failure.* Fairfax, VA: George Mason Univ. Press.

Cramer, C. And J. Tschurhart. 1983. "Power Pooling: An Exercise in Industrial Coordination." *Land Economics* 59 (1): 24–34.

Crampes, C., and J.J. Laffont. "Transfers and Incentives in the Spanish Electricity Sector" Gramaq, Idei, Universite des Sciences Sociales de Toulouse.

Crane, C.M., S.S. Leonard, J.J. Russo, and C.R.E. Bolton. 1991. Pacific Gas and Electric Company Marketing Services Department.

Crew, M. and P.Kleindorfer. 1978. "Reality and Public Utility Pricing." *American Economic Review* 68 (March): 31–40.

Crew, Michael A. ed., 1980. *Issues in Public Utility Economics and Regulation.* Lexington, MA: Lexington Books.

Crew, M.A. ed., 1982. *Regulatory Reform and Public Utilities.* Lexington, MA.: The MIT Press.

Crew, M.A. and P.R. Kleindorfer. 1986. The *Economics of Public Utility Regulation.* Cambridge, MA: The MIT Press.

Crew, M.A., and C.K. Rowley. 1987. A Public Choice Theory of Monopoly Regulation. Presented at the Advanced Workshop in Public Utility Economics and Regulation: Sixth Annual Conference, New York.

Crocker, K. 1983. "Vertical Integration and the Strategic Use of Private Information." *Bell Journal of Economics.* 14: 236–248.

Cross, L. 1989. "A Primer on Recent FERC Actions Final Orders No. 451 and 380." *Gas Energy Review* 15 (7): 2–7.

Cudahy, R.D. 1989. "Return on Investment and Fairness in Regulation." *Public Utilities Fortnightly* 123 (3): 19–23.

Cudahy, R. and T. Dreessen 1996. *A Review of the Energy Services Company (ESCO) Industry in the United States.* Prepared by NAESCO for World Bank Industry and Energy Department.

(December).

Cudahy, R.D. and J.R. Malko. 1976. "Electric Peak Load Pricing: Madison Gas and Beyond." *Wisconsin Law Review*. (3):47.

Culbreath, H.L. 1981. An Overview of the Financial Difficulties of the Electric Utility Industry. Testimony before the House Subcommittee on Energy Conservation and Power.

Curtis, C.B. 1992. "Maintaining a Proper Balance Between Federal and State Authority—Is There a Place for Regional Regulation?" *The Electricity Journal* (January/February).

Curtis, C.B. 1992. "A Modest Proposal." *The Electricity Journal* (6 January).

Danielsen, A.L. and D.R. Kamerschen. eds. 1983. *Current Issues in Public Utility Economics: Essay in Honor of James Bonbright*. Lexington, MA: Lexington Books.

Danner, C. 1983. "Common Ground." *Public Utilities Fortnightly* (1 May).

Danner, C. 1986. Prepared Direct Testimony of Dr. Carl Danner. Illinois Commerce Commission Docket 86-0256. Illinois Citizens Utility Board.

Dasovich, J., W. Meyer and V.A. Coe. 1993. "California's Electric Services Industry: Perspectives on the Past, Strategies for the Future." Division of Strategic Planning, California Public Utilities Commission, San Francisco, CA.

David, P.A. 1970. "Learning by Doing and Tariff Protection: A Reconsideration of the Case of the Ante-Bellum United States Cotton Textile Industry." *Journal of Economic History* (30): 521–601.

Davidow, W.H. and M.S. Malone. 1992. *The Virtual Corporation: Structuring and Revitalizing the Corporation for the Twenty-First Century*. New York: Harper Business.

Davis, R. 1996. Stranded Cost Recovery: How Much Is Justified? Presented at the National Leadership Summit on: Information Technology and Energy Policy. The Alliance For Competitive Electricity.

Deloitte & Touche, LLP. 1996."Federal, State and Local Tax Implication of Electric Utility Industry Restructuring: The Electric Industry Restructuring Series." The National Council on Competition and the Electric Industry (October).

Demsetz, H. 1968. "Why Regulate Utilities?" *Journal of Law and Economics* 11 (April): 55–65.

Dixit, A. 1992. "Investment and Hysteresis." *Journal of Economics Perspectives* (1): 107–32.

Douglas, John. 1996. "Custom Power: Optimizing Distribution Services". *EPRI Journal*. (May/June).

Dowlatabadi, H. and M. Toman. 1989. "Changes in Electricity Markets and Implications for Generation Technologies." Washington, DC: Resources for the Future, Energy and Natural Resources Division. (January).

DPA Group Inc. and Charles River Associates Inc. 1989. *BC Hydro: Cost Risk Areas and Management Options from Purchased Power*. Vancouver, British Columbia: The DPA Group.

Dunn, Jr., W.H. 1996. "Practical Aspects of Electrical Restructuring." *The Electricity Journal* 9 (8).

Dworzak, D. and J. Long. 1990. *Regulation of State Bidding Programs*. v.II. Washington, D.C.: Edison Electric Institute.

ECO Northwest. 1993. "Environmental Externalities and Electric Utility Regulation." National Association of Regulatory Utility Commissioners (September).

Edison Electric Institute. 1989b. "Customer Wheeling: A Fiction, Contrary to the Public Interest." *Transmission Issues Monograph* (3).

Edison Electric Institute. 1996. "Performance Based Regulation: Design and Implementation Strategies." Washington, DC. #40-96-30.

Edison Electric Institute. 1995. "Restructuring of the Natural Gas Pipeline Industry to Assure Equality of Gas Transportation: Application to Open Access Electric Transportation." Washington, DC. #40-95-15.

Edison Electric Institute. 1995b. "A Price Cap Designer's Handbook." Washington, DC. #04-94-24.

Edison Electric Institute. 1994. "Integrated Resource Planning in the States: 1994 Sourcebook." Washington, DC.

Edison Electric Institute. 1994a. "Environmental Externalities: An Issue Under Critical Review." Washington, DC.

Edison Electric Institute. 1994b. "Electrotechnology: An Environmental Solution." Washington, DC.

Edison Electric Institute. 1993. "Types of Incentive Regulation: A Primer for the Electric Utility Industry. Washington, DC: Edison Electric Institute Finance, Regulation and Power Supply Policy Group.

Edison Electric Institute. 1992. "The Case Against Retail Wheeling: A Response to Advocates of Retail Wheeling." *Transmission Issues Monograph* (5).

Edison Electric Institute. 1991a. "Restructuring And Sale of the Electricity Supply Industry in England and Wales." Washington, DC.

Edison Electric Institute. 1991b. "Recovery of Indirect Costs of Transmission Services." Washington DC. #04-91-13.

Edison Electric Institute. 1989. "Capacity and Generation of Non-Utility Sources of Energy." Washington, D.C.

Edison Electric Institute. 1988a. "Engineering and Reliability Effects of Increased Wheeling and Transmission Access: Factors for Consideration." *Transmission Issues Monograph* (2).

Edison Electric Institute. 1988b. Motion of Edison Electric Institute on Rehearing. Docket EC88-2-000. United States of America before the Federal Energy Regulatory Commission.

Edison Electric Institute. 1986. "Restructuring in the Electric Utility Industry: The Legal and Regulatory Considerations." Report prepared by Nixon, Hargrave, Devans & Doyle.

Edison Electric Institute. 1983. *Electric Utility Diversification: A Guide to the Strategic Issues and Options.* Handbook, Volume 1. Washington, DC.

Edison Electric Institute. 1982. "Alternative Models of Electric Power Deregulation." Washington, DC.

Edison Electric Institute. 1974. *Historical Statistics of the Electric Utility Industry Through 1970.* Washington, DC.

Edison Electric Institute. 1972. "Ten Year Report on Load Diversity Based on 1962-71 Load Data." Washington, DC.

Edison Electric Institute 1970. "Major Electric Power Facilities and the Environment." Washington, DC: Plant Siting Task Force.

Einhorn, M.A. 1990. "Electricity Wheeling and Incentive Regulation." *Journal of Regulatory Economics* (2):173–189.

ELCON (the Electricity Consumers Council) 1987. *Efforts to Encourage Competition Without Transmission Access are Fatally Flawed. Memorandum to Federal Energy Regulatory Commission.* (7 September).

Electric Power Research Institute. 1995. "Fielding A Real-time Pricing Program: Pennsylvania Power & Light Case Study." TR-105042. Final Report August. Madison, Wisconsin: Prepared by Christensen Associates.

Electric Power Research Institute. 1996. "Interconnected Operations (Ancillary) Services: Workshop to Review Final Reports on Definitions and Requirements for Managing Unbundled IOS." (August).

Electric Power Research Institute. 1996. *Distributed Resources: A Market Assessment.* Palo Alto, CA: TR-106055.

Electric Power Research Institute. 1992. *Distributed Utility Valuation Project* Monograph. Palo Alto, CA: (October).

Electric Power Research Institute. 1988. *Technical Assessment Guide.* Palo Alto, CA.

Electric Power Research Institute. 1983. *Study of Power Plant Construction Lead Times.* Palo Alto, CA.

Electrical World. 1996. "Competition, Deregulation: Is the US Rushing into the Dark?" *Electrical World T&D Edition* 210 (10).

Electricity Journal. 1990. "Wheeling, Dealing and Data." (Jan/Feb).

Energetics, 1995. *Mechanisms to Address Stranded Benefits: A Compilation of Recent Proposals.* Prepared for National Renewable Energy Laboratory and Competitive Resource Strategies Office of Utility Technologies. (October).

Energetics, 1995. *Making Electricity Restructuring Work for Small Customers: A Preliminary Review of Fourteen "Load Aggregation" Proposals.* Prepared for The National Renewable Energy Laboratory. Washington, DC.

Energy, Economics and Climate Change. 1994. *Policy Update.* Cutter Information Corp.

Energy and Environmental Economics, 1996. "Summary of Area and Time-Specific Cost Studies." Manuscript. San Francisco, CA.

Energy and Environmental Economics. 1994. "Designing Profitable Rate Options Using Area and Time-Specific Costs." Prepared by Central Power and Light Company. TC3629 Project Results: (November).

Energy Information Administration (EIA). 1996. *Emissions of Greenhouse Gases in the United States* 1995. DOE/EIA-0573 (95).

Energy Information Administration (EIA). 1995. *Annual Energy Outlook 1995—With Projection To 2010.* (January).

Energy Information Administration (EIA). 1995. *World Nuclear Outlook.* DOE/EIA-0436(95).

Energy Information Administration (EIA). 1995. *U.S. Electric Utility Demand-Side Management 1993.* DOE/EIA-0589(93).

Energy Information Administration (EIA). 1995a. *International Energy Outlook.* DOE/EIA-0484(95).

Energy Information Administration (EIA). 1995b. *Electric Power Annual,* 1994. DOE/EIA-0348(94).

Energy Information Administration (EIA). 1995c. *Electric Sales and Revenue 1993.* DOE/EIA 0540 (93).

Energy Information Administration (EIA). 1995d. *Financial Statistics of Major Publicly Owned Electric Utilities, 1994.* DOE/EIA-0437(94)2.

Energy Information Administration (EIA). 1994. *Financial Impacts of Nonutility Power Purchases on Investor-Owned Electric Utilities.* DOE/EIA-0580.

Energy Information Administration (EIA). 1993a. *Financial Statistics of Major U.S. Investor-Owned Utilities 1992.* DOE/EIA-0437(92)1.

Energy Information Administration (EIA). 1993b. *The Changing Structure of the Electric Power Industry, 1970–91.* DOE/EIA-0562.

Energy Information Administration (EIA). 1993c. *The Public Utility Holding Company Act of 1935.* DOE/EIA-0563.

Environmental Protection Agency (EPA) 1995. *Acid Rain Program Emissions Scorecard 1994.* EPA 430/R-95-012.

Environmental Protection Agency (EPA). 1995b. *National Air Quality and Emissions Trend Report, 1995.*

Environmental Protection Agency (EPA). 1994. *National Air Pollution Trends Report. 2.0 Summary of 1994 Emissions.*

Environmental Protection Agency (EPA). 1994. "Summary of National Emission Trends and Economic, Demographic and Regulatory Influences on Historic Trends in Emissions." *National Air Pollution Trends Report.*

Environmental Protection Agency (EPA). 1994. "Appendix A: National Emissions (1970 to 1994) by Subcategory." *National Air Pollution Trends Report.*

Environmental Protection Agency (EPA). 1993. *The Plain English Guide to The Clean Air Act.* (April) EPA 400-K-93-001.

Environmental Protection Agency (EPA). 1992. *Implementing the 1990 Clean Air Act: The First Two Years.* Washington, DC: Office of Air and Radiation. (November) EPA 400-R-92-013.

Eto, J., C. Goldman, and M.S. Kito. 1996. "Ratepayer-Funded DMS After Restructuring." *The Electricity Journal* 9(7).

Eto, J., S. Kito, L. Shown and R. Sonnenblick. 1995. "Where did the Money Go? The Cost and Performance of the Largest Commercial Sector DSM Programs." Lawrence Berkeley Laboratory. (December).

Eto, J., E. Vine, L. Shown, R. Sonneblick and C. Payne. "The Total Cost and Measured Performance of Utility-Sponsored Energy Efficiency Programs." *Energy Journal* 17: 31–52.

Evans, D.S., ed. 1989. *Breaking Up Bell: Essays on Industrial Organization and Regulation.* New York: North-Holland.

Evans, L. and S. Garber. 1985. "Public Utility Regulators Are Only Human: A Positive Theory of Rational Constraints." *American Economic Review* (June).

Evans, N. 1995. "Unbundling of Services in a Competitive Wholesale Market: Lessons from the UK." In Proceedings: 1996 EPRI Conference on Innovative Approaches to Electricity Pricing. Palo Alto, CA: Electric Power Research Institute. (March) EPRI TR-106232.

Fairman, J.F., and J.C. Scott. 1977. "Transmission, Power Pools, and Competition in the Electric Utility Industry." *Hastings Law Journal* 28: 1159–1207.

Fairman, J.F. 1995 "The Franchise Bottleneck." *The Electricity Journal* (May).

Fanara, P. Jr., J. E. Suelflow and R.A. Draba. 1980. "Energy and Competition: The Saga of Electricity Power." *Antitrust Bulletin* (Spring): 125–142.

Fang, J.M. and P.S. Galen. 1996. *Electric Industry Restructuring and Environmental Issues: A Comparative Analysis of the Experience in California, New York and Wisconsin.* Golden, CO: National Renewable Energy Laboratory. NREL/TP-461-20478.

Farhar, B.C. 1993. *Trends in Public Perceptions and Preferences on Energy and Environmental Policy.* Report prepared for and sponsored by the Office of Planning and Assessment, Office of Conservation and Renewable Energy, U.S. Department of Energy. Golden, Colorado: National Renewable Energy Laboratory. (February).

Faruqui, A. and J.R. Malko. 1983. "The Residential Demand for Electricity by Time-of-Use: A Survey of Evidence from Twelve Experiments with Peak Load Pricing." *Energy: The International Journal.* (4): 781–795.

Faulhaber, G.R. 1987. The FCC's Path to Deregulation: Turnpike or Quagmire?" *Public Utilities Fortnightly* 120 (5): 22–26.

Federal Energy Administration, 1975. "Survey of Electric Utility Powerplant Delays." FEA/G-75/495.

Federal Energy Regulatory Commission (FERC). 1983. "Electric Rate Handbook." Prepared by Michael E. Small, Special Assistant to Deputy General Counsel for Litigation and Enforcement. (October).

Federal Energy Regulatory Commission (FERC). 1993. Staff Discussion Paper, Transmission Pricing Issues. Docket RM 93-19-000 (30 June).

Federal Energy Regulatory Commission (FERC). 1989. *The Transmission Task Force's Report to the Commission: Electricity Transmission: Realities, Theory, and Policy Alternatives.* Washington, DC.

Federal Energy Regulatory Commission (FERC). 1987. *Regulating Independent Power Producers: A Policy Analysis.* Washington, DC: Office of Economic Policy.

Federal Energy Regulatory Commission (FERC). 1984. *A Review of Selected Transmission Agreements.* Washington, DC: Office of Electric Power Regulation.

Federal Energy Regulatory Commission (FERC). 1983. "Electric Rate Handbook." Prepared by Michael E. Small, Special Assistant to Deputy General Counsel for Litigation and Enforcement. (October).

Federal Energy Regulatory Commission (FERC). 1981. *Power Pooling in the United States.* FERC-0049. Washington, DC: Office of Electric Power Regulation.

Federal Power Commission. 1967. "Prevention of Power Failures." Volume I-Report of the Commission (July).

Feiler, T. and C.Seiple. 1995. "Electric Stranded Investment: Not as Much as You Think." *Public Utilities Fornightly* (15 January).

Feist, J., R. Farhang, J. Erickson, E. Sterakos, P. Brodie and P. Liepe. 1994. "Super-Efficient Refrigerators: The Golden Carrot from Concept to Reality." Summer Study on Energy Efficiency.

Feldman, D.L.,ed. 1996. *The Energy Crisis: Unresolved Issues and Enduring Legacies.* Baltimore: The Johns Hopkins Univ. Press.

Ferguson, J.S. 1996. "Evolution or Revolution: Dismantling the FASB Standard on Decommissioning Costs." *Public Utilities Fortnightly* 134 (10): 26–31.

Ferguson, J.S. 1995. "Fossil Plant Decommissioning: Tracking Deferred Cost in a Competitive Market." *Public Utilities Fortnightly* 133 (12).

Fernando, C., P. Kleindorfer, R.D. Tabors, F. Pickel and S.J. Robinson. 1995. *Unbundling the US Electric Power Industry: A Blueprint for Change.* Tabors Caramanis & Associates. Cambridge, MA. (March).

Fessler, D.W. 1996. "Is a Social Perspective Sustainable in the Changing Dynamic of the Electric Service Industry?" California Public Utilities Commission (26 March).

Finance, Regulation & Power Supply Policy Group. 1996. "Utilities and Energy Efficiency Services Under Increasing Competition." Washington, DC: Edison Electric Institute.

Finsinger J. and I. Vogelsang. 1981. "Alternative Institutional Frameworks for Price Incentive Mechanisms." *Kyklos* 34: 388–404.

Fisher, C.F., Jr., S. Paik, and W.R. Schriver. 1986. *Power Plant Economy of Scale and Cost Trends—Further Analyses and Review of Empirical Studies.* Report Prepared by Construction Resources Analysis, The University of Tennessee, for Oak Ridge National Laboratory.

Flaim, Dr. T.A. 1994. "Methods for Dealing with Transition Costs for the Electric Utility Industry". In *The Electric Industry in Transition.* Arlington, VA: Public Utilities Reports, Inc. & the New York State Energy Research and Development Authority.

Flanigan, J. 1988. Utilities Light Up With Mergers. *Boston Business Journal* (15 August): 20.

Ford Foundation, Energy Policy Project. 1974. *A Time to Choose: America's Energy Future.* Final Report. Consumers Union ed. Cambridge, MA: Ballinger.

Fox-Penner, P.S. 1996. "The Impacts of PURPA, Retrospect and Prospect." Article available from the author.

Fox-Penner, P.S. 1994. "Critical Trends in State Utility Regulation." *Natural Resources and the Environment.* (8): 17–20.

Fox-Penner, P.S. 1993. "Efficiency and the Public Interest: QF Transmission and the Energy Policy Act." *Energy Law Journal* (14): 51–73.

Fox-Penner, P.S. 1992. "Quality Maintenance and Monitoring in Price-Cap Regulation." Article available from the author.

Fox-Penner, P.S., P. O'Rourke, and P.J. Spinney. 1990. *Competetive Procurement of Electric Utility Resources.* Palo Alto, CA: Electric Power Research Institute. CU-6898s.

Fox-Penner, P.S. 1990a. "Regulating Independent Power Producers: Lessons of the PURPA Approach." *Resources and Energy* (12): 117–141.

Fox-Penner, P.S. 1990b. "Cogeneration After PURPA: Energy Conservation and Industry Structure." *Journal of Law and Economics* 33 (October): 517–552.

Fox-Penner, P.S. 1990c. *Electric Power Transmission and Wheeling: A Technical Primer.* Washington, DC: The Edison Electric Institute.

Fox-Penner, P.S. 1990d. "Is Deintegrated Generation Efficient? A Proposed Empirical Research Framework." Proceedings of 13th Annual International Conference of the International Association of Energy Economists. Copenhagen, Denmark. (June).

Fox-Penner, P.S. 1990e. "Self-Dealing and Utility Generation Purchases: Precedents and Options." Boston, MA: Charles River Associates.

Fox-Penner, P.S. 1989. IPP Bidding: The View from Today's Utilities. Presented at the American Cogeneration Association Annual Conference.

Frame, R. 1993. Characteristics of a "Good" Retail Wheeling System. Presented at the Electric Utility Business Environment Conference, March 16. Electric Utility Consultants, Inc., Denver, Colorado.

Freeman, D. 1994. "Competition in the Electric Industry: An Unguided Missile?" In *The Electric Industry in Transition,* 15–20. Arlington, VA: Public Utilities Reports, Inc. & the New York State Energy Research and Development Authority.

French, R.X., and S.Z. Haddad. 1981. "The Economics of Reliability and Scale in Generating Unit Size Selection." *Public Utilities Fortnightly* 107 (8): 33–38.

French, K., and R. McCormick. 1984. "Sealed Bids, Sunk Costs and the Process of Competition." *Journal of Business* 57: 417–443.

Fromm, G., ed. 1981. *Studies in Public Regulation.* Cambridge, MA: MIT Press.

Fungiello, P.J. 1973. *Toward a National Power Policy.* Pittsburgh: Univ. of Pittsburgh Press.

Galbraith, J.K. 1954. *The Great Crash of 1929.* Boston: Houghton-Mifflin.

Garfield, P.J. and W. Lovejoy. 1964. *Public Utility Economics.* Englewood Cliffs, N.J.: Prentice-Hall.

Gegax, D. and K. Nowotny. 1993. "Competition in the Electric Utility Industry." *Yale Journal of Regulation* 10 (Winter): 63–88.

Gegax, D. and J. Tschirhart. "An Analysis of Interfirm Cooperation: Theory and Evidence from Electric Power Pools." *Southern Economic Journal* 50 (April): 1077–1098.

Gellings, C.W. and J. S. McMenamin. 1993. "DSM: What's Real, What's Not, and How We Can Tell the Difference." Electric Power Research Institute and Regional Economic Research, Inc. (10 July).

General Electric Company. 1979. *Load Modeling, Generation Expansion, Production Costing, Investment Costing.* Schenectady, NY: Electric Utility Engineering.

George, S.S., and M.M. Schnitzer. 1989. Multi-Attribute Bidding for Future Energy Resources: Results from a Simulated Auction. Presented at Utility Opportunities for New Generation, sponsored by the Edison Electric Institute and Electric Power Research Institute.

Gilbert, R.J. and E.P. Kahn. *International Comparisons of Electricity Regulation.* Cambridge: Cambridge Univ. Press.

Gilbert, J.S. 1989. "Cogeneration—Competition or Resource?" In *Utility Opportunities for New Generation.* Synergic Resources Corp, ed. Palo Alto, CA: Electric Power Research Institute. CU-6605.

Gilsdorf, K. 1988. *Do Cost Economies Arise from Vertical Integration in the Electric Utility Industry?* St. Paul, MN: Energy Issues Intervention Office, Minnesota Department of Public Service.

Gilsdorf, K. 1995. "Testing for Subadditivity of Vertically Integrated Electric Utilities." *Southern Economic Journal* 62 (July): 126–138.

Gjengedal, T., C. Marney and S. Johansen. 1993. Reducing Environmental Impacts from Electric Power Production through Environmental Dispatch. Presented at the ECEEE 1993 Summer Study. (Vol 2 of Proceedings): 169–179.

Goldberg, V., ed. 1989. *Readings in the Economics of Contract Law.* Cambridge: Cambridge Univ. Press.

Goldman, C.A. and J.F. Busch. 1992. "DSM Bidding—The Next Generation: Early Experience with DSM Bidding suggests Nonutility Service Providers can Absorb Substantial Risk of Performance, Although Bid Prices are Sometimes High. ESCOs and Utilities are Moving Toward Partnerships that Make Good Use of the Strengths of Both Groups." *The Electricity Journal.* (May): 34–43.

Goldman, C. and D. Dayton. 1996. "Future Prospects for ESCOs in a Restructured Electric Industry." Proceedings, ACEEE 1996 Summer Study on Energy Efficiency in Buildings. Washington, DC: American Council for an Energy Department. (December).

Goldman, C., M. Kito and M. Moezzi. 1995. "Evaluation of Public Service Electric & Gas Company's Standard Offer Program." Berkeley, CA: Lawrence Berkeley National Laboratory Report (July).

Goldstone, S. 1996. "The Changing Nature of the Public Interest in Energy Efficiency Due to Restructuring: Implications for the Two Track Approach." Prepared for the California Public Utility Commission Energy Service Working Group (19 April).

Gollop, F.M. "Environmental Regulations and Productivity Growth: The Case of Fossil-Fueled Generation," *Journal of Political Economy* 91 (August): 654–74.

Golub, B.W., Bohn, R., Schweppe, F. and R. Tabors. 1989. "An Approach for Deregulating the Generation of Electricity" In *Electric Power Strategic Issues.* Plummer, J., ed. Arlington, VA: Public Utilities Reports, Inc.

Golub, B.W., and L.S. Hyman. 1983. "The Financial Difficulties and Consequences of Deregulation Through Divestiture." *Public Utilities Fortnightly* 111 (4): 19–25.

Goodwin, L.M. 1990. *Avoiding the Minefields in Power Sales Contracts. Cogeneration Project Handbook.* Fairfield, CT: Pequot Publishing.

Goodwin, L.M. 1989a. "Negotiating Power Contracts." *Independent Energy* (September): 57–60.

Goodwin, L.M. 1989b. Current Issues in Power Sale Contract Negotiations. Presented at the Third Annual Meeting and Exposition of the American Cogeneration Association and Cogeneration and Independent Power Coalition of America.

Gordon, R.J. 1982. "The Productivity Slowdown in the Steam-Electric Generating Industry." Evanston, IL: Northwestern University.

Gordon, R.L. 1990. "Timidity in Electric Utility Deregulation." *Resources and Energy* (Netherlands) 12(1): 17–32.

Gordon, R. 1990. "Deregulating Electric Utilities." *Resources and Energy* 12.

Graves, F.C. *Capacity Prices in a Competitive Power Market.* Technical report prepared by The Brattle Group. Cambridge, MA.

Graves, F. 1995. Prices and Procedures of an ISO in Supporting a Competitive Power Market. Presented at the Restructuring Electric Transmission Conference, September 27.

Gray, H. M. 1940. "The Passing of the Public Utility Concept." *Journal of Land and Public Utility Economics* (February).

Green, R.J. and D.M. Newbery. 1996. "Competition in the British Electricity Spot Market." *Journal of Political Economy.* Univ. of Chicago Press.

Green, D.G. 1990. The Antitrust Laws And Transmission Access In The Electric Utility Industry: An Outline of Basic Principles. Presented at American Bar Association Annual Meeting, August, Washington, D.C.

Greenwade, J.D. 1987. "Access by Cogenerators to Utility Transmission Systems." *Cogeneration Journal* 2 (3): 52–62.

Griffes, P.H. 1990. *Institutions Surrounding Jointly-Owned Electricity Generating Plants.* Preliminary. Chicago: U.S. General Accounting Office and University of Chicago.

Gross, G., Dr. and Dr. N. Balu. 1996. *A Synthesis Paper on Wholesale Markets, Power Pool Proposals.* Prepared for The National Council on Competition and the Electric Industry.

Gruening, E. 1964. *The Public Pays: A Study of Power Propaganda.* New York: Vanguard Press.

Guyant, A. 1995. "The Issues and Time Remaining Before Super Hardships Arrive in Low-Income Electric Utility Service." Public Service Commission of Wisconsin (June).

Haase, P. 1996. "Stability Assessment." *EPRI Journal* (July/August).

Hadley, S. and E. Hirst. 1995. "Utility DSM Programs from 1989 Through 1998: Continuation or Cross Roads?" Oak Ridge National Laboratory. ORLN/CON-40.

Hagerman, R.L. and B.T. Ratchford. 1978. "Some Determinants of Allowed Rates of Return on Equity to Electric Utilities." *Bell Journal of Economics* (Spring): 46–55.

Haman-Guild, R. and J.L. Pfeffer. 1987. "Competitive Bidding for New Electric Power Supplies: Deregulation or Reregulation?" *Public Utilities Fortnightly* 120 (6): 9–20.

Hamilton, N. and C.L. Bros. 1985. "The Need for Standard Contracts and Prices for Small Power Producers." *Public Utilities Fortnightly* 115 (11).

Hamrin J., W. Marcus, F. Morse and C. Weinberg. 1994. "Affected with the Public Interest— Electric Utility Restructuring in an Era of Competition." The National Association of Regulatory Utility Commissioners Renewable Subcommittee of the Conservation Committee (September).

Hanser, P., J. Wharton and P. Fox-Penner. 1997. "Real-Time Pricing—Restructuring's Big Bang". *Public Utilities Fortnightly* 135 (5).

Hart, O. and B. Holmstrom. 1986. "The Theory of Contracts." In *Advances in Economic Theory.* T. Bewley, ed. Cambridge, MA: Cambridge Univ. Press.

Harte, J. 1992. "Acid Rain". In *The Energy-Environment Connection,* 50–74. J.M. Hollander, ed. Washington, DC: Island Press.

Hartman, 1991. "A Critical Analysis of the Proposed Merger Between Kansas Power and Light Company and Kansas Gas and Electric Company." Berkeley, CA: Law and Economics Consulting Group, Inc. (April).

Hassett, K. and G. Metcalf. 1991. Residential Energy Tax Credits and Home Improvement Behavior. Presented at Universities Research Conference: Economics of the Environment. National Bureau of Economic Research, Inc. Cambridge, MA.

Hassett, K. and G. Metcalf. 1993. "Energy Conservation Investment: Do Consumers Discount the Future Correctly?" *Energy Policy,* 21 (6): 710–716.

Hawes, D.W. 1995. "Electric Utility Mergers and Aquisitions Seen in a Larger Perspective." *The Electricity Journal* 8 (8): 11.

Hazlett, T. 1985. "Private Contracting versus Public Regulation as a Solution to the Natural Monopoly Problem." In *Unnatural Monopolies: The Case for Deregulating Public Utilities.* Poole, R.W., Jr., ed. Lexington, MA: Lexington Books.

Heimann, F. 1991. "The Electric Policy Study." *Research in Law and Economics.* 13: 1–6.

Hellman, R. 1972. *Government Competition in the Electric Utility Industry.* NY: Praeger.

Hempling, S. and N. Rader. 1996. "State Implementatiiom of Renewable Portfolio Standards: A Review of Federal Law Issues." Prepared by American Wind Energy Association Under Contract to U.S. Department of Energy (January).

Henderson, J.S. 1988. "There Are No Distortions of Short-Term Generation Choices if Electricity Transmission is Priced Flexibly." Proceedings of the Fifth Biennial Regulatory Information Conference. Columbus, OH: National Association of Regulatory Utility Commissioners.

Henderson, J.S. 1985. "Cost Estimation for Vertically Integrated Firms: The Case of Electricity." In *Analyzing the Impact of Regulatory Change in Public Utilities.* Michael Crew, ed. Lexington, MA: D.C. Heath & Co.

Henney, A. 1996. "Competition, Confusion, and Chaos: The Metering Muddle." *Public Utilities Fortnightly* 134 (20): 26–29.

Herriott, S. 1995. "The Organizational Economics of Power Brokers and Centrally Dispatched Power Pools." *Land Economics* 61 (3): 308–313.

Hewlett, J.G. 1995. Stranded Nuclear Power Plant Retirement Cost: Is There a Problem? Presented at the Restructuring Electric Transmission Conference, September 27.

Hill, L.J. 1995. "Perfomanced Based Regulation in the Electric Industry," National Conference of State Legislatures 20 (22).

Hill, L.J. and M.H. Brown. 1995. "Changing Electric Markets." National Council of State Legislatures: vol. 2 of the Utility Series (December).

Hines, V. 1989. "Changing Times at the Power Pool." *Energy User News* 14 (2): 11.

Hirchberg, J. and D. Aiger. 1983. "An Analysis of Commercial and Industrial Customer Response to Time-of Use Rates." *The Energy Journal* 4: 103–126.

Hirsh, R.F. 1995. Consensus, Confrontation and Control in the American Electric Utility System: An Interpretive Framework for the Virtual Utility Conference. Presented at the Symposium on the Virtual Utility, Rensselaer Polytechnic Institute, March 31.

Hirst, E. and J. Reed. 1991. *Handbook of Evaluation of Utility DSM Programs.* Oak Ridge National Laboratory.

Hirst, E. and J. Eto. 1995. "Justification for Electric-Utility Energy-Efficiency Programs." U.S. Department of Energy under contract number DE-AC05-84OR21400 (August).

Hirst, E. and B. Tonn. 1996. "Social Goals Electric-Utility Deregulation." *Perspectives* (Spring).

Hirst, E. and L. Baxter. 1995. "How Stranded Will Electric Utilities Be?" *Public Utilities Fortnightly* 133 (4): 30–32.

Hirst, E. and J. Reed. 1991. *Handbook of Evaluation of Utility DSM Programs.* Oak Ridge National Laboratory.

Hirst, E. 1988. "Integrated Resource Planning: The Role of Regulatory Commissions." *Public Utilities Fortnightly* 122 (6): 34–42.

Hirst, E. 1994. "Electricity-Utility DSM Programs in a Competitive Market." Sponsored by Office of Energy and Renewable Energy (April).

Hirst, E. 1994. "Cost and Effects of Electric-Utility DSM Programs: 1989 through 1997." Oak Ridge National Laboratory.

Hirst, E., S. Hadley and L. Baxter. 1996a. "Methods to Estimate Stranded Commitments for a Restructuring U.S. Electricity Industry." Oak Ridge Labortory: (January 1996).

Hirst, E., S. Hadley and L. Baxter. 1996b. "Factors that Affect Electric-Utility Stranded Commitments." Oak Ridge National Labortory (July).

Hirst, E., R. Cavanagh and P. Miller. 1996. "The Future of DSM in a Restructured US Electricity Industry." *Energy Policy* 24 (4): 303–315.

Hobbs, B.F. 1992. "Environmental Adders and Emissions Trading: Oil and Water?" *The Electricity Journal.* (August/September): 26–34.

Hoff, T.E. and C. Herig. 1996. *Managing Risk Using Renewable Energy Technologies.* Managing Risk. Draft, 1/19/96.

Hoffman, A.R. 1996. The Role of Renewables in a Restructured Utility Environment. Presented at Workshop on Energy Services in a Restructured Electric Industry, Association of Energy Services Professionals. Presented by Office of Energy Efficiency and Renewable Energy: U.S Department of Energy.

Hogan, W.W. 1995. "A Wholesale Pool Spot Market Must be Administered by the Independent Operator: Avoiding the Separation Fallacy." Harvard University (25 October).

Hogan, W.W. 1996. "Coordination for Competition in an Electricity Market." Response to an Inquiry Concerning Alternative Power Pooling Institutions Under the Federal Power Act-Docket No. RM94-20-000 (2 March).

Hogan, W. 1995. "A Wholesale Pool Spot Market Must Be Administered by the Independent System Operator: Avoiding the Separation Fallacy." *The Electricity Journal* (December): 26–37.

Hogan, W.W. 1994. "Order Instituting Rulemaking and Order Instituting Investigation: on the Commission's Proposed Policies Governing Restructuring California's Electric Services Industry and Reforming Regulation" California Public Utilities Commission (15 June).

Hogan, W. 1994a. "Markets in Real Electric Networks Require Reactive Prices." *Energy Journal* 14(3): 171–200.

Hogan, W.W. 1994b. "Retail Wheeling Fiction vs. Efficient Direct Access Fact." National Association of Regulatory Utility Commissioners Committee on Electricity (25 July).

Hogan, W. 1994c. "Efficient Direct Access: Comments on the California Blue Book Proposal." *The Electricity Journal* 7 (7).

Hogan, W.W. 1989. "Approximating Efficient Short-Run Prices for Electric Power Transmission." Cambridge, MA: Harvard University, John F. Kennedy School of Government, Energy and Environmental Policy Center.

Hohenemser, C., R.L. Goble and P. Slovic. 1992. "Nuclear Power". In *The Energy-Environment Connection* 133–175. J.M. Hollander, ed. Washington, DC: Island Press.

Holden, M., Jr. 1981. "Problems for Utility Deregulation." *Wall Street Journal* (10 June): 31.

Hollander, J.M. and D. Brown. 1992. "Air Pollution". In *The Energy Environment Connection,* 15–49. J.M. Hollander, ed. Washington, DC: Island Press.

Holmes, A. 1992. *Electricity in Europe: Present Status and Prospects for the 1990s.* London: Financial Times Information Service.

Holmes, A. 1992a. "Evolution and De-evolution of a European Power Grid." *Electricity Journal* (October).

Houghton, J.T., L.G. Meira Filho, B.A. Callander, N. Harris, A. Kattenberg and K. Maskell, eds. 1995. *Climate Change 1995: The Science of Climate Change.* Published for the Intergovernmental Panel on Climate Change. Cambridge: Cambridge Univ. Press.

House Science Committee, Democratic Membership. 1996. "Brown Releases GAO Report Documenting Decline in Electricity R&D." *Science & Technology News.* (16 September).

Houthakker, H.S. 1951. "Electricity Tariffs in Theory and Practice." *The Economic Journal* (March 1951.) 61: 1–25.

Hughes, W.R. and F. M. Fisher. 1971. "Would Increasing Residential Electric Rates Help Preserve Environmental Quality?" *Public Utilities Fortnightly* (1 April): 22–31.

Huettner, D.A. and J.H. Landon. "Electric Utilities: Scale Economies and Diseconomies." *Southern Economic Journal* 44 (4): 883–912.

Hughes, W.R. 1971. "Scale Frontiers in Electric Power." In *Technological Change in Regulated Industry,* W.W. Capron, ed. Washington, DC: Brookings Institution.

Hughes, T. 1983. *Networks of Power.* Baltimore: Johns Hopkins Univ. Press.

Hung-po C., S. Oren, S. Smith and R.B. Wilson. 1990. "Service Design in the Electricity Power Industry." Report P-6543. Palo Alto, CA: Electric Power Research Institute.

Hung-po C. and R.B. Wilson. 1989. "Priority Service: Pricing, investment, and Market Organization." *American Economist Review* 77: 899–916.

Hunt, G.L. 1990. "Is It Time To Rethink Nepool?". In *The Transmission Symposium,* 24–28. NY: Public Utilities Reports, Inc. and The Management Exchange.

Hunt, S. and G. Shuttleworth. 1996. *Competition and Choice in Electricity.* Great Britain: Bookcraft (Bath) Ltd., Avon.

Hyman, L.S. 1992. *America's Electric Utilities: Past, Present, and Future.* Arlington, VA: Public Utilities Reports, Inc.

Hyman, L.S. 1981. Financial Aspects of Electric Utility Deregulation. Presented to the California Public Utility Workshop on Deregulation.

Ilić, M., F.C. Graves, H. Fink and A.M. DiCaprio. 1996. "A Framework for Operations in Competitive Open Access Environment." *The Electricity Journal* (April).

Illinois Commerce Commission. 1987. *Electric Wheeling in Illinois.* Report of the Illinois Commerce Commission to the Illinois General Assembly.

Intergovernmental Panel on Climate Change. 1996. Summary for Policymakers: Scientific-Technical Analyses of Impacts, Adaptations, and Mitigation of Climate Change. Cambridge, MA: Cambridge Univ. Press

International Brotherhood of Electrical Workers. 1996. "Competition in the Electric Utility Industry: Should Washington Solve a Problem That Doesn't Exist?"

International Energy Agency. 1994. *Climate Change Policy Initiatives: 1994 Update.* Volume 1 OECD Countries. Paris: OECD/IEA, 1994.

International Energy Agency. 1991. *Utility Pricing and Access: Competition for Monopolies.* Paris: IEA.

Interstate Natural Gas Association of America. 1996. *Comparison of Gas and Electric Industry Restructuring Costs.* Washington, DC. (August).

Isaac, R.M. 1991. "Price Cap Regulation: A Case Study of Some Pitfalls of Implementation." *Journal of Regulatory Economics.*

Jaffe, A.B. and F.A. Felder. 1996. "Should Electricity Markets Have a Capacity Requirement? If so, How Should It Be Priced?" *The Electricity Journal* (December): 52–60.

Jaffe, A. and R. Stavins. 1991. The Energy Paradox and the Diffusion of Conservation Technology. Presented at Universities Research Conference: Economics of the Environment. National Bureau of Economic Research, Inc., Cambridge, MA.

Jarell, G.A. 1978. "The Demand for State Regulation of the Electric Utility Industry." *Journal of Law and Economics* (October): 269–296.

Jensen, V. and M. Scott. 1988. *An Empirical Analysis of Electricity Wheeling in Illinois.* ILENR/_RE-SP-87/23. Springfield, IL: Illinois Department of Energy and Natural Resources.

Johansson, T.B., H. Kelly, A.K.N. Reddy, R.H. Williams and L. Burnham, eds. 1993. *Renewable Energy: Sources for Fuels and Electricity.* Washington, DC: Island Press.

Johnson, B. 1980. "Cost Allocations: Limits, Problems, and Alternatives." *Public Utilities Fortnightly* (4 December): 33–36.

Johnson, B. 1993. Improved Models of Individual Energy Technology Choice: Their Implication for Energy Technology Markets Participants and Policy-Makers. Presented at the Energy Modeling Forum, Terman Engineering Center, September.

Johnson, L. 1989. *Price Caps in Telecommunications Regulatory Reform.* Rand Northeastern University Library. Prepared for The John and Mary R. Markle Foundation. N-2894-MF/RC.

Johnson, L. 1978. "Boundaries to Monopoly and Regulation in Modern Telecommunications." In *Communications for Tomorrow.* G.O. Robinson, ed. New York: Praeger.

Johnson, R.B., S.S. Oren and A.J. Svoboda. 1996. *Equity and Efficiency of Unit Commitment in Competitive Electricity Markets.* Berkeley, CA: Pacific Gas and Electric Co. and Univ. of California.

Johnson, R.B., S.S. Oren and A.J. Svoboda. 1996. Equity and Efficiency of Unit Commitment in Competitive Electricity Markets. Presented at the Power Conference, March 15.

Jones, P.S.M. 1989. *Nuclear Power: Policy and Prospects.* New York: John Wiley & Sons.

Jones, L.P., P.Tandon and I. Vogelsang. 1990. *Selling Public Enterprises.* Cambridge, MA: The MIT Press.

Joskow, P.L. and R. Schmalensee. 1986. "Incentive Regulation for Electric Utilities." *Yale Journal on Regulation* 4 (1): 1–49.

Joskow, P.L. and R. Schmalensee. 1983. *Markets for Power: An Analysis of Electric Utility Deregulation.* Cambridge, MA: The MIT Press.

Joskow, P.L. 1992. "Dealing With Environmental Externalities: Let's Do It Right!" Washington, DC: Edison Electric Institute.

Joskow, P.L. 1989. "Regulatory Failure, Regulatory Reform, and Structural Change in the Electrical Power Industry." In Brookings Papers on Microeconomic Activity, 125–200.

Joskow, P.L. 1989. "The Effects of Economic Regulation." In *Handbook of Industrial Organization,* Willig and Schmalensee, eds. Amsterdam: North-Holland/Elsevier.

Joskow, P.L. 1987. Competition and Deregulation in the Electric Utility Industry. MIT Center for Energy Policy Research Policy Discussion Paper. Cambridge, MA: Massachusetts Institute of Technology Energy Laboratory.

Joskow, P.L. 1985. "Mixing Regulatory and Antitrust Policies in the Electric Power Industry: The Price Squeeze and Retail Market Competition." In *Antitrust and Regulation: Essays in Memory of John J. McGowan.* F.M. Fisher, ed. Cambridge, MA: MIT Press.

Joskow, P.L. 1975. "Applying Economic Principles to Public Utility Rate Structures: The Case of Electricity." In *Studies in Electric Utility Regulation,* 40–41. C.J. Cicchetti and J.L. Jurewitz, eds. Cambridge, MA: Ballinger Publishing Co.

Joskow, P.L. 1995. "Apendix A." In *Horizontal Market Power In Wholesale Power Market.* Cambridge, MA: Massachusetts Institute of Technology (August).

Joskow, P.L. and N.L. Rose. 1989. "The Effects of Economic Regulation." In *Handbook of Industrial Organization.* Schmalensee and Willing, eds. Amsterdam, North-Holland//Elsevier.

Joskow, P.L. 1996. "Technical Conference Concerning Independent System Operators and Reform of Power Pools Under the Federal Law Act." Federal Energy Regulatory Commission (24 January).

Joskow, P.L. 1973. "Pricing Decision of Regulatory Firm: a Behavioral Approach." *Bell Journal of Economics and Management Science* (Spring): 118–140.

Joskow, P.L., and N.L. Rose. 1985. "The Effects of Technological Change, Experience, and Environmental Regulation on the Construction of Coal-Burning Generating Units." *Rand Journal of Economics.* 16: 19.

Joskow, P.L. and R.G. Noll. 1981. "Regulation in Theory and Practice: An Overview." In *Studies in Public Regulation.* G. Fromm, ed., Cambridge, MA: MIT Press.

Jurewitz, J.L. 1988. "Deregulation of Electricity: A View from Utility Management." *Contemporary Policy Issues* 6 (3): 25–41.

Jurewitz, J.L. 1994. "Retail Wheeling: Why the Proponents Must Bear the Burden of Proof." *The Electricity Journal.* (April).

Kahn, A.E. 1994. "Competition in the Electric Industry Is Inevitable and Desirable." In *The Electric Industry in Transition,* 21–31. Arlington, VA: Public Utilities Reports, Inc. & the New York State Energy Research and Development Authority.

Kahn, A. 1988. *The Economics of Regulation.* Cambridge, MA: The MIT Press.

Kahn, A.E. 1978. "Applications of Economics to Utility Rate Structures." *Public Utilities Fortnightly* 101(19): 13–17.

Kahn, E.P. 1994. "Preparing for the Inevitable: The Nationalization of the U.S. Nuclear Industry in a Competitive Electricity Market". In *The Electric Industry in Transition,* 199–212. Arlington, VA: Public Utilities Reports, Inc. & The New York State Energy Research and Development Authority.

Kahn, E.P., C.A. Goldman, S. Stoft and D. Berman. 1989. "Evaluation Methods in Competitive Bidding for Electric Power." Lawrence Berkeley Laboratory LBL-26924.

Kahn, E.P, S.Stoft, C.Marnay, D. Berman. 1990. "Contracts for Dispatchable Power: Economic Implications for the Competitive Bidding Market." Lawrence Berkeley Laboratory LBL-29447.

Kahn, E.P., and R. Gilbert. 1993. *Competition and Institutional Change in U.S. Electric Power Regulation.* University of California Program on Workable Energy Regulation, Report PWP-011 (May).

Kahn, E.P. 1991. "Risks in Independent Power Contracts: An Empirical Survey." *Electricity Journal* (November): 30.

Kamerschen, D.R., J.B. Kau and C.W. Paul. 1982. "Potential Determinants of Rates in Electric Utilities." *Policy Studies Review:* (February): 277–290.

Kamerschen, D.R. and D.C. Keenan. 1983. "Caveats on Applying Ramsey Pricing." *In Current Issues in Public Utility Economics: Essays in Honor of James C. Bonbright.* Danielsen and Kamerschen, eds. Lexington, MA: Lexington Books.

Kamerschen, D.R. and R.L. Wallace. 1973. "The Simple Analytics and Some Evidence on the Opportunity Cost and the Capital Attraction Criterion for Utility Regulation." *Public Utilities Fortnightly* (15 February): 43–50.

Kaplow, L. 1985. "Extension of Monopoly Power Through Leverage." *Columbia Law Review* 85: 515–556.

Kassler, P. 1994. *Energy for Development.* London: Group Public Affairs, Shell Interntional Petroleum Company Ltd.

Kaufman, A. and D.P. Dulchinos. 1986. *The Federal Power Marketing Administrations: To Privatize Or Not to Privatize.* Report No. 86-90 S.

Kelly, K., ed. 1987. "Non-Technical Impediments to Power Transfers." Columbus, OH: National Regulatory Research Institute, NRRI 87-8.

Kelly, K., N. Simmons, Jr. and T. Prior. 1979. "Fuel Adjustment Clause Design." Columbus, Ohio: National Regulatory Research Institute.

Kelly, K. 1990. "Impediments to Increased Access: Three Perspectives: Electric & Gas Research." In *The Transmission Symposium,* 45–49. NY: Public Utilities Reports, Inc. and The Management Exchange.

Kemeny, J.G. 1980. "Saving American Democracy: The Lessons of Three Mile Island." *Technology Review* 83 (June–July): 64–75.

Kennedy, J.B. and R.A. Baudino. 1991. "Retail Wheeling: Expanding Competition in the Electric Industry." Sponsored by The Ad Hoc Committee For A Competitive Electric Supply System. (April).

Kennedy, J.B. 1978. "Peak Load Pricing and Industrial Consumers' Response." *Assessing New Pricing Concepts in Public Utilities.* East Lansing, MI: MSU Institute of Public Utilities.

Khanna, S.K. 1982. "Economic Regulation and Technical Change: A Review of the Literature." *Public Utilities Fortnightly* 110: 35–44.

Kihlstrom, R.E. and D. Levhari. 1977. "Quality, Regulation and Efficiency." *Kyklos.* 30 (2): 214–234.

Kihm, S.G. and D.W. York. 1994. A Critical Review of Retail Wheeling. Presented at The NARUC-DOE Fifth National Conference on Integrated Resource Planning. Kalispell, Montana: May 15–18.

King, C.L. 1996. "Competition at the Meter: Lessons From the U.K." *Public Utilities Fortnightly* (1 November).

King, M.J., G.A. Heffner, S. Johansen and B. Kick. 1996. "Public Purpose Energy Efficiency Programs and Utilities in Restructured Markets." *The Electricity Journal* (July): 14–25.

King, Owen, Gill and Michael. 1994. "Energy Services Market: Will Competition Be Left to Chance?" Report of a Research Project Undertaken for the Energy Saving Trust Fund and the Gas Consumers Council. London (September).

Kirsch, L.D. and H. Singh. 1995. "Pricing Ancillary Electric Power Services," *The Electricity Journal* 8 (8): 28.

Kleeman, S. 1989. "Deregulating the Electric Utility Industry." Illinois Department of Energy and Natural Resources.

Klein, B., R.G. Crawford, and A.A. Alchian. 1978. "Vertical Integration, Appropriable Rents, and the Competitive Contracting Process." *Journal of Law and Economics* 21 (2): 297–326.

Knutson, K. 1996. "As PURPA Plays Out…Advantage, NUGs." *Public Utilities Fortnightly* (8 May): 8.

Koger, R.K. 1979. "Is There Economic Justification for a So-Called Lifeline Rate?" *Public Utilities Fortnightly* (10 May): 11–17.

Kolbe, A. L. And L.S. Borucki. 1996. "The Impact of Stranded-Cost Risk on Required Rates of Return for Electric Utilities: Theory and an Example." Cambridge, MA: The Brattle Group. (June).

Kolbe, A. L. and J.A. Read, Jr. (with G.R. Hall.) 1984. "The Cost of Capital, Estimating the Rate for Public Utilities." Cambridge, MA: MIT Press.

Kolbe, A. L., S. Johnson and Johannes P. Pfeifenberger. 1993. "Purchased Power—Risk And Rewards." Prepared for Edison Electric Institute (December).

Komanoff, C. 1981. *Power Plant Cost Escalation: Nuclear and Coal Capital Costs, Regulation, and Economics.* New York: Komanoff Energy Associates.

Korn, D.H. 1996. "Nuclear Decommissioning Trust Funds: Rethinking the Approach." *Public Utilities Fortnightly* (15 November).

Krapels, E. and V. Stagliano. 1996. "A Milestone Year: Power in the Commodity Markets." *Public Utilities Fortnightly* (15 May).

Kriz, M. 1996. "Power Brokers." *National Journal* (30 November): 2594–2599.

Lagassa, G.K. 1980. "Implementing the Soft Path in a Hard World: Decentralization and the Problem of Electric Power Grids." In *Energy Policy and Public Administration.* G.A. Daneke and G.K. Lagassa, eds. Lexington, MA: Lexington Books.

Lancaster, K. 1979. "The Problem of Second Best in Regulation to Electricity Pricing." *Electricity Utility Rate Design Study* (7 August).

Landon, J.H., and D.A. Huettner. 1976. "Restructuring the Electric Utility Industry: A Modest Proposal." In *Electric Power Reform: The Alternatives for Michigan,* 217–229. W.H. Shaker and Wilbert Steffy, eds. Ann Arbor, MI: Institute of Science and Technology, University of Michigan.

Lee, T.H. and F.J. Ellert. 1991. "Technical Implication of Our Regulatory Structure." In Research In *Law and Economics,* Vol. 13, 191–212. R.O. Zerbe, Jr. and V.P. Goldberg, eds., Greenwich: JAI Press.

Lee, H. and N. Darani. 1996. "Electricity Restructuring and the Environment." *The Electricity Journal* (December): 10–15.

Lee, H. and N. Darani. 1995. "Electricity Restructuring And The Environment." Environment and National Resources Program, Center for Science and International Affairs, Kennedy Center School of Government, Harvard University (22 November).

Leigland, J. and R. Lamb. 1986. *WPP$$: Who is to Blame for the WPPSS Disaster.* Cambridge, MA: Ballinger Publishing.

Lessels, D.J. 1980. "The Economic Distribution System Cost and Investments." *Public Utilities Fortnightly:* 106 (4 December): 37–40.

Lesser, J. and M.Ainspan. 1996. "Using Markets to Value Stranded Costs." *The Electricity Journal* 9(8).

Lewis, T.R. and D.E.M. Sappington. 1992. "Incentives for Conservation and Quality-Improvement by Public Utilities." *The American Economic Review.* (December).

Lewis, W.A. 1949. *Overhead Costs: Some Essays in Economic Analysis.* New York: Rinehart.

Lindsley, E.F. 1983. "Planning Practically for a Decentralized Electric System: How Past Experience Can Guide Us." In *Decentralized Electricity Production.* Brown, H., ed. New Haven: Yale Univ. Press.

Lobsenz, G. 1996. "State Regulators Question Independence Of System Operators," *The Energy Daily:* (23 October).

Lock, R. and M. Stein. 1996. "Electricity Transmission." In *Energy Law and Transactions.* Muchow, D. and Mogel, W. New York: Matthew Bender.

Loeb, M., and W.A. Magat. 1979. "A Decentralized Method for Utility Regulation." *Journal of Law and Economics* 22: 399–404.

Loube, R. 1995. "Price Cap Regulation: Problems and Solutions." *Land Economics* (August).

Lovins, A.B. 1977. *Soft Energy Paths.* San Francisco: Ballinger.

Lovins, A.B. 1976. "Energy Strategy: The Road Not Taken?" *Foreign Affairs* 55 (1): 87–92.

Macaulay, S. 1963. "Non-Contractual Relations in Business: A Preliminary Study." Reprinted In *Readings in the Economics of Contract Law.* Goldberg, V., ed. Cambridge: Cambridge Univ. Press.

MacAvoy, P.W. ed., 1970. *The Crisis of Regulatory Commissions: An Introduction to the Current Issue of Public Policy.* New York: W.W. Norton and Company, Inc.

MacKerron, Gordon. 1996. "Problems of Regulation and Competition in the England and Wales Electricity System." Presented at a meeting of the Harvard Electricity Policy Group in Dallas, TX, January 25–26.

Malko, J.R. and T.B. Nicolai. 1980. "Using Accounting Cost and Marginal Cost in Electricity: Some Current Challenges and Activities. In *Issues in Public Utility Economics and Regulation.* Crew, ed. Lexington, MA: Lexington Books.

Malko, R.J. and J. Simpson. 1978. "Time-of-Use Pricing in Practice: Some Recent Regulatory Action." *Assessing New Pricing Concepts in Public Utilities.* East Lansing, MI: MSU Institute of Public Utilities. 399–419.

Malko, J.R. and A. Faruqui. 1980. "Implementing Time-of-Day Pricing of Electricity: Some Current Challenges and Activities. In *Issues in Public Utility Economis and Regulation,* 399–419. Crew, ed. Lexington, MA: Lexinton Books.

Malko, J.R. and D. Smith, R.G. Uhler. 1981. "Costing For Ratemaking, Topic 2." A report to the National Association of Regulatory Utility Commissioners, Electric Utility Rate Design Study. Report No. 85 Palo Alto, CA: Electric Power Research Institute.

Malone, T.W. and J.F. Rockart. 1991 "Computers, Networks, and the Corporation." *Scientific American* 265 (3): 128.

Maloney, M.T., R.E. McCormick and R.D. Sauer. 1996. "Customer Choice, Consumer Value: An Analysis of Retail Competition in America's Electric Industry." Volume 1. Washington, DC: Citizens for a Sound Economy Foundation.

Mann, P.C. 1977. "User Power and Electricity Rates." *Journal of Law and Economics.* 17 (2).

Mann, P.C. 1977. "Rate Structure Alternatives for Electricity." *Public Utilities Fortnightly* 113 (20 January): 19–24.

MAPP. 1996. Press Release announcing the "Membership's Vote on the Reliability Council's New Functions". In Can Regional Reliability Councils Continue to Operate as Voluntary Organizations? How Might the Changes in a Restructured Industry Affect the Role of Reliability Councils?

Marchand, M. 1968. "A Note on Optimal Tolls in an Imperfect Environment." *Econometrica* 36 (3–4): 575–581.

Marritz, R.O. 1988. "Investigating in Efficiency." *The Electricity Journal* 1 (2): 22–35.

Marritz, R.O., and G.C. Culp. 1979. *Governmental Impediments to Electric System Efficiency Through Integration. The National Power Grid Study: Technical Study Reports.* Volume II. Washington, DC: U.S. Department of Energy, Economic Regulatory Administration, Office of Utility Systems, 313–351.

Massachusetts Electric Company. 1989. *Alternate Energy Negotiation-Bidding Experiment: Report.*

Massella, I.M. 1984. "Rate Moderation Plans—Cushioning Rate Shock." *Public Utilities Fortnightly* 113 (16 February): 52–56.

Massey, W.L., C.A. Patrzia, J.J. Rice and G.L. Wortham. 1992. "Regional Bodies Are They Lawful?" *The Electricity Journal* 5 (1).

Masten, S.E., J.W. Meehan, and E.A. Snyder. 1988. *The Costs of Organization.* Draft paper prepared for the University of Michigan and Colby College.

Mauldin, M. 1989. *Integrating Environmental Issues into Electric Utility Planning.* Washington, DC: Barakat, Howard & Chamberlin, Inc. P1910/EI-Paper/8-14-89.

McDonald, F. 1962. *Insull.* Chicago: Univ. of Chicago Press.

McGraw-Hill. 1996. "155 Independent Power Companies: Profiles of Industry Players and Projects." New York: The McGraw-Hill Companies.

McGuire, B. 1995. *Price-Driven Coordination in a Lossy Power Grid.* University of California Energy Institute, Berkeley, CA.

Meeks, J.E. 1972. "Concentration in the Electric Power Industry: The Impact of Antitrust Policy." *Columbia Law Review* (January): 64–130.

Messenger, M. 1996. "California Retrospective." *The Electricity Journal* 9 (6).

Messing, M., H.P. Friesema, and D. Morell. 1979. *Centralized Power: The Politics of Scale in Electricity Generation.* Cambridge, MA: Oelgeschlager, Gunn & Hain.

Metague, S.J. 1991. "Current Access and Pricing Proposals: Pacific Gas & Electric." In *The Transmission Symposium,* 110–114. NY: Public Utilities Reports Inc. and The Management Exchange.

Michaels, R.J. 1996. "Markets of the Future, Utilities of the Past." *The Electricity Journal* 9(8).

Michaels, R.J. 1996. "Stranded Investments, Stranded Intellectuals." *Regulation* 1 (8 April): 47–51.

Miller, E.S. 1994. "Economic Regulation and the Social Contract: An Appraisal of Recent Developments in the Social Control of Telecommunications." *Journal of Economic Issues* (September).

Mills, McCarthy & Associates. 1994. A C.L.E.A.N.E.R. ECONOMY: The Economic and Environmental Impacts of Electrotechnologies: Clean Electrotechnologies Advancing Net Energy Reductions. Chevy Chase, MD: The Edison Electric Institute. (June).

Mills, McCarthy & Associates, Inc. 1995. " Does Price Really Matter? The Importance of Cheap Electricity for the Economy." Western Fuels Association Inc. (January).

Mitchell, B.M. and P.R. Kleindorfer. 1980. *Regulation Industries and Public Enterprise.* Cambridge: Lexington Press.

Mitchell, B.M., W.S. Manning and J.P. Acton. 1978. *Peak Load Pricing: European Lesson for U.S. Energy Policy.* Cambridge, MA: Ballinger, 1978.

Mitnik, B. 1980. *The Political Economy of Regulation.* New York: Columbia University Press.

Monteverde, K., and D.J. Teece. 1982. "Supplier Switching Costs and Vertical Integration in the Automobile Industry." *Bell Journal of Economics* 13 (1): 206–213.

Moody's Investor Service. 1991. "Credit Quality Trends in Privatization of Britain's Electricity Supply Industry." Moody's Investor Service, New York, N.Y. (February).

Moody's Investors Service. 1995. "Stranded Cost Will Threaten Credit Quality of U.S. Electrics." *Global Credit Research* (August).

Moore, T. 1986. "Network Access and the Future of Power Transmission." *EPRI Journal* 11 (3): 4–13.

Moore, T. G. 1970. "The Effectiveness of Regulation of Electric Utility Prices." *Southern Economic Journal:* 36 (April): 365–375.

Moorhouse, J. C. 1986. "Electric Power: Deregulation and the Public Interest." San Francisco: Pacific Studies in Public Policy, Pacific Research Institute for Public Policy.

Moot, J.S. 1991. "Electric Utility Mergers: Uncertainty Looms Over Regulation Approvals at the FERC." *Energy Law Journal* 12 (1).

Morin, R.A. 1984. *Utilities' Cost of Capital.* Arlington, VA: Public Utilities Reports, Inc.

Morgan, G. M. 1993. "What Would It Take to Revive Nuclear Power in the United States?" *Environment* (March).

Morrison, S.A., and C. Winston. "The Dymanics of Airline Pricing and Competition." *The American Economic Review* 80 (2): 389–393.

Moscovitz, D. 1989. *Profits and Progress Through Least-Cost Planning.* Washington, DC: National Association of Regulatory Utility Commissioners.

Moscovitz, D. 1988. Will Least-Cost Planning Work Without Significant Regulatory Reform? Presented at NARUC Least-Cost Planning Seminar, April.

Moscovitz, D. 1982. *An Overview of Reliability Criteria.* Princeton, NJ: NERC.

Mosher, W.E. and F. Crawford, 1933. *Public Utility Regulation.* New York: Harper Brothers.

Moskowitz, D., T. Austin, C. Harrington and C. Weinberg. 1993."Future Utility and Regulatory Structures." Regulatory Assistance Project, Gardiner, ME. (December).

Mueller, M. and J.R. Schement. 1995. "A Profile of Telecommunication Access in Camden, New Jersey." New Brunswick, NJ: Rutgers Univ. School of Communication.

Muller, F. 1996. "Mitigating Climate Change: The Case for Energy Taxes." *Environment* (March): 13–43.

Mulligan, J. 1985. "An Assessment of Cost Savings from Coordination in U.S. Electric Power Generation: Comment." *Land Economics* 61 (2): 205–207.

Munasinghe, M. and J.J. Warford. 1985. "Electricity Pricing: Theory and Case Studies" Published for The World Bank. Baltimore, MD: The Johns Hopkins Univ. Press.

Munson, R. 1984. *The Power Makers.* Emmaus, PA: The Rodale Press.

Muntzing, L.M 1975. "The New NRC—Its Mandate and Challenge." *Public Utilities Fortnightly* (13 March).

Nadel, Jordan, Eto, and Kelly. 1993. *Gas DSM—Opportunities and Experiences. Summary of U.S. DSM Activities.* Palo Alto, CA: Electric Power Research Institute. EPRI TR-102021.

Nadel, S. M., M.W. Reid and D.R. Wolcott, 1992. *Regulatory Incentives for Demand-Side Management.* Washington, DC: American Council for an Energy Efficient Economy

Naill, R. and S. Belanger. 1989. "Impacts of Deregulation on U.S. Electric Utilities." *Public Utilities Fortnightly* (12 October).

Nakarado, G. 1994. "Substituting Precision Design and Information Technologies for Traditional Energy Use Patterns". Renewable Energy and Other Information-rich Tools in the Brave New World. Chicago: National Renewable Energy Laboratory.

National Academy of Sciences. 1996. "Electricity in Economic Growth."

National Association of Regulatory Utility Commissioners, 1988. "Least Cost Planning Handbook for Public Utility Commissioners." Washington, DC.

National Association of Regulatory Utility Commissioners. 1973. "Electric Utility Cost Allocation Manual by J.J. Doran, F.M. Hoppe, R.Koger, and W.W. Lindsay." Washington, DC.

National Council. 1993. "Profiles of Regulatory Agencies of the United States and Canada." Washington, DC: National Association of Regulatory Utility Commissioners.

National Governor's Association. 1986. "Moving Power: Flexibility for the Future." Washington, DC: Committee on Energy and the Environment.

National Governor's Association. 1983. "An Analysis of Options for Structural Reform in Electic Utility Regulation." Washington, DC: NGA Task Force on Electric Utility Regulation.

National Independent Energy Producers. 1989. *Independent Energy Producers: The New Electric Generating Sector.* Working Paper Number One. Washington, DC: Vanguard Communications, Inc.

National Independent Energy Producers. 1990. *Bidding for Power: The Emergence of Competetive Bidding in Electric Generation.* Working Paper Number Two. Washington, DC.

National Independent Energy Producers. 1995. *Is Competition Here?* Washington, DC.

National Resource Council Committee on Nuclear and Alternative Energy Systems. 1979. *Energy in Transition*. Washington, DC: National Research Council, National Academy of Sciences.

Navarro, P. 1996. "Ten Questions for the Restructuring Regulator." *The Electricity Journal* 9 (7).

Navarro, P. 1982. "Our Stake in the Electric Utility Dilemma, *Harvard Business Review* (May–June): 87–97.

Navarro, P. 1996. "Seven Basic Rules for the PBR Regulator." *The Electricity Journal:* (April): 24–30.

Nelson, C., S. Peck, and Uhler, R. 1989. "The NERC Fan in Retrospect and Lessons for the Future." *Energy Journal* 10(2): 91–107.

Nelson, R.A. and M.E. Wohar. 1983. "Regulation, Scale Economies, and Productivity in Steam Electric Generation." *International Economic Review* 24: 57.

New England River Basin Commission, 1972. "Laws and Procedures of Power Plant Siting." Boston, MA.

Newbery, D.M. and M.G. Pollitt. 1996. "The Restructuring and Privatisation of the CEGB—Was it Worth it?" Cambridge, UK: Univ. of Cambridge.

Newbery, D.M. 1995. *Electricity Power Sector Restructuring: England and Wales*. Cambridge, UK.

Newbery, D.M. 1995a. Electricity Power Sector Restructuring: England and Wales. Presented at the POWER Conference, October 19.

Newbery, D.M. 1995b. *Power Markets and Market Power*. Prepared for the EEE conference on Competitive Restructuring of the Electricity Supply Industries, London, April 16.

Newcomb, J. 1994. "The Future of Energy Efficiency Services in a Competitive Environment". Strategic Issues Paper. (May).

Nichols, E. 1928. *Public Utility Service and Discrimination*. Washington, DC: Public Utilities Reports, Inc.

Niehans, J. 1987. "Transactions Costs" *New Palgrave Dictionary of Economics*. London: MacMillan.

Nissel, Hans E. 1983. "Basic Concepts and Design of Interruptible Power Rates." *Public Utilities Fortnightly* 111 (9 June): 36–41.

Noll, R.G. ed. 1985. "Regulatory Policy and the Social Sciences." Berkeley: University of California Press.

Noll, R. 1989. "Economic Perspectives on the Politics of Regulation." In *Handbook of Industrial Organization*. R. Schmalensee and R. Willig,. Amsterdam: North-Holland.

Norello, D.P. 1992. "New Nox Requirements Under Title I of the Clean Air Act." *Environmental Reporter*. Washington, DC. (July): 938–941.

North American Electric Reliability Council. 1996. Future Role of NERC—II White Paper. Agenda Item 10. BOT Meeting. September 16–17.

North American Electric Reliability Council. 1992. *Control Area Concepts and Obligations*. Princeton: North American Electric Reliability Council.

North American Electric Reliability Council Engineering and Operating Committees. 1987. *Reliability Considerations for Integrating Non-Utility Generating Facilities with the Bulk Electric Systems*. Reference document prepared by NERC.

North American Electric Reliability Council. 1989. *1989 Electricity Supply & Demand for 1989–1998*. Annual Data Summary. Princeton: North American Electric Reliability Council.

North American Electric Reliability Council. 1988. *1988 Electricity Supply & Demand for 1988–1997*. Annual Data Summary. Princeton: North American Electric Reliability Council.

Norton, F.L., IV, and M.B. Early. 1984. "Access to Electric Transmission and Distribution Lines." *Energy Law Journal* 5 (1): 47–76.

Nowak, G.P. and S.L. Taylor. 1996. "The Law of Public Utilities" In *Energy Law and Transactions*. Muchow, D.J. and W.A. Mogel, eds. New York: Matthew Bender.

Nowotny, K. 1988. "Transmission Technology and Electric Utility Regulation." *Journal of Economic Issues* 22 (June): 555–514.

O'Connor, P.R. 1987. "On Deregulation." *Electric Potential* 3 (2): 3–6.

OECD. 1985. *Environmental Effects of Electricity Generation*. Paris: OECD Publications Office.

Office of Technology Assessment. 1995. "Renewing Our Energy Future." Congress of the United States.

Office of Technology Assessment. 1994. *Studies of the Environmental Costs of Electricity*. Background Paper. Washington, DC.

Office of Technology Assessment. 1984. *Nuclear Power in an Age of Uncertainty*. Washington, DC.

Olson, C.E. 1970. *Cost Considerations for Efficient Electricity Supply*. East Lansing, MI: Michigan State University, Graduate School of Business Administration, Division of Research, Institute of Public Utilities, MSU Public Utilities Studies.

Olson, W.P. and K.W. Costello. 1995. "Electricity Matters: A New Incentives Approach for a Changing Electric Industry." *The Electricity Journal:* (January/February): 28–40.

Oppenheim, J. 1989. "Potential Cost of Competition: A Customers Prospective Brownouts, Death Spirals, and Alternatives and The Existence of Alternative Suppliers Does Not Necessarily Mean Competition Exists." In *Utility Opportunities for New Generation*. Edison Electric Institute and Electric Power Research Institute. EPRI CU-6605.

Oren, S., P.T. Spiller, P. Varaiya and F. Wu. 1995. "Nodal Prices and Transmission Rights: A Critical Appraisal." *The Electricity Journal* 8 (3).

Organization for Economic Co-Operation and Development. 1989. "Projected Cost of Generating Electricity From Stations for Commissioning in the Period 1995–2000."

Osterberg and Sheehan. 1994. "On the Brink of Disaster: A State-by-State Analysis of Low-Income Natural Gas Winter Heating Bills." Scappoose, OR: Flying Pencil Publications.

Ottinger, R.L., D.R. Wooley, N.A. Robinson, D.R. Hodas, S.E. Babb, S.C. Buchanan, P.L. Chernick, E. Cverhiss, A. Krupnick, W. Harrington, S. Radin and U. Fritsche. 1991. *Environmental Costs of Electricity*. Prepared by Pace University Center for Environmental Legal Studies. New York: Oceana Publications, Inc.

Pace, J.D. 1987. "Wheeling and the Obligation to Serve." *Energy Law Journal* 8: 265–302.

Pace, J.D. 1981. Deregulating Electric Generation: An Economist's Perspective. Presented before the International Association of Energy Economists.

Palmer, K., P. Fox-Penner, D. Simpson and M. Toman. 1992. *Contracting Incentives in Electricity Generation Fuel Markets*. Arlington, VA: Public Utilities Reports, Inc..

Palmer, K. and D. Burtraw. 1996. "Electricity Restructuring and Regional Air Pollution." Washington, DC: Resources for the Future (July) Discussion Paper 96-17-REV2.

Panzar, J.C., and R.D. Willig. 1981. "Economics of Scope." *American Economic Review* 71: 268–272.

Panzar, J.C., and R.D. Willig. 1977. "Free Entry and the Sustainability of Natural Monopoly." *Bell Journal of Economics* 8: 1–22.

Panzar, J.C. 1979. Sustainability, Efficiency, and Vertical Integration. In *Regulated Industries and Public Enterprise*. B.M. Mitchell and P.R. Kleindorfer, eds., Lexington, MA: Lexington Books.

Park, R.E. and J.P. Acton. 1984. *Response to Time-of Day Electricity Rates by Large Business Customers.* Rand Corporation.

Peck, S.C. and J.P. Weyant. 1985. "Electricity Growth in the Future." *The Electricity Journal* 6 (1).

Peltzman, S. 1978. "The Public Power Alternative." In *Electric Power Reform: The Alternatives for Michigan.* Shaker, W.H., W. Steffy, and P.W. McCracken, eds. Ann Arbor, MI: Univ. of Michigan Press.

Peltzman, S. 1971. "Pricing in Public and Private Enterprises: Electric Utilities in the United States." *Journal of Law and Economics.* Volume 14 (1).

Pender, R.B. 1991. "Decade of Change Begins for Maturing Independent Energy Industry." *Cogeneration Journal* 6(2): 8–11.

Penn, D.W. 1995. "How Likely Is Long-Run Competition in Electricity Generation? Warning Signs." *The Electricity Journal* 8 (1).

Penn, D.W. 1994. "Change: The Old and New World of Competition." Transmission Access, the National Energy Policy Act of 1992, and FERC Implementation (August).

Penn, D.W. 1986. "A Municipal Perspective on Electric Transmission Access Questions." *Public Utilities Fortnightly* (6 February): 15–22.

Perry, M.K. 1989. "Vertical Integration: Determinants and Effects." In *Handbook of Industrial Organization.* Schmalensee, R. and R.D. Willig, eds. Amsterdam: Elsevier Science Publishers.

Pfannenstiel, J. 1981."Implementing Marginal Cost Pricing in the Electric Utility Industry," In *Applications of Economic Principles in Public Utility Industries,* 53–72. Sichel, W. and T.G. Gies, eds. Michigan Business Studies, Vol. 2, No. 3.

Pfeifenberger, J.P. and W.B. Tye. 1995. "Handle With Care: A Primer on Incentive Regulation." *Energy Policy* 23 (9): 769–779.

Pfiefenberger, J.P., P.R. Amman, G.A. Taylor and The Brattle Group. 1996. "Distributed Generaton Technology in a Newly Competitive Electric Power Industry." Proceedings of the 1996 American Power Conference, Chicago, April 9.

Phillips, A., ed. 1975. *Promoting Competition in Regulated Markets.* Washington, DC: The Brookings Institution.

Phillips, A. 1980. "Ramsey Pricing and Sustainability with Interdependent Demands." In *Regulated Industries and Public Enterprise: European and United States Perspectives,* 187–204. B.M. Mitchell and P.R. Kleindorfer, eds. Lexington, MA: Lexington Books.

Phillips, Jr., C. F. 1993. *The Regulation of Public Utilities: Theory and Practice.* Arlington, VA: Public Utilities Reports, Inc.

Phillips, C.F., Jr. 1969. *The Economics of Regulation,* rev. ed. (Homewood.Ill.: Richard D. Irwin) Chapter 2, "The Economic Concepts of Regulation," 19–44; Chapter 3, "The Legal Concept of Public Regulation," 45–82.

Pierce, R.J. 1991. "Using the Gas Industry As A Guide to Reconstituting the Electricity Industry". *Research in Law and Economics.* 13: 7–56.

Pierce, R.J. 1983. "Reconsidering the Roles of Regulation and Competition in the Natural Gas Industry." *Harvard Law Review* 97 (2): 345–385.

Pierce, R.J. 1986. "A Proposal to Deregulate the Market for Bulk Power." *Virginia Law Review* 72(7): 1183–1235.

Pindyck, R. S. 1991. "Irreversibility, Uncertainty and Investments." *Journal of Economic Literature* 39: 1110–1148.

Plant Siting Task Force. 1970. *Major Electric Power Facilities and the Environment.* Prepared at the Request of the Edison Electric Institute Committee on Environment.

Platt, H. 1991. *The Electric City.* Chicago: Univ. of Chicago Press.

Plum, L.V. 1938. "A Critique of the Federal Power Act." *Journal of Land and Public Utility Economics* 14 (May): 147.

Plummer, J. and S. Troppmann. 1990. *Competition in Electricity: New Markets and Structures.* Arlington, VA.: Public Utilities Reports, Inc.

Poirer, J.L. and M. Rollor. 1989. The Need for New Capacity: Upcoming QF/IPP Markets. Presented at the Third Annual Meeting and Exposition of the American Cogeneration Association and Cogeneration and Independent Power Coalition of America.

Pollit, M.G. 1996. "Ownership and Performance in Electricity Utilities: The International Evidence on Privatization and Efficiency." Oxford Institute for Energy Studies (July).

Poole, Robert W., Jr. ed., 1985. *Unnatural Monopolies: The Case For Deregulating Public Utilities.* Lexington, MA: Lexington Books.

Porter, K. 1996. "Numbers in Schaefer Legislation". National Renewable Energy Laboratory. (16 September).

Porter, K. 1995. "A Summary of the California Public Commission's Two Competing Electric Utility Restructuring Proposals" NREL (October).

Posner, R.A. 1971. "Taxation By Regulation." *Stanford Law Review.* (February): 548–643.

Posner, R.A. 1969. "Natural Monopoly and its Regulation." *Stanford Law Review:* (February): 548–643.

President's Commission on Environment Quality. 1992. "Energy Efficiency Resource Directory: A Guide to Utility Programs." American Gas Association et. al: (September).

Price, T. *Political Electricity.* Oxford: Oxford Univ. Press.

Price Waterhouse. 1987. "Competition: Pressures for Change." Palo Alto. CA: Electric Power Research Institute, EM-5226.

Primeaux, W.J., Jr. 1986. *Direct Electric Utility Competition.* New York: Praeger.

Primeaux, W.J., Jr. 1978. "Rate Base Methods and Realized Rates of Return." Economic Inquiry. 16 (January): 95–107.

Primeaux, W.J., Jr. 1975. *A Reexamination of the Monopoly Market Structure for Electric Utilities. In Promoting Competition in Regulated Markets.* A. Phillips, ed., Washington, DC: The Brookings Institution.

Public Service Commission of Wisconsin. 1995. "Report of the Advisory Committee on Electric Utility Restructuring. PSCW Docket No. 05-EI-114: (October).

Public Utilities Reports, Inc. And the New York State Energy Research and Development Authority. 1994. *The Electric Industry in Transition.* Arlington, VA.: Public Utilities Reports, Inc.

Public Utilities Reports, Inc. 1981. *Utility Diversification, Strategies & Issues.* Prepared from proceedings of the Management Conference. (June).

Puget. 1996. Letter to FERC explaining Puget Power & Light Company's Decision to Withdraw from WSCC and the Company's Continuing Involvement with WSCC. (22 March) In Can Regional Reliability Councils Continue to Operate as Voluntary Organizations?: How Might the Changes in a Restructured Industry Affect the Role of Reliability Councils?

Quarles, J. And W. H. Lewis, Jr. "The New Clean Air Act: A Guide to the Clean Air Program As Amended in 1990." Washington, DC: Morgan, Lewis, & Bockius.

Rader, N.A. and R.B. Norgaard. 1996. "Efficiency and Sustainability in Restructured Electricity Markets: The Renewables Portfolio Standard." *The Electricity Journal.* (July): 37–49.

Ramsey, F.P. 1927."A Contribution to the Theory of Taxation," *The Economic Journal.* 27 (145): 47–61.

Ramsey, M.L. 1937. *Pyramids of Power: The Story of Roosevelt, Insull, and the Utility Wars.* New York: Bobbs Merrill

Rau, N. 1988. "The Evaluation of Transactions in Interconnected Systems." Columbus, OH: National Regulatory Research Institute.

Reed, Daniel J. 1978. "Utility Rates Under the National Energy Act, Quo Vadis?" *Public Utilities Fortnightly* (20 July 20).

Reed, M., J.Brown and J. Deem. 1983. "Incentives for Demand-Side Management." Washington, DC: National Association of Regulatory Utility Commissioners. 2nd ed.

Regulatory Assistance Project. 1995. "System Benefits Charge." *Issuesletter* (September).

Regulatory Assistance Project. 1984. *Issuesletter* (September).

Reid, M.W., J.B. Brown and J.C. Deem. 1993. "Incentive for Demand-Side Management." National Association of Regulatory Utility Commissioners (March).

Reid, E. and C. Droffner. 1995. "EIS: Digests of Environmental Impact Statements." Bethesda, MD: Cambridge Scientific Abstracts 19(2).

Resource Data Interntional, Inc. 1995. *Energy Choices in a Competitive Era: The Role of Renewable and Traditional Energy Resources in America's Electric Generation Mix.* Center for Energy and Economic Development.

Rhodes, R. 1993. *Nuclear Renewal.* New York: Viking Press, 1993.

Ridley, S. 1995. "Seeing the Forest From the Trees: Emergences of the Competitive Franchise." *The Electricity Journal* (May).

Riordan, M. and D. Sappington. 1987. "Information, Incentives, and Organizational Modes." *Quarterly Journal of Economics* 102: 243–263.

Ripley, W. 1927. *Main Street and Wall Street.* Boston: Little Brown and Company.

Roll, J.B. and E.B. Lande. 1980. "Lifeline Rates: Impacts and Significance." *Public Utilities Fortnightly* (31 July).

Rose, K. 1996. " An Economic and Legal Perspective on Electric Utility Transition Costs." The National Regulatory Research Institute (July).

Rosek, R., 1989. "Competitive Bidding In Electricity Markets: A Survey." *Energy Journal* 10(4): 117–138.

Rosen, Harvey S. 1988. "Public Finance." Department of Economics, Princeton University.

Rosenthal, B. 1989. *Regulation of State Bidding Programs.* Vol. I. Washington, DC: Edison Electric Institute.

Rosenthal, B. 1988. "Summaries of State Laws and Regulations Regarding Utility Power Wheeling." Draft Report to Edison Electric Institute, Washington, DC (October).

Rosso, D.J., B.J. Springer, and K.B. Anderson. 1988. "The Raiders are Coming! (Maybe)." *Public Utilities Fortnightly* 121 (12): 9–17.

Roth, G.J. 1967. *Paying for Roads: The Economics of Traffic Congestion.* Harmondsworth, England: Penguin Books.

Rowe, J. and P. Graeing. 1996. "Constitutional Aspects." *The Electricity Journal* 9 (7).

Rudolph, R. and S. Ridley. 1987. *Power Struggle: The Hundred-Year War over Electricity.* New York: Harper & Row Publishers.

Rustbakke, H.M. ed. 1983. *Electric Utility Systems and Practices.* New York: Wiley.

Salop, S.C. and J.J. Simons. 1984. "A Practical Guide To Merger Analysis." *The Antitrust Bulletin.* 29(4).

Sanstad, A., C. Blumstein and S. Stoft 1995. "How High are Option Values in Energy-Efficiency Investments?" *Energy Policy* 23 (9): 739–744.

Santa, D.F. and C.S. Sikora. 1994. "Open Access and Transition Costs: Will the Electric Industry Transition Track the Natural Gas Industry Restructuring?" *Energy Law Journal* 15: 273–321.

Sappington, D.E.M., and J. Stiglitz. 1986. "Information and Regulation." In *Public Regulation: New Perspectives in Institutions and Policies,* E.E. Bailey, ed.. Cambridge, MA: MIT Press.

Sappington, D.E.M., and D. L. Weissman. 1994. "Designing Superior Incentive Regulation: Accounting For All." *Public Utilities Fortnightly* (15 February).

Sawhill, J.C. and L. P. Silverman. 1983. "Build Flexibility, Not Power Plants." *Public Utilities Fortnightly* (26 May).

Sawhill, J. and Braverman. 1985. "Transformed Utilities: More Power to You." *Harvard Business Review* (July–August).

Schmalensee, R. 1986. "Good Regulatory Regimes." *The Rand Journal of Economics:* (Fall).

Schmalensee, R. and B.W. Golub. 1983. "Estimating Effective Concentration in Deregulated Wholehouse Electricity Markets." Working Paper MIT-EL 83-001WP. MIT Energy Laboratory (January).

Schmalensee, R. 1979. *The Control of Natural Monopolies.* Lexington: Lexington Books.

Schmalensee, R. 1974. "Estimating the Costs and Benefits of Utility Regulation." *Quarterly Review of Economics and Business.* 14 (Summer): 51–64.

Schoengold, D. 1994. *Competition in the Electric Industry: The Non-Jurisdictional Impacts of Restructuring.* Prepared by MSB Energy Associates. Middleton, WI. (July).

Schroedeer, C.H., L.Wiggins, D.T. Wormhoudt. 1981. "Flexibility of Scale in Large Conventional Coal-Fired Power Plants." *Energy Policy* 9 (June): 127–135.

Schuler, J.F., Jr. 1996. "Generation: Big or Small?" *Public Utilities Fortnightly* (15 September): 30–34.

Schultz, D.K. 1996. "Access to Information: Crucial for Competition." *The Electricity Journal* 9 (6).

Schulz, W. 1980. Conditions for Effective Franchise Bidding in the West German Electricity Sector. In *Regulated Industries and Public Enterprise: European and United States Perspectives.* B.M. Mitchell and P.R. Kleindorf, eds. Lexington, MA: Lexington Books.

Schumacher, E.F. 1973. *Small is Beautiful.* New York: Harper & Row.

Schwarz, J. *The New Dealers.* New York: Random House, 1994.

Schweitzer, M. 1996. "The Muni Option." *The Electricity Journal* 9 (8).

Schweitzer, M. and T.R. Young. 1994. "State Regulation and its Effects on Electric-Utility Use of DSM Resources." Oak Ridge Laboratory (August).

Schweppe F.C., et. al. 1988. *Spot Pricing of Electricity.* Boston, MA: Kluwer Academic Publishers.

Schweppe, F.C., M.C. Caramanis and R.E. Bohn. 1982. *Optimal Spot Pricing: Practice and Theory. IEEE Transaction on Apparatus and Systems.* Vol. PAS-101. Washington, DC.

Shapiro, M.E. 1986. *Utility Diversification: A Detour with Serious Risks for Ratepayers.* Sacramento, CA: Senate Office of Research. HD2766 .S47.

Sharkey, W.W. 1982. *The Theory of Natural Monopoly.* Cambridge, England: Press Syndicate of the Univ. of Cambridge.

Shepard, W.G. 1985. *Public Policies Towards Business.* Homewood, Illinois: Richard Irwin, Inc.

Shepard, W.G. 1983. "Current Issues in Public-Utility Economics: Price Structures in Electricity." Essay in Honor of James C. Bonbright.

Shepard, W. 1983. Price Structure in "Electricity." *Current Issues in Public-Utility Economics.*

Shepherd, W.G. 1996a. Monopoly and AntiTrust Policies In Network-Based Markets Such As Electricity: Session III: Industrial Organization, Technocolical Change and Strategic Response to Deregulation. Presented for the Symposium on the Virtual Utility, Saratoga Springs, New York.

Shepherd, W.G. and T.G. Gies. 1974. *Regulation in Further Perspective: The Little Engine that Might.* MA: A Subsidiary of J.B. Lippincott Company.

Shugar, D., R. Orans, et. al. 1992. "Distributed Photovoltaic Generation: A Comparison of System Costs vs. Benefits for Cocopah Substation." Arizona Public Service Company (November).

Sidak, J.G. and D.F. Spulber. 1996. "Deregulatory Takings And Breach of the Regulatory Contract." *New York University of Law Reveiw* (October).

Siddiqi, Riaz and J. Woodley. 1994. "Real-Time Pricing's Hidden Surprise." *Fortnightly* (1 March).

Sioshansi, F. 1990. "Pricing and Marketing Electricity: New Directions in Power Supply." Special Report No. 2016, The Economist Intelligence Unit, 40 Duke St. London W1A1DW.

Skoog, Ronald A., ed. 1980. *The Design and Cost Characteristics of Telecommunications Networks.* Hollmdel. NJ: Bell Laboratories.

Smith, E. M. 1994. " Business and Technology Change Towards the Year 2000: Redefining the Electricity Industry's Customer Markets, Products and Services." In *The Electric Industry in Transition.* Arlington, VA: Public Utiliites Reports, Inc. & the New York State Energy Research and Development Authority.

Smith, J.M. and K.F. Skousen. 1977. *Intermediate Accounting: Comprehensive Volume.* Cincinnati: South-Western Publishing Co.

Smith, V.L. 1988. "Electric Power Deregulation Background and Prospects." *Contemporary Policy Issues* 6 (3): 14–24.

Sorenson, B. 1983. "The Grid as an Energy Absorber and Redistributor." In *Decentralized Electricity Production.* Brown, H., ed. New Haven: Yale Univ. Press.

Southern States Energy Board, 1985. "Energy Facility Siting: Legislative Directory." Atlanta, GA.

Southern States Energy Board, 1978. "Energy Facility Siting in the U.S.: Topical Monograph." Atlanta, GA.

Spann, R.M. 1976. "Restructuring the Electric Utility Industry: A Framework for Evaluating Public Policy Options in Electric Utilities." In *Electric Power Reform: The Alternatives for Michigan.* W. Shaker, ed. East Lansing, MI: Michigan State Univ. Press.

Spiewak, S.A. 1987. *Cogeneration and Small Power Production Manual.* Lilburn, GA: The Fairmont Press, Inc.

Sponseller, D. 1987. "An Overview of Utility Reorganization Activity." *Public Utilities Fortnightly* 120 (8): 42–46.

Sporn, Phillip. 1971. *The Social Organization of Electric Power Supply in Modern Societies.* Cambridge: MIT Press: 1971.

Spulber, D.F. *Regulation and Markets.* Cambridge: MIT Press, 1989.

Stalon, C.G. and E.C. Woychik. 1995. "What Model for Restructuring? The Competitive Power Market Working Group Debate," *The Electricity Journal* 8(6): 63.

Stalon, C.G. 1994. "Stranded Investments Costs: Desirable and Less Desirable Solutions". In *The Electric Industry in Transition,* 185–198. Arlington, VA: Public Utilities Reports, Inc. & the New York State Energy Reserch and Development Authority.

Stalon, Charles G. 1992. " Focusing on QUANGOs." *The Electricity Journal* (6 January).

Stalon, C.G. 1991. "The Significance of the FERC's Transmission Task Force Report in the Evolution of the Electric Industry." In *Research In Law and Economics* Vol. 13, 105–149 R.O. Zerbe, Jr., and V.P. Goldberg, eds. Greenwich: JAI Press.

Stalon, C.G. 1988. Interdependence of Pricing and Transmission Access. Paper presented to The Second Annual ACA/CIPCA Conference and Exhibit, Chicago.

Stigler, G.J. 1968. *The Organization of Industry.* Chicago: The University of Chicago Press.

Stigler, G.J. and C. Friedland. 1962. "What Can the Regulators Regulate? The Case of Electricity." *Journal of Law and Economics.* 5 (October): 1–16.

Stoft, S. 1996. "Analysis of the California WEPEX Applications to FERC," POWER-Program On Workable Energy Regulation (15 October).

Stoft, S. 1996. "California's ISO: Why Not Clear the Market?" *The Electricity Journal* (December): 38–43.

Stoll, H.G., ed. 1989. *Least-Cost Electric Utility Planning*. New York: Wiley-Interscience.

Studness, C.M. 1996. "Stranded Cost Recovery: It's UnAmerican." *Public Utilities Fortnightly* (15 July): 43.

Studness, C.M. 1995. "Price-cap Regulation: Will it Survive in the U.K.?" *Public Utilities Fortnightly* (15 June).

Studness, C. 1991. "The Failure of Utility Regulation and the Case Deregulation." *Public Utilities Fortnightly* (15 September).

Stuntz, L.G. 1992. "Is it Time to Consider Regional Solutions to Power Planning Problems? One Federal View." *The Electricity Journal* (January/February).

Stuntz, L. 1992. "A Federal Review." *The Electricity Journal* 5 (1).

Subbakrishna, Nagendra. 1994. Long Term Rate and Bill Impacts of DSM in New York State. Presented at the NARUA-DOE Fifth National Integrated Resource Planning Conference, May.

Swidler, J.C. 1991. "An Unthinkably Horrible Situation." *Public Utilities Fortnightly.* (15 September).

Symons Jr., W. 1978. "California Rate Experiments: Lifeline or Leadweight?" *Public Utilities Fortnightly* (16 October).

Synergic Resources Corp, ed. 1989. *Utility Opportunities for New Generation*. Palo Alto: Electric Power Research Institute, CU-6605.

Taylor, T.L. and P.M. Schwartz. 1996. "Near RTP: Advanced Notice and Price Risk—A Customer/Utility Model." Department of Energy, Grant No.DEFG48-95R810582 A000: (1996.)

Technology Futures, Inc., and Scientific Foresight, Inc. 1984. *Principles for Electric Power Policy*. Westport, CT: Quorum Books.

Telson, M. L. 1975. "The Economics of Levels of Reliability for Electric Power Generation Systems." *Bell Journal of Economics* 6 (Autumn): 679–694.

Tenenbaum, B., Lock, R. and J. Barker, Jr. 1992. "Electricity Privatization: Structural, Competitive, and Regulatory Options." *Energy Policy* (December).

Tenenbaum, B. and Henderson, J.S. 1991. "The History of Market-Based Pricing." *Electricity Journal* (December): 30–46.

The Electricity Journal. 1988. The Great PacifiCorp-Utah P&L Merger Case. *The Electricity Journal* 11 (5): 19–25.

The Union of Concerned Scientists. 1975. *The Nuclear Fuel Cycle: A Survey of the Public Health Environmental, and National Security Effects of Nuclear Power*. Cambridge: The MIT Press.

Tonn, B., E. Hirst and D. Bauer. 1995. "Public-Policy Responsibilities in a Restructured Electricity Industry." Oak Ridge Labortory (June).

"Trabandt Outlines Strong Opposition to Retail Wheeling of Power." *Energy Report* 20 (3), 1980.

Transmission Task Force. 1989. The Report of the Transmission Task Force. Washington: Federal Energy Regulatory Commission.

Troxel, E. 1957. "Power and Public Utilities Problems: Discussion" *American Economic Review.* Papers and Proceeding. 47 (May): 403–415.

Troxel, E. 1947. *Economics of Utilities.* New York: Rinehart, 1947. Chapter I, "The Legal Concept of Public Utility," 3–24; Chapter 2, "Economic Characteristics of Public Utilities," 25–48.

Tye, W.B. and F. Graves. 1996. "The Economics of Negative Barriers to Entry: How to Recover Stranded Cost and Achieve Competition on Equal Terms in the Electricity Industry." The Brattle Group (14 May).

Tye, W.B. and C. Lapuerta. 1996. "The Economics of Pricing Network Interconnection: Theory and Application to the Market for Telecommunications in New Zealand." *The Yale Journal on Regulation.* 13: 2–88.

Tye, W.B. and H. Leonard. 1983. "On the Problems of Applying Ramsey Pricing to the Railroad Industry with Uncertain Demand Elasticities". *Transportation Research.* 17A.(6): 439–450.

Tye, W.B. 1991. *The Transition to Deregulation.* NY: Quorum Books.

Tye, W.B. 1987. *Encouraging Competition Among Competitors: The Case of Motor Carrier Deregulation And Collective Ratemaking.* New York: Quorum Books.

U.K. Secretary of State for Energy. 1988. *Privatising Electricity: The Government's proposals for the privatisation of the electricity supply industry in England and Wales.* Cmd. 322.

U.S. Department of Commerce. 1987. National Telecommnications and Information Administration. *NTLA Regulatory Alternatives Report.* Washington, DC.

U.S. D.O.E./E.I.A. 1993. "Center For Clean Air Policy New York Case Study Electricity Restructuring and the Environment." *Electric Power Monthly.*

U.S. Department of Energy. 1987. "Designing PURPA Power Purchase Auctions: Theory and Practice." Office of Policy, Planning, and Analysis, DOE/SF/00098-H1.

U.S. Department of Energy. 1986. "Emerging Policy Issues in PURPA Implementation." DOE/PE-70404-H1.

U.S. Department of Energy. 1981. National Electric Reliability Study: Final Report. DOE/EP-004.

U.S. Department of Energy. 1980a. "The National Power Grid Study. Vol I." Washington, D.C. DOE/_ERA/0056/3

U.S. Department of Energy. 1980b. "Power Pooling: Issues and Approaches." DOE/ERA/6385-1.

U.S. Department of Energy. 1979. "The National Power Grid Study. Vol I. Report; Vol. II. Techinical Study Reports." DOE/ERA/0066-2.

U.S. Department of Energy. 1987. "Energy Security: A Report to the President of the United States." (March).

U.S. Department of Energy. 1996. "The Electric Power Outages in the Western United States, July 2–3, 1996. Report to the President." (August).

U.S. General Accounting Office. 1984. "Analysis of the Financial Health of the Electric Utility Industry." GAO/RCED 84–22, (11 June).

U.S. Congress Office of Technology Assessment. 1989. "Electric Power Wheeling and Dealing: Technological Considerations for Increasing Competition." Washington, DC: U.S. Congress Office of Technology Assessment.

U.S. National Acid Precipitation Assessment Program. 1996. "Tracking and Analysis Framework."

United Nations Environment Programme. 1993. *The Impact of Climate Change.* Nairobi, UNEP, 1993. (UNEP/GEMS Environment Library No 10).

Verleger, Jr., Philip K. 1993. "Adjusting to Volatile Energy Prices." Institute for International Economics (November).

Vickers, J., and G. Yarrow. 1988. *Privatization: An Economic Analysis.* Cambridge, MA: MIT Press.

Vickrey, W. 1967. "Optimization of Traffic and Facilities." *Journal of Transport Economics and Policy* 1 (2): 123–136.

Vince, C.A. and J.C. Fogel. 1995. "Franchise Competition in Electric Utility Industry." *The Electricity Journal* (May).

Vine, E., D.Crawley and P. Centolella. 1991. "Energy Efficiency and the Environment: Forging the Link." Washington, DC: American Council Energy for an Energy-Efficient Economy.

Vogelsang, I., and J. Finsinger. 1979. "A Regulatory Adjustment Process for Optimal Pricing by Multiproduct Monopoly Firms." *Bell Journal of Economics* 10: 157–171.

Wall, C. 1995. "Market Power and Power Markets: Voice In Current Debate." Harvard Electricity Policy Group Kennedy School of Government, Harvard University (4 April).

Watson, R.T., Marufu C.Z. and R.H. Moss, eds. 1995. *Climate Change 1995: Impacts, Adaptations and Mitigation of Climate Change: Scientific-Technical Analyses.* Published for the Intergovernmental Panel on Climate Change. Cambridge: Cambridge Univ. Press.

Waverman, L. 1975. "The Regulation of Intercity Telecommunications." In *Promoting Competition in Regulated Markets.* A. Phillips, ed. Washington, DC: The Brookings Institution.

Weinberg, C. J. 1993. *The Restructuring of the Electric Utiliity Technology Forces, R & D and Sustainability.* Royal Institute of International Affairs. London (19 November).

Weinberg, C., J. Iannucci, and M. M. Reading. 1991. "The Distributed Utility: Technology, Customer, and Public Policy Changes Shaping the Electrical Utility of Tomorrow." *Energy Systems and Policy* 15 (4): 307–322.

Weiner, M., N. Nohria, A. Hickman and H. Smith. 1996. "The Future Structure of the North American Utility Industry." In Proceedings of the Sympsium on the Virtual Utility. Awerbuch, S., ed., Saratoga Springs, N.Y., April 1.

Weisman, D.L. 1989. "Competitive Markets and Carriers of Last Resort." *Public Utilities Fortnightly* 124 (1): 17–23.

Weiss, E.R. and J. Salzman. 1989. "The Greening of American Energy Policy." *St. John's Law Review,* 63 (Summer): 691–714.

Weiss, L. and S. Spiewak, eds. 1989. *The Wheeling and Transmission Manual.* Washington, DC: Wheeling & Transmission Monthly Publications.

Weiss, L.W. 1975. "Antitrust in the Electric Power Industry." In *Promoting Competition in Regulated Markets.* A. Phillips, ed. Washington, DC: The Brookings Institution.

Wellisz, S.H. 1963. "Regulation of Natural Gas Pipeline Companies: An Economic Analysis." *Journal of Political Economy* 71 (February): 30–43.

Werden, G.J. 1982. "Market Delineation and the Justice Department's Merger Guidelines." *Duke Law Journal.* 1983: 514–578.

Werden, G.J. 1996. "Identifying Market Power in Electric Generation." *Public Utilities Fortnightly.* (15 February): 16–21.

Western Interstate Energy Board. 1996. Meeting of the Committee on Regional Electric Power Cooperation. Portland, Oregon. April 11–12.

White, W.S., Jr. 1982. "A Closer Look at Electric Utility Deregulation." *Public Utilities Fortnightly* 109: 19–23.

White, L.J. 1972. "Quality Variation When Prices are Regulated." *Bell Journal of Economics* (Fall).

Wibberly, P. and A. Green. 1995. "Pricing of Constraints—The England and Wales Experience." In Proceedings: 1996 EPRI Conference on Innovative Approaches to Electricity Pricing. Palo Alto, CA: Electric Power Research Institute. (March) EPRI TR-106232.

Wilkinson, M. 1989. "Power Monopolies and the Challenge of the Market: American Theory and British Practice." Discussion Paper E-89-12, Energy and Environmental Policy Center, Kennedy School of Government, Harvard University (September).

Williamson, O.E. 1985. "Assessing Contracts." *Journal of Law, Economics, and Organization* 1 (1): 177–208.

Williamson, O.E. 1979. "Transaction Cost Economics: The Governance of Contractual Relations." *Journal of Law and Economics* 22: 223–261.

Williamson, O.E. 1976. "Franchise Bidding for Natural Monopolies: In General and With Respect to CATV." *Bell Journal of Economics* 7(Spring) : 73–104.

Williamson, O.E. 1975. *Markets and Hierarchies, Analysis and Antitrust Implications.* New York: The Free Press.

Winston, C.E. 1950. *Economics and Public Utilities.* New York: Appleton-Century-Crofts.

Wiser, R., S. Pickle and C. Goldman. 1996. California Renewable Energy Policy and Implementation Issues: An Overview of Recent Regulatory and Legislative Action. Energy & Environment Division: Ernest Orlando Lawrence Berkeley National Laboratory. University of California. Berkeley, CA: LBNL-39247: UC-1321.

Wolfram, C. D. 1995. *Measuring Duopoly Power in the British Electricity Spot Market.* Cambridge, MA: MIT Department of Economics.

Woolf T. and J. Michals. 1995. "Performanced-Based Ratemaking: Opportunities and Risk in a Competitive Electricity Industry." *The Electricity Journal* (October): 64–73.

WSCC. 1996. Executive Policy Issues Task Force Report to the Board of Trustees. March 12, 1996 and a white paper entitled "Why Control Areas Must Be Members of a Reliability Council". In Can Regional Reliability Councils Continue to Operate as Voluntary Organizations? How Might the Changes in a Restructured Industry Affect the Role of Reliability Councils?

Wu, F.F. and P. Varaiya. 1995. *Coordinated Multilateral Trades for Electric Power Networks: Theory and Implementation.* Berkeley, CA: University of California.

Xavier, P. 1995. "Price Cap Regulation for Telecommunications: How has it performed in practice?" *Telecommunications Policy.* UMI Article Ref. No.: TCP-2002-4.

Zajac, E.E. 1978. *Fairness or Efficiency: An Introduction to Public Utility Pricing.* Cambridge, MA.: Ballinger Publishing.

Zardkoohi, A. 1986. "Competition in the Production of Electricity." In *Electric Power: Deregulation and the Public Interest.* J.C. Moorhouse, ed. San Francisco: Pacific Research Institute for Public Policy.

Ziegler, B.T. 1995. "Affiliate Transactions and Electric Industry Restructuring." *The Electricity Journal* 8 (8): 20.

Zimmer, M.J. 1989. Independent Power Producers: From a Developer's Perspective or Where Goes a Pro-Competitive National Electric Policy? Presented at Utility Opportunities for New Generation, sponsored by Edison Electric Institute and Electric Power Research Institute

SUBJECT INDEX